Advancing Agricultural Production in Africa

The Commonwealth Agricultural Bureaux (CAB) provides a wide range of specialist services to agriculture and related fields world-wide. For further information on CAB's information, identification, and biocontrol services contact Clive Hemming, CAB, Farnham Royal, Slough SL2 3BN, UK. See also Chapters 78-81 in this volume.

Advancing Agricultural Production in Africa

Proceedings of CAB's First Scientific Conference
Arusha, Tanzania, 12–18 February 1984

Edited by D. L. HAWKSWORTH

Organized in collaboration with the Government of Tanzania

 COMMONWEALTH AGRICULTURAL BUREAUX

Commonwealth Agricultural Bureaux
Farnham Royal
Slough SL2 3BN
UK
Tel. (02814) 2662
Telex 847964

ISBN 0 85198 537 8

Printed and bound in Great Britain by Kingprint of Richmond.

Contents

(E) *Crop Improvement Strategies*

Chairman: J.M. HIRST

(F) *Symposium Report*

III. ANIMAL PRODUCTIVITY AND HEALTH: INTEGRATION WITH MARKETING AND
CONSUMER PREFERENCE

Convenor and Rapporteur: P.R. ELLIS

(A) *Introduction*

Chairman: S. BEKURE

(B) *The Problems of Marketing*

Chairman: A.A. MAJOK

(C) *The Adjustment and Servicing of Small-Scale Production Systems*

Chairman: I.S. MPELUMBE

(B) *Terrain, Land and Water*

Chairman: D.J. GREENLAND

(C) *Biological Resources*

Chairman; A.C. MASCARENHAS

(D) *Symposium Report*

V. COMMONWEALTH AGRICULTURAL BUREAUX SERVICES

VI. INFORMATION SERVICES

Chairman: E.E. KAUNGAMNO

I. Extension level rice publications in African languages (R. HARGROVE). II. Agricultural user population and their information needs: a case study of Badeku Pilot Rural Development Project in Nigeria (C.E. WILLIAMS and S.K.T. WILLIAMS). III.Recent developments in the transfer of agricultural information in India (D.B.E. REDDY). IV. Training programmes in agricultural bibliographical and information services (M. BELLAMY). V. A computerized agricultural information system in Africa (M. HAILU).

VII. CULTURAL AND SOCIAL ASPECTS

VIII. REVIEW OF THE CONFERENCE

Chairman: A.S. MSANGI

Close of Conference: The Hon. Prof. J.B. MACHUNDA (Minister of Agriculture, Tanzania)

Preface

The Arusha Conference confronted what is certainly one of the most pressing problems of this decade, that of *Advancing Agricultural Production in Africa*. It was therefore a most appropriate topic for CAB's first international scientific conference. The interest generated by the Conference can be gauged by the fact that it attracted some 360 delegates and speakers from 36 countries, considerably in excess of the expected attendance.

CAB was particularly honoured that the President of Tanzania, Julius Nyerere, agreed to present the Opening Address of the Conference. His speech, which stressed the importance of the small farmer and low-input systems, was warmly received.

The concept of a CAB Conference arose at the 1980 CAB Review Conference in London. The Hon. Dr J.S. Malecela, then Minister of Agriculture for Tanzania, invited CAB to hold a conference in his country, an offer CAB was pleased to accept. CAB was fortunate in persuading Professor A.H. Bunting, University of Reading, to chair the CAB Organizing Committee. He decided on the topic and framework in consultation with the Tanzanian Organizing Committee, which was chaired first by Professor H.Y. Kayumbo and later by Professor A.S. Msangi, successively the Director of the Tanzanian National Scientific Research Council, UTAFITI.

The substance of the Conference was addressed to administrators, directors and senior professional scientists in African countries responsible for research and related services for development in agriculture. The aim was to support and strengthen efforts to align such work to the needs, resources and constraints of development in these nations. Thus, specialists with African experience were invited for discussion of the requirements of particular products, constraints to their production, and recent advances which might be more widely applied to better the situation.

The overall result of the Conference was a comprehensive survey of the current status and constraints to *Advancing Agricultural Production in Africa*. It is hoped that the proceedings of the Conference will prove of value to all concerned. I am sure that it will be understood and appreciated that opinions presented in the contributions are those of the authors and do not necessarily reflect those of the Commonwealth Agricultural Bureaux or the Tanzanian National Scientific Research Council, or those of the authors' own organizations or institutions.

D.G. CROSBY
Chairman
CAB Executive Council

Canadian High Commission
London
22 May 1984

Acknowledgements

In addition to the Tanzanian National Scientific Research Council, the Conference of which this publication is a record was held with the support of the Food and Agricultural Organization of the United Nations, the Commonwealth Secretariat, the Commonwealth Foundation, the Commonwealth Fund for Technical Co-operation, the International Agricultural Research Centres, the Australian Development Assistance Bureau, the British Council, the Overseas Development Agency, the Ford Foundation, the Government of Tanzania, the Irish Department of Foreign Affairs, the Netherlands Ministry of Foreign Affairs, the Rockefeller Foundation, and the Swedish International Development Authority.

Financial or other support is gratefully acknowledged from Massey Ferguson (World Export Operations), Userlink Systems Ltd., 3M (UK) PLC Business Products Division, and May & Baker Animal Health Ltd. (UK).

The Editor would also like to record his gratitude to the staff at the Commonwealth Mycological Institute, Kew, for assistance in the checking of proofs, and to Mr F.S. Dobson and Mr R. O'Shea of Kingprint Ltd. for expediating the publication of this volume.

Organizing Committees

COMMONWEALTH AGRICULTURAL BUREAUX

Prof. A.H. Bunting (Chairman)

D.G.R. Belshaw
E. Birichi
Dr J. Bridge
P.R. Ellis
D.C.P. Evans
W. Finlayson
P.S. Gooch

Dr D.J. Greathead
Prof. P.T. Haskell
Dr D.L. Hawksworth
Prof. J.M. Hirst
Mrs L.E. Howell
Dr N.W. Hussey

J.E. Johnston
Dr M.N.G.A. Khan
C. Parker
A.J. Smyth
Dr J.M. Waller
Prof. M.J. Way
Dr E.K. Woodford

TANZANIA

Prof. A.S. Msangi (Chairman)

W.S. Abeli
A.S. Hamad
Dr J.N.R. Kasembe
Prof. H.Y. Kayumbo

Dr J.M. Liwenga
Mrs V.F. Malima
Dr B.J.N. Ndungura

Conference Officers

Chairman: Prof. A.H. Bunting

General Secretary: Dr D.L. Hawksworth
Programme Secretary: P.S. Gooch
Information Officer: Ms M. Bellamy
Publications Officer: G.P. Rimington
Secretariat: Mrs A. Atkinson and Mrs S. Eames

Dr D.G. Crosby (*Chairman, CAB Executive Council*) presenting a rose bowl to Mwalimu Dr Julius K. Nyerere (*President of Tanzania*) during the Opening Ceremony of the Conference. The Hon. Prof. J.B. Machunda (*Minister of Agriculture, Tanzania*) is in the centre.

Chapter 1

Opening Address

MWALIMU DR JULIUS K. NYERERE
President of Tanzania

On behalf of the Government and People of Tanzania, and on my own behalf, I welcome to this country and to Arusha all our guests from African and Commonwealth states. We are pleased to have you with us. You have come here to participate in the first Commonwealth Agricultural Bureaux Scientific Conference to be held in Africa, and we are honoured to be your host. I hope that the facilities we are able to offer will contribute to the usefulness of your meeting. But I also hope that you are able to enjoy your stay with us, and that you will have an opportunity to see something of the surrounding countryside and its wildlife.

The subject of your Conference is a challenging and important one. The peoples and the nations of Africa depend upon African agriculture now and will do so for the foreseeable future. Our agricultural production has to expand if we are to survive; it has to expand faster if we are to lift ourselves out of our present poverty and provide for all our people the basic material requirements of human dignity and real freedom.

Yet that expansion has to be achieved without destroying the physical environment of our countries; without polluting our land, or our water; and in a manner which does not cause soil erosion or the destruction of our forests. Equally important, the agricultural expansion has to be directed at meeting our needs — the needs of Africa and Africans. It has to be balanced between increased output of the foodstuffs our people need to eat, and the output of commercial crops which are necessary for our clothing, our housing, and our exports. At the same time, the agricultural expansion has to be achieved in a manner which is consistent with the total development of the African people and their own societies. For the purpose of agricultural production in this continent must be, and must be recognised as being, the welfare and advancement of the people who now live, and will in the future live, on our continent. Agricultural production is not desired for its own sake, but as a means of improving the lives of men, women and children.

It is on that basis that we have to transform our traditional agricultural methods and production techniques. This is essential if we are to meet the challenges of an expanding population which demands public and private consumption goods in ever-increasing quantities. But we have to start from where we are now, without trying to pretend that we have knowledge and capital which we do not have. Also we have to advance in full recognition that we do not know all the answers to the problems of expanding agricultural output in Africa, and realising that we could do more harm than good by over-enthusiastic application of scientific theories in what are really unknown circumstances.

There is much which we in Africa can learn from other parts of the world, from experience in other countries, and from techniques which succeed in different societies. But we would be well-advised to pause and think before assuming that because a method of increasing output succeeds elsewhere it will therefore be directly applicable in Africa — or in all parts of Africa. The production problems in tropical areas are different, and in many ways greater, than those which prevail in temperate climates.

Africa's physical environment is very fragile. There is much evidence of the speed with which erosion can carry away our top soil when traditional small-scale farming is replaced by large and capital intensive methods of cultivation, and when tropical forests are cut down in order to expand the arable acreage. The delicate balance of nature in tropical conditions gets easily upset, water supplies are interrupted or poisoned; and human, animal and plant diseases spread only too quickly when we seek to apply systems of agricultural production which have apparently succeeded in easier climatic circumstances. There is, in other words, a very clear danger that we can make things worse in the medium- and long-term when we try to achieve a quick improvement of our agricultural productivity and our output.

None of this means that we can avoid making changes. Change is inevitable; nothing stands still in nature, or society. All that it means is that we have to move with great care, just as our peasants have learned over the generations to move cautiously – for their so-called conservatism is often little more than a realisation from experience that changed practices have costs as well as benefits. The importance of agricultural research is thus very clear. But unfortunately research into the problems of tropical agriculture is comparatively recent, and very small in quantity when compared with the amount of research done by the more developed areas of the world for their own conditions.

Further, such research into tropical farming which has been done has mostly been premised on the assumption that the future lies in plantations or large farm systems – that only on such a basis can production be greatly increased. This is a reflection of orthodox theories about the advantage of mass production methods; it also reflects the interests of farm machinery manufacturers and the belief that because something is bigger and faster it is therefore better. We in Africa have only in the last few years begun to question these assumptions, and to ask whether they are always and in all circumstances valid.

Also such research as has been done on tropical crop production, and soil regeneration, has usually – in fact almost always – presupposed the ready availability of chemicals such as fertiliser and insecticide, of adequate and regular water supplies, and of a basic scientific and technical knowledge on the part of the farmer. And sociological questions have been regarded as irrelevant both to the research and its beneficial outcome.

These assumptions are valid only rarely in African countries. In most of our states agricultural production is overwhelmingly peasant production, and consists for the most part of farms which a single peasant family can cultivate without extra labour or mechanised implements. Even if – as happens in a few countries in eastern and southern Africa – estates do produce a sizeable proportion of the marketed food or export crops, the importance of peasant production remains. Peasants at least feed themselves, and they are the majority of the population virtually everywhere. For the most part even the urban workers maintain links with their rural family, and many of them look to the family shamba to provide security for their old age or their sickness.

Therefore, without denying the contribution which can be made by large farms, I do stress that we cannot over-emphasize the importance of the sociological preponderance of peasants in Africa. It has macro-economic implications; for the most part they use by far the largest proportion of the arable land, with an obvious link between their output per acre and the total agricultural production of the country. It has social implications; unless the rains fail the peasants at least feed the majority of the population and provide some minimum of life for infants, the elderly and the disabled. In addition, this structure of production underlies the whole traditional way of life, the relations between individuals, families, and groups, as well as the ethical base of society and people's sense of belonging. Finally, it is the possession of land, with all that that implies, which prevents – or at least slows – the drift to the urban areas. In other words, the proletarianization of the peasants, especially if this took place over a short period, would cause social, economic, and political chaos in African states at the same time as it exaccerbated their national economic problems.

Thus, it would not constitute an advance for Africa in human terms if the peasants were squeezed out and large-scale private or public farming systems were put in their place. This is true even if it could be proved that the result would be a greater total production in terms of tonnes of crops or kilos of meat. For human and social stability would have been destroyed and total consumption may not be greater; landless and unemployed labourers would not have access to that increased production. And in any case there is considerable evidence to suggest that good husbandry on small farms gives a higher average yield per acre than capital-intensive farming on a large scale.

Advancing agriculture, even in terms of output, thus becomes a question of what you are aiming at. If there is a shortage of labour, but plenty of land and capital, advancing agriculture implies a different kind of farming, and different techniques, from that in circumstances where capital is short, and alternative employment is not available.

In Africa now, mechanised equipment, chemicals, and technical or scientific knowledge are not readily available to the farmers. Nor are they likely to be in the near future. Those facts have to be taken as a starting point for any consideration about how to advance agricultural production in this continent. For Third World countries are — almost by definition — short of capital, of economic infrastructure, and of technical and scientific knowledge in the rural areas. Thus, for example, even when tractors or combine harvesters have been acquired, they sometimes have to be sent hundreds of miles for servicing, and minor breakdowns can result in their being out of action for months at a time. And even when fertilisers are produced within the state, the transport problems of poor countries sometimes mean that they are not available at the right time in the right place. The long-term effect of their continual application on particular African soils is also a point to be watched in countries which do not have reliable and locally available soil-testing laboratories.

Further, most African countries still have to import advanced agricultural equipment, chemicals, and even spare parts. Increasingly, the imports cannot be paid for from the earnings of exports because the terms of trade constantly move against commodity producers. Few if any African states are in balance in their foreign transactions — most of us have the kind of deficit which makes it impossible to import sufficient goods even to maintain machinery and equipment at the level attained in the 1970s.

Basing African agricultural advance exclusively, or even mainly, on industrial and capital intensive techniques of production is thus a contradiction in terms. A tractor can plough hundreds of acres; waiting for one which does not exist, or is not working, means that you get less output than if you had cultivated with a hand-hoe, and much less than if you had based your production plans on the use of animal-drawn implements. Putting total reliance upon chemical fertilisers which may not arrive in time, while ignoring the benefits of crop-rotation, and the use of manure and compost, is like trying to cross a river at the rapids, instead of walking a mile to the ford. If inter-cropping and crop rotation can reduce insect and disease infestation it is not an advance in agriculture to abandon them in favour of monoculture and artificial insecticides.

In saying these things I have not forgotten that this is a scientific conference; on the contrary I stress them because the application of scientific knowledge to our problems could make a major contribution to the increased well-being of our people. But it must be science directed at our needs, our objectives, and tailored to fit our circumstances.

Policy-makers here come from different countries, and the policies they will be seeking to implement — and therefore the help they need from scientists and economists — will differ. But they will differ in emphasis more than in nature. All of us are concerned about future soil fertility, about pollution of our land and water, about getting a greater output per acre and per hour of labour. All of us are concerned about the economic security of our peasants, about social and political stability in changing conditions. And all of us are now very conscious of the need to expand agriculture in a manner which increases national — and even local — self-reliance.

I have to emphasize this last point. For your meeting is taking place at what is for many African countries a time of economic crisis. We need to expand our production of food and commercial crops as quickly as possible, and we need to improve their quality. But methods of doing this which demand more imports — even short term — are in practice not useful to us now however apparently superior they may be to anything we can do with local resources. We just do not have the foreign exchange which would be needed. Nor can we rely on foreign aid for it. Apart from the fact that foreign aid is being reduced by many countries, and that the resources of international institutions are being cut, we have learned from experience that basing your plans on import dependence is to walk up a blind alley.

Self-reliance is the policy of Tanzania and has been for many years, but all the countries of Africa are now realising its importance — even the oil producing nations are being driven in that direction. And Tanzania itself has in too many respects thought of achieving self-reliance in the future by increasing import-dependence in the short-run. We are now realising that while some degree of increased import dependence is inevitable when you begin to develop, your development plans are more secure the greater the extent to which they are themselves based on self-reliant methods. Now we have no choice. Such foreign assistance as we can get through the continued generosity of friends must be used only for projects and imports for which there is no conceivable local substitute. However urgent your need to arrive at your destination you cannot travel by Concord if you do not have one, nor by plane if you cannot afford the fare. Africa's need for agricultural advance is indeed great and urgent -- people's lives depend upon it. But wishes are not horses. Our task is to use our local resources in the most efficient way possible.

Your task in this Conference is to help us to do that. You have to help Africa to meet the needs of its people, where they are now, with the resources of knowledge, skill, and material which are currently available to African states. From that base you have to help us by indicating some roads forward, and showing how they can be traversed.

I am confident that the experienced people from the Commonwealth Agricultural Bureaux, from African and other Third World Governments and Research Institutes, and from Tanzania itself, will use this coming week for that purpose. I believe that the Papers you consider, and the discussion which takes place on them, will be directed at meeting practical problems experienced by our farmers large and small. And I hope that in your discussions you will take full account of the economic and sociological constraints and opportunities which arise from the nature of our societies, as well as the need to use modern knowledge for the benefit of Africa's rural people. For I repeat; Africa's farmers are cautious more than conservative -- and they have cause to be. We need to learn from their experience; it is an essential input in any agricultural research. Peasants and scientists working together can advance agricultural production in Africa. I am sure that during the coming week all those who participate in your Conference will improve their capacity to play a constructive part in the work which has to be done.

You have my good wishes.

Chapter 2

The International, National and Sectoral Contexts of Falling Agricultural Productivity in Sub-Saharan Africa

D. G. R. BELSHAW

Dean of the School of Development Studies and Reader in Agricultural Economics, University of East Anglia, University Plain, Norwich, NR4 7TJ, UK.

The Nature of the African Agrarian Crisis

The purpose to this contribution is to identify and examine a wider set of perceptions of the causes of the present agrarian crisis in Africa, as it has been termed, than are usually discussed in the increasing number of meetings and publications which are devoted to it. First, though, what exactly *is* the nature of the African agrarian crisis? This itself is open to different interpretations. Some stress a decline in total or per capita food production over the long-term, after smoothing out the effects of short-run natural catastrophes and other variations. For example the United States Department of Agriculture, the World Bank and the Organisation of African Unity are unexpected bedfellows in sharing this view. Whether the situation is quite so clear cut as they believe, however, is contestable. One must agree that strongly rising food imports *are* observable in a number of African countries, and that the prices of food supplied to the urban populations have tended to rise at least as fast as the general rate of inflation. But these rising import figures are consistent with urban populations eating more food, and especially more preferred foods, as their total incomes have risen. In economic terms market demand may have risen faster than supply, rather than domestic supply contracting.

Further, the production data which suggest that average per capita food output has fallen significantly are themselves notoriously unreliable. There are two main reasons for this: a large proportion of food output is consumed on the farm or sold in the rural areas. In Africa these amounts are rarely recorded and measured accurately. Secondly, with increasing inflation a rising proportion of the marketed food surplus is sold outside normal channels at price levels considerably above those which are officially prescribed. By their very nature these parallel or hidden economy transactions are not reflected in the official statistics which, therefore, tend to understate both the production and the domestic marketed surplus of foodstuffs.

What can be more generally agreed, however, is that there have been marked declines in total production of African agricultural exports. A recent study (Ghai 1983), for example, showed that between 1969-71 and 1977-79:

> "there were absolute declines in the volume of exports of no less than 16 commodites, including such important crops as cocoa, coffee, maize, cotton, groundnuts, oil seed cake, bananas and rubber. Consequently the African share in exports from developing countries has fallen sharply for most commodities between 1961-63 and 1977-79"

The statistics for agricultural exports are more reliable, of course, than for agricultural production, although even here transborder smuggling will lead to under-recording. Independent evidence, however, suggests that prices for food sold for domestic consumption improved relative to the prices of export commodities. The numerous studies which have demonstrated the keen responsiveness of African farmers to price incentives also support the view that average food availablility for the domestic population is greater than the official statistics suggest. In summary, the agrarian crisis manifests itself clearly on the export commodity side, but less certainly and generally on the food output side (cf. Eicher 1982).

Returning to the task of identifying the main *causes* of the agrarian crisis, such a review should help us understand why several different strategies have been advocated to resolve the problem. It is true that many of the factors involved cannot be usefully expressed through a few simple statistics, both because of lack of data and because situations change with time and vary within and between countries. It is the intention of this conference, however, to study these causes, and identify their remedies, which are general to much or all of the continent. We are bound to employ, therefore, a largely qualitative and generalised analysis; even so, it should be possible to pick out the main issues which each country has to consider.

There are, I suggest, two broad kinds of approach to this task. The first concentrates on the *internal* difficulties facing the agricultural sector. These usually involve the amount and quality of natural resources and other production factors, labour, capital, and the available production technologies relative to the numbers of rural people, their aspirations and their basic needs. Problems of this type must be viewed as long-term and deep seated ones. The second kind of perspective or focus emphasizes the shocks and constraints imposed on the agricultural sector from outside by both international and domestic macro-economic developments and decisions. This approach sees the origins of the agrarian crisis as more recent and temporary and their effects as more readily reversible, given the political will to act.

In many African countries, but by no means all of them, it would appear that both sets of factors are working simultaneously and that their mutual interaction causes further twists so that the crisis is spiralling deeper. Two consequences for policy-making follow from this. First, the design of strategies which will effectively attack situations of this type must be agreed upon at the national and, often, the international level as well as within the agricultural sector itself. Secondly, such strategies must incorporate both short-term *and* long-term measures in the form of a carefully phased sequence of policies, projects and programmes which are effective, mutually reinforcing and capable of adjustment to rapid changes in external circumstances.

The Agricultural Resources and Technology Focus: Resulting Diagnoses and Prescriptions

Let us examine the first approach in more detail. Views of the cause of the agrarian crisis which focus on the scale and quality of the resources for agricultural production usually stress several interacting factors:

(1) High rates of rural population growth which put increasing pressure on land, water and vegetation resources. Indigenous technical changes and diversification of activity can absorb more population up to a point but past success in this area cannot be extrapolated into the future. Livingstone (1983) has used the analogy of the sponge: the absorptive limit of the agricultural sector – in terms of numbers of people held within it – will be reached suddenly and with severe consequences.

(2) The physical quality of the agricultural resource-base will deteriorate, if it has not begun to do so already, and open rural destitution will become apparent. The break-down of traditional shifting cultivation methods and the private enclosure of communal lands are also adduced as related factors.

(3) Adverse weather has an increasingly devastating impact. This is partly because with population pressure more people are exposed in high-risk enviroments, partly because they have fewer reserves of grain, cash, livestock, reciprocal relationships and so on to fall back on in adverse times, and also, it is argued, because natural resource degradation turns minor climatic events into major ones. Whether long-term climatic changes are also to be observed generates less argeement; professional meteorologists seem unable to substantiate statistically the evidence for such a view.

(4) Massive rural-to-urban migration flows remove the majority of adult males from the agricultural sector, so that critical labour and cash inputs are missing from the male side of the typically dualistic household/farm enterprise (Ghai 1983). This view, though, seems to underplay the possible importance of the flows remittance income from the urban to the rural sector, the entrepreneurial capacities of women and the possible beneficial effects on skills and attitudes of periods of employment in large-scale agriculture or urban enterprises.

In any case, we must note that these four kinds of experience are by no means universal in sub-Saharan Africa. If they were the primary cause of the agrarian crisis, we should expect to find that countries with under-utilised natural resources, like Zaïre, the Sudan and Zambia, are exempted from its presence. The comparative performance data in Table 1, however, do not lend support to this view. Nor does the presence in the upper segment of performance of resource scarce (high man:land ratio) countries like Rwanda and Burundi. Further, these processes are almost always gradual and operative over the long-term. They do not convincingly explain why such a large proportion of African countries experienced rapidly falling agricultural productivity over the same period, i.e. from the early 1970s up to the present time.

Diagnoses of this first type often emphasize the need for the intensification of agricultural production through the widespread adoption of modern technology. Those policy-makers concerned about the drain on foreign exchange of food imports or with doubts about the security of food surpluses for the urban population often favour the expansion of a large-scale modern farming sector. Alternatively, concern about the consequences for income distribution and welfare within the agricultural sector leads to the advocacy of appropriate "Green Revolution" technology based on bio-chemical innovations (e.g. Eicher 1982). This usually means high-input/high-output

Table 1   Average agricultural sector performance, countries in sub-Saharan Africa 1970-81.

Country	Value of agricultural output per capita change 1970-81 (%) (a)	Total value of agricultural output: change 1970-81 (%) (b)	Food nutrients per capita: change 1970-81 (%) (c)	Total food nutrients change 1970-81 (%) (d)	Rank order by change in food (%) (e)	Per capita dietary energy + nutritional satisfaction (1978-80) (f)	Value added per capita in agriculture in 1970 (1975 US $) (g)
(a) Countries with rising agricultural productivity							
Ivory Coast	+ 10	+ 71	+ 13	+ 76	1	114	164
Rwanda	+ 7	+ 48	+ 5	+ 45	4	95	74
Benin	+ 7	+ 47	+ 8	+ 49	3	100	55
Zimbabwe	+ 6	+ 34	+ 11	+ 61	2	80	70
Burundi	+ 1	+ 26	0	+ 24	7	92	68
Cameroon	+ 1	+ 28	+ 2	+ 30	6	106	107
(b) Countries with falling agricultural productivity							
Ethiopia	−2	+ 7	−16	+ 7	23	74	44
Upper Volta	−2	−2	−3	+ 27	8	85	40
Zambia	−5	+ 16	−4	+ 35	9	95	65
Kenya	−6	+ 43	−14	+ 31	20	89	62
Mali	−7	+ 23	−10	+ 20	13	85	48
Madagascar	−7	+ 23	−8	+ 22	11	107	101
Nigeria	−9	+ 28	−9	+ 29	12	99	160
Chad	−9	+ 14	−5	+ 18	10	76	67
Guinea	−10	+ 18	−10	+ 18	13	84	112
Sudan	−12	+ 18	+ 4	+ 39	5	101	101
Togo	−12	+ 19	−11	+ 19	16	92	71
Tanzania	−14	+ 20	−11	+ 24	16	87	63
Zaïre	−14	+ 29	−13	+ 17	19	96	18
Niger	−14	+ 17	−14	+ 18	20	92	109
Liberia	−15	+ 24	−12	+ 28	18	98	101
Ghana	−15	0	−28	0	27	88	264
Congo	−16	+ 10	−17	+ 9	24	99	77
Lesotho	−16	+ 9	−10	+ 16	13	107	32
Mauritania	−22	+ 6	−22	+ 6	25	89	93
Sierra Leone	−23	+ 2	−23	+ 3	26	92	71
Uganda	−30	−2	−15	+ 18	22	80	196
Somalia	−39	+ 7	−39	+ 7	28	92	128

Sources: (a) FAO (1982) *Production Year-book*, 81; (b) *Ibid.*, 77; (c) *Ibid.*, 79; (d) *Ibid.*, 75; (e) Ranked according to change in food availability per capita, 1970-1981. (f) FAO (1981) *The State of Food and Agriculture*, table 16. (g)(i) FAO (1982) *Production Year-book* and (ii) World Bank (1983) *World Development Report*, Washington DC.

Notes: (1) Data were only partially available for the following countries: Angola, Botswana, Central African Republic, Gabon, The Gambia, Guineau Bissau, Malawi, Mozambique, Senegal and Swaziland. (2) No data were available for Djibouti or Rio Uni. (3) Mauritius, Réunion and the Seychelles have been excluded.

seed and fertilizer packages; although they are highly divisible they are often inapproriate to the financial positions of the mass of small-scale African producers, as Lipton (1983) has argued recently. The use of composite rather than hybrid seed, for example, and greater emphasis on inter-cropping with legumes, green-manuring, composting, etc., rather than on the exclusive application of purchased artificial fertilizers, may represent important policy options which will affect the total numbers of farms adopting new technologies and thus the impact on average productivity. Obviously, this is less likely to improve significantly if the majority of rural producers are excluded from the process.

A common alternative progression of analysis is to view the long-term solutions to such deep-seated processes within the agricultural sector as lying outside it, in the form of the structural transformation of the economy (i.e. the relative and absolute growth of industry, if not absolute decline of agriculture) accompanied by the transfer of an increasing proportion of population from the countryside to the towns. The emphasis in the Organization for African Unity's Lagos Plan of Action, for example, has been summarized in the following terms (Lipton 1983):

> "The OAU's 'Lagos Plan of Action' .... is orientated to the replacement of food imports, but despite much stress upon agriculture's 'priority' seeks aid above all for new publicly-owned transport links and industrial producers; such emphasis might help farmers indirectly, but both are geared primarily towards greater political independence, and towards securing Africa's share in 'targets' set by UNCTAD for growing Third World proportions of industrial production".

As a short-term remedy, the OAU strategy faces the problem that unless industrial export-led growth is possible (which is unlikely in the short-term) growth in industrial output and employment is limited by the size of the domestic market. This is dominated by the low (and possibly declining) real incomes of the rural majority. Lipton himself favours an emphasis on an 'Asian-style' labour intensive agricultural technology, but this too is difficult to envisage as an *immediately* applicable strategy. In parentheses one could say, however, that the agrarian successes achieved in India do suggest that they may have considerable merit for incorporation in longer-term 'balanced growth with equity' strategies in Africa.

In summary, this analytical approach does focus on important problems and their solutions. But the question remains whether success can be achieved without the presence of certain pre-conditions in the external environment of the agricultural sector.

International and Domestic Macro-economic Variables: Diagnoses and Prescriptions

An alternative school of thought has seen the current problems of most African agricultural sectors as having two causes: the first is the combination of a series of shocks administered by the international economic system since the first half of the 1970s. The second is the form in which these changes have been mediated to the agricultural sector through the macro-economic policies of domestic Governments. On the *international* side, several factors are usually identified as combining together (see e.g. Khan & Knight 1983):

(1) The impact of the dramatic increases in the price of non-renewable energy products in 1973/74 and again in 1979/80. These acted both *directly* upon agriculture through the rising costs of petro-chemical imports and, especially, of rural transport and also through subsequent *indirect* effects in the form of increases in the prices of domestic industrial products and the consequences of deterioration in the balance of payments positions (see further below).

(2) The "export of inflation" from industrial market economies which were able to some degree to pass on their higher energy costs in the form of higher prices for their industrial goods exports. This process further exacerbated the balance of payments positions of non-oil developing countries. Where protected domestic industries were able to pass on to the consumer the price increases of intermediate and capital goods imports agriculture's terms of trade tended to worsen further (because in Africa the majority of consumers are agricultural producers).

(3) The further impact of the world economic recession from 1979-83, which sharply depressed the prices of most primary commodities in real terms.

(4) The added pressure on the balance of payments, via the debt service ratio, of high levels of interest rates in the international capital markets which have prevailed in recent years.

To these factors must be added the further effects on African agriculture of intensified agricultural protectionism in Western Europe. This operates most obviously to eliminate markets in Europe for actual or potential African exports of specific commodities, such as sugar and meat. More importantly, in terms of political possibilites for reform, subsidised exports of European agricultural products also reduce the prices which African exports can obtain in Third World country markets, such as the Gulf States and the Middle East generally.

To these international effects must be added, in many African countries, the adverse consequences for agriculture of the domestic response in the shape of the chosen macro-economic policies. One or more of the following responses are found:

(1) Inflationary demand management, that is running deficit budgets to maintain the levels of public recurrent services, with the result that the internal rate of inflation accelerates further. This also serves to put additional strains on confidence in the domestic currency, especially when:

(2) Rigid exchange rate policies are followed. A fixed nominal rate in a situation of a decling real rate of exchange protects the purchasers of imported goods, especially those whose money incomes can maintain parity with the rate of inflation. But it penalises the producers of export goods. This is because they receive payment in a depreciating domestic currency but face rising prices for their purchases from non-export producers, e.g. domestic manufacturers and foodstuffs. In the typical African situation, industrial firms and urban consumers of imported goods gain while agricultural producers of exports lose.

(3) Extensive price controls and other forms of market intervention make the problem worse, since parallel economy and smuggling opportunities are increased. Consequently, the tax collected by Government diminishes still further, as does government control of foreign exchange, when primary exports increasingly "leak" across international frontiers.

The consequences of these policies are sharp declines in (a) the prices received by agriculture for its products compared with its purchases* (i.e. the domestic net barter terms of trade facing the agricultural sector), and (b) a relative price swing in favour of foodstuffs for domestic sale rather than exports (except where these can be remuneratively smuggled). Any short-term progress, according to this perspective, requires a reversal of these processes. This requires that a series of essentially painful and possibly politically risky measures are implemented. Such measures would usually include the progressive devaluation or unpegging of the domestic currency and the restructuring of internal prices, or their freeing altogether, so that agriculture's terms of trade are improved. Another important component, often, is the progressive elimination of deficit financing by reducing the rates of growth of public recurrent expenditures.

Conclusion: Towards Effective Agricultural Development Strategies

It seems unfortunately to be the case that the short-term strategy measures advocated by the second school of thought will become more necessary and more painful the longer any kind of adjustment is deferred. On the other hand, an emphasis on purely macro-economic readjustment runs the risk that agricultural supply-side constraints are not attacked. In consequence the strategy may fail, the pain notwithstanding.

One major issue concerns the maintenance of the *rural* transport and marketing network (as distinct from the international and inter-city networks) and its operation at reasonable cost. Another relates to the delivery and cost of purchased farm inputs; some animals, more so than

---

*These purchases are dominated by the consumer goods bought by rural producers and their families. Farm production inputs are much less important, reflecting the dominance of land and family labour resources in peasant agriculture.

crops are particularly susceptible to the absence of these. Also, the capacity of publicly-run agricultural service institutions may be cut back, with disastrous effects, as part of the overall economy campaign unless the micro-level effects on agricultural supply are taken fully into account.

Table 2   Domestic terms of trade of export crops for selected African countries 1971-79. Source, World Bank (1981)

	1971	1972	1973	1974	1975	1976	1977	1978	1979
			(1970 = 100 unless otherwise specified)						
Cameroon									
Barter terms of trade	98.2	90.5	85.2	81.2	74.5	73.4	77.5	88.2	--
Income terms of trade	89.8	96.9	84.0	81.8	81.1	66.4	67.2	91.7	--
(cocoa, coffee, cotton)									
Ghana									
Barter terms of trade	91.2	82.8	88.0	89.4	86.0	58.8	34.0	35.8	46.3
Income terms of trade (cocoa)	85.9	92.4	88.4	75.2	78.0	56.1	26.0	22.6	27.7
Ivory Coast									
Barter terms of trade	111.2	109.3	97.8	99.7	126.5	111.8	97.3	119.3	101.6
Income terms of trade	113.4	134.1	111.6	134.3	171.5	178.0	144.3	170.6	131.2
(cocoa, coffee, cotton, palmoli)									
Kenya									
Barter terms of trade	98.3	93.1	79.3	83.9	123.4	93.6	55.0	49.5	58.5
Income terms of trade	129.2	157.0	177.9	198.5	170.1	280.0	449.6	263.5	218.0
(coffee, tea, pyrethrum,									
cotton, maize, wheat, sisal)									
Malawi									
Barter terms of trade	105.9	129.1	123.0	100.1	94.3	115.4	119.0	116.7	--
Income terms of trade	108.3	122.2	139.4	104.5	104.6	98.6	120.5	124.6	--
(tobacco, groundnuts, cotton,									
maize)									
Mali									
Barter terms of trade	81.2	80.6	69.0	61.2	83.1	97.4	65.3	60.0	50.0
Income terms of trade	99.2	98.8	76.8	55.7	94.1	135.8	123.5	90.1	76.4
(cotton, groundnuts)									
Nigeria									
Barter terms of trade	--	--	96.5	119.2	125.2	95.9	119.1	93.0	109.0
Income terms of trade	--	--	75.0	102.5	104.6	79.4	80.0	43.6	58.1
(cocoa, cotton, palm kernels)									
Sengel									
Barter terms for trade	97.3	111.9	104.2	114.1	120.6	115.1	103.4	101.5	91.3
Income terms for trade	77.3	146.1	81.4	84.5	141.1	204.2	148.7	72.2	104.7
(groundnuts, cotton)									
Tanzania									
Barter terms of trade	96.9	95.3	88.5	76.0	63.5	90.7	110.0	79.5	67.4
Income terms of trade	101.8	102.8	96.3	78.8	68.8	84.5	102.5	73.7	62.7
(coffee, tobacco, cashews, cotton)									
Togo									
Barter terms of trade	98.5	92.8	88.5	80.0	79.2	76.2	68.4	80.6	90.1
Income terms of trade	110.5	92.8	64.1	57.2	57.0	62.0	49.4	48.8	57.8
(cocoa, coffee, cotton)									
Upper Volta									
Barter terms of trade	99.0	101.7	102.8	108.0	91.9	101.7	105.5	101.7	92.6
Income terms of trade	106.9	130.7	110.0	140.2	181.6	214.0	156.0	229.5	245.8
(cotton, sesame)									
Zambia (1971 = 100)									
Barter terms of trade	100.0	82.1	113.6	104.8	91.8	84.7	104.6	98.1	127.2
Income terms of trade	100.0	142.6	97.9	126.1	125.6	169.6	133.1	92.9	90.4
(maize, groundnuts, tobacco)									

Other important short-run strategy components on the supply side of the agricultural sector should probably include:

(1) Improvements in agricultural marketing efficiency by removing price controls, cutting out statutorily protected monopoly control by parastatals or co-operatives and improving the quality of market information.

(2) Assisting farmers to switch resources into relatively profitable enterprises and vice-versa, out of less profitable ones, through orienting production credit, subsidies, processing capacity, etc. accordingly. Table 2 indicates the way in which productivity gains from a combination of improved technology and shifting resources into higher value activities can be, and have been, secured (e.g. Malawi, Kenya, Ivory Coast, Upper Volta), more than offsetting the effects of declining international net barter terms of trade (relative price changes).

(3) Assuming that the macro-economic preconditions for successful agricultural development are restored, then decentralised planning to broaden the base of participation in productivity-raising technology, production infrastructure, etc. should be resumed. Lagging regions, excluded minorities, women and other groups can be targeted to achieve efficiency *and* equity objectives at the same time. This would be in line with the precepts of the latest versions of rural development planning which had been evolving before the macro-economic preconditions for their success were so often destroyed around the end of the 1970s.

(4) Finally, resources should be transferred into under-utilised high pay-off uses, such as small-scale irrigation and multiple land-use schemes in forested areas, and taken out of low return activities such as management and capital intensive parasatal production projects.

A related broad issue, concerns the extent to which the poor countries of Africa can successfully readjust on their own. Clearly a favourable international economic climate will reduce the costs of adjustment and make its implementation more likely. The heralded climb of western industrial market economies out of recession will give them (as aid donors and trade partners) and African Governments the opportunity to co-operate to redress the balance in favour of the neglected agricultural sectors and rural populations.

Short-term measures of the type discussed, introduced in a period of expanding international trade, should then secure the preconditions for the successful investment of resource in the *medium and long-term* activities needed by African agriculture. These focus on devising more productive small-scale agricultural technology and further deepening the rural production infrastructure activities which are the main concern of the first school of thought and of this conference. The preceeding discussion, however, may have reinforced the view that a successful resolution of the African agrarian crisis is likely to require not only successful measures within the agricultural sector but political will and managerial competence at the national and international levels, both within Africa and outside it.

References

Eicher, C. K. (1982) Facing up to Africa's food crisis. *Foreign Affairs* 61, 151-174.

FAO (1982) *Production Yearbook, 1981.* Rome; FAO.

FAO (1982) *The State of Food and Agriculture, 1981.* Rome; FAO.

Ghai, D. P. (1983) Stagnation and inequality in African agriculture. In *Growth and Equity in Agricultural Development* (A. Maunder; K. Ohkawa, eds) 63-78, Aldershot; Gower.

Henn, J. K. (1983) Feeding the cities and feeding the peasants: what role for Africa's women farmers? *World Development* II, 1043-1055.

Khan, M. S.; M. Knight (1983) Sources of payments problems in LDCs. *Finance and Development* 20, 4.

Lipton, M. (1983) African agricultural development: the EEC's new role. *Institute of Development Studies Bulletin* 14, 21-23.

Livingstone, I. (1983) *The "Sponge Effect": Population, Employment and Incomes in Kenya.* [Development Studies Discussion Paper 152.] Norwich; University of East Anglia.

World Bank (1981) *Accelerated Development in Sub-Saharan Africa.* Washington, DC; World Bank.

World Bank (1983) *World Development Report, 1983.* Washington DC; World Bank.

Chapter 3

The Structure and Performance of the Agricultural Sector in some African Countries

D. G. R. BELSHAW

Dean of the School of Development Studies and Reader in Agricultural Economics, University of East Anglia, University Plain, Norwich NR4 7TJ, UK.

Introduction

On the first day of the Conference, a number of country delegations presented statements which described the present structure and recent performance of the agricultural sectors of their respective countries. For each of those African countries which produced a document, a summary of the main points is provided in the following sections. These are presented in the sequence Tanzania, Kenya, Ethiopia, Malawi, Zimbabwe, Botswana and Mauritius.

Written country statements were also produced by the following non-African Commonwealth countries: Bangladesh, Canada and Cyprus. Verbal statements were also made at the Conference by representatives of the Governments of Lesotho and Zambia. All country documents prepared for the Conference are listed in the Appendix to this paper.

The State of Agriculture in Tanzania[1]

The Structure of the Agricultural Sector

In Tanzania, as in many developing countries, agriculture is the mainstay of the economy. It is the major source of employment and hence national income. The agricultural sector supports about 90% of the population and accounts for over 40% of the GDP and 75% of the foreign exchange earnings. Furthermore, it is the major source of food for the people and raw materials for the expanding industrial sector as well as a vital market for domestic industrial goods.

Smallholder village farming employing very little capital is the predominant form of agriculture practised in Tanzania. With the exception of a few commodities, the smallholders account for the bulk of agricultural output. In view of this the Government puts a very high priority on the development of small holder farming.

Performance of the Agricultural Sector

In recent years the performance of the agricultural sector has not been commensurate with its role in the national economy. While growth in agricultural production was satisfactory in the 1960s, it has declined substantially in the 1970s. Production has continued to decline for both food and export crops resulting in increased food imports and diminished foreign exchange earning capacity.

The observed poor performance in agricultural production is due to many inter-related factors both external and internal. With regard to the uncontrollable exogenous factors, agricultural production has been seriously hampered by the escalation in international inflation in the 1970s caused by the rapid increase in oil prices, coupled with the continued decline in terms of trade. Bad weather, particularly droughts and floods, has also reduced significantly the output of the

[1]This section is based on the summary of the Tanzanian document (Appendix, reference 1) which was verbally presented to the Conference by the Principal Secretary of the Ministry of Agriculture, Government of Tanzania.

predominantly rainfed agriculture. In the case of internal factors, agriculture has not received the priority it deserves in terms of resource deployment.

While the importance of the agricultural sector was recognised since the early 1960s, available empirical evidence indicates that it has been stifled of badly needed manpower and finance. In recent years, this problem was aggravated by the acute shortage of foreign exchange which seriously affected the availability of inputs for agriculture, as well as raw materials and equipment for the supporting sectors, particularly transport and industries. Other factors include policy deficiencies particularly in research, extension and prices; inadequate infrastructure especially feeder roads; and the low level of technology applied in agriculture.

While these factors are equally applicable to the livestock sector, the problems of land use management practices, animal nutrition, diseases and breeding are of particular importance. Despite the overall understocking in many areas, there is severe overgrazing in the semi-arid regions where mixed farming is undertaken, resulting in low productivity and land degradation. The skewed distribution of water, inadequate provision of basic socio-economic infrastructure, and pursuance of uncoordinated and conflicting objectives on land use cause further problems.

Fish catches have also been on the decline, mainly because of inadequate fishing gear and other equipment. This is again due to shortage of foreign exchange.

In the forestry sector, the major problem is the serious threat of deforestation resulting from unbalanced utilization of forest resources, particularly the increased demand for fuelwood and expansion of agricultural land. This has caused serious problems in soil erosion and siltation.

The Government has recognised since the early 1960s the role of game resources in ecological systems and the need to protect them against depletion. To date about 30% of the total mainland area has been set aside for wildlife.

Prospects and Strategies for the Future

So far a rather gloomy picture has emerged concerning the status of agriculture in Tanzania. Therefore at this juncture some mention of the bright side is warranted.

Tanzania has a big potential in agriculture as reflected in its favourable man/arable land ratios, diverse agro-climates, substantial untapped and under-utilised resources in livestock production and important fishing resources in its lakes and in the ocean.

It is estimated that only about 6.2 M ha (15.7%) of the 39.4 M ha of potential rainfed agriculture land in Mainland Tanzania is currently under cultivation. This has led to pronounced extensive land use, a tendency that is now diminishing due to emphasis on permanent settlement in village communities. Due to the constraints already mentioned, about 60% of the livestock is kept on only about 10% of the land which is also used for crop production.

In terms of fisheries development Tanzania has about 50 000 km² of lake area, and 800 km of coastline. In addition, there is a country-wide network of rivers, ponds and swamps that are exploitable for both subsistence and internal commercial fishing purposes.

With regard to forestry, about 50% of the land area of mainland Tanzania is earmarked for forest and to date about 13 M ha have been gazetted as forest reserve. Plantation forests cover some 69 000 ha. Also under the village afforestation programme which aims at establishing fuel wood lots, about 494 000 ha have been planted.

Taking into consideration this big potential in agricultural production and unsatisfactory performance of the sector in recent years, the Government has taken a number of measures aimed at raising production. Since 1982 it has embarked on a three year Structural Adjustment Programme which aims *inter alia* at promoting agricultural production through instituting appropriate pricing policies and provision of inputs and incentive goods to the farmers. The sectors that support agriculture will also be rehabilitated and consolidated. Also in the same year the Government has adopted the National Agricultural and Livestock Policies that provide for short- and long-term corrective measures that are needed to revive production.

Tanzania also has a National Food Strategy Programme which aims at attaining self-sufficiency in food and expanding export earnings. The programme incorporates 24 crops (17 food and 7 cash), livestock and poultry products (meat, egg and milk) and fish. It is estimated to cost about Shs 70 000 M phased out into Shs 14 000 M in the short-term (1982–85), Shs 17 000 M in the medium-term (1986–1990), and Shs 39 000 M in the long-term (1991–2000).

In the short-term, efforts to improve output will focus on rehabilitation and consolidation of the physical infrastructure and existing production capacity, improvement of capacity utilisation, streamlining the institutional setup with a view of raising operating efficiency, introduction of a package of producer incentives, and strengthening the extension services.

In the medium-term, attention will be given to maximizing capacity utilisation, evolving a price structure in favour of comparative advantage in production, intensifying research, particularly in the development of input packages suitable for different agro-ecological zones, and establishment of village-level irrigation schemes. Emphasis will also be placed on augmenting farm power particularly through the more diversified use of oxen and improvement of the operations of the rural credit system.

In the long-term, concentration will be mainly on expansion of the productive capacity together with protection of the environment against irreversible damage. Some of the major environmental concerns in Tanzania include loss of soil fertility due to over-use, inappropriate cropping rotations and fertilizer use, waste of water resources, excessive use of energy, deforestation, desertification and depletion of game resources.

Programmes for livestock production will aim essentially at raising animal productivity in the long-term through encouraging proper stocking levels, range management and pasture improvement.

In the fishing industry, immediate attention will be given to the rehabilitation of fishing gear and improvement of handling and processing facilities. Out of the projected total increase in agricultural output under the National Food Strategy Programme, area expansion will account for about 53 % and the yield increases for 47 %. The contribution of yield increases to output growth is higher for cereals (56 %) than either root starches (37 %) or cash crops (50 %). In the case of maize, rice, beans, and sisal, the contribution of yield increase to total output growth will be more than 60 %.

The scope for raising agricultural production through the application of appropriate scientific and technical knowledge is enormous. This fact under-scores, then, the importance of this Conference. The contribution of science and technology to agricultural output in other regions of the world is well known.

What is now needed is a case-by-case development and application of improved technical packages suitable for different socio-economic and political environments. This Conference draws together participants with a wide experience in the problems of agricultural and rural development.

In this regard the Conference offers an opportune time for the exchange of ideas and experiences. It should deliberate on the constraints on research programmes in agriculture. It is hoped, therefore, that the Conference will produce concrete resolutions aimed at charting out a pragmatic recovery programme for agricultural production.

Finally, the contribution of national Governments in ameliorating the problems of agriculture and rural development cannot be over-emphasized. There must be a political willingness and commitment to respond positively to the needs of the peasants who constitute the majority of the rural poor. At the same time, Governments have to pursue policies that support and reinforce the contribution of science and technology in expanding agricultural production and improving the welfare of that same group – the rural poor.

The State of Agriculture in Kenya[1]

The Structure of the Agricultural Sector

The Republic of Kenya covers an area of 582 644 km² or approximately 58 264 400 ha. Out of this total area, 6 785 000 h (11.6 %) is classified as high potential land, which receives an annual rainfall of 857.5 mm and above. Another 3 157 000 ha (5.4 %) is classified as medium potential area and receives an annual rainfall of between 735 mm to 857.5 mm y^{-1}.

Approximately 72 % of the total land area (an estimated 42 105 000 ha) is classified as low potential land. The rainfall within this region is below 612 mm y^{-1} and it is also usually very

[1]This section is drawn from material presented in the document submitted by Kenya (Appendix, reference 2).

erratic. Another 4 867 000 ha (8 %) is unclassified land, usually moorlands, while the balance of about 3 % is under water. Most of the agricultural activities are concentrated in the high potential and medium potential areas which occupy about 20 % of the total country area. However, due to population increase in these regions, agriculture is spilling over into low potential areas where possibilities of obtaining a crop in any given season carry a high risk.

A dualistic farming pattern evolved out of the land tenure system. The small farm sector comprises some 1.5 M small farms and about 250 000 pastoral holdings. In addition, the small farm sector includes settlement schemes (400 00 ha) with about 35 000 homesteads, and areas of legal settlement in large farm sub-division or in forest land or any other publicly owned land. In total the small farm sector comprises over 1.7 M holdings. Most farms have 2 ha or less; few are more than five ha in size (see Table 1).

Table 1   Farm Size Distribution in the Small Farm Sector of Kenya 1974/75

Farm size Group (ha)	Less than 0.5	0.5 to 0.9	1.0 to 1.9	2.0 to 2.9	3.0 to 3.9	4.0 to 4.9	5.0 to 5.9	8.0 to more
Frequency %	14	18	27	15	9	7	7	4

Source: Statistical Abstract, 1978.

The large farm sector which evolved on alienated Government land is made up over 3000 holdings, of which between 1540 and 1800 are mixed farms. The remainder are ranches, plantations and Government farms.

Since Independence in 1963 there has been a dramatic change in land ownership, resulting in the sub-division of many large farms into smaller units which are now managed as small-scale farms.

Table 2   Farm Size Distribution in the Large Farm Sector of Kenya, 1976

Farm size Group (ha)	Less than 100	100 to 199	200 to 299	300 to 399	400 to 499	500 to 599	1000 to 1999	2000 to 3999	4000 to 19 999	20 000 and more	Total
No. of holdings	1133	384	345	258	219	492	211	111	107	13	3273
Frequency %	35	12	11	8	7	15	6	3	3	—	

Source: Statistical Abstract, 1978.

The farms described above have adopted cropping patterns and farming techniques which suit the natural, economic and socio-political conditions for each location, and as a result of this adoption, more or less distinct farming systems have evolved. In Kenya, six distinct farming systems can be identified as follows: shifting cultivation, fallow systems, ley and dairy systems, arable irrigation farming, perennial crops systems and grazing systems.

The Performance of the Agricultural Sector

The role of agriculture in the Kenyan economy can be assessed in two main ways, i.e. its contributions to national product and to employment. Firstly, agriculture plays a big role in helping the country to sustain the other sectors of the economy. This is clearly shown by the fact that most of its processing industries rely on raw materials from the agricultural sector. These include meat and dairy products, beverages and tobacco, canned vegetables, fish, oils and fats, bakery products, furniture and leather products. Besides, the top foreign exchange earners are mainly agricultural products such as coffee, tea, pyrethrum, meat and meat products, hides and skins etc. From 1975 the contribution of agriculture to total GNP (at factor cost) has constantly been more than 30%, as compared to a contribution of about 11 to 13% from the manufacturing sector and about 14% from public services.

Secondly, the agricultural sector is a major source of employment for the majority of the people. Roughly 80% of Kenya's population lives in rural areas, and the bulk of this population — approximately 14 million people — is in one way or another involved in agriculture. Though concrete figures are difficult to obtain, it is estimated that about 8 million people are employed directly or indirectly in the agricultural sector, with many more employed indirectly in other agricultural supportive institutions.

Kenya aims at self-sufficiency in food production and has done fairly well in maintaining this position except for 1980 and 1981 when the country was forced to import large quantities of maize to fill the gap caused by crop failures as a result of drought in 1979 and 1980. Otherwise imports are usually limited to luxury items such as refined sugar, temperate fruits, chocolate products etc. However, even imports of luxury food items are now being reduced drastically due to shortages of foreign exchange. Exports of food items have so far been negligible except for horticultural crops which are being developed for the export market (see below).

The major export crops are coffee, tea, pyrethrum and horticultural products. The value of agricultural products exported from 1972 to 1980 averages about 55% of the value of all commodities exported during the same period.

Coffee is the major export earner, with exports lying in the 75 000–100 000 t y^{-1} range since 1975 compared with 50 000–55 000 t in the early 1970s. Currently the smallholder sector accounts for about 65% of the total coffee produced as well as about 75% of total area under coffee.

The total area planted to tea in the smallholder sector has risen from around 30 000 ha in 1972 to over 51 000 ha in 1980, with an additional 26 000 ha in the large-scale estates.

Pyrethrum is mainly cultivated by small scale farmers who account for about 62% of its production. With the increase in production from 327 t in 1935 to 17 560 t in 1981, Kenya has become the major world producer of this crop, accounting for roughly 80% of total world output.

The major horticultural exports are pineapples, both fresh and tinned, cut flowers, mangoes and french beans with many other products they are exported to high-price European markets. Export values exceeded K£ one million for the first time in 1973 and by 1981 they had exceeded K£12 million.

Apart from the leading export activities, several minor ones have made significant advances in recent years. Sisal production has picked up again since the 1970s and in 1982 Kenya produced 50 028 t as compared to average production of about 35 000 t in the period 1975–80. Local industries consume about 20% while the remaining 80% is exported.

Kenya was a net importer of sugar until 1979 when she attained self-sufficiency and produced exportable surplus of some 95 000 t in 1980. In 1980 the country also produced 132 800 t of molasses, of which about 56% was exported and the rest consumed locally for animal feeds and alcohol production for both local and export markets.

Turning to the staple food crops, maize is the most important food crop in Kenya; it forms the staple diet of the majority of Kenyans. Maize production in Kenya has fluctuated considerably, with some years producing exportable surpluses, whereas others are marked by shortages, e.g. 1965, 1979 and 1980 when importations of maize were necessary. In 1983 the domestic requirement for maize was estimated at 2 777 000 t, increasing to 3 154 000 t by 1989. Thus production must increase at a rate of 9% per annum between 1980 and 1989 if the country is to be self-sufficient in maize.

It is apparent from the above that since the shortages in 1979/80, production has caught up and has increased even beyond the high levels achieved in 1977. Much progress has been achieved over the years in maize research. Hybrids suitable for cultivation in the various agroecological zones have been developed.

On the other hand, with wheat the most important goal is to achieve the self-sufficiency which has eluded the country since 1976. The area under wheat has slowly decreased. During the late 1960s over 160 000 ha were planted with wheat, compared to some 138 000 ha in 1977 and 97 000 ha in 1981. This has meant that the country has had to import considerable amounts of wheat every year.

In the livestock sector, smallholders produced 75 % of the total milk output, 65 % of total beef production, and substantial amounts of the eggs, poultry meat, mutton and goat meat produced. About 50 % of these are marketed to the urban areas and the rest is home consumed. During the last two decades, smallholder livestock farming has developed rapidly from a basically traditional, subsistence-oriented activities into modern commercial farming enterprises. Visible signs of this development are the successful introduction of high-yielding grade cattle, numbering about 1.5 M, which are virtually replacing the low-yielding Zebu breeds through A.I. services. The adoption of high-yielding arable fodder crops like Napier/Bana grass to replace natural pasture is also significant. In addition, commercialised poultry varieties have been successfully introduced.

Development of the forestry resource is needed to ensure a sustained and growing supply of timber for these traditional domestic needs and to satisfy the expanding industrial demand for forest products. But there are three severe constraints that affect forestry development namely: (1) serious depletion of forest and tree resources through uncontrolled felling of trees, especially for charcoal burning, in particular on private land and in local authority forests; (2) the encroachment of agricultural settlements onto forest land; and (3) poor management of the sawmilling industry that has led to inefficient and wasteful exploitation of scarce forestry resources.

Kenya possesses over 10 000 km² of lakes of mostly fresh water, 3200 km of rivers plus numerous small highland streams. There are also several fishponds and 640 km of marine coast on the western Indian Ocean. These resources and the fish stock they support, which has a potential yield of about 150 000 t, as yet make a very small contribution to GDP. They also still have only a small impact on employment and nutrition.

Prospects and Strategies for the Future

Some of the major constraints faced by agriculture are land tenure, unavailability of labour for the agricultural sector, lack of agricultural machinery, lack of agricultural credit, lack of markets for some agricultural produce and fluctuating world commodity prices. The Government is aware of these problems and has several programmes and projects aimed at solving them.

Land tenure problems are being tackled and solved successfully through land adjudication in trust land areas. As regards credit for agricultural production the Government has established bodies such as the Agricultural Finance Corporation and the Co-operative Bank of Kenya which provide seasonal loans for agricultural production. Farmers can also raise seasonal credit from commercial banks and other farmers, organizations such as the Kenya Farmers Association.

The marketing of agricultural produce is done by farmers directly within the country or through parastatal organizations such as National Cereals and Produce Board, Kenya Tea Development Authority, Kenya Meat Commission, Coffee Board of Kenya etc. which provide for both domestic and export markets.

The Government also supports agricultural research, provides agricultural extension services and regularly reviews agricultural prices. It trains both medium-level and high level staff for its agricultural industries who provide their services either through the public or private sector. To ensure that these requirements for agriculture and other sectors of the economy are met the Government works on the basis of development plans which normally cover a five year period. The Government has also decided to base development efforts at District level which is expected to go a long way towards boosting agricultural development.

The trends in the value of cash export crops have been described in the previous section. It is clear that agriculture is making a strong contribution both to national economic growth and to the incomes of many small-scale and poor farmers.

In order to achieve self sufficiency in food, the Government in 1981 developed a National Policy which has the following overall objectives:—

(1)   to maintain a position of broad self-sufficiency in the main foodstuffs in order to enable the nation to be fed without using scarce foreign exchange on food imports;
(2)   to achieve a calculated degree of security of food supply for each area of the country;
(3)   to ensure that these foodstuffs are distributed in such a manner that every member of the population has a nutritionally adequate diet.

This policy is being implemented actively and is producing good results.

The major objective of forest development in Kenya is to promote the establishment and maintenance of vegetative cover on both private and public lands. Such cover is not limited to forest trees but also includes other productive multipurpose tree or shrub species. The promotion of this objective is not restricted to high potential areas but covers both low potential and more arid parts of the country where avoidance of desertification is of the utmost priority.

The following steps are being take to achieve these objectives:

(1)   Review of currently designated catchment forests in terms of their management and extent. From this review will arise a comprehensive programme of action which will direct the Forest Department's role in the nation's resource conservation programme;
(2)   A special campaign has already been mounted to popularise among the rural population tree planting and nurturing and the establishment and correct management of farm wood lots. For this campaign to succeed, correct varieties of seedlings are provided free of charge or at nominal prices by the Forest Department in every administrative division throughout the country;
(3)   In addition to a forest research programme that has emphasized the needs of plantation forests in high potential areas in the past, the special need to establish and maintain tree and shrub species for arid and semi-arid zones of the country is also being looked into;
(4)   Management of small-scale sawmills is to be improved through programmes involving extension, training and credit to ensure that they adopt recommended practices and continue to provide employment opportunities in rural areas.

The main objective of the fisheries sector is to promote the maximum exploitation, on a suitable basis, of fishery resources in order to generate additional employment opportunities and income and increase the availability of animal protein. The main steps being taken are:

(1)   The development of mechanized trawling fleets on Lake Victoria and at the coast to harvest hitherto unexploited fish resources;
(2)   The improvement of traditional fishing methods by provision of motorised boats, improved fishing gear and improved fishing practices;
(3)   The improvement and provision of on-shore facilities such as landing beaches or stations, ice plants and markets;
(4)   The promotion of fish farming in inland areas.

Kenya's main objective of tourism and wildlife development is to maximize net returns, subject to important social, cultural and environmental constraints. Thus while every effort is being made to increase tourism and the wildlife's sector's contribution to foreign exchange earnings, incomes, government revenue and employment, careful planning of development must ensure that adverse effects accompanying growth in tourism are kept to a minimum.

The State of Agriculture in Ethiopia[1]

The Structure of the Agricultural Sector

Ethiopia's principal natural resource is the rich endowment of agricultural land. Out of the total area of 122.3 M ha, agricultural land is estimated to comprise about 81 967 700 ha or

[1]This section is based on information contained in the paper submitted to the Conference by the Government of Ethiopia (Appendix reference 3).

67% of the total land area; this consists of about 66 M ha or 54.1% of cropped and fallow land. The area of Ethiopia suitable for rainfed cultivation is estimated to be about one quarter of the total or a little more than 30 M ha. At present, only about 6 M ha are used i.e. only about 20% of the cultivable land. It is estimated that the potential irrigable area is about 2.25 M ha of which only 89 000 ha or about 4% is currently being irrigated. Thus the large area of arable land not yet fully exploited combined with suitable climate, fertile soil and water resources from many rivers indicates a good potential for agricultural development.

About 15.8 M ha (or 13% of the total land area) is or has been under crops. Out of the total cropped land some 94% of it is fragmented and under private holdings. Surveys by the Ministry of Agriculture indicate that about two-thirds of the farmers own holdings are less than one hectare of cropland. Smallholder production in the main cropping areas, which are in the highlands, is characterised by rainfed cultivation on small, scattered, irregular plots, making extensive use of land with low cultivation standards and yield levels, using little or no fertilizer and suffering high field and storage crop losses. Thus, agricultural productivity is generally low. The farmers have traditional skills but lack adequate supporting factors which promote optimum production such as good marketing systems, access roads, credit, extension and veternary services, improved seeds, fertilizer and other farm inputs. It is estimated that only about 25% of the total agricultural production is marketed. The absence of secondary and tertiary roads raises transport costs to such high levels as to discourage expanded production for the market.

Erosion, however, is an acute problem. The heavy rain which falls on the bare, steeply-sloping land easily loosens the top soil and washes it away to be carried as the silt load of muddy Ethiopian rivers at one billion tonnes per year.

The total area under the cultivation of food crops in 1981/82 was 6.2 M ha. Of this, the greater part amounting to 4.4 M ha were devoted to cereals — teff, barley, maize, wheat sorghum and millet—; about 785 630 ha were under the cultivation of various pulses, and about 27 080 ha were under oil crops. The production of food crops is by and large in the hands of individual peasants. At present out of the total area under cultivation for food crops and out of the total output more than 95 and 92% respectively are in the hands of individual peasants. The share of producers co-operatives is 0.4% of the area and 0.6% of the yield. These percentage figures indicate that the state farms and farmers producers co-operatives play a very limited role in this area. They do, however, account for about 20–25% of marketed output.

The main industrial crops in Etihopia are coffee, oilseeds (sesame and cotton seed), cotton, sisal, tobacco, fruits, pepper and sugar cane. The area under the cultivation of these crops other than coffee is estimated at 33 000 ha. The area under coffee which is the major export crop is estimated at about 500 000 ha making Ethiopia the largest arabica coffee producer in Africa. Production is around 200 000 to 250 000 t or about 4–5% of total world production. It is a small holders' crop as only 2–3% is produced in large plantations. Some coffee comes from systematically grown plots while the rest comes from coffee forests. Coffee berry disease has reduced production by about 20% since 1971. Domestic consumption of coffee is estimated to be around 110 000 t or about 55% of the total production. This implies a per capita consumption of about 3.75 kg y^{-1} which is higher than in most South American producer countries. It is estimated that about 25% of the employed population is involved directly or indirectly in coffee production, processing and marketing. Coffee also provides the main source of government revenue: 23% in 1977/78.

Ethiopia is first in Africa and tenth in the world in the size of its livestock resources. The latest estimates of the country's livestock resources include 29 M cattle, 23 M sheep, 19 M goats, and 52 M poultry. About one-half of the total land area is permanent pasture devoted to livestock production. It is estimated that around 60% of all livestock is owned by highland farmers and the balance by the nomadic tribes of the lowland areas. In certain areas there are too many animals in relation to pasture and there is overgrazing. The overall grazing situation indicates that the highland area is overstocked by 8.4 M animals while the lowland area is understocked by 4.4 M animals. Livestock is a major source of draught power and transportation and provides about 10% of the protein intake in rural areas. While animal husbandry contributes about 20–25% to the total agricultural output, in terms of contribution to export earnings only hides and skins are a major item. Thus, its contribution to export earnings is disproportionate to its potential.

Agriculture is the foundation of the Ethiopian economy and it is likely to maintain this position for very many years to come. It contributes around 50% to GDP, 90% to foreign trade and provides employment opportunities to about 86% of the population. This is currently estimated at 33 M and is growing at 2.8% y^{-1}.

Crops account for about 80% of the gross value of agricultural production and livestock products for the balance 20%. Coffee is the single most important crop contributing over half the agricultural GDP. Coffee yields, however, are on the low side, varying from about 250 kg ha^{-1}, for forest coffee to 700 kg ha^{-1}, for plantations. A substantial increase of coffee production would necessitate large scale programmes of disease control, rehabilitation, replanting with higher-yielding varieties and improvements in processing.

Ethiopia was a net exporter of cereals in the post-World War II period but it has now become a grain deficit country partly due to the failure of local production to meet the increasing demand and partly due to periodic droughts particularly since 1973.

The droughts affected the lowlands of the south and southeast in 1974/75 and caused the death of about 80% of the cattle while the Ogaden war in 1977/78 resulted in further livestock casualities. However, with its large potential, the livestock industry, with products ranging from meat, meat products, milk, dairy products, to hides and skins, remains the single most important industry that can diversify in a relatively short time the mono-crop export economy. Endemic diseases, shortage of animal feeds, uncontrolled grazing and breeding and underdeveloped marketing systems are the major constraints to increased output.

Natural forest reserves of the country have for long been depleted as a result of human habitation, population pressure and exploitation for timber and firewood. Out of an estimated 8 M ha of natural forests, only 3 to 4 M ha, or about 3% of the total land area, are estimated to be under forest today. The depletion of forest resources and the neglect of reafforestation have led to an extensive and alarming loss of top soil by erosion and shortage of forest products. Some of the cities like Addis Ababa are facing near crisis situations with regard to fuel-wood and charcoal while the construction industry is facing a timber shortage.

The catches from marine fishery have fallen from over 18 000 t in 1970/71 to 954 t in 1975/76. Most of this catch is exported on an irregular basis to neighbouring countries. Sea fish consumption in the interior of the country is limited because it is much higher in price than beef. In the case of fresh water fish, it is estimated that only 600 to 1200 t y^{-1} are being currently exploited in the fresh water lakes. These figures compare badly with estimates of fish catch potential: the estimated annual potential production of the Red Sea coast is about 66 000 t of fish; the Rift Valley lakes have an estimated potential yield of about 20 000–26 000 t of freshwater fish a year and the network of rivers and streams could yield a further additional annual output of 7000–10 000 t.

The main constraints to be overcome in exploiting the identified potential of the agricultural sector have been identified by the agricultural task force include the following:

(1)   Backwardness of agricultural technology;
(2)   Fragmented holdings and scattered settlement;
(3)   Population pressure on land resources in the highland areas;
(4)   Shortage of capital, reflecting low savings rates and inefficient use of existing capital;
(5)   Shortage of trained agricultural manpower for managing large farms, agricultural research and extension, rural administration, etc.
(6)   Poor rural physical infrastructure – roads, crop stores, irrigation works, etc.
(7)   Lack of development of alternative rural energy sources to fuelwood;
(8)   Lack of reliable information about key features of the natural resource base;
(9)   Deteriorating terms of trade facing agricultural producers;
(10) Inadequate priority given to investment of scarce funds in the agricultural sector;
(11) Disparities in development between different regions of the country, reflecting the past effort in only a few areas;
(12) Limited agricultural planning capacity exacerbated by lack of trained manpower.

The State of Agriculture in Malawi[1]

**The Structure of the
Agricultural Sector**

The quantity and quality of the natural resource base imposes important constraints on agricultural productivity in Malawi. The land area is some 9.4 M ha. The population was estimated to be 5.6 M in 1977, increasing at a rate of 2.9% y^{-1}. The average population density was 60 persons km^{-2}. Some 76% of the land area is available for agicultural use.

The arable soils of Malawi may suffer from physical and chemical limitations. Among the physical soil properties, structure is one of the most important for agriculture. Of the upland soils, the luvisols have a good structure which is quite stable under proper cultural practices. However, under unimproved agriculture, continuous use of the soil is bound to destroy the structure. Ferralsols, acrisols and nitosols all tend to have weak soil structure. Of the alluvial soils, vertisols have very good structure, especially within the topsoil. However, the structure of fluvisols depends on the texture of the different strata; sandy strata are structureless while clayey strata may have good structure. If drained, gleysols have good structure. However, the structure appears to be poor under wet conditions. Regosols (sandy soils) are structureless and solonetz have weak structure or are structureless. Improvement of soil structure depends on good soil management practices. Fallow can help restore structure. However, pressure on land does not allow resting land for long periods. So, for a long term solution, good cultural practices are emphasised.

This summary of the nutrient status of the soils indicates that most levels are low. Considering the three main agricultural soils, nitrogen, phosphorous and potassium levels range from very low to medium. Thus, the nutrient status is limiting as regards these nutrients. Calcium and magnesium levels are also low. Organic matter tends to be adequate, but it is notable that the level is low in the xanthic ferrasols which are widely prevalent and widely used for arable farming in Malawi. Nitosols which occur in high rainfall areas, are mostly used for tree crops and tend to have adequate levels of nutrients and organic matter. However, they are highly leached; hence both calcium and magnesium are low. The cation exchange capacity is also low.

On the whole, alluvial soils have adequate levels of the nutrients, although low levels occur. Particularly phosphorous is very low in the gleysols, while nitrogen is low in the solonetz.

Malawi experiences rainfall of high intensities, most soils have poor structure and steep slopes are common in most parts of the country. Hence the problem of soil erosion is widespread. Soils experiencing serious erosion problems are dystric nitrosols, rhodic ferralsols and some vertisols. Areas in which nitosols and rhodic ferralsols occur are of higher elevation, have moderate or steep slopes and have high rainfall. This combination of factors has caused these areas to be dissected.

The importance of the agricultural sector to Malawi's economy cannot be overemphasized, since it accounts for about 40% of the total population. The sector is divided into two sectors, namely the estate sector which accounts for 15% of Gross Domestic Product and approximately 70% of total agricultural exports, the smallholder sub-sector accounting for the difference.

Maize is Malawi's main staple food crop. Annual per capita consumption lies in the range 225 to 250 kg of shelled maize. At least 90% is produced by smallholders, the remainder being grown by estates. The crop occupies almost 75% of smallholder cultivated land. Only about 10% of total production enters official marketing channels. Malawi's policy of self sufficiency in foodstuffs has necessitated increased maize production using improved cultural production.

The National Sample Survey of Agriculture, 1980/81, indicated that some 43 000 t of rice were produced on 22 000 ha with a further 3200 ha on 16 irrigation schemes.

Until 1980 virtually all wheat produced in Malawi was grown by smallholders under rainfed conditions as a winter crop in the cool highland areas of the Northern, Central, and Southern Regions. The last three years have seen a rapid expansion of irrigated wheat production by a few estates.

Groundnuts are an important smallholder crop because they contribute significantly towards dietary requirements in most parts of the country, improve soil fertility and structure, provide cash income and are an excellent livestock feed, both haulms and cake.

[1]This section is summarised from the very detailed report submitted to the Conference by the Government of Malawi (Appendix reference 4).

Malawi grows both confectionery (Chalimbana, Malimba and RG1) and oil groundnuts (Manipintar) the former for home consumption and export and the latter for home consumption only. The Chalimbana is grown throughout the country and accounts for about 90 % of sales to the State Marketing Corporation (ADMARC).

Sorghum, millet, cassava, guar beans, pulses and spices are also grown, mainly by small-holders.

Tobacco is the major export earner within Malawi's economy, and during 1982 it accounted for 58.9 % of total exports. The crop, which consists of five types, occupies some 70 000 ha of arable land providing employment to about half of the total agricultural labour force. Of the five tobacco types produced in Malawi, Virginian flue-cured and burley tobacco was produced by commercial estates, while the western fire-cured, sun/air cured and oriental tobacco is produced in the smallholder sector. In total, Malawi's tobacco accounts for about 1 % of a total world production 5.5M t, the largest market share taken by flue-cured of which Malawi's contribution is 1.1 %.

Tea has been grown in Malawi since the early 1900s. It has consistently ranked second in importance other than tobacco as a foreign exchange earner. During the period 1977–81, tea export earnings have averaged 17 % of total exports, compared with 51 % for tobacco and 14 % for sugar. Now tea is in third place after sugar. Of the approximately 19 000 ha planted to tea, just over 2300 ha are owned by some 4800 smallholders, ranging in size from 0.25 to 1.0 ha. Over 80 % of the tea is owned by 26 estates.

Sugarcane is produced by two large estates (total 16 000 ha) and by the Sugar Smallholder Authority (700 ha). Total commercial Malawi production increased from 107 600 t in 1979 to 166 600 t in 1981. Sugar exports have also increased. Smaller quantities of cotton and coffee are also produced.

An estimated 3.2 M ha of grazing land is at present available for livestock. A large area (over 600 000 ha) consists of dambo land which provides useful dry season grazing. The average carrying capacity of the natural pasture is estimated to be 12 ha LU^{-1} for upland grazing, and 2.4 ha LU^{-1} for dambo grazing. As the total number of Livestock Units in the country including goats and sheep is estimated to be in the region of 650 000, the whole grazing area provides sufficient grazing for about 9 months per annum. However, it should be noted that the distribution of the grazing area does not always coincide with the distribution of livestock. The most severe grazing problems seem to exist in the Lower Shire Valley and Lilongwe district, where both human and livestock populations is high. Little or no grazing control is practised, which leads to the deterioration of the sward while indiscriminate burning of the grazing lands, when on a large scale, aggravates the grazing problems even further.

The fisheries of Malawi are divided into three categories, the traditional, semi-mechanized and fully mechanized traditional group; these fishermen use either dugout canoes or boats without any mechanical power. They are completely dependent on manual power for the operations of the different gear in use.

The Performance of the Agricultural Sector

Agriculture is the most important sector of the Malawian economy, as it employs about 90 % of the population and in 1980 50 % of all paid employees were in agriculture. In the same year it accounted for 37 % of the Gross Domestic Product and contributed 94.5 % of all export earnings. These export earnings were derived from mainly tobacco (46.2 %), tea (13.7 %), sugar (17.8 %) and groundnuts (7.3 %).

It is not possible to measure smallholder production and growth with accuracy because of lack of data. However, ADMARC purchase figures from smallholders are taken to represent 10 % of total smallholder food crops production; they show an increase in aggregate production from 183 391 t in 1972 to 302 452 t in 1982.

However, the bulk of the country's agriculture exports comes from the estate subsector which has to-date functioned as a principal earner of foreign exchange and engine of growth for Malawi's development. The estate subsector contributes 15 % of total agricultural production but accounts for nearly 70 % of all agricultural exports. Estate production is mainly centred on flue-cured tobacco, tea and sugar. Over the past 10 to 15 years this sector has been more dynamic, its output rising at about 15 % per annum, its share in exports rising from about one-third to two-thirds and its employment generation accounting for one half of the new formal sector jobs.

Over the 14 years since 1964 agriculture's share in the Gross Domestic Product declined from 58 to 41 %, but in absolute terms agriculture was growing. The real growth of the agricultural sector is estimated at 3 % y^{-1}. In fact production of some of the crops has been declining in recent years (since 1980), while others have been fluctuating up and down. For example, cotton production reached the peak of 24 218 t in 1979 but has since been falling. Purchases in groundnuts have been falling since 1979. Several reasons are responsible for this poor performance such as (1) poor (series of droughts) weather conditions which have been hitting the country since 1979/80, (2) less renumerative producer prices set by ADMARC, especially prices for cotton, tobacco and groundnuts; the latter together with rice are competed for by private traders who offer higher prices than ADMARC; and (3) the rapid expansion of estates through alienation of customary land.

Malawi is generally self sufficient in food supply except in very bad years. The main staple food of the country is maize, but there are many other food stuffs of importance which supplement the food supply of the people in the rural areas. These crops include cassava, sorghum, potatoes, cocoyams, plantains, etc. The total import of foodstuffs increased over the period 1978 to 1980. But the increase was not considerable. Also, the composition of these food stuffs indicates they are for a specialised market, not the bulk of the population.

Generally earnings in agriculture are lower than those in industry. In 1980 the average monthly earnings for all industries were K43.66 while that of agriculture were K15.91. By the second quarter of 1982 the situation had improved by 50 % but the earnings were still below half of the national average.

Prospects and Strategies for the future

In 1968/69, the Malawi Government, with external financial assistance, launched four very expensive and intensively managed Integrated Rural Development Projects. The Projects were sited in some of the most density populated area of the country with the highest agricultural potential. For all of them, the aims were to sustain self-sufficiency in food staples, expand agricultural exports and improve rural incomes. The outputs of these projects had common features:

(1)   Provision of infrastructure (offices, staff houses, rural roads, input/produce markets, water and health facilities).
(2)   Provision of credit facilities for farm inputs.
(3)   Improvement of agricultural extension.

During the period 1968/76, a total of about K126 M from external sources had been invested. At that level of investment, it was realised that it would cost far too much and take far too long to spread the benefits to the entire population. Hence a change in approach brought The National Rural Development Programme (NRDP).

The "birth" of this programme was the "fruit" of very lengthy negotiations between the Malawi Government and the World Bank. A five-year Phase I of the Programme was launched in 1977/78 with financial aid from the World Bank and the Canadian Government (CIDA). The basic object of NRDP was not different from the earlier agricultural development programmes but emphasis was on investing in those activities with a more immediate impact on agricultural production, e.g. extension, credit, input supply and input/produce markets. Emphasis was to be placed on improving productivity per unit area, soil conservation, water-shed management and afforestation.

Under this programme, the country was divided up into 8 Agricultural Development Divisions and about 40 Rural Development Projects (RDPS) and the policy was that all projects would come under "development financing" on the basis of four phases of five years each. At the planning stage, a target period of 20 years was set during which all the 40 planned RDPS would be financed. It was anticipated that 3–4 RDPS would be joining the "race" every year. Since its inception (1978) this objective has not been achieved due to problems with funding and other reasons.

The State of Agriculture in Zimbabwe

The Structure of the Agricultural Sector

Agriculture is critically important to the Zimbabwean economy; it generates 20 % of the Gross National Product, 35 % of export earnings and 35 % of all employment opportunities. Some 75 % of the population is engaged in agriculture as a primary source of economic activity.

Within the agricultural sector there are approximately 5000 large-scale commercial farmers. 10 000 small-scale and 860 000 communal or peasant farmers. Their respective contribution to gross agricultural output is 70 %, 5 % and 25 %. Approximately 50 % of dry land arable potential has been exploited and approximately 20 % of the irrigation potential.

The Performance of the Agricultural Sector

Over the past 20 years Zimbabwe has been basically self-sufficient in food. A traditional maize grain exporter, self-sufficiency in wheat was achieved in 1972. Self-sufficiency in oilseeds was achieved in 1969. The only commodities now imported are tropical products which cannot be grown profitably in Zimbabwe, such as cocoa, rubber and rice. In the livestock sector, the country is self-sufficient in all products and is a significant exporter of beef and animal by-products such as hides and skins. It is also a significant producer of cotton and tobacco. Over the past three years serious drought conditions have been experienced and at the present time a critical situation in respect of water and grain supplies exists. Imports of both maize and wheat will be necessary in 1984/85; the former because of the widespread failure of rainfed crops and the latter because there is insufficient water for winter irrigation. However, outputs of cotton, tobacco, oilseeds, beef and milk are all expected to be greater than last year. The quality of the tobacco crop is especially good.

From the output point of view, the main agricultural commodities are shown by descending order of value in Table 3.

Table 3   Zimbabwe: 1984 Output at Farm Prices.

Tobacco	Z$ 230 M	Dairy	Z$ 70 M
Beef	Z$ 180 M	Sugar	Z$ 60 M
Cotton	Z$ 125 M	Tea/coffee	Z$ 38 M
Maize	Z$ 90 M	Wheat	Z$ 25 M

Strategies for the Future

For the future the priorities are as follows:
(1)   To maintain self-sufficiency in food.
(2)   To expand the production and sales of products where Zimbabwe has a comparative advantage (tobacco, beef, maize and cotton).
(3)   To increase irrigation capacity in order to stabilise total output in dry years.
(4)   To raise the productivity of peasant farmers.
(5)   To halt and reverse the process of land degradation in peasant farming districts.
(6)   To assist other States in the region to raise their agricultural production.

The State of Agriculture in Botswana[1]
(with particular reference to crop production)

The Structure of the Agricultural Sector

Botswana has an area of about 582 000 km², and a population of approximately 800 000. It is estimated that about 84 % of the total land area consists of Khalahari sands and that less than 5 % of Botswana's land area is suitable for arable agriculture.

The climate of Botswana is predominantly semi-arid, with rainfall that ranges from 650 mm in the north-east to a low of 250 mm in the extreme south-west. Effective rains fall between October and March, in the form of showers which are limited in amount and distribution. Crop evapotranspiration rates are high, ranging from 1.5 metres to 2.0 metres per annum. Apart from the Okavango river system, there are no perennial rivers in Botswana. With the exception of the northern part of the country above latitude 20°S, no area is free from the risk of frost.

The soils of Botswana are predominantly sandy. In the eastern hardveld which forms the catchment of the Limpopo, soils are loamy sand, with small areas of heavier soils in depressions.

Most of these soils are hard, clod forming and have a tendency of capping. Natural fertility of these soils is low especially in phosphorous.

Although cattle account for the bulk of the agricultural production, many more people are dependent on arable agriculture.

Mixtures of crops are common and pulses are normally grown together with the main grain crop. The predominant crops include sorghum, maize, millet, cowpeas and groundnuts. However, a limited amount of cash crops (sunflower, cotton and citrus) are exported from the freehold farming areas.

Drought and *Quelea quelea* are probably the two most serious factors influencing farmers' choice of crops. Sorghum and millet require the greatest amount of labour in respect of bird scaring during the grain formation period. Farmers are aware that under poor rainfall conditions, sorghum will perform better than maize, but they still grow maize because the latter crop is not subject to bird damage. At the moment, the sorghum varieties grown are mainly food types and the most popular are Segaolane and Town. Segaolane is a local selection which takes about 130 days to physiological maturity. Under good rainfall conditions average yield on the experiment station is approximately 4 t ha^{-1}. But it is also susceptible to *Striga*, aphid and various grain moulds.

Cultivation is done by ox-drawn or tractor-drawn mould-board plough. The majority of the farmers broadcast seed prior to planting and weeding is done by hand.

The performance of the Agricultural Sector

Adverse climatic conditions frequently limit the yield and in years of extremely low rainfall crops may fail completely. Average annual yield of sorghum is 260 kg ha^{-1}; maize 266 kg ha^{-1}; millet 153 kg ha^{-1}. Other constraints to crop production in the traditional sector include (1) lack of draught power − most of the farming households depend upon borrowed draught animals to plough; (2) late planting resulting in poor crop establishment; (3) lack of weeding.

The estimated annual demand for sorghum and maize is about 100 000 tonnes. The gap between crop production and demand is still very high. Only in the 1974 and 1976 seasons has self-sufficiency been approached. Approximately half of Botswana's basic grain needs have to be met from imports.

Strategies for the Future: The Role of Research

Current programmes in crop research include the following;

(1) Sorghum and Millet Improvement: Research is directed mainly to sorghum and millet breeding. Introduction and testing of new varieties of these crops have resulted in better crop performance. However, current varieties of sorghum are rather susceptible to foliar diseases, various head moulds and parasitic weeds such as *Striga*. Improved early maturing varieties are needed to replace the existing ones which are generally late maturing. Sorghum breeding work is now going on to develop suitable varieties with good grain quality and disease resistance.

(2) Cowpea Improvement Programme: This programme is mainly concerned with evaluation and selection of suitable varieties. A short season variety ER7 has been identified and this variety will be multiplied and distributed for commercial use.

(3) Oilseed Crops: Major emphasis has been put on the evaluation of sunflower and groundnut varieties. A few promising varieties have been released for commercial use.

(4) Sorghum/legume rotations: This section is responsible for investigating the yield response of sorghum following cowpea or groundnuts compared to continuous sorghum with and without fertilizer and to determine yield advantages of intercrop versus monocrop. Preliminary findings have indicated that legumes in the rotation can benefit the performance of the following sorghum crop.

(5) Other research projects: These include work on fertilizer, plant protection and various systems evaluation studies.

[1]This section is based on the paper submitted to the Conference by the Government of Botswana (Appendix reference 6).

The State of Agriculture in Mauritius[1]

The Structure of the Agricultural Sector

Maritius is situated in the Indian Ocean, some 800 km east of the Malagasy Republic. It is a small island of 1856 km^2 (186 000 ha), densely populated (516 km^{-2}), with a population of 959 905 by the end of 1982 and growing at a rate of about 1 % annually. The island has a central plateau 610 m above sea level and enjoys a subtropical maritime climate with no extreme seasonal variation. Summer months are wet, extending from November to April and winter months are dry and extend from May to October. The overall island temperature means are 20°C in July and 25°C in January. The mean annual rainfall is 5000 mm on the central plateau and 1000 mm on the drier western coast.

Agriculturally important soils in Mauritius fall into two main groups classified under the Hawaiian system:

(1)   The typical mature ferralitic soils or latosols of the tropics whose parent rock has decomposed to such an extent as to leave no undecomposed minerals. Such mature soils consist of the Low Humic Latosols, Humic Latosols and Humic Ferrugenous Latosols;

(2)   The typical immature soils in which the minerals are still in process of weathering. This immature series are classified as Latosolic Reddish Prairie and Latosolic Brown Forest soils.

About 52.7 % (98 000 ha) of the total area of the island is cultivated with sugarcane, some 6000 ha is under tea and about 2000 ha is cultivated with tobacco, vegetable and other foodcrops. Other important agricultural products are fruit and horticultural crops and livestock products (meat, milk and eggs). There are over 30 000 small sugarcane planters, 3000 foodcrop growers and 12 000 livestock breeders. The high proportion of small sized plantations, the high degree of part-time farming (over 60 % of cane planters) and the nature of the subsidiary occupations of farmers present special problems. A Livestock Census conducted this year has revealed that the small breeders possess 16 011 head of cattle, 6 818 pigs, 76 175 goats and 595 sheep. The cattle population owned by large breeders has been increasing over the past few years to reach 33 % of the total cattle population of about 24 000 head.

There are about 57 000 ha of forest lands which include forest plantations (11 600 ha), partially degraded native forests (4585 ha) and the rest privately owned forest lands. Forest plantations are situated mostly on the uplands.

The exploitation of the native forest has been so extensive in the past that today only about 1 % of the original forest of the island is left. This is confined to the steep mountain slopes and gorges and to regions which are not easily accessible. The native forest of the island contains a unique flora and fauna which is not found anywhere else in the world.

The Performance of the Agricultural Sector

The share of agriculture in the gross domestic product had been increasing slowly after independence to reach 55 % in 1974 following the international sugar boom but it has decreased steadily since to reach 30 % in 1979. Agricultural production contributed about 15 % to the Gross Domestic Product (GDP) in 1982.

Foodcrop production did not reach the level usually forecast, i.e. 80 000 t; only about 50 000 t of foodcrops (potatoes, maize, beans and peas, groundnuts, vegetables and fruits) were produced during 1981. Livestock production amounted to 1274 t of meat comprised as follows: cattle: 528 t, goat and sheep: 107 t, pig: 639 t. Fish production amounted to 3104 t during 1981.

As domestic food production has not been able to satisfy demand, the country is still heavily dependent upon imports for its food requirements. These consist mainly of the staple foods, rice and wheat flour (26 % of total food imports), dairy produce, meat and meat preparations. About 25 % of the total import expenditure in 1981 was spent on food. Looking at the individual food items, rice, which is the staple food of the country, is all imported; some 80 000 t are imported annually. Local production of rice has never attained any significant level because of the high costs of production. In 1981, the country imported some 13 000 t of maize, whereas the domestic production for that year was 1081 t. On the other hand the country produced about 16 000 t of

---

[1]This section is drawn from material contained in the paper cited in the Appendix as reference 7.

table potatoes in 1981, which was in excess of the country's requirement by 1500 t. The annual production of vegetables has fluctuated around 30 000 t. The major problems encountered by producers are the unavailability of land, inadequate irrigation and also a lack of limited marketing and credit facilities.

Mauritius relies heavily on imports for its requirements of milk and meat. Local production is only catering for 12.5 % of the country's requirements of milk and 15 % for beef. The total import bill of milk and meat represented 20 % of the total food imports. Mauritius imported 4115 t of beef in 1981 while local production for the same year was 528 t, accounting for only 9.5 % of local consumption. Similarly, local milk production represented only 4.0 % of total milk consumption.

As far as poultry is concerned, considerable progress has been recorded over the past few years. A remarkable preference has been observed for the consumption of chicken meat, especially for cost reasons. The local production has steadily increased from 4929 t in 1979 to 6000 t in 1982. As a result, the country is presently self-sufficient in poultry meat and eggs. The annual egg production is roughly 35M units.

Goat rearing for meat is essentially a part-time activity in the coastal rural areas. They are either allowed to graze freely or tethered on marginal and sub-marginal land or are permanently confined in sheds. The three main breeds of goat on the island are the Barbari which is the local breed, the Jumna Pari and the Anglo-Nubian. The last two breeders face the same problem of availability of fodder, absence of a proper marketing system, shortage of breeding animals, lack of knowledge as regards management techniques and good husbandry. In 1982, the local production of goat meat and mutton reached 108 t but 3340 t were imported for a value of 28.3M rupees. There is also a modest augmentation of domestic meat supplies from sheep, pigs and deer, resulting in a slow decline in imports of meat in these categories.

In the foreseeable future exports will continue to remain the most important source of foreign exchange earnings. Sugar (including molasses) has been the major export commodity while tea and fish have been less significant. Agriculture contributed about 67 % of the total foreign exchange earnings of the country in 1982. Mauritius produces about 650 000 t of sugar. Its marketing is governed by the Lomé II Convention under which Mauritius is committed to supply 500 000 t of sugar annually to the European Economic Community at a guaranteed price which is in line with that paid to European beet sugar producers and well above the average world market price.

Besides meeting the EEC quota and the local consumption which amounts to 35 000 t, the surplus sugar is being sold on the world market.

Mauritius is the smallest producer of the International Tea Promotion Association member countries, with an annual production of about 4600 t of black CTC tea in 1982 of which 4000 t are exported. This amounts to less than 0.5 % of the total world production. The tea industry employs 15 000 people on a full-time or part-time basis − i.e. 5 % of the labour force employed in the country − and the tea industry accounts for about 3 % of the country's total foreign exchange earnings. In 1971, with the help of the World Bank, Government created the Tea Development Authority in an effort to intensify tea cultivation and create employment. The Tea Development Authority leases land to small holders and by 1982 there were 1737 small holders cultivating an area of about 1650 ha.

The production of tobacco is regulated from year to year by the Tobacco Board which is the body responsible for the purchase of all tobacco produced locally under permit and for any importation of foreign tobacco used for blending by the sole manufacturer operating in the country. The local production of tobacco has increased over the years with a reduction in the quantity of imported tobacco. The total production of tobacco for 1981/82 was about 1258 t. In the last few years, with the increase in the price of cigarettes, it appears that the demand for the cheaper brands of cigarettes is on the increase. Consequently, it is expected that the amarello air-cured tobacco will increase substantially. The production of tobacco does not present much difficulty provided that the tobacco producers obtain suitable rotational lands with adequate irrigation. Pests and diseases are well controlled and do not constitute a major constraint in tobacco production.

The local production of timber satisfies only 20−40 % of the domestic requirement. The local production varies depending upon the climatic conditions prevailing in the country. In a cyclonic year, more timber is put on the market as a result of salvaging operations in the forest plantations.

With the increasing prices of fossil fuel, more people are turning to fuelwood. The production of fuelwood is of the order 20 000 m³ annually. In addition, a large volume of wood in the form of dry privet and chinese guava are removed from the forest by members of the public.

Fish is consumed by every section of the population. The consumption per head has increased from 6.5 kg in 1969 to 18 kg in 1982. Production lags far behind consumption and imports have remained at a high level – around 14 000 t y⁻¹. Mauritius has a fairly extensive "marine economic zone" but the prospects of bank fishing can only be determined in the light of assessments of fish stocks in the area. The output from the fishing banks has dropped considerably from 3700 t in 1977 to 1700 t in 1981.

Tuna fishing is presently being carried out by foreign fishing vessels for processing and sale in the Far Eastern markets. Some 3000–4000 t of skipjack fished in the Western Indian Ocean are sold to the Tuna Canning Factory in Mauritius. The European Economic Community countries and the USA provide the main markets for canned skipjack. The quantity of canned tuna sold on the local market is 30 t. Negotiations of licensing agreements with neighbouring countries within the context of regional co-operation are expected to open new fishing grounds. The culture of the fresh water prawn introduced into Mauritius in 1972, is well established and commerical production has started. The importation of frozen crustaceans is about 50 t annually and the annual production of the fresh water prawn has reached 20 t. However, the production potential of the existing ponds is about 50 t and this can be achieved with better feed and improved pond management.

Prospects and Strategies for the Future

The overall food strategy for Mauritius is the promotion of self-sufficiency in specific food items, to reduce the import food bill and to improve the nutritional status of the population.

The objectives to be attained in the foodcrop sector over the period 1983–87 are:
(1)   to attain self-sufficiency in maize, onions, garlic, ginger, tumeric, coriander and tamarind;
(2)   to increase the local production of beans, peas, coconuts, citrus fruits etc.;
(3)   to encourage the establishment of soya bean, vanilla and edible mushrooms;
(4)   to determine whether rice cultivation can be commercially established.

To attain these objectives, Government is speeding up the rate of agricultural development of:
(1)   direct public investment in agricultural infrastructure, e.g. irrigation projects
(2)   measures and policies to increase production (e.g. by ensuring the availability of land for foodcrop production, price support schemes, credit facilities, tax and other incentives)
(3)   provision of storage and better marketing facilities.

Possible effective measures can be illustrated by reference to two major food crops – maize and soya bean. It is expected that by 1987, domestic production of maize should satisfy the average annual requirement of 15 000 t, through utilising the potential of cultivating maize in the interlines of virgin and ratoon cane. Drying units are being established in different parts of the island. The small farmers will be grouped into co-operatives to enable them to have access to farm machinery for the mechanization of maize production. In the case of soya, Government proposes to start commercial pilot projects to study all aspects of production as there exists a high demand for imported soya products.

The role of agriculture as a major foreign exchange earner will be extended through the production of high-value agricultural crops suitable for export. Sugar being the main foreign exchange earner, fiscal incentives in the form of reduced export tax on sugar have already been granted to the sugar industry. At present Government tax revenues from agriculture come mainly form the export duty on sugar. It seems unlikely that there will be any further major investment in the sugar industry, but it is the objective of Government to group the 30 000 small farmers into co-operatives so that better cash facilities might be provided to them and they may increase their productivity by following advice tendered to them by the Extension Service.

In the context of the diversification of agricultural exports, the role of tea as a foreign exchange earner will be strenghthened. Government will initiate a series of administrative and technical measures at the Tea Authority to raise productivity in this sector and to increase competeveness of local tea on the export markets. Other measures include improved cultural practices, mechanical plucking, quality control and blending of local tea with high quality imported tea.

It is Government policy to boost the production of beef, milk, goat milk, venison and pork. Cattle breeding is currently being undertaken by large and small breeders. Large scale breeding based on modern techniques is carried out on the sugar estates. The small breeders in most cases are owners of one or two head of milk cattle and the majority of them undertake goat raising at the same time. The overiding constraints to the development of the cattle population owned by small cowkeepers are the scarcity of fodder as a result of urbanisation, the high cost of feed and concentrates and low marketing efficiency. Government has recently established a Livestock Holding Centre run by the Mauritius Meat Authority to improve the efficiency of marketing.

To alleviate the problem of fodder availability, Government is reviewing its policy regarding agro-forestry whereby it is intended to increase the fodder area in the forests.

The Role of Research

Major policies for agricultural research are detailed below:

Research on crops will be specifically oriented towards the removal of major constraints to production, identification and testing of high yielding crop varieties, studies on storage, processing and appropriate food technology, development of a sound national seed multiplication programme for the production of high quality seeds, and the design of light agricultural implements suitable to local systems of production. Sugarcane breeding is being done with a view to produce higher yielding varieties with higher sucrose content and with desirable characteristics to satisfy the local agro-climatic conditions and to possess resistance to pests and diseases. Considerable research is presently being conducted regarding the response of varieties to drip irrigation, fertilizer response, the use of ripeners for crop maturity, mechanical harvesting and loading of sugarcane. Research in improved methods of sugar technology is actively being undertaken for more efficient sugar manufacture and better energy conservation to reduce the cost of manufacture.

As for livestock research, special attention will be focused on improving the productivity of livestock adaptable to local conditions through selection and breeding, improving the present disease surveillance systems and optimising the utilisation of agricultural, industrial and domestic by-products as alternative sources of feeds for livestock. There is also an extensive range of strongly applied research projects for inland, lagoon and oceanic fisheries.

Appendix

Country Statements on the Structure and Performance of the Agricultural Sector prepared for the CAB/Government of Tanzania Conference on "Advancing Agricultural Production in Africa", Arusha, Tanzania 1984.

(1)   The Status and Prospects of Agricultural Production in Tanzania (41 pp.).
(2)   Kenya Agricultural and other Related Development (compiled by W.E. Adero) (47 pp. plus 12 appendices).
(3)   Agriculture in the Ethiopian Economy (prepared by L. Birke) (15 pp.).
(4)   Malawi: Position Paper (123 pp. plus 4 maps).
(5)   Zimbabwe: Country Statement (2 pp.).
(6)   Crop Production in Botswana (6 pp. plus 2 tables).
(7)   An Outlook on Agricultural Production in Mauritius (prepared by G.M. Lallmahomod and A.L. Owadally) (21 pp.).
(8)   The Commonwealth Agricultural Bureaux and Agricultural Development in Bangladesh (prepared by M.M. Rashid) (7 pp.).
(9)   Canadian Overseas Development Aid for Agriculture in Africa (prepared by J.S. Clark) (6 pp.).
(10)   Assessment of the Performance of the Agricultural Sector of Cyprus (69 pp. plus 10 tables).
(11)   Delivery of Animal Health and Production Services – General Aspects – A Case of Lesotho (prepared by M. Moteane) (4 pp.).

Chapter 4

Relevance of Agricultural Research in Australia to the Semi-Arid Tropics of West and Eastern Africa

B. K. FILSHIE

CSIRO Centre for International Research Co-operation, Canberra Saving Centre, City Walk, Canberra, A.C.T. 2608, Australia

Introduction

Socio-economic Constraints to Technology Transfer from Australia to Sub-Saharan Africa

Socio-economic differences between the two regions impose just as important constraints as do physical differences in technology transfer. Agricultural production in north-west Australia uses low inputs of labour, high inputs of land and (for crop production) high capital inputs and exports in products to other regions. In the subsistence farming operations of populous developing countries, the picture is completely reversed with a high input of labour, and low inputs of land and capital. Technology developed for one socio-economic environment is unlikely to be directly transferable to another. Thus the tall-growing forms of grain sorghum developed for hand-harvesting in Africa have proved unsuitable, despite their good climatic adaptation, for the mechanised farming practices of north-west Australia. Security of land tenure can also affect decisions to adopt new technology. This all points to the need for location-specific research in order to adapt present technologies to particular socio-economic and technical environments.

Rationale for Australian Involvement in Co-operative Agricultural Research in Sub-Saharan Africa

There are three primary reasons why collaboration between Australian agricultural research institutions and their counterparts in Africa is desirable. The major one is that sub-Saharan Africa (SSA) is the only developing region of the world where per capita food supplies have been consistently declining since 1970 (USDA 1980). This trend must be reversed if immense human suffering is to be avoided. The second reason is the marked similarities in the agroclimatic characteristics of the northern region of Australia and many parts of SSA. This is especially so in the semi-arid tropical zones and will be elaborated upon later. Finally, agricultural research in developing countries has a high payoff, both in terms of the return on the investment, and in terms of equity, although research investments remain well below those required to fully exploit the opportunities for increased agricultural production and the enhancement of economic development and human welfare.

The Australian Centre for International Agricultural Research (ACIAR)

ACIAR was established in June 1982 as an organization devoted entirely to the mobilisation of Australia's agricultural research capacity for the benefit of developing countries. ACIAR does not itself conduct research but achieves it objectives of identifying agricultural research problems in developing countries and finding solutions to such problems by commissioning research by Australian institutions and individuals in partnership with developing country research groups. The work may be carried out both in Australia and the developing country.

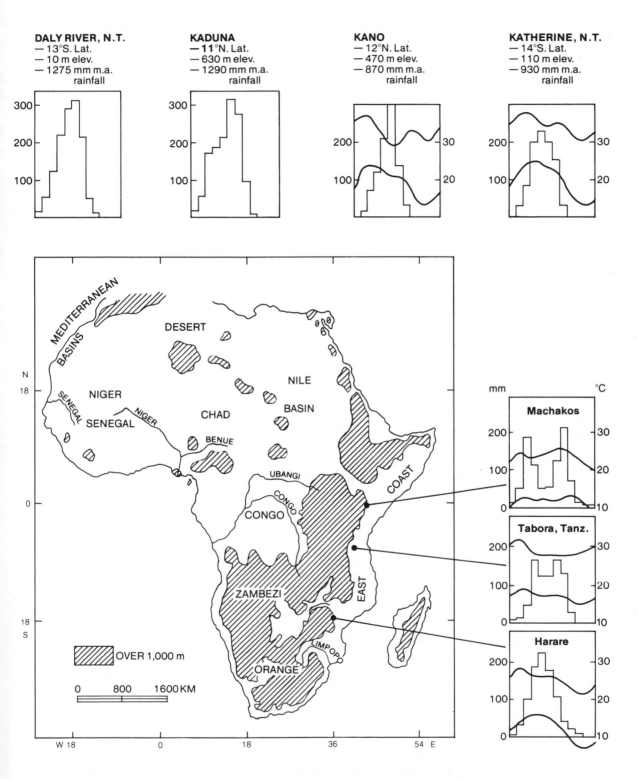

Fig. 1   Map of Africa showing the large proportion of eastern and southern Africa comprised of land >
1000 m elevation. Inserts show monthly mean rainfall (mm) and mean maximum and minimum
temperatures for two west and three eastern African stations and for Katherine and Daly River.

Agroclimatic Similarities Between Northern Australia and Africa

The term "semi-arid tropics" (SAT) has no generally accepted precise meaning. Here it is taken as the broad zone accepted by the International Crops Research Institute for the Semi-Arid Tropics (ICRISAT) as falling within its mandate. That zone, as mapped, includes much of northern Australia and the major part of sub-Saharan Africa. However there is a small but significant humid tropical zone in North Queensland (7-9.5 months growing season). The North Queensland tablelands have subtropical climates.

The distribution and use of arable soils in the SAT has recently been reviewed by Swindale (1982). The most common soils of the lower rainfall regions of sub-Saharan Africa are various sandy soils with little clay increase with depth. Similar sandy soils in the Australian SAT occur in north-west Australia and far north Queensland. In higher rainfall areas of Africa soils similar to the red earths (Alfisols) of northern Australia occur, particularly on siliceous parent materials. Soil structural problems due to high intensity rainfall are common in both continents (Lal 1984).

The native grasslands of the far north of Australia are very similar to those of the savanna zone of Africa with their inherent high sensitivity to overgrazing (Andrew et al. 1984). As in Africa, the dry savannas are much more productive than the moist savannas. Extensive grazing in northern Australia is based on exotic animals since, unlike the situation in Africa, northern Australia has no native or indigenous ungulate herbivores. Also in contrast to northern Australia, there is in Africa a diversity of large predators, and serious animal and human diseases such as trypanosomiasis and malaria, are present (malaria was eliminated from northern Australia in the 1960s but dingoes can still be a problem for young stock (McCosker et al. 1984)).

Mean climatic data for the northern regions of West Africa and northern Australia are similar, although the consequences of slightly higher elevations in West Africa are not clear. For example, Kaduna in Nigeria (11°N) has a mean annual rainfall of 1290 mm compared with 1275 mm at Daly River in the Northern Territory (13°S) and the distribution of mean monthly totals during the respective wet seasons is almost identical (R.L. McCown personal communication, Fig. 1). Also seasonal trends in rainfall and temperature are similar for Kano (12°N) and Katherine (14°S). However most of the SAT in Eastern Africa occurs at elevations above 1000 m (Fig. 1); much higher and cooler than Australia's highest altitude region, the Atherton Tableland. The seasonal rainfall pattern in the south of Eastern Africa is similar (e.g. Harare) to that in northern Australia, but at lower latitudes (e.g. Tanzania, Kenya, etc.) it becomes progressively bi-modal.

Using pattern analysis methods, Cook & Russell (1983) determined that the main homoclimates for Katherine and Kimberley in northern Australia were in Africa particularly Sudan, Upper Volta, Malagasy and Mozambique (Fig. 2). Ouagadougou in Upper Volta and El Roseires in the Sudan were similar to both Kimberley and Katherine. This work, while helpful, is somewhat unsatisfactory in that the attributes of climate selected in the analysis and which indicate similarity may not be those that impose/determine/constitute the climatic constraint to a particular agricultural system (Williams et al. 1984). The main limiting factor for crop production in the SAT is the highly irregular rainfall pattern and the use of mean data can often produce misleading results. In both Australia and Africa rainfall is erratic in the SAT, and droughts of varying duration within the wet season are frequent. Estimates of variability and climatic risk based not on mean data but on an examination of the actual weekly sequence of rainfall and evaporation are given by Williams et al. (1984) for selected northern Australian and African (Senegal, Upper Volta) stations. For regions having similar amounts of annual rainfall, the effective rain period is generally longer, and the variability of the effective rain period (both its onset and cessation) is larger in the Australian than in the West African stations. The percentage of crop failure years in Australia is consequently larger than for the African stations examined. This work of Williams et al. (1984) is limited in extent and further water balance studies are needed to fully understand the climatic similarities or differences between different regions in the SAT. Furthermore much of the precipitation occurs in high intensity storms and a major part of this rainfall runs off the surface or infiltrates beyond the root zone of crops. Present estimates of the rainfall runoff are far from precise, again making the comparison of agroclimatic regions difficult.

Much of eastern and southern tropical Africa falls in a mid-altitude climatic zone that has

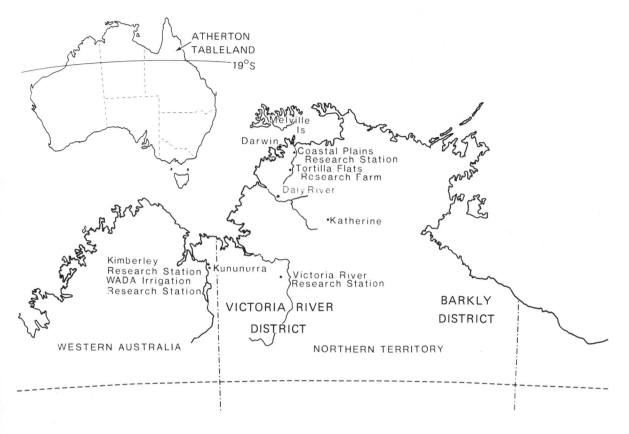

Fig. 2   Map of north-west Australia.

scarcely any analogue in Australia. Therefore, although the principles of agricultural technology can be exchanged between the two continents, it will be necessary to adapt them to the relevant local climate/soil conditions.

Research in Northern Australia Relevant to Sub-Saharan Africa

Plant Adaptation

Current research aimed at obtaining an understanding of the influence of environmental factors (particularly daylength, temperature and water deficits) on crop growth and development is highly relevant to crop improvement in the lowland SAT of Africa, but the cooler conditions at higher altitudes would prevent direct application of such knowledge to parts of eastern Africa. An excellent example has been studies on the genotype × environment interaction over a wide range of genetic diversity within soybean that has led to the devlopment and/or selection of crop types for both irrigated and rainfed wet season conditions and irrigated dry season conditions (Garside *et al.* 1984). Similar studies on mungbeans are in progress and preliminary studies on a range of grain legumes including guar, cowpea, sesame and pigeon pea have commenced. Comparative studies on the tropical cereals (sorghum, maize and bullrush millet) are attempting to identify those plant attributes favourable to growth and yield under conditions of both short term intermittent water deficits during the wet season and sustained water deficits at the end of the wet season (R.C. Muchow & A.A. Done, pers. comm.). Reseach aimed at understanding the phenology and floral biology and their effect on yield of horticultural species with commercial potential in northern Australia, particularly mango and cashew nut (Scholefield & Blackburn 1984), is highly relevant to other regions in the semi-arid tropics. Similarly studies on the improvement of quality of the harvested product are equally important in both the Australian and African SAT. Current research on seed quality and weathering resistance in mungbean (Imrie & Putland 1982, Williams *et al.* 1982), head mould resistance in sorghum (Done *et al.* 1984) and postharvest handling of tropical fruits (Chaplin 1981, Scholefield & Blackburn 1984) are relevant examples.

The attributes upon which the selection of crop and horticultural species is based in northern Australia may not necessarily match those relevant elsewhere in the world's SAT. However, there are many common attributes which are relevant throughout semi-arid tropics of the world, including establishment ability, ability to withstand water shortage, and disease resistance. Even in improved subsistence systems, attributes such as even maturity and lodging resistance may assume importance. Nonetheless, although the final selection of cultivars for the African SAT must be done *in situ* the efficiency of selection will be improved if it is based on a thorough understanding of the crop response to environmental factors.

In contrast to crop and horticultural species, pasture species suitable for the Australian SAT are likely to be directly useful in similar rainfall and temperature regimes in the African SAT, since common extensive grazing systems prevail. Currently nine introductions are considered to be persistent and productive in northern Australia, namely *Cenchrus ciliaris*, *Urochloa mosambicensis*, *Digitaria decumbens*, *Brachiaria mutica*, *Cenchrus setigeus*, *Hymenachne acutigluma*, *Stylosanthes hamata*, *Calopogonium muconoides* and *Alysicarpus vaginalis* (Winter *et al.* 1984a). This list includes several grass species that originated and are successful in the African SAT. The priority in Africa is to obtain well-adapted legume species. Physiological and ecological information has assisted the understanding of adaptation but its use in the selection or breeding of better plants is still under study (Winter *et al.* 1984a). For the forseeable future, plant introduction and evaluation programs will be the main source of improved cultivars for northern Australia, although plant breeding may assist in the production of disease resistant cultivars, e.g. anthracnose resistance in *Stylosanthes*.

Crop Management

Seeding through crop residue mulch or pasture sod by the no-till system is a potentially useful soil management system in the SAT to conserve soil and water, reduce the maximum soil temperature, prevent raindrop impact causing surface crusting and sealing, and maintain organic matter at adequate levels (Lal 1984). Current research in north-west Australia (McCown *et al.* 1984) has relevance to the situation in west and eastern Africa. Australian experience to date using no-till technology indicates that: (1) the lower maximum soil temperatures which are a consequence of mulch retention benefit the establishment of seedlings; (2) good control of weeds is obtained by the application of glyphosate (Roundup) at moderate rates similar to those used in temperate regions; and (3) the most successful planter, in terms of seedling emergence over a wide range of conditions, has been a narrow tyne, preceded by a rolling coulter to cut surface mulch, and followed by a narrow in-furrow press wheel.

Most of the research into weed control in northwest Australia has involved herbicide evaluation (Peake *et al.* 1984) and would need adaptation to take account of the labour intensive nature of agriculture in the African SAT. Current research on alternative methods of weed control such as crop rotation and smothering by dense crops (A.L. Garside pers. comm.) could be relevant, but much more information is needed about weed biology and the nature of crop-weed competition before more use can be made of alternative methods in an integrated weed management system. Some basic research is currently being conducted into insect and disease control in north-west Australia (Allwood *et al.* 1984, Irwin *et al.* 1984). Nevertheless an integrated approach to pest control to minimize crop losses is necessary and the Australian experiences should generally be relevant between regions in the SAT.

Studies in nutrient management in cropping systems in northern Australia should have relevance to similar soil types in sub-Saharan Africa. Current research (Jones *et al.* 1984) includes: (1) assessing the significance of inputs of nitrogen from legumes when grown in rotation with crops; (2) assessing the losses of nitrogen and phosphorus from the soil/plant system-losses of nitrogen by erosion, leaching denitrification, and volatilization are of particular significance; and (3) assessing the efficiency of fertilizer use as determined by form of fertilizer, placement, timing of application, use of amendments (e.g. inhibitors and coatings), and crop husbandry practices.

Pasture Management and Beef Production

The beef industry in northern Australia is based primarily on grazing native grasslands. Current research deals with both the management of native pastures and improvement of cattle management and production (Winter *et al*. 1984b). It includes studies of the influence of soil characteristics, grazing and fire on the ecology of the existing grasslands. Native pastures commonly do not provide a diet adequate in nitrogen, phosphorus, sulphur and sodium for much of the year. These deficiencies can be overcome by feeding non-protein nitrogen (together with a carbohydrate source such as mollasses) and mineral supplements, but productivity from native pastures is ultimately limited by their low digestibility and susceptibility to overgrazing. The main emphasis of research on improved pastures is on the use of annual, semi-perennial and perennial legumes sown into native pastures alone or with exotic annual or perennial grasses. Stocking rates can be increased several-fold by oversowing a legume into native pastures but animal production and the stability and persistence of these pastures are related to the fertility level. With high applications of fertilizer and with annual grasses present, the pastures inevitably become dominated by the latter while at low fertility levels the legumes have predominated and mineral supplements have been required for good animal production. Further research is required into systems for using these pastures for year round grazing, strategic grazing or in combination with native pastures.

Integration of Crop and Beef Production

A research program in northern Australia is currently in progress which uses a systems approach (Dillon & Virmani 1984) in evaluating the biological feasibility of a no-till tropical legume ley-farming system for this climatic zone. The system includes the following features (McCown *et al*. 1984): (1) self-regenerating legume ley pastures of 1-3 years duration grown in rotation with maize or sorghum; (2) cattle grazes on native grass pastures during the wet season and leguminous pastures and crop residues in the dry season; (3) crops are sown directly into the pasture which is chemically killed at, or shortly before, planting; and (4) the pasture legume sward which volunteers from hard seed is allowed to form an understory in the main crop. Important research findings to date include: (a) one year of legume ley has generally provided a succeeding crop with 50-75 kg fertilizer N. Longer leys have had greater effects and a longer residual effect; (b) mulch retention due to no-tillage has generally resulted in increases to maize grain yield of over 20%; (c) herbicide and planting technologies have been developed which allow on-farm evaluation to proceed; (d) cattle have gained weight during the dry season when grazing legume leys and maize stover. Clearly the transfer of this new technology to the African SAT will depend on its successful adaptation to the socio-economic circumstances existing in sub-Saharan Africa.

Animal Health and Ruminant Production

The major part of the SAT in Australia is used for ruminant production and many of the problems are similar to those encountered in Africa. However, there are social and economic differences in the reasons for having stock. The Australian producers are looking for maximum production (or turnoff) whereas the African owner is often looking for the maintenance of the greatest possible numbers of stock and also stock for draught purposes. Basic nutritional research in Australia has increased the understanding of the constraints to livestock/ruminant production on low quality pastures characteristic of the SAT (reviewed by Jones 1983). While in Australia nutrition is the basic problem for ruminant production in the SAT, Africa also has major problems of disease and ecto-and endo-parasites.

Disease is less of a problem in the Australian SAT partly as a result of Australia's isolation and also through eradication campaigns. The disease of bovine pleuropneumonia has been eradicated and the campaigns against bovine brucellosis and tuberculosis are in their final phases, with large parts of Australia already being free from the two disease. The experience gained from these campaigns and the constant risk of introduction of other exotic pathogens has meant that Australian scientists maintain an active and effective understanding of the major exotic diseases.

The tick is a major disease problem of cattle in northern Australia with economic effects resulting from the spread of infectious diseases and the depression of cattle growth rates. Control has relied heavily on acaricides but chemical resistance has occurred. Reseach activities in Australia include computer modelling of tick population dynamics to improve understanding and control, stock breeding for tick resistance, and the development of vaccines against the cattle tick.

Australian scientists have long been aware of the effect of internal parasites on stock production and much research has gone into the control of internal parasites through antihelminthic drugs. With the increasing cost of drugs and the development of parasite drug resistance, techniques of biological control are being investigated such as manipulation of susceptible stock and of the infective stage larval population.

Programmes of eradication of diseases require specific tests for the detection of infected stock. The ELISA (Enzyme-linked immuno-sorbent assay) has many advantages over present disease tests – it can detect either the anti-body as in a serological test, or the antigen, i.e. the disease causing agent. This test is suitable for use in the field and does not require facilities of a diagnostic laboratory. At present research in Australia is being directed to establishing suitable tests of this system for diseases in a range of livestock.

Social Forestry

A serious problem facing sub-Saharan Africa is the growing shortage of fuelwood which is a major source of domestic energy for over 90% of the population. The loss of tree cover is also having a detrimental effect on the hydrological characteristics of catchments, soil erosion and the general stability and quality of the environment.

Wood, either harvested directly or as charcoal, accounts for up to 90% of the total energy consumed in Eastern Africa. In Tanzania, for example, fuelwood accounts for 97% of the total wood products consumed in the country and virtually all of this is harvested from natural forests (Kasembe et al. 1983). In the semi-arid regions of the country, because of over-exploitation up to 300 man-days per family are devoted to firewood collection which represents a serious inroad into scarce labour resources (Kasembe et al. 1983). Because of this rapid exploitation of the existing natural hardwood forests, there is an urgent need to establish fuelwood plantations throughout the areas of greatest demand and especially close to major urban centres. The area of planted forest required to satisfy this need in Eastern Africa by the year 2000 has been calculated by the World Bank as follows:

Country	M ha
Ethiopia	2.0
Sudan	2.3
Somalia	0.2
Kenya	1.0
Tanzania	1.2
Uganda	0.5

The current rate of fuelwood establishment falls far short of these targets and a major planting effort will be required in the next two decades to achieve a minimum level of fuelwood self-sufficiency in the region. In addition to wood lots for fuel production, there is considerable scope in Africa for the development of agro-forestry systems involving the integration of trees and food crops. Trees, especially legumes and other nitrogen-fixing species, when used in this way complement the cropping enterprise through the provision of biological nitrogen and through stabilisation of fragile tropical soils following clearing and cultivation for crop production. Nitrogen-fixing tree species such as *Leucaena, Acacia* and *Casuarina* also provide valuable fuelwood and can be regularly cut to provide high-value fodder to supplement the low quality diets of domestic ruminants.

In the past the main Australian genus used in Africa, *Eucalyptus,* has proved very valuable because of its rapid growth and the general suitability of the wood for a variety of purposes. At the same time the species commonly in use have been pressed into service for purposes for which they are not ideally suited and sometimes without appropriate management techniques for their production.

There exists still in Australia a large suite of tree species which so far has scarcely been tapped for use in agroforestry including not only *Eucalyptus* but several other genera, especially *Casuarina* and *Acacia*. While the potential of such species as well as additional *Eucalyptus* species is evident, their full employment in Africa will be realised only if there is a considerable research effort to assess their full value and establish the precise means for introducing and managing them specially in small scale fuelwood production and agroforestry systems. Both *Casuarina* and the newly described *Allocasuarina* as well as *Acacia* are nitrogen-fixing plants which will contribute significantly to the nitrogen balance in the systems in which they are grown, but considerable research is still necessary to determine the best combinations of plant and symbiotic micro-organisms. In *Casuarina* and the related genus there are at least 15 species which merit close attention while only three are at all commonly seen in cultivation and their use has been restricted. Among the Australian Acacias, of which there are some 700, there are some 50 or more species of tree form and most of these remain to be screened for silvicultural use.

Precise field collection is needed of seed of the little known species in all genera including *Eucalyptus*. They should be experimentally evaluated in Africa in comparison with indigenous species as rapidly as possible, by active research co-operation between scientists in the African countries and Australia. This co-operation should encompass fields such as systematic and dendrological research, vegetative propagation, symbiont selection and inoculation, provenance and gene/environment interactions, assessment of tree breeding and improvement techniques, to provide an experimental and information base on which active planting programmes can be based.

This contribution was compiled from contributions from the following Australian research workers: Dr R.C. Muchow (CSIRO Division of Tropical Crops and Pastures, Darwin), Mr A.G. Eyles (CSIRO Division of Tropical Crops and Pastures, St. Lucia), Dr J.G. Ryan (Deputy Director, ACIAR, Canberra), Prof. L.D. Pryor (Forestry Consultant to ACIAR, Canberra), Prof. J.R. McWilliam (Director, ACIAR, Canberra), Dr B.S. Harrap (CSIRO, Centre for International Research Cooperation (CIRC, Canberra), and Mr D.J. Brett (CSIRO, CIRC, Canberra).

References

Allwood, A.J.; Strickland, G.R.; Learmonth, S.E.; Evenson, J.P. (1984) Insects. In *Agro-research for the Semi-Arid Tropics: north-west Australia* (R.C. Muchow, ed), in press. Brisbane; University of Queensland Press.

Andrew, M.H.; Gowland, P.N.; Holt, J.A.; Mott, J.J.; Strickland, G.R. (1984) Vegetation and Fauna. In *Agro-research for the Semi-Arid Tropics: north-west Australia* (R.C. Muchow, ed), in press. Brisbane; University of Queensland Press.

Chaplin, G.R. (1981) Post-harvest handling of mangoes. *Northern Territory Department of Primary Production, Australia, Technical Bulletin* 55.

Cook, S.J.; Russell, J.S. (1983) Climate characteristics of CSIRO Division of Tropical Crops and Pastures Field Stations in Northern Australia. *CSIRO Division of Tropical Crops and Pastures, Technical Paper*, in press.

Dillon, J.L.; Virmani, S.M. (1984) The farming systems approach. In *Agro-research for the Semi-Arid Tropics: north-west Australia* (R.C. Muchow, ed), in press. Brisbane; University of Queensland Press.

Done, A.A.; Muchow, R.C.; Warren, J.D.; Kernot, I.I. (1984) Maize and sorghum. In *Agro-research for the Semi-Arid Tropics: north-west Australia* (R.C. Muchow, ed), in press. Brisbane; University of Queensland Press.

Garside, A.L.; Beech, D.F.; Putland, P.S. (1984) Grain legumes and oilseed crops. In *Agro-research for the Semi-Arid Tropics: north-west Australia* (R.C. Muchow, ed), in press. Brisbane; University of Queensland Press.

Imrie, B.C.; Putland, P.S. (1982) Breeding mungbeans for the monsoonal tropics. In *CSIRO Division of Tropical Crops and Pastures Annual Report 1981-82* (G.T. Adams, ed), 41-2. Melbourne; CSIRO.

Irwin, J.A.G.; Persley, G.J.; Conde, B.C.; Pitkethley, R.N. (1984) Diseases. In *Agro-research for the Semi-Arid Tropics: north-west Australia* (R.C. Muchow, ed), in press. Brisbane; University of Queensland Press.

Jones, R.J. (1983) Improving the nutrition of grazing animals using legumes, fertilizer and mineral supplements. In *Proceedings of the Eastern Africa/ACIAR Consultation on Agricultural Research, 18-22 July 1983, Nairobi, Kenya.*

Jones, R.K.; Myers, R.J.K.; Wright, G.C.; Day, K.J.; Mayers, B.A. (1984) Fertilizers. In *Agro-research for the Semi-Arid Tropics: north-west Australia* (R.C. Muchow, ed), in press. Brisbane; University of Queensland Press.

Kasembe, J.N.R.; Macha, A.M.; Mphuru, A.N. (1983) Country paper on agricultural research in Tanzania. In *Proceedings of the Eastern Africa/ACIAR Consultation on Agricultural Research, 18-22 July 1983, Nairobi, Kenya*

Lal, R. (1984) Soil surface management. In *Agro-research for the Semi-Arid Tropics: north-west Australia* (R.C. Muchow, ed), in press. Brisbane; University of Queensland Press.

McCosker, T.H.; Eggington, A.R.; Doyle, F.W. (1984) Observations on post-weaning performance of Brahman cross animals in the Darwin district of the Northern Territory. *Proceedings of the Australian Society for Animal Production* 15, 452-455.

McCown, R.L.; Jones, R.K.; Peake, D.C.I. (1984) Evaluation of a no-till, tropical legume ley-farming strategy. In *Agro-research for the Semi-Arid Tropics: north-west Australia* (R.C. Muchow, ed), in press. Brisbane; University of Queensland Press.

Muchow, R.C.; Yule, D.F. (1984) Irrigation. In *Agro-research for the Semi-Arid Tropics: north-west Austalia* (R.C. Muchow, ed), in press. Brisbane; University of Queensland Press.

Pcakc, D.C.I.; Fulton, M.C.; Putland, P.S. (1984) Weeds. In *Agro-research for the Semi-Arid Tropics: north-west Austalia* (R.C. Muchow, ed), in press. Brisbane; University of Queensland Press.

Scholefield, P.B.; Blackburn, K.H. (1984) Horticultural crops. In *Agro-research for the Semi-Arid Tropics: north-west Australia* (R.C. Muchow, ed), in press. Brisbane; University of Queensland Press.

Swindale, L.D. (1982) Distribution and use of arable soils in the semi-arid tropics. In *Managing Soil Resources. Plenary Session Papers, 12th International Congress of Soil Science, New Delhi,* 67-100.

USDA (1980) *Food Problems and Prospects in Sub-Saharan Africa: The Decade of the 1980's,* United States Department of Agriculture.

Williams, J.; Day, K.J.; Isbell, R.F.; Reddy, S.J. (1984) Soils and climate. In *Agro-research for the Semi-Arid Tropics: north-west Australia* (R.C. Muchow, ed), in press. Brisbane; University of Queensland Press.

Williams, R.W.; Lawn, R.J.; Imrie, B.C.; Byth, D.E. (1982) Weathering resitance in mungbean. In *CSIRO Division of Tropical Crops and Pastures Animal Report 1981-82* (G.T. Adams, ed), 41. Melbourne; CSIRO.

Winter, W.H.; Cameron, A.G.; Reid, R.; Stockwell, T.G.; Page, M.C. (1984a) Improved pasture plants. In *Agro-research for the Semi-Arid Tropics: north-west Australia* (R.C. Muchow, ed), in press. Brisbane; University of Queensland Press.

Winter, W.H.; McCosker, T.H.; Pratchett, D.; Austin, J.D.A. (1984b) Intensification of beef production. In *Agro-research for the Semi-Arid Tropics: north-west Australia* (R.C. Muchow, ed), in press. Brisbane; University of Queensland Press.

Chapter 5

The Experience of Commercial Banking in Agricultural and Rural Development

P.M. BOLAM

Head of Group Agricultural Services, Barclays Bank plc, Juxon House, 94 St Paul's Churchyard, London EC4M 8EH, UK.

It is a privilege to spend a week with agriculturalists and policy makers from so many countries and to address you for a short time on the "experiences of commercial banking in agricultural and rural development". We are drawn together with one central theme, that of finding ways in which the population of developing countries can feed themselves economically and consistently. The task has an urgency with few parallels in the modern world, for it is not only seeking to supply the increasing needs of the present population, but catering for the future as well. The increase will be 1.6% – 1.7% per annum, so that in 1984 there will be another 70-80 million more mouths to feed; each month a town equivalent to London will grow up. With better nutrition and the wonders of modern medicine, more children are born, more survive and on average, people live longer.

It is not part of my paper to talk on the problems of population growth but it has a direct bearing on our general theme. This enormous task would be made the easier and simpler with more active steps in population control – at conception rather than denying people full potential or full lifespan through malnutrition.

I have four credentials to stand on this platform: (1) I have been an agriculturalist for many years and latterly gained a great love of Africa. (2) I have been an agricultural banker for the last eleven years. (3) I am a Trustee of the Plunkett Foundation and (4) I am a very active member of the Royal Agricultural Society of the Commonwealth. So I have lived amongst crops and livestock all my life dealing with both large and small farming businesses and latterly have been involved in helping to lend money to them and to all the ancillary Industries as Head of Group Agricultural Services Department.

The Plunkett Foundation must be known to almost everyone at this Conference for it has gained so much respect by guiding and training those in farming co-operatives in Commonwealth countries. The RASC visits some Commonwealth countries every second year. We had a memorable time last year in Zambia and Zimbabwe studying the farming, but above all, having fruitful discussions on the problems and possibilities of Africa feeding itself. Contributions made by the African member countries were notable. It is a particular pleasure to meet them again at this Conference.

For most developing countries, the food producing sector is the most important of any in their national economy. For the thirty four or so countries classified by the World Bank as "low-income" countries, agriculture has made up 37% of their gross domestic product in 1981, compared with 14% in "middle-income" countries. Population growth within the developing countries is still increasing to the extent that during the next decade a 35% increase in food supplies will be needed just to match this rate of increase. When population growth is added to the increasing demand of the present population, the enormity of the problem is obvious. The food producing sector is an income-earner, engages the largest proportion of the population and produces the greater part of the GNP. Some feel that growing food has been a routine task for thousands of years. Since today lots of farmers are still on small plots of land, the job lacks glamour, importance and satisfaction compared with posts in administration and modern industry. So many of the rural population have the objective of getting away from farming to a job with more opportunity, when the urgent need is for them to remain in some aspect of food production, marketing and processing. The actual production of more food, both for human and animal consumption must have top priority, if economic growth in developing countries is to be achieved. In countries short of food, or where food is a vital export of the economy, the whole chain of providing resources, distributing them to the farmer on time, ensuring inputs are properly used and making sure that the crops are of marketable quality, are all together important ingredients of successful food production. They cannot therefore succeed unless the price structure and the financial funding and management skills are not available. It is significant that countries which make a good success of their agriculture are making the best job of developing their industry.

Table 1   Growth rates of agricultural output

Country Group	Agricultural Output		Per Capita	
	1960/70	1970/80	1960/70	1970/80
Developing Countries	2.8	2.7	0.3	0.3
Africa	2.7	1.3	0.2	1.4
Industrial Market economies	2.1	2.0	1.1	1.2
South East Asia	2.9	3.8	0.3	1.4

To date, Africa has failed to increase its food output faster than its population. How far then has lack of finance contributed to the African food problem?

Table 2   Total net resource receipts of developing countries from all sources.

	1970	%	1975	%	1980	%	1981	%
							$ Bn at 1981 Prices	
Official develop-ment assistance	20.83	42.5	30.72	36.9	35.27	38.0	35.51	34.1
Non-concessional flows	28.14	57.5	52.51	63.1	57.53	62.0	68.49	65.9
Total	48.97	100.00	83.23	100.00	92.80	100.00	104.00	100.00

The total financial resource receipts from all sources in 1981 reached $104Bn ($92.8Bn in 1980). Total flows accounted for between 4% and 5% of the developing economies' gross domestic product (GDP) and grew by 12% between 1980 and 1981. The growing importance of non-concessional flows is clearly evident, having increased from 57.5% to 62% during the 1970s. 1981 however, showed a reduction of official development assistance, largely as a result of a drop in OPEC support and fairly static levels of support from OECD in the form of bi-lateral aid and through their contributions to multilateral aid funded through international agencies.

Table 3   Total net resource receipts of developing countries from all sources.

Bilateral	1970	%	1975	%	1980	%	1981	%
							$ Bn at 1981 Prices	
			Official development assistance					
OECD	5.66	70.1	9.81	48.8	18.02	49.6	18.28	51.5
OPEC	0.39	4.8	5.68	28.3	8.26	22.7	6.91	19.5
Other	0.96	11.9	0.76	3.8	2.34	6.4	2.32	6.5
Multilateral agencies	1.07	13.2	3.84	19.1	7.74	21.3	8.00	22.5
Total	8.08	100.00	20.09	100.00	36.36	100.00	35.51	100.00

Table 4   Total net resource receipts of developing countries from all sources

	1970	%	1975	%	1980	%	1981	%
							$ Bn at 1981 Prices	
			Non-concessional flows					
Direct Investment	3.69	33.8	11.51	33.5	10.36	17.5	14.64	21.4
Bank Sector	3.00	27.5	12.0	35.00	19.00	32.00	25.00	36.5
Other Bilateral	3.54	32.4	8.25	24.00	25.10	42.3	23.85	34.8
Multilateral	0.69	6.3	2.58	7.5	4.85	8.2	5.00	7.3
Total	10.92	100	34.34	100	59.31	100	68.49	100

Bank lending to developing countries between 1970 and 1981 grew from $3Bn to $25Bn, or about 19 % per annum in cash terms, or around 8 % in real terms. What needs to be stressed is that the banking sector, and direct investment, made a sizeable contribution to the 12 % growth in aid between 1980 and 1981.

With the pressures on OECD countries, which account for about 90 % of the official development assistance, making it difficult to raise their aid: GNP ratios above the 0.36 % achieved during 1981, the non-concessional flow, particularly the commercial bank lending and direct investment, takes on a new significance. To quote OECD, "The outlook for international bank lendings suggests that the constraints are not primarily the ability of the banks to fund increased lending. The question is whether current developments, including recession conditions and high interest rates, "sharply reduce" all receipts and an increased awareness of political risk might lead to a deceleration in the growth of international bank lending to the developing country borrowers". Having repeated that quotation, let me make it plain that the major problems lie elsewhere in the world and not in this great continent of Africa. Historically, financial aid may not have been as high as the levels requested but the growth pattern has not been sluggish. Commercial lending has in part substituted for the lack of growth in institutional support, but it is interesting to note that within the levels of institutional aid, Africa's share of official development assistance has been 33.3 % against a 22.3 % share of the population within the less developed countries – a larger share per head.

In spite of this higher level of financial support, it is still evident from the decline in African food production that success is not solely a function of the aggregate levels of finance and aid. In other words more money is unlikely to provide a solution, in the absence of changes in economic policy.

Clearly, it will not be possible to meet the total requirement of the developing economies in terms of aid, and it may be that the present economic problems will slow down the rate of growth in total aid sources. It does not, however, follow from that that agricultural development will suffer as a result. A large number of countries now recognise that the drafting of "food strategies" provides a suitably agreed focal point on which aid recipient countries and multi-national and national aid agencies can accelerate agricultural production and economic growth. The "food strategy" has won the support of the United Nations General Assembly, the Organisation for African Unity, as well as the major donor countries of OECD. In fact, in 1982, the World Food Council opened the implementation phase for the campaign for food strategies.

The emphasis on policies and priority in food production will help to provide the momentum for continued development of agriculture but, oh dear, how weather can upset our planning! You of all people, know the effects of drought and the risks of rain-fed crops, but don't make it the excuse for a poor operation.

Within the last decade or so, to meet the demand for more food, there have been two main strategies. One is the large-scale project, mainly financed by outside money, growing stock and crop on land where often there has been no recent food production. These large-scale projects are an immense challenge to management and husbandry skills since often the seed, fertilisers, machinery and breeding stock have had to be transferred hundreds of miles to the site which has taken an immense planning. Some of the schemes which may have involved buildings and irrigation have been successful whilst others, alas, have been partial or total failures. We must know why, so that others don't make the same mistakes. In the light of long experience, it must be possible in 1984 to produce a format and timetable as a pro-forma and guide to future schemes, for there are many lessons to be learned and all resources in large projects, money, husbandry skills, seed fertilisers and management ability are in short supply, though some don't believe it. Aid comes from Government or Institutions, from taxes paid by ordinary people like you and me. There needs to be more accountability and assessment of value. What are the tangible results say per £100 000 spent. A whole industry has grown up in this area.

All my experience has taught me to have the greatest respect for the opinion of the farmers and the population who live in the countryside. They know the soil, the climate – what is possible and what is not. Some of these people should be involved in the planning at the earliest stage. Believe me, this experience is not just confined to your countries, but applies to anywhere in the world. The question for this Conference to answer is surely "When is a large-scale project justifiable and at what level of financial risk?"

The other method is to start producing more food by working with the small farmer and training him in better cultivation methods and understanding the simplest recording systems. I wonder if the Conference would consider that this perhaps provides a more permanent and long-term solution. As an agriculturalist of some experience, I have a feeling that starting with the small farmer and building upwards will, in the long run, create a more permanent source of food for the country involved, but the incentives must be adequate. The subsistence farmer of today must be the commercial farmer of tomorrow. If he isn't, we have wasted £Ms and many will starve to death.

There are many good examples of this approach in all the countries represented at this Conference. Barclays Zambian Lima Loan Scheme is one which readily comes to my mind – I will return to this later. In short, I believe there is a far greater possibility of success by involving more of the farmers in producing food for their country.

Many authorities, including the World Bank, believe that agricultural production is best stimulated by income incentives for producers, and have stressed the need to "improve the formulation and implementation of agricultural pricing policies". In short, pay the right price and streamline the system.

It is all too easy to be critical when developing countries face a choice of priorities and the demands of different sectors of the population, but it is worth recording the research of Professor Peterson. He compared real prices received by farmers in fifty three countries, industrialized and developing, in the period 1968/70 and showed that farmers in industrialized countries received more favourable prices than their counterparts in the Third World. He demonstrated that in those countries where the difference between product prices and costs was smallest, growth in agricultural output had been penalised by 40 – 60% with a corresponding loss of 3% in national income. Unfavourable ratios could reflect low product prices or high input costs, but where these ratios have been favourable, as in the EEC, increases in agricultural production have been stimulated. I must stress that I am talking particularly about prices for domestic consumption which includes raw materials for industry. Is it surprising therefore, that farmers in some developing countries feel that they have inadequate incentives to make money and improve their lot and that of their families?

Time does not permit me to elaborate on several aspects of economic policy which have a bearing on the national economy and food production. However, end product price distortion can occur when developing countries wish to build up domestic industry in a protected market. Control of inflation makes it essential to maintain high exchange rates but these artificially high rates militate against agricultural exports and hold back agricultural development by encouraging the importation of food stuffs paid for in an over-valued currency.

Authorities must ensure that sufficient public expenditure is allocated to improve infrastructure to ensure the adequate distribution of resources needed to produce crop and stock and to get the farmers' produce to market in prime condition. If adequate attention is not paid to these priorities, you will not get the food. Where these disincentives are such that output prices make agricultural production unprofitable, it is fairly evident that increasing the level of domestic credit to such businesses will simply add to the problems of the sector. There is little point in criticising lack of support from commercial credit institutions, if the ability to finance the cost of that credit, through agricultural production, does not exist.

In reviewing the role that the commercial banks have played, we must not ignore the part that the co-operatives and similar savings institutions for the small farmer have played. He will ultimately use this money (and others) to borrow, for crop and stock production, playing a major part in rural agriculture to date. People tend to overlook the fact that commercial banks, publicy and privately owned, are not simply lending bodies but have a dual function, for they also operate as savings institutions. The latter function is particularly important in developing countries since credit is a scarce resource and commercial banks, by providing banking facilities for co-operatives and other parties, are able to mobilise savings to finance internal development of a major kind.

It is precisely because credit is such a scarce resource that commercial banks have a duty not to waste a country's rare assets. The banks must account to their investors, depositors and shareholders for their actions and are therefore not to be considered by any borrower as charitable institutions. The banks are governed by the profit motive to enable them to expand their lending activities, but they deal with other people's money and consequently cannot afford to take unacceptable risks. Whilst some feel that this is a hard commercial line, if we look at the development of African agriculture over the next fifty years, the increase in food is dependent upon the farmers looking at their operation as a commercial venture and not as subsistence farming.

Our experience, particularly lending to small farmers through crop finance schemes, is that if they are visited regularly by bank advisors, in this case by young local men qualified in agriculture, and realise that with help to buy good seed, correct fertilisers and sprays, growing crops properly increases their chance of heavy crops which will allow them to repay their borrowings, then they can borrow again the following year to advance their business. The same applies to livestock of course. This method of elementary education surely leads to the development of the small subsistence farmer into one who will be commercially minded and therefore ultimately the backbone of food production. In Kenya 50% of both tea and coffee production is from small farmers and most of this is exported.

Commercial banks have traditionally been involved in short-term lending to the agricultural sector abroad. Funds have generally been made available to the industry by way of advances to marketing boards and co-operatives. This type of finance meets a vital need in less developed countries by ensuring that co-operatives and agricultural merchants have sufficient supplies of inputs available to producers for timely application. For example, Barclays Bank of Kenya Limited is involved with financing leading merchants and co-operatives of agricultural inputs, farm machinery and tools, to assure supplies to individual farms. The meat industry in Botswana is another example (see Chapter 44).

The success of these loan facilities in indirectly providing finance to farmer producers will ensure the continuance of such schemes, especially where these organisations provide a ready made infrastructure equivalent to a commercial bank's domestic network.

As long ago as 1945, Barclays recognised the gap which existed in the supply of medium term finance in developing countries and established the world's first specialist vehicle for the provision of term finance and equity investment for development projects, BODC in fact. This was in some way a "pump primer". Much of the work was ultimately taken over by other international agencies.

However, Barclays continued to provide support through their International Development Fund, which has a relatively small supply of finance to support projects in developing nations by way of loans, grants, guarantees and equity participation. The main aim of the Fund is to bring all projects up to commercial viability, since this is vital to a developing nation for sustained growth. We have supported many projects in African countries. Let me quote one that is exceptional for us which others might find helpful.

The Lima Loan Scheme in Zambia

This was a scheme which was part of a major food production drive in Zambia, spearheaded by President Kuanda in 1980, involving many foreign countries. Barclays established a domestic branch banking system in Zambia in 1916 and have been closely involved in farming ever since. Traditionally, we were involved with crop finance and term loans for livestock, but the involvement with very small-scale farmers had been somewhat haphazard, for lending small amounts with no ancillary earnings, often at high risk, takes considerable time and effort. Yet Barclays took a policy decision to devise a properly structured scheme to assist in financing production and educating farmers, both financially and technically, in association with Government bodies.

The Operation

We employed young credit supervisors, locally trained agriculturalists, each one looking after 200 farmers and answerable to the local bank managers. Each travelled by motorcycle and was housed near the bank. Each identified suitable small farmers lying within an acceptable distance of the local bank and recommended the advance, with the branch manager undertaking ultimate responsibility. Security available was very limited and documentation was kept to a minimum but supervisors maintained careful crop and credit histories, advising the farmer and teaching him the elements of credit management, whilst ensuring that the loan was used for its intended purpose and the agreed acreage had been planted or livestock bought. The loans had to be recovered for the scheme to progress because there was a specific amount of capital put aside for this purpose.

The scheme provided speedy decision making, so that inputs could be purchased and planting completed at the *proper time*. When applications were approved, payments could start immediately, maximising the opportunity for good crops. In the 1982/83 season we lent US $2.4M to almost 2500 farming families, financing 53 000 acres, mainly maize, but with a proportion of sunflower and cotton. The repayment performance is very encouraging – well over 90%.

What have we learned?

(1)   Proper management and organisation was *absolutely vital*, needing continual review to ensure that the scheme was as efficient as possible.

(2)   Regular visits by supervisors were essential and there was a correlation between the number of visits, the crop yield and successful loan recovery.

(3)   In effect the scheme encouraged farmers to increase their acreage and total production though there was a tendency for reduced output per unit area since management demands are more easily met over a wider acreage rather than making each acre more intensive.

(4)   Farmers' attitudes and inducements may vary according to area, size and type of farm. Initially, the scheme was for farms between 20 and 50 acres. Interestingly, our most successful areas in terms of productivity were where farms seemed to be between 15 and 20 acres. The evidence suggests we are stimulating the emergent commercial farmer.

(5)   It was necessary to kill the misconception that small-scale farming loans were "handouts" and repayment was unnecessary! We are not in the welfare business. Remember our aim is to develop good local commercial farmers. The farmer was fundamentally honest but needed to be reminded of his obligations. Having said this however, there are some remarkable success stories and many people increased their crop yeilds per acre by 200 – 300%.

(6)   We hope that the rise in income will encourage rural savings for old age and other envisaged needs, whilst providing more employment in the countryside. There is some evidence that this is happening.

This scheme has aroused interest within the European Commission and FAO.

Conclusions

(1)   Africa is a vast continent, three times the size of the USA or Australia.

(2)   Africa is short of food. The problem must be solved by those who live in Africa. Massive imports of food, whatever the price, is no solution.

(3)   Africa has the sun, the giver of all life and energy, many fertile soils and access to vast quantities of water for irrigation.

(4)   Africa has the capacity to be the bread basket of the world, to feed itself and export enormous quantities of food for foreign currency and import substitution.

(5)   But there must be the will and sheer determination to succeed. The goals are achievable. It is up to the leaders, the urban and rural peoples.

(6)   Farmers are the key – they are the same the world over – most of them will produce the food you need if you pay them to do so. There are plenty of success stories.

(7)   But when he presents his crop to the market, he must be paid the right price for the quality and quantity at the time of delivery, in cash. That may involve organisational changes in co-operatives and security.

(8)   The changes are not of the farmers making. The countries want more food urgently. You provide him with better seed, hybrid varieties that he must buy because retained seed is not suitable. To get yield potential, he needs more fertilisers of the right type, accessible to the farm before he needs it and agrochemicals to combat pest and disease. That needs cash, or short-term crop finance. After all the merchant wants his money too. Otherwise he has a cash flow problem. He becomes a reluctant banker forced to lend money, through credit to the purchaser. If he cannot be paid, it is the responsibility of marketing organisations to get their business conducted properly. Besides, if he has to sell to one buyer, there is a moral obligation to pay. Competitive markets are different.

(9)   The alternative is for the farmer to revert to subsistence levels. His family will be fed. The townspeople will suffer.

(10)   Then take care, the next generation, the potential larger commercial farmer, better educated, will not necessarily accept the way of life of their parents for their expectations are higher. They may leave the land. It has happened.

(11)   If producer prices are the victim of national and international manipulation, or constraints once more that vast increase will not be forthcoming to meet the continually increasing demand.

(12)   There is an urgent need for better business management skill, to ensure resources are at hand for the farmer at the right time – delay is a waste to everyone – and at harvest, to store and market the crop. Management must be unfettered and isolated from political pressures to achieve their stated business objectives.

There is no doubt whatsoever that Africa can feed itself if this is the prime objective. As a representation of the banking systems operating in this great continent of yours, we will play our part in ensuring that finance is not the stumbling block, providing the right economic environment exists. I hope the Conference will accept my forthright comments in the spirit in which I make them, namely a very deep concern for the food problem which will become increasingly severe and will not, in my view, be solved by massive gifts of food from others.

Chapter 6

The Effects of Pest and Non-pest Constraints on the Production and Utilization of Crops

T.R. ODHIAMBO

Director, The International Centre of Insect Physiology and Ecology (ICIPE), P.O. Box 30772, Nairobi, Kenya.

Introduction

Ten years ago the World Food Conference in Rome called world attention to the seriousness of hunger, and ushered in new initiatives to improve food security, particularly for the developing countries. The subsequent governmental actions have focussed primarily on two areas: firstly, in increasing food production at the national level; and, secondly, in reducing price fluctuations in world markets. The *Lagos Plan of Action for the Economic Development of Africa, 1980–2000*, formulated by the Heads of State and Government meeting in Lagos in April 1980 dwelt on these two themes, and recognized pest management as one of the crucial elements in Africa's fight for food security.

These actions, however, have still to make a major break in the declining trend in agricultural production, and the worsening food deficit situation in Africa. While Asia and Latin America have, over the last two decades, achieved considerable Green Revolution production levels in their staple food crops (rice, maize and wheat), and have shown modest increases of exports of these commodities (2.5% for Asia and 3.6% for Latin America over the period 1961/65 – 73/77), Africa has registered a worsening crisis in food production and in agricultural production as a whole. Over the period 1961/65 – 76/80, output of basic food staples in sub-Saharan Africa increased at the rate of only 1.7% y^{-1}. When this is related to the human population growth rate, there was in fact a decline of food production per capita of the order of 1.1% y^{-1} (Paulino 1983).

Yet, agriculture is the dominant economic sector in Africa, and seems destined to remain so for the foreseeable future. In eastern and central Africa, more than half of the economically active population is engaged in agriculture, and this level usually exceeds 70%, and in oil-rich Nigeria, agriculture is still the dominant economic sector in terms of its contribution to the national Gross Domestic Product (GDP). Agricultural production, then, is a truly central policy issue for the continent. Its mandate is five-fold (West 1976): (1) responsibility for producing food for domestic consumption; (2) responsibility for earning foreign exchange through world markets (to pay for consumer goods, as well as for the importation of capital goods and raw materials for the domestic manufacturing industry); (3) provision of rising income for the rural people; (4) provision of savings for the development of the domestic non-agricultural sector; and (5) creation of gainfull employment.

Risks in Farming in Africa

Risks in farming in Africa are higher than in any other part of the developing world, when key factors are taken together: (1) Most soils are of low natural fertility, are extremely fragile and rapidly lose their productive capacity under intensive cultivation, and are poorly known in scientific terms; (2) the physical environment is harsh, and exerts environmental stress in most croplands: the weather is highly variable, the rainfall is often badly distributed and comes in convective storms, high soil temperatures are frequent, and water stress is common; (3) pests, diseases and weeds are diverse and abundant, and we know little of their detailed biology, host relationships, host defences against them, and the actual losses due to their predations under different cropping systems; and (4) productivity of agricultural labour seems to be low, and is becoming scarce at a time when it is most needed.

A crucial element permeating this risk complex is the weakness of the knowledge base upon which we are structuring agricultural production technology in Africa. It is not helpful to talk of technology transfer to Africa from the developed temperate world when there is little to transfer relevant to African conditions. Nor is it perceptive of African decision-makers when they regard mission-orientated basic research, which would provide this essential knowledge base, as a luxury item on the national or regional budgets. Further, when the people who intimately know about farming in Africa and desperately depend on it are the traditional farmers, it is short-sighted to ignore them and their traditional bank of knowledge.

The risks involved in farming in Africa translate into the chronic problem of wide fluctuations in food consumption. This is due principally to the degree of variability in crop production, and the extent to which this variability is compensated for by changes in imports or strategic reserves. Variability in food production is high; in the Sahel it ranged from $9.56 - 22.07\%$ coefficient of variation in the period 1961–1977. Thus, every Sahelian country has a high probability that annual cereal production variation will be more than 10% of the trend (McIntire 1981). Similarly, although smaller, fluctuations are common in other parts of Africa south of the Sahara. In North Africa they are exceeded.

A priority in agricultural policy should therefore be the establishment of means for: (1) increasing food production through mission-orientated bio-agronomic research (including the rationalisation and upgrading of traditional crop production systems), followed by investment in implementation mechanisms for these upgraded crop management practices; and (2) creating buffer stocks that can absorb the surplus from high production years to act as a hedge for years of poor harvests. However, the establishment of buffer stocks is not really cost-effective (Adams 1983); a more efficacious and cheaper mechanism is to purchase needed stocks from the market place.

The vital question for African scientists, farmers, and decision-makers to address is the development of locale-specific technologies that permit continuous crop production over many years. This is the most important problem to solve in Africa.

Major Constraints on Crop Production

Soil poverty and fragility

Oxisols and ultisols occupy extensive areas in the tropics; they are characterized by a poverty in nutrients, high acidity, and deficiencies in phosphorous and some micro-nutrients. Consequently, even though physical characteristics may favour crop production, they can sustain only restricted utilization. These soils are strongly acidic and leached, and are therefore poor and easily degradable. The soils of the lowland humid tropics are fragile, as seen when heavy equipment is used to clear a forest area. In the Cameroons, removal of a mere 2.5 cm layer of litter and humus decreased the yield of maize by approximately 50%; if 7.5 cm of the top soil was removed, the remaining soil became completely infertile (Tributh 1980).

The west African semi-arid tropics (WASAT), in the same geographical area as the humid tropical region, show an equally severe low natural fertility. These soils have a low structural porosity, which tends to reduce root penetration and water translocation, show a tendency to compaction and hardening during the dry season, and have a high and increasing susceptibility to erosion in circumstances of continuous cultivation and up to 60% loss of rain-water due to run-off. The WASAT soils are consequently shallow and highly unstable, exhibiting rapidly declining productivity under continuous cultivation.

We are now starting to appreciate the vitally important place that traditional systems of soil fertility management have played, namely the evolution of the bush fallow system as an adaptation to this constraint (Getahun et al. 1982). The challenge is to upgrade this knowledge and develop a package which will permit continuous and sustainable crop production in the same plot over many years. Emerging technology components include land development methods and forest clearing techniques that disturb the soil humus and top-soils as little as possible, minimum or "zero tillage" combined with mulching techniques, agroforestry systems (especially alley cropping), and mixed or relay cropping methods (ter Kuile 1983).

Harsh or unpredictable physical environment

A feature of rainfall in Africa is that although it may be adequate for crop production in aggregate, its distribution is so variable and unpredictable that it often causes multiple replantings, or the failure of grain-filling in cereals or proper fruiting, or even spoilation of the mature crop. In East and Central Africa rainfall is the dominant, constraining natural resource in respect of the agricultural potential (Collinson 1983), only 3–6% of the area has an adequate amount for permanent agriculture (over 1250 mm y^{-1}), it is highly variable during the cropping season; and shows a variable start and finish. The rainfall patterns limit the full potential of the abundant solar energy being exploited.

High soil temperatures found in Africa usually discourage the exploitation of microbial nitrogen-fixation legumes. The potential for improving nitrogen-fixation in legumes is being explored. IITA (1982) has screened 750 cowpea rhizobial isolates from Nigeria and Niger. Many rhizobia from the hot, dry environment of Maradi grew as well at over 37–44°C as at 30°C, but few of those from cooler environments tolerate 37°C. Thirteen isolates were effective on several high-yielding cowpea cultivars; grain yield of VITA-7 inoculated with 3 of the 13 isolates was markedly greater than that of uninoculated cowpea. One question to be pursued further is the survival of high temperature-tolerant rhizobial strains in the absence of the host. Identifying or developing cultivars tolerant to environmental stress is an important avenue to follow in assuring output in often harsh and variable climates. This needs to be coupled with a more effective agrometeorological service in all parts of Africa (see Chapter 62).

Pressure of pests, diseases, and weeds

Crop pests, plant diseases, and arthropod-borne diseases of livestock, are probably the greatest barrier to the introduction of higher-yielding crop varieties and animal breeds. Research to enable us to understand host defences to pests and disease vectors, the development of resistant plant cultivars, and the development of animal resistance (or immunity) to disease vectors, is crucial as a foundation for the design of integrated management systems.

Weeds also substantially limit crop yields in tropical Africa. Weeds seem to be better adapted to the harsh climate and nutrient-poor soils of Africa than food crops (ter Kuile 1983); the abundance and diversity of plant life in the tropics simply translates into a more abundant and diversified weed population which competes with crops in the largely fragile and rapidly depleted tropical soils (Odhiambo 1984).

These sources of potential yield losses add up to about 42% in Africa (Table 1). These losses, inflicted on an already unassured agicultural production and low yields, aggravate an already poor productivity level, often the lowest in the world. The traditional farmer's primary incentive in Africa is therefore to seek stability in his agricultural output.

Table 1   Losses of potential crop production by pests, diseases and weeds together (Wortman & Cummings 1978).

Region	Value (US$ M)	Loss (% potential value)
South America	9 276	33.0
Europe	35 842	25.0
Africa	10 843	41.6
Asia	35 715	43.3
World	137 439	33.8

Pesticides (insecticides, acaricides, fungicides, herbicides, etc.) are not the simple and final answer to pest and disease problems. They are expensive, and seem to be destined to be (1) an emergency measure (such as the control of an outbreak of a migrant pest), or (2) in a selective and species-specific manner within an integrated pest management system (IPM).

Mixed cropping (or intercropping) systems, the predominant technology of crop production in Africa, and a longstanding traditional practice, seems to offer a technology that can stabilize crop production without appreciable loss to soil fertility, and contain pest losses within a sub-economic level (Odhiambo 1979). After a long period of hesitation there are now serious investigations into the complex relations represented in the intercrop situation in progress. This type of basic research is relevant, important, and crucial to the African farmer, not only to understand what is going on, but also to provide a knowledge base on which to build up rational, productive intercropping and agroforestry systems which, in turn, give stable yields on a sustainable basis.

Socio-economic constraints   The performance linkage between agricultural contributions to the national GDP, investment in research and development activities, and the articulation of technological inovations to actual implementation by the farmer, are of vital importance in a productive agricultural enterprise. While the agricultural sector is the leading contributor to the GDP of Nigeria, the entire expenditure on this by Government only amounts to 5.5 % of total public sector expenditure. In the period 1975–1980, the Nigerian Development Plan allocated only 0.3 % of its federal expenditures to agricultural research, whereas the agricultural sector was expected to contribute 21 % to the GDP (Idachaba 1980).

Lack of an effective delivery system of research results (and of new technological developments) to farmers is prevalent and acute in most of the continent. Further, while extension officers are trained in formal agriculture (usually largely based on temperate patterns) the farmer feels that he is being talked at rather than talked with, and becomes reinforced in his isolation.

Future Potentials

It is a good sign that many scientists and decision-makers in Africa have not cried out for a Green Revolution in the continent, for this requires, as it has done in Asia and Latin America, the selection and development of high-yielding crop varieties that require high soil fertility (or fertilizers), good soil moisture (or irrigation), good pest control (or a pesticidal umbrella), and good agronomic management to give optimum performance. It seems to me that we should consider a different alternative, one that fits more closely to the agro-ecological framework of Africa, and its socio-cultural knowledge of agricultural production. It eschews the achievement of high yields as a first goal. The alternative *modular approach* to progressive crop improvement requires:–

(1) Development of locale-specific agricultural production methods which assure a stable and continuous productivity under the existing resource-poor farmer knowledge base, but rationalised and upgraded to yield predictable results in a more technologically ordered manner. Stably-performing crop varieties and animal breeds would be expected to have a sufficient range of productivity potential to respond with higher outputs when environmental and other constraints are significantly reduced.

(2) Development of an integrated pest management system (IPM) for pests, diseases, and weeds which will reduce yield losses to sub-economic levels on a continuing basis, without making conventional pesticides the centre-piece of the scheme. Components would include host resistance to pests, cultural methods, the use of biological agents (parasites, predators, and pathogens), utilization of genetic techniques for modifying pest populations, and the development of better forecasting of pest and disease outbreaks. Pest management is more of a regional problem in Africa than in other areas in the world, and must therefore be addressed with all the talent and resources we can muster in Africa.

(3) Use of the full potential of our enormous endowment of germ-plasm (cereals, legumes, vegetables, pasture plants, tree-crops, wild ungulates, etc.) to develop, modify, or create cultivars or breeds having a wide range of characteristics to fit more effectively into the new, upgraded cropping and husbandry systems that will arise from mission-oriented research on successful traditional systems (intercropping, agro-forestry, and mixed farming). Such a development will naturally lead to a more intensive use of land, and to a more commercial type of farming, thus gradually phasing out the resource-poor farmer.

The accent here is on a partnership between the natural scientist, the technologist, the social scientist, the farmer, and the state. This may remind one of the way Thomas (1973) analogised the ways of the honeybee and the way meaningful science can happen:

> "If you want a bee to make honey, you do not issue protocols on solar navigation or carbohydrate chemistry, you put it together with other bees (and you'd better do this quickly, for solitary bees do not stay alive) and you do what you can to arrange the general environment around the hive. If the air is right, the science will come in its season, like pure honey."

References

Adams, R.H. (1983) The role of research in policy development: the creation of the IMF Cereal Import Facility. *World Development* 11, 549-563.

Collinson, M. (1983) Technological potentials for food production: East and Central Africa. In *Conference on Accelerating Agricultural Growth in Sub-Saharan Africa*. Harare; International Food Policy Research Institute and the Department of Land Management, University of Zimbabwe.

Getahun, A.; Wilson, G.F.; Kang, B.T. (1982) The role of trees in farming systems in the humid tropics. In *Agro-forestry in the African Humid Tropics* (L.H. MacDonald, ed.), 28-35. Tokyo; United Nations University.

Idachaba, F.S. (1980) *Agricultural research policy in Nigeria*. [Research Report No. 17.] Washington, DC; International Food Policy Research Institute, 70 pp.

International Institute of Tropical Agriculture (IITA) (1982) *Research Highlights for 1982*. Ibadan; IITA.

McIntire, J. (1981) *Food Security in the Sahel: variable import levy, grain reserves, and foreign exchange assistance*. [Research Report No. 26.] Washington, DC; International Food Policy Research Institute, 70 pp.

Odhiambo, T.R. (1979) Linking basic research to crop improvement programs for the less-developed countries: biological control of insects. In *Linking research to crop production* (Staples, R.C.; Kuhr, R.J. eds), 153-198. New York; Plenum Press.

Odhiambo, T.R. (1984) International aspects of crop protection: the needs of tropical developing countries. *Insect Science and Its Application*, in press.

Paulino, L.A. (1983) *The Evolving Food Situation in Sub-Saharan Africa*. Washington, DC; International Food Policy Research Institute, 81 pp.

ter Kuile, C.H.H. (1983) Technological potentials for food production in the humid and sub-humid tropics of Africa. In *Conference on Accelerated Agricultural Growth in Sub-Saharan Africa*. Harare; International Food Policy Research Institute and the Department of Land Management, University of Zimbabwe.

Thomas, L. (1973) *Lives of a Cell.* Boston; Massachusetts Medical Society.
Tributh, H. (1980) Problems of soil science in connection with cultivation measures in Cameroon. *Natural Resources and Development* 12, 82-90.
West, Q.M. (1976) Food and cash crop competition. In *Nutrition and agricultural development* (Scrimshaw, N.S.; Behar, M. eds), 163-169. New York; Plenum Press.
Wortman, S.; Cummings, R.W. (1978) *To Feed this World.* Baltimore; John Hopkins University Press, 440 pp.

Chapter 7

Cereal Production Needs and Achievements

H. DOGGETT

15 Bandon Road, Girton, Cambridge CB3 0LU, UK.

Introduction

The major cereals are well served, so far as germplasm and breeding are concerned, by the International Agricultural Research Centres. There are two cereals deserving of more attention: finger millet, and *Setaria* millet. Finger millet is one of the oldest African cereals, perhaps the oldest; sorghum and finger millet must have been developed around the same time. Finger millet is very drought-resistant, but also tolerates waterlogging: like sorghum it may be transplanted. *Setaria* millet has a low water requirement, and a good quality grain. It is likely to be valuable as an additional crop in a cropping sequence, as well as a main crop in very drought-prone areas.

Maize, sorghum, and bullrush millet are well served by the Centres. The major problems have been identified, such as grain quality, and a start has been made towards their solution. Wheat also can draw on good stocks of germplasm. The potential of rice has not been explored properly in Africa. This is likely to become an important crop as excellent rice can be grown in Tanzania and Kenya.

On the breeding side, recurrent selection in populations is likely to become most important, we still need to learn how to handle populations both economically and effectively. They are especially useful in situations where there is a lack of continuity of professional staff. A population can be maintained by technicians, and expanded into line selections later when the staff situation improves.

Production needs

We must consider advancing agricultural production against a background of falling per capita production in Africa, certainly of food crops. I propose to leave the achievements to others and concentrate on the needs for production. Having spent 26 years in agricultural research on rather isolated stations in East Africa, I wonder why there is so little to show in farmers' fields for the research carried out on those stations?

The small producer

The cereals producer in many parts of Africa is the small farmer. He is often a man with only a few hand tools, a family to assist him as his sole labour force, and a plot of land. The cropping season is a period of unremitting hard toil, with labour peaks in which priority has to go to food and basic needs for the family. His limited financial resources are swamped by his great needs. The answer to advancing agricultural production in Africa lies in making it worth the small farmer's while to produce more, and to produce it well. If it is sufficiently worthwhile to use inputs and changed technology he will do it but only if it is fully worth his while to do so.

Prices and marketing

Prices are a key issue: attractive, consistent farm-gate prices will bring the increased production needed. Good farm-gate prices can only be paid when there is an efficient buying and marketing system reaching right to the farmer's village. Payment must be cash on delivery; the minimum purchase price must be advertised before the time of land preparation and must be adhered to even if circumstances change. The requirement to provide firm support prices for essential crops cannot easily be fulfilled by most African Governments. In a bumper harvest the surpluses purchased can neither be stored nor disposed of without severe financial loss. Donor agencies should consider providing further assistance in this area.

Taxation systems yielding sufficient Government revenue are difficult to design and implement; inevitably, the difference in the price obtained in the export market for cash crops and that paid to the producer is used to meet Government expenditures, so reducing the price paid to the small farmer. Another factor operates in the case of cereals. For realistic, practical reasons, food prices in towns need to be kept as low as possible. Where money to subsidise those prices is not available, less money can be paid to the producer. It ceases to be worth the farmer's while to produce surplus grain. In the 1950s in Sukumaland, pearl millet (mawele) was grown on the hill sands, sorghum on heavier land, and there was some maize; two years ago, the cereals were gone and in their place were expanses of cassava. Without diminishing in any way the credit due to Indian scientists and plant breeders at International Centres, there can be no doubt that a major factor in the achievement of Indian grain production targets has been the implementation of the Government's Agricultural Price Support policy. The strength of the Indian economy makes it possible to operate these price supports.

In the Teso district of Uganda last year I found farming to be booming. Kampala is a wonderful market for all kinds of produce. Cassava is being grown in large quantities, and is used in three ways. There, it is not only a subsistence crop; lorry loads are sent for sale in Kampala and some is used to feed pigs. Substantial numbers of pigs are sent to Kampala and fetch an excellent price. The Teso farmer does not appreciate that pigs cannot be raised in Uganda because of disease problems and just keeps sending more to market. A good range of basic goods is available in the shops but these are too expensive for people living on salaries; however, a lorry load of pigs will buy a great deal. This boom did not result from planning but farmer initiative.

Perhaps the main agroclimatological zones should be designated and the most reliable, best adapted cereal named for each zone. A support price might then be promulgated for each cereal, at least within its designated zone. In the past, maize is the only cereal which has received price support, to the detriment of the other cereals whose prices were uncertain.

Technology

The cereals must form a component of a cropping or farming system. The best rotational system for a viable mix of food and commercial crops needs to be devised. Particular attention must be given to the interaction of crops in sequences so that the best available "natural" level of weed and pest control is achieved. This will be very location-specific, and must be determined through national research programmes. National crop improvement programmes need to be strong, with good continuity of staff. Improved varieties often provide the key to the use of new technologies in developing cropping systems as they are usually much more responsive to inputs than are traditional types. The potential of new cropping systems will not be realised until every crop in a system is represented by an improved cultivar. Several African crops have received little or no attention from plant breeders, and even those which are included in the programme of an International Centre also require breeding work in national programmes to adapt material to local conditions. For example, in Sri Lanka the national rice research station has taken new International Rice Research Institute (IRRI) lines and crossed them with local material to produce a whole range of well adapted rice; soil and water conditions differ from place to place in Sri Lanka. Some of these varieties have also proved successful in South China and parts of South East Africa.

There is already a body of knowledge available in most African countries on cropping practices that result in consistently good yields. President Nyerere in his address to the "Resource Efficient Farming Methods" Workshop on 16 May 1983 noted the value of using farmyard manure (FYM) or compost. The traditional method of agriculture that moved southwards from Ethiopia used manure. Results all confirm its value; Ukiriguru annual reports for 1948-57 show that 3 tons of FYM still gave a residual benefit after 3 years, 5–7 tons gave significant responses in the fifth residual year on both cotton and millet. The use of cattle manure was developed by African farmers in Africa, and used by their descendants for many years, but today the practice has largely disappeared. Why? The prospects for adoption of new technologies and research advances must yield an adequate return for the small farmer's hard work.

Involving the Small Farmer

There is often a gulf between the small farmer and extension workers and scientists of a Government agricultural department. The small farmer carries out long, hard manual labour and possesses the transmitted knowledge and wisdom of experience, but seldom has much education. Departmental officials are well educated, but seldom have had to make a success of farming themselves. Failure to establish mutual respect can result. There can be a failure to recognize the extent of the farmer's widsom and knowledge. The farmer's traditional methodology may be just brushed aside, and he may be advised, or sometimes even commanded, to implement a changed system which has been devised on an experimental station by scientists. Each group has much to learn from the other. Much has been written about Farming or Cropping Systems Research. Under the heading "A philosophy of collaborative research" Harwood (1979) wrote:

> "Small farmers in the developing countries, however, need research that is aimed directly at the practical problems of agricultural development, and attuned to the actual circumstances of their lives. The method proposed includes a certain amount of basic research in varietal improvement, disease and pest management, plant physiology, and soil fertility. But the major emphasis is on production research, planned and carried out by and with the farmers on their own fields. This fresh approach is not a substitute for either basic research or continued technological development. Rather, it is a way of making sure that the fruits of knowledge and technology are shared with the smaller farmers, who are often excluded from agricultural improvement programs."

The Asian Cropping Systems Network (ACSN), led by Harwood for some years, has shown how successful research of this type can be. It consists of a network of projects in national programmes, linked with IRRI and each other by a network co-ordinator. We must learn how to implement in Africa "production research planned and carried out by and with the farmers on their own fields" (Harwood loc.cit). Work at Ahmado Bello University in Nigeria has been moving towards this for some years and CIMMYT is now trying to initiate this approach in Eastern Africa.

Attitudes towards Agriculture

President Nyerere reminded us last May that 80 % of the people of Tanzania obtain their living from the land, and that almost all the remaining 20 % are the children of peasant farmers. The same must be true of many African countries, yet a sense of pride in being farmers is often lacking.

Agriculture is often regarded as an inferior occupation, and even extension workers may feel positively despised. Yet without agriculture, those in the more admired and respected occupations would either starve or be obliged to leave the country. The economically successful western countries all have a strong agricultural sector which is carefully nurtured. The building of a sense of pride in farmers and agricultural achievements is essential to advance agricultural production. Farmers, extension workers, and agricultural scientists must become highly respected members of the community.

National Research and Extension Services

Strong national agricultural services are the essential foundation for research together with the farmer, which will lead to improved farming and so to increased agricultural production, including production of cereals.

A strong developing agriculture requires able and experienced national scientists (including social scientists) and extension workers, as well as good farmers. It is hard to attract such people, and to keep them in agriculture, for three reasons:−

(1)   The problems of returning to one's own country after several years of postgraduate training in the West: the difference in the standard of living, the isolation of some of the agricultural stations, the problems of good schooling for one's children, medical services, of finding a job for an educated wife, and the companionship of other well-educated ladies for her.

(2)   The attractions of life in the cities, and the enticement of other posts which offer better prospects of financial reward, rapid promotion, and political or social influence.

(3)   Agricultural projects are popular with donor agencies, and are liberally staffed with foreigners. Local governments are reluctant to spare their best men to double-up for foreigners when they are needed so badly elsewhere, and the best local scientists can find more challenging and congenial jobs than understudying a foreign "expert". Too few foreign "experts" have a thorough understanding of tropical small farmer subsistence agriculture, or are young and flexible enough to learn and adapt rapidly.

Possible Solutions

Postgraduate Training

It is unrealistic to expect men and women who have lived and worked for 3-5 years in a Western university to return to their own countries and settle down easily on isolated stations. More postgraduate training could be done in Africa, or elsewhere in the developing world, with more senior scientists exchanging with their peers in the West. Donor agencies should be invited to support more postgraduate institutes of agriculture as training centres.

Rewards

The agricultural sector needs most of the able, best qualified men, because a stong national agricultural service provides the foundation of the basic economy. The attractions of the agricultural sector need to be improved to a level where people working in other sectors of the economy are trying to get jobs in the agricultural sector. Salary, "perks", and prestige are all important.

Donor agencies might consider "topping up" local salaries of agriculturalists nominated by national Governments who are qualified for posts in the international market. As an economy strengthens, the level of "topping up" would be expected to decline.

Foreign experts may be used as network co-ordinators for groups of projects, so that local project leaders have to carry the responsibility themselves, but are visited regularly, and have group workshops and exchange visits between projects organised for them. The foreign experts can then act as guides and friends, but leave actual managment and control in local hands. They would then be a part of the training programme, helping to guide and "coach" young agriculturalists as they gain practical experience and self-confidence.

Conclusion

In summary, I suggest that the following are important for increased crop production:– (1) Attractive, stable farm prices for produce. (2) Farmer orientated, farmer involved research. (3) Greatly increased respect for agriculture and agriculturalists. And (4) strong national research programmes, requiring (a) more graduate and postgraduate training done by locally based institutions, (b) inducements for nationals to stay in their own countries although qualified to work abroad, and (c) foreign advisers deployed as network co-ordinators rather than as project scientists or leaders.

Reference

Harwood, R.R. (1979) *Small Farm Development.* Boulder, Colo; Westview Press.

Chapter 8

Environmental Constraints to Rainfed Cereals Production in Africa: a country assessment

A.H. KASSAM[1], G.M. HIGGINS[2] and H.T. VAN VELTHUIZEN[3]

[1]Echemess Development Services, 5/7 Singer Street, London EC2A 4QA, UK; [2]Soil Resources, Management and Conservation Service, Land and Water Development Division, FAO, Rome; [3]FAO Consultant.

Introduction

Environmental constraints to rainfed crop production are imposed by factors of climate, soil and landform, and in association with factors of technology and management set agronomically attainable limits to production potential. Food self-sufficiency is one of the prime objectives of national agricultural policies of African countries, and in the formulation of development plans and strategies, a knowledge of the environmental land resource endowment and its potential is an essential prerequisite. Planning for an accelerated expansion in cereal production calls for a comprehensive knowledge of technology-specific environmental constraints for the different cereal crops concerned. Information on environmental constraints has to indicate their location, geographical extents and effects on response to production inputs, and thus on crop-specific suitability of land utilization types (describing the associated farming systems circumstances). This reference information is generally formulated as part of the national level data base routinely used by national and district planners.

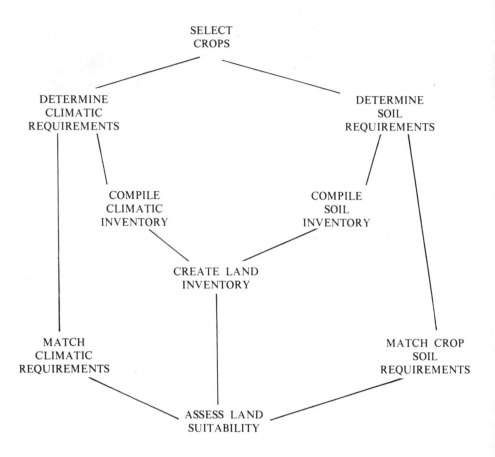

Fig. 1   Schematic outline of the methodology

This paper describes an assessment of environmental constraints to rainfed production potential of four major cereals undertaken in Mozambique as part of a larger national assessment of potential land use. It outlines the methodolgy used in the assessment of the environmental resources inventory of climate and soil compiled to rate environmental constraints, their extents and effects on rainfed production potential of the four example cereal crops. Such an approach can contribute towards institutional and farm level development in the agricultural sector, in a relatively short period of time, given the incentive and financial resources for investment. The methodology is suited for adaptation to studies covering a wider range of crops at sub-national and project levels, in most parts of the African continent. A first approximation assessment for all the countries in Africa for the four rainfed cereal crops was undertaken by the FAO's Agro-ecological Zones Project (FAO 1978) using the approach described in this paper, and generalized results for the African region are presented as an overview.

Methodology

The methodology used for assessing environmental land suitability of rainfed crops including cereals is schematically summarized in Fig. 1.

The following sections describe the various activities that are necessary to determine crop requirements, to prepare soil and climatic inventories, assess agro-climatic and agro-edaphic suitabilities and finally to prepare an environmental land suitability classification.

Crops and Land Utilization Types

The four selected rainfed cereal crops are: maize, sorghum, pearl millet and wheat. Each crop is evaluated at two levels of inputs, a low and a high. Combined description of produce, inputs, technical know-how, etc. form the basis of the definition of the land utilization types employed in the assessment. The assumptions used for the two levels of input circumstances, to describe the conditions under which the crops are to be produced, are as follows (Kassam *et al.* 1982a).

Under the low level of inputs assumptions, cereal production is geared towards subsistence farming, and is based on the use of traditional local cultivars, manual labour, no or negligible use of fertilizer and biocides, and use of fallow periods to maintain long-term soil fertility.

Under the high level of input assumptions, cereal production is geared towards commercial farming and is based on the use of improved cultivars with high yield potential, mechanized labour, adequate fertilizer and biocides, and fallow periods to conserve moisture.

Pertinent in the definition of the land utilization types are the usages, growth cycles and other characteristics of the cereal species and cultivars to be grown.

Climatic and Soil Requirements

Climatic requirements

To enable the cereal crops to be matched to a climatic resources inventory, climatic requirements of crops were based on their agro-climatic adaptability characteristics as related to productivity (Kassam *et al.* 1977). Moisture and temperature are specially important in relation to aspects such as photosynthesis, growth habit, phenology, yield formation, pests and diseases and cultural operations (Kassam *et al.* 1982a).

Photosynthetic and phenological characteristics of the crops bear a relationship to yield. The rate of growth is directly related to crop photosynthesis, which is conditioned by the carbon assimilation pathway and its response to temperature and radiation.

Within any suitable length of moisture growing period, the temperature regime (and photoperiodic regime when the available cultivars are day-length sensitive) governs which crop cultivars can be considered for production. When the climate of an area is phenologically suitable for a crop, it is possible to relate the temperature and radiation regimes to crop productivity from the knowledge of the photosynthetic response to radiation and temperature.

Accordingly, a complete inventory of crop climatic requirements was prepared using a first order grouping based on phototsynthetic characteristics. For cereals, four main groups are recognized, namely:

Group I – (e.g. wheat) C_3 photosynthesis pathway; optimum temperature range for photosynthesis 15-20°C; adapted to moderately cool and cool conditions (mean daily temperature 10-20°C). *Group II* – (e.g. rice) C_3 photosynthesis pathway; optimum temperature range for photosynthesis 25-30°C; adapted to moderately warm and warm conditions (mean daily temperature > 20°C). *Group III* – (e.g. pearl millet, lowland sorghum, lowland maize) C_4 photosynthesis pathway; optimum temperature range for photosynthesis 30-35°C; adapted to moderately warm and warm conditions (mean daily temperature > 20°C). *Group IV*– (e.g. highland sorghum, highland maize) C_3 photosynthesis pathway; optimum temperature range for photosynthesis 20-30°C; adapted to moderately cool conditions (mean daily temperature 15-20°C).

Soil requirements

To enable the cereal crops to be matched to soil resources, edaphic requirements of the crops were determined, based on their edaphic adaptability characteristics as related to productivity (Kassam *et al.* 1982a). These requirements must be understood within the context of limitations imposed by landform and other external features (phases) which may have a significant influence on the use that can be made of the soil.

Soil requirements were defined in terms of internal and external requirements, and for each crop, soil characteristics considered meaningful to crop production were listed e.g. soil depth, soil texture, salinity, stoniness, etc. For each crop, each property was then quantitatively sub-divided into those for optimum conditions and for a range of conditions e.g. pearl millet: optimum soil pH

5.5-7.5, range of soil pH 5.2-8.2. Beyond the critical range the crop cannot be expected to yield satisfactorily unless special precautionary management measures are taken.

Inventory of Environmental Land Resources

Climatic resources inventory

The usefulness of a climatic inventory, for predicting agroclimatic suitability of crops, is dependent on how far the climatic requirements of the crops can be matched with the climatic parameters used in the inventory. Crop requirements are therefore a prerequisite for determining these parameters. A special climatic data bank, comprising 134 stations each with up to 31 years of rainfall records, was compiled for the study (Kassam *et al.* 1981) while the details of the climatic resources inventory compiled for the assessment are given in Kassam *et al.* (1982b).

Providing temperature requirements are met, the degree of success in the growth and production of a crop is largely dependent on how well its optimum length of growth cycle fits with the period when water is available for growth. Curtailment of the growth cycle is naturally reflected by decreased yield and the same is true for enforced extended growth cycles.

The climatic resources inventory characterizes both heat and moisture conditions through the concept of the length of growing period, being defined as the duration (in days) when both water and temperature permit crop growth. A reference moisture supply (from rainfall and soil) of more than half potential evapotranspiration has been considered to permit crop growth, while reference mean daily temperatures greater than 5°C have been considered as being conducive to growth.

Quantification of the *heat attributes* during the growing period is achieved by defining thermal zones representing the actual temperature regime during the growing period. The temperature thresholds used in these definitions accord with those differentiating the four groups of crops previously described, and therefore allow matching of the temperature requirements of the crops with the temperature parameters used in the climatic resources inventory.

Quantification of *moisture conditions* in the growing period is based on a water balance model comparing precipitation (P) with potential evapotranspiration (PET).

The data utilized for the calculation of the water balance and for further climate-related calculations comprises meteorological station records where extended data on rainfall, maximum and minimum temperatures, vapour pressure, wind speed and sunshine duration are available on a monthly basis.

The growing period is the time when moisture supply exceeds half potential evapotranspiration; it includes the time required to evaportranspire up to 100 mm of soil moisture storage. A normal growing period has a humid phase when moisture supply is greater than potential evapotranspiration. When there is no humid phase, the growing period is defined as intermediate.

The quantification of moisture regimes is based on the analysis of length of growing period for each year separately and the computation of: the number of separate growing periods per year; length of each growing period and its various moisture periods; the quality of moisture conditions during the growing period and its various moisture periods; and year-to-year variability.

The climatic resources inventory was compiled by plotting, on a 1:2M base map, the individual station values of: (1) temperature regimes during the growing period, and identifying areas with a similar temperature regimes and delineating these regions (thermal zones); (2) historical profiles of patterns of numbers of growing period, and identifying areas with a similar pattern and delineating these regions (pattern zones); and (3) mean total dominant length of growing period, and identifying areas with similar lengths and delineating these by constructing isolines of growing periods with values 60, 75, 90, 120, 180, 210, 270 and 300 days, delineating lengths of growing period zones of 30-59 days, 60-74 days, 75-89 days, 90-119 days, etc.

Soil resources inventory

The soil map of Mozambique at 1:2M scale (Voortman & Spiers 1981) has served as the inventory of soil resources, providing the essential soil, slope, texture and phase data. The legend of the soil map is in accordance with the terminology of the FAO/UNESCO Soil Map of the World (FAO 1971-81), and in Mozambique 34 different soil units of the FAO/UNESCO system occur (16 major soil units) (Kassam *et al.* 1982c).

The soil units have been defined in terms of measurable and observable properties of the soil itself and are combined into 32 "diagnostic horizons". These soil properties and diagnostic horizons are relevant to soil use and production potential, and therefore have a practical application value for predicting possible optimum uses of soils.

Three textural classes (coarse, medium, fine) and three slope classes (0.8 %, 8-30, > 30) were distinguished. Phases indicate land characteristics which are not considered in the definition of the soil units. For Mozambique four phases (stony, petric, lithic, dune) were recognized.

Land resources inventory

Superimposition of the climatic inventory on the soil map allows the creation of unique land units within which soil and climatic conditions are known and quantified. The resultant map output is original and creates the ecological land units of the study, wherein areas with similar soils and climates are delineated (Higgins & Kassam, 1980).

After compilation of the land resources inventory in the map-form, area measurement of each soil mapping unit was effected as it occurred in each thermal zone, in each pattern zone and in each length of growing period, in each province. The data resulting from the above described measurements is used to create the computerized land resources inventory. Information on both mapping unit composition and extent is converted by a computer programme into extents (km²) of all individual soil units in each map unit, as they occur in each thermal zone, pattern of growing period zone and length of growing period zone by province (Kassam *et al.* 1982c)

Agro-Climatic Suitability Assessment

Thermal zone matching and constraint-free yields

Comparison of the climatic requirements of the crops, with the climatic conditions of land units, is the basis of the agro-climatic suitability assessment. Firstly the temperature requirements of the crops are compared with the temperature conditions of the identified thermal zones. This step indicates which crops should be considered, from a temperature viewpoint.

The next step is to compute the constraint-free yield of a suitable crop in the various lengths of growing period zones inventoried in that thermal zone. This is calculated from the net biomass production (total plant dry matter), taking into account the biomass production capacity of the crops (of assumed growth cycles and leaf area indexes) as influenced by the response of photosynthesis to radiation and temperature (Kassam 1977, Kassam *et al.* 1982d).

Agro-climatic constraints

The constraint-free crop yields reflect yield potentials in the prevailing temperature and radiation regimes on crop photosynthesis and growth in the various lengths of growing period zones.

However, yield reductions due to agro-climatic constraints of rainfall variability and moisture stress, pests, diseases and weeds, and workability need to be taken into account to quantify agronomically attainable yields.

Four groups of constraints and three severity ratings have been used in the inventory of agro-climatic constraints (Kassam *et al.* 1982d). They are: *group "a"* resulting from moisture stress during the growing period; *group "b"* due to pests, diseases and weeds, directly affecting the physical growth of the crop; *group "c"* due to various factors, including pests and diseases, and temperatures, affecting yield formation and quality; and *group "d"* arising from difficulties of workability and produce handling.

The three severity ratings are: 0: none or only slight constraints, resulting in no significant yield losses; 1: moderate constraints, resulting in yield losses of the order of 25 %; and 2: severe constraints, resulting in yield losses of the order of 50 %. The severity of the four groups of constraints were assessed by crop, thermal zone, length of growing period zone and level of inputs (Kassam *et al.* 1982d).

In general, with increasing length of growing period and wetness, constraints due to pests and diseases (groups "b" and "c") become severe earlier for the low input cultivator. However, as the length of growing period continues to extend, even the high input level cultivator cannot keep them under control and, in the longer growing periods, these two groups of constraints become severe yield reducing factors at both levels of inputs. Other factors, such as poor grain quality in very short lengths of growing period zones, are of similar severity for both levels of inputs.

The attainable crop yields in each length of growing period (for each year separately) are derived by application of the appropriate yield losses, defined by the agro-climatic constraints ratings, to the constraint-free yields.

The reductions are made consecutively according to the presence (or absence) of the agro-climatic constraints in each thermal zone, in each length of growing period zone and at each level of inputs.

The agro-climatic suitability classification (Kassam *et al.* 1982d) is based on computed values of agronomically attainable yields, taking into account the year-to-year variability in the *number* of growing periods per year and in the *lengths* of growing periods. This gives the average (long-term baseline) agronomically attainable yields for each crop by each mean total dominant length of growing period, by each pattern of growing period, by each thermal zone for each input level.

The agro-climatic suitability classification for each crop at both input levels, was achieved by considering the whole agronomically attainable yield range and classifying each individual (inventoried) mean total dominant growing period zone yield into one of the following five classes. If the yield of a crop from a particular zone is 80% or more of the maximum attainable, that zone is assessed as agro-climatically very suitable (VS) for that crop. Zones with yields of 60 to less than 80% are classified as "suitable" (S); 40 to less than 60% as "moderately suitable" (MS); 20 to less than 40% as "marginally suitable" (mS); and less than 20% as "not suitable" (NS).

An exception to the general methodology for agro-climatic suitability assessment applies to fluvisols because the climatic inventory does not fully reflect their particular moisture regime and, to some extent, temperature regime. Accordingly, a separate classification for fluvisols has been considered.

Agro-edaphic Suitability Classification

The agro-edaphic suitability classification is based on (1) the soil unit evaluation and (2) modification of the soil unit evaluation by limitations imposed by slope, texture and phase conditions (Kassam *et al.* 1982d).

The soil unit evaluation is based on how far the conditions of a soil unit meet crop requirements under a specified level of inputs. There are five basic classes for each crop and level of inputs. If the soil unit largely meets the crop's requirements and yield suppressions if any would be less than 20% it was adjudged S1 (very suitable). If the soil unit only partly met the crop's requirements, it was adjudged S2 (suitable), or S3 (moderately suitable) or S4 (marginally suitable), corresponding to yield suppressions of 20-40, 40-60 and 60-80% respectively. If the soil unit failed to meet the crop's minimum requirements, it is graded as N (not suitable). A rating of N would override any agro-climatic suitability, and the agro-climatically attainable yields are not possible. Similar rules were made for the various slope and texture classes and soil phases.

Environmental Land Suitability Classification

Both the agro-climatic suitability classification and the agro-edaphic ratings are necessary to arrive at the environmental land suitability classification. This takes account of all the inventoried attributes of land and compares them with crop requirements, to give an easily understood picture of the suitability of land for the production of the crop.

In essence, the land suitability classification has been computed by modifying the extents of lands in the five agro-climatic suitability classes by the ratings of the various soils inventoried in those areas. The area of each growing period zone, its agro-climatic suitability, and the extent and degree of soil limitations to crop production, is used to compute the area of land variously suited to the production of the particular crops at the two levels of inputs considered.

This has been achieved by applying the programme illustrated in Fig. 2. The assessment is carried out separately for each crop and level of inputs.

Firstly, the crop's temperature requirements with regard to photosynthesis and phenology are compared with the prevailing temperature conditions of each thermal zone. If they do not accord,

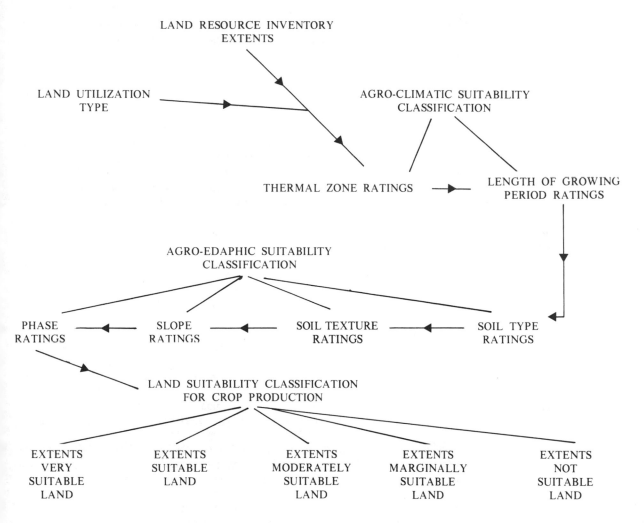

Fig. 2   Schematic outline of the land suitability assessment programme

that thermal zone is classified as not suitable. If they do accord with crop requirements, all growing period zones in that thermal zone are considered for further suitability assessment.

This consists of applying the length of growing period ratings of the agro-climatic suitability classification, to the computed areas of the various growing period zones. Thus, if a particular growing period zone is agro-climatically "very suitable" (VS) for the production of the crop, then all areas of this zone are classified, in the first instance, as "very suitable" from the agro-climatic viewpoint. If half the areas of a growing period zone are agro-climatically "very suitable" (VS) and half "suitable" (S), then half of the extent of that growing period zone is computed as "very suitable" and half as "suitable".

The next step is an appraisal of the soil units present in each growing period zone. The rating of soil units, for the crop and level of inputs under consideration, is applied to the computed area of the growing period zone occupied by each soil unit. This leads to appropriate modifications of the agro-climatic suitability assessment. Subsequently, the ratings for different soil texture and slope classes, and soil phases are applied consecutively to arrive at the final land suitability appraisal for the crop, for the level of inputs under consideration.

The five classes of environmental land suitabilities are similar to those used for the agro-climatic suitability classification (very suitable to not suitable) and are related to attainable yield as a percentage of the maximum attainable under optimum agro-climatic and agro-edaphic conditions; these provide the necessary data for calculation of the rainfed production potential of any given area in Mozambique.

Assessment of the Africa Region

Using the approach described, the FAO's Agro-ecological Zones Project undertook a first approximation assessment of environmental constraints to rainfed production and land suitability of a number of crops, including cereals, for all the countries in Africa (FAO 1978). In this regional assessment, 11 thermal zones (characterizing the rainy season thermal regimes), and 19 moisture zones (characterizing the rainfed lengths of growing periods) were inventoried. These climatic mapping units in combination with 89 FAO soil units, 3 slope and texture classes and 11 phases inventoried for Africa led to an inventory of environmental land resources (by individual countries) comprising a total of 36 868 climate-soil land units over a total area of 2 878 $\times$ 10^6 ha (Higgins & Kassam 1980).

Climate resources inventory

The thermal zones inventoried in the tropical and sub-tropical Africa correspond to areas with mean temperatures (during the rainfed growing period) of: more than 20°C (warm), 15-20°C (moderately cool), 5-15°C (cool) and less than 5°C (cold). The temperature thresholds used here accord with those differentiating the four major cereal crop groups described earlier. These are subsequently matched to the 19 moisture regimes inventoried. Areas with year-round humid growing period (365 + days) were inventoried separately from areas with year-round growing period with some rainfall deficit (365 − days).

The data shows that 29.4 % of the Africa region has dry desert conditions, 17 % has inadequate growing periods (1-74 days), and 0.3 % is cold with severe temperature constraints. The other 53.3 % is climatically suitable for rainfed cereal production (including rice) of some kind with 19 % having shorter (75-179 days), 33.6 % long (180-365 days) and 0.7 % year round growing periods.

Soil resources inventory

The soil inventory of the Africa region was compiled from the FAO/UNESCO Soil Map of the World (FAO 1971-1981). A generalized interpretation of this data reveals the following overall major limitations: 18.6 % is occupied by cambisols, fluvisols, luvisols and nitosols, which do not have inherent fertility limitations, 14.6 % by acrisols and ferralsols which do have severe fertility limitations, 3.4 % by vertisols which are heavy cracking clay soils, 2.2 % by solonchaks and solonets that are salt-affected soils, 5.3 % by gleysols which are poorly drained soil, 13.1 % by lithosols and other shallow soils, 19.7 % by arenosols and regosols which are coarse textured soils and 16 % by xerosols and yermosols which are semi-desert and desert soils.

Environmental land suitability classification

This is derived from the agro-climatic suitability classification and the agro-edaphic suitability classification and estimates the extent of land variously suited to the rainfed production of each crop, by input level, and classified by thermal zones and by lengths of growing periods.

Conclusions

Comparison of the individual country estimates with data on area already being harvested reveals that there is a large overall land reserve for arable crops including cereals in Africa. However, in some countries such as Kenya, Rwanda, Somalia, Niger and Reunion, these reserves are very limited particularly when considered in relation to the cereal food requirements of their future (year 1990 and 2000) projected populations (Higgins *et al.* 1981).

Any large scale national level increase in cereal production may of necessity have to come from intensified production from the presently cultivated areas instead of from deliberate expansion of area under cultivation with a low level of technology and under a subsistence setting.

In addition to providing data on extent, location and attainable potential of rainfed arable and cereal areas, and the extent of land reserve, the country assessment such as that for Mozambique also provides data for answers to questions related to effective planning of land use for cereal production. Critical areas where land resources are insufficient to meet cereal food needs of the population now (or projected to be) living on them (Higgins *et al.* 1983), can be revealed.

The assessment can also indicate which areas and which cereal crops will give the highest returns to increased inputs. Areas giving maximum productivity returns from an increased level of inputs are located in the areas of maximum suitability for each cereal. In the case of rainfed lowland maize, the area of maximum suitability is located in the 150-210 days lengths of growing period zones.

Country assessments also allow limitations to production and relevant research priorities to be identified together with the location and extent to which they apply. The land inventory reveals areas affected by environmental constraints to production, and allows formulation of research and development priorities to overcome them and to meet the most important needs of an area as a whole.

The need for adequate inventories of environmental resources (climate and soil) to meet such basic information needs of planners, research workers and development agencies, is clear. It is hoped that this paper also emphasizes the need for an interdisciplinary and interinstitutional approach in such work, for environmental resource inventories which cannot be matched to land use and development requirements will be of limited value.

References

FAO (1971-1981) *FAO/UNESCO Soil Map of the World, 1:5 000 000*, 10 vols. Paris; UNESCO.

FAO (1978) Reports of the Agro-ecological Zones Project. *World Soil Resources Report* 48 (1) – *Africa*. Rome; FAO.

Higgins, G.M.; Kassam, A.H. (1980) The agro-ecological zones land inventory. Working Sheet 10. In *Report on the second FAO/ UNFPA expert consultation on land resources for populations of the future*. [Project INT/75/ P13.] Rome; FAO.

Higgins, G.M.; Kassam, A.H.; Naiken, L.; Shah, M.M. (1981) Africa's agricultural potential. *Ceres* 14(5), 13-21.

Higgins, G.M.; Kassam, A.H.; Naiken, L.; Shah, M.M.; Fischer, G. (1983) Potential population supporting capacities of lands in the developing world. *Technical Report of Project FPA/INT/513 – Land Resources for Populations of the Future*. Rome; FAO/UNFPA/IISA.

Kassam, A.H. (1977) Net biomass and yield of crops. *Consultant's Report. Agro-ecological Zones Project*. Rome; AGLS, FAO.

Kassam, A.H.; Kowal, J.M.; Saraf, S. (1977) Climatic adaptability of crops. *Consultants' Report. Agro-ecological Zones Project*. Rome; AGLS, FAO.

Kassam, A.H.; van Velthuizen, H.; Higgins, G.M.; Christoforides, A.; Voortman, R.L.; Spiers, B. (1981) Data bank and analysis of growing period. *Project MOZ/75/011 Land and Water Use Planning, Mozambique* [Field Document no. 33.]

Kassam, A.H.; van Velthuizen, H.; Higgins, G.M.; Christoforides, A.; Voortman, R.L.; Spiers, B. (1982a) Land utilization types and ecological adaptability of crops. *Project MOZ/75/011, Land and Water Use Planning, Mozambique*. [Field Document no. 32.]

Kassam, A.H.; van Velthuizen, H.; Higgins, G.M.; Christoforides, A.; Voortman, R.L.; Spiers, B. (1982b) Climatic resources inventory of Mozambique. *Project MOZ/75/011, Land and Water Use Planning, Mozambique*. [Field Document no. 34.]

Kassam, A.H.; van Velthuizen, H.; Higgins, G.M.; Christoforides, A.; Voortman, R.L.; Spiers, B. (1982c) Land resources inventory of Mozambique. *Project MOZ/75/011, Land and Water Use Planning, Mozambique*. [Field Document no. 35.]

Kassam, A.H.; van Velthuizen H.; Higgins, G.M.; Christoforides, A.; Voortman, R.L.; Spiers, B. (1982d) Agro-climatic and agro-edaphic suitabilities for rainfed crop production in Mozambique. *Project MOZ/75/011, Land and Water Use Planning, Mozambique*. [Field Document no. 36.]

Voortman, R.L.; Spiers, B. (1981) A national soil map of Mozambique at scale 1:2 000 000. *FAO/UNDP Project MOZ/75/011, Land and Water Use Planning Mozambique*. [Field Document.]

Chapter 9

Rice – Production Problems in Africa

KAUNG ZAN, V.T. JOHN and M.S. ALAM

International Institute of Tropical Agriculture (IITA), P.M.B. 5320 Ibadan, Nigeria.

Introduction

Rice is a strategic component and a crucial element in the staple food economies of several African countries. Demand for rice in sub-Saharan Africa is becoming more acute as a result of a general dietary shift from conventional foods, and is likely to continue to rise. Production therefore has to be stepped up to avert a serious economic drain on the foreign exchange reserves in most countries.

Africa had about 3.5% of the world's total rice area in 1982 and nearly 2.2% of total production. About 50% of the rice area is classified as upland, which in many respects is unfavourable for rice production. The remainder is subjected to varied water regimes and nutrient imbalances, while only 11% of the entire area is under irrigated ecology.

In order to increase production, the World Bank loaned the various rice projects in Africa US $199M (FAO 1983) in 1982. The estimated area of rice in Africa for 1985 is 5.7M ha and for 1990 6.8M ha. During the next 15 years, an additional 3M ha are expected to be under irrigation.

Rice situation in Africa

Rice Area

The African rice area in general has increased from 1961-65 to 1982 by 62.7%. West African rice land forms nearly 50% of the entire rice area in Africa and shows a 28% increase from 1970 to 1980. On the other hand, the northern region shows a decrease of 15% and central, east and south Africa, an increase of 11.0% over the 1970 figure. The total rice production in Africa is obviously affected by the contribution from West Africa.

Eighteen out of a total of 31 African rice growing countries increased their rice area from 1970 to 1980 with a total area of 3 789 000 ha in 1980. Seven countries with a total area of 1 011 000 ha in 1980 showed a decrease in area, prominent among these being Guinea and Egypt. Six countries with a total area of 75 000 ha did not show any change in area.

Production

Rice production in Africa registered a steady increase of 64.6% from 1961-65 to 1982 simulating the world trend.

The increased production from 1970 to 1980 was contributed by two regions – West Africa, and central and eastern Africa. On the other hand, the decreased production of northern Africa is symbolized by the drastic fall in hectarage of rice land in Egypt.

Per hectare yields in paddy production in the major rice growing countries in tropical Africa, excluding Egypt, remained low in general from 1969-71 to 1980. The major reason for the increase in total rice production in Africa is the increased area under cultivation; the yields per hectare in individual countries need increasing.

Based on the per hectare productivity compared between the figures of 1970 and 1980, fourteen countries have registered an increase in per hectare yields of rice, whereas 11 countries show a decrease and 5 no change.

**Self-sufficiency –
Supply/Demand Situation**

Next to Far Eastern countries, Africa is the major importer of rice but unlike in the Far East imports in Africa have steadily risen from 1979 to 1982. Although the main rice exporting country in Africa is Egypt, the quantity exported has been steadily decreasing from a total of 178 000 t in 1976-78 to 25 000 t in 1982 (FAO 1983).

There is an almost constant or sometimes increasing trend in rice imports by several rice growing countries of Africa. The increased demand for importation stems from the fact that the self-sufficiency level has been decreasing while the per capita consumption has been increasing. The local supply for consumption remained almost constant while the per capita consumption has increased from 1975 and, probably as a result of increased population, has lead to higher import levels and lowered self-sufficiency.

Problems and Prospects of Rice Production

Systems of rice culture

It is generally agreed that rice growing systems (eco-systems) in the region should be grouped as follows: (1) *Dryland rice* grown on naturally freely drained soils, are entirely dependent upon rainfall. (2) *Hydromorphic rice* is grown on the soils where rice roots are periodically saturated by fluctuating water table. (3) *Mangrove swamp rice* is grown in swamp along the coastal regions with tidal intrusion. (4) *Inland swamp rice* is grown on flat or V-shaped valley bottoms and flooded to different degrees, which sometimes leads to floating conditions. (4) *Paddy rice* is grown on bunded paddies, either under rainfed or irrigated conditions.

Although reliable statistics on the areas under different ecosystems are difficult to obtain, reasonable estimates have been made for West Africa. Recently estimates of area in upland, lowland rainfed, deep water and irrigated ecologies for all of sub-Saharan Africa show that of a total rice area in Africa of 4.1M ha nearly 50 % is upland (including hydromorphic), mostly in West Africa, 30 % to lowland rainfed, mostly in central and east Africa, 11 % is irrigated and 8 % is deep water rice all quantities to West Africa.

Climatic resources and constraints

Climatic factors such as rainfall, temperature, solar radiation and photoperiod influence the physiological expression of the genetic potential of the variety. Climate also influences indirectly the incidence of pests and diseases, which in turn affect the grain yield.

The quantity, reliability, duration and frequency of rains affect the suitability of an area for rice cultivation. In the high rainfall humid zone of West Africa, where upland rice is grown extensively, drought is not a major problem but moisture supply becomes the most limiting factor for upland rice in transition and savanna regions of West Africa and elsewhere particularly on soils having a low water holding capacity. This situation justifies drought resistance research in rice varieties.

Rainfall regimes in major parts of central Africa are more favourable for rice production although the full potential or rainfed rice production there is yet to be fully assessed.

Temperature greatly influences the growth pattern, duration and finally the yield of rice plants. Air temperatures normally below 20°C and above 30°C markedly affect the growth and yield of rice (Yoshida 1978). The mean minimum temperatures in several rice areas of central and eastern Africa and Malagasy during the rice growing season drop below 25°C and at times much lower.

The effect of high temperature is more pronounced at the heading stage, when the fertility of spikelets is adversely affected by temperatures above 35°C, as in Sahelian countries.

Solar radiation in the tropics is higher in the dry than wet season. Consequently, the dry season yield is higher. Low solar radiation, particularly during the reproductive stage, has a pronounced effect on spikelet number. This situation occurs in rainfed rice in the forest belt of West Africa and seems to limit the possibility of obtaining high yields (Posner 1978). Even under good management, light can easily become a limiting factor in the humid zone of Africa.

Soil resources and constraints

The world's most productive rice land consists mostly of young alluvial soils in river deltas and inland valleys. Soils in such rice areas are predominantly hydromorphic and fertile, but their distribution in the wetter regions of tropical Africa nevertheless limits the rapid expansion of rice production.

(1) *Well-drained upland soils* are marginally suitable for upland rice because of their low

available water holding capacity and even relatively short dry spells are detrimental to the performance of upland rice which depends largely on available soil moisture reserves and tolerance of the rice variety to drought stress. Nitrogen is the most limiting plant nutrient for upland rice. The behaviour of available nitrogen in soils with a seasonally fluctuating ground-water table may cause severe nitrogen deficiency. Phosphorous and sulphur deficiency have been observed in rainfed upland rice grown in the savanna and iron deficiency may be observed with continuous cultivation or after burning following land clearing. Leaching losses, soil acidity and multiple nutrient deficiencies in high rainfall regions may constitute severe production problems on strongly leached soils.

(2) *Hydromorphic and swamp soils.* Iron toxicity in rice grown in the inland valleys or swamps is common, especially in Liberia, Sierra Leone and Nigeria. Poor drainage caused by the physical conditions of the terrain may result in unfavourable reduced soil conditions.

(3) *Lowland coastal soils.* The majority of the mangrove swamp soils along the West African coast are potential or actual acid sulphate soils (Moormann & Veldkamp 1977) containing variable amounts of pyrite (FeS) which, upon aeration of the soil, will oxidize, leading to soil acidification. This also leads to formation of high levels of Fe^{2+} and Al^{3+}, which can affect the growth and production of rice. Marine salinity derived directly from sea water intrusion in some lowland coastal areas of West Africa is a limiting factor. During the wet season, with the higher flow volume of the river, salinity can be reduced by gradual washing.

Assessment and Evaluation of Potential Rice land in Africa

Systematic assessments are yet to be carried out on the soil conditions and land characteristics for the various types of rice cultivation in Africa. Lack of more detailed soil and land use surveys in most African countries is among the major constraints limiting rapid expansion of rice areas in Africa. In Malagasy and Tanzania (Zanzibar and Pemba), most of the rice land is irrigated and includes a wide range of soils and land types. Even where rainfall is adequate in mainland Tanzania, rice is less likely to be grown than any other food crop. However, the country offers good opportunities for rapid expansion of rice cultivation in valleys and swamp land. The possibility and potential of utilising the three large hydromorphic lowlands, namely the lower Niger river delta, the Congo river basin, and the upper White Nile swamps, for rice cultivation are yet to be explored but appear to offer a high potential.

Biological Constraints

Among biological constraints, lack of high yielding improved varieties, diseases and insect pests, weeds and vertebrate pests are the major ones. For raising the yield per unit area, the small farmers of Africa need varieties with reasonable but stable yield levels, appropriate growth duration, and tolerance or resistance to environmental stresses and devastating diseases and insects. Cramer (1967) estimated the world's average yield loss at about one-third of the value of the rice crop; 13.8 % is caused by insect pests, 11.6 % by diseases and 9.5 % by weeds. According to Barr *et al.* (1975), the yield loss attributable to insects, disease and weeds in Africa was 33.7 % of the potential production.

(a) Superior varieties

High yielding semi-dwarf varieties developed by the International Rice Research Institute (IRRI) in the Philippines are grown in several African countries for lowland rainfed or irrigated conditions, but varieties that are well adapted to the complex African ecologies are still to be identified. The rice research program of the International Institute of Tropical Agriculture (IITA) is developing rice varieties adapted to free drained upland, hydromorphic, irrigated and shallow swamp ecologies of Africa. The strategy is genetic improvement by breeding specific adaptational attributes of local African varieties into more productive genotypes from IRRI and elsewhere. As a result, the upland rice varieties developed have resistance to drought, blast, grain discoloration and lodging and hence respond positively to higher management. Similar achievements have been made by the Institut de Recherches Agronomiques Tropicales (IRAT) in the Ivory Coast. The West Africa Rice Development Association (WARDA), an inter-governmental organization of 15 West African countries undertakes regional trials and at Rokupr has the only centre in Africa working on the improvement of mangrove rice.

(b) Insect Pests

Insect pests are one of the major constraints of rice production in Africa. Between 10 and 15 species are considered to be pests of major importance (Akinsola 1982). Among the insect pests, stalk-eyed fly is indigenous to Africa and is more prevalent in irrigated and hydromorphic ecologies. Species of lepidopterous stemborers and gallmidge are different from those in Asia. Insect control by the use of resistant varieties is inexpensive and compatible with other control methods in the integrated pest management approach. Several rice varieties resistant to insects have been developed by IRRI, but some insect pest species of rice in Africa are different from those found in Asia. At IITA, major efforts have been made to identify the sources of resistance to stemborers, stalk-eyed fly, gallmidge, caseworm and whorl maggot.

(c) Diseases

Rice blast (*Pyricularia oryzae*) is the most widespread and devastating disease in Africa and is highly variable in pathogenicity. Many lines from crosses involving African land varieties, with high yielding exotic semi-dwarf varieties at IITA have been found to possess a high level of resistance to leaf blast but the relationship with neck blast is not yet well understood. Rice yellow mottle virus (RYMV) is considered a potential threat to rice production and occurs in West Africa and Kenya. Investigation at IITA resulted in the identification of several resistant sources and a breeding programme has already been started to develop varieties resistant/tolerant to this virus. Other diseases, including leaf scald and glume discolouration are being studied at IITA. Bacterial leaf blight (*Xanthomonas campestris* pv. *oryzae*) and leaf streak (*X. campestris* pv. *oryzitale*) occur in some African countries, but their extent is still limited.

(d) Weeds

Weed control is one of the most labour intensive operations in rice production. The yield reduction due to weeds in Nigerian rice production ranges from 33-100%. In Africa weed control is mainly carried out by manual labour. For upland, hydromorphic, irrigated and deepwater rice ecosystem, removal of weeds before seeding is largely determined by the area of rice grown and availability of family labour. Use of herbicides in rice culture is still limited for a number of reasons, among which the cost-benefit ratio is crucial. In several countries of Asia and Africa, hand weeding is the most common practice among small farmers; the time of first weeding is important. Poor land preparation at planting time is a major factor affecting the severity of weed infestation. In paddy rice water management is another factor in weed control.

(e) Land preparation

Proper land preparation is essential for good stand establishment and subsequent crop growth. In Africa, land preparation is mainly done by manual labour. In tsetse-free areas draught animals are being introduced, and their use is increasing slowly. Both the cultivated area and yields could be increased by means of more draught animal power and improved animal draught implements and tools. Mechanization will help the area and quality of tillage only if the infrastructure for maintenance and repairs, including availability of spare parts, is well established. For the fragile upland soils of West Africa minimal or zero tillage, using a rolling injection planter developed by IITA, is a viable option for minimising soil erosion.

(f) Fertiliser management

Fertiliser is one of the most effective inputs for production of rice but is also expensive and its use in African rice culture is still very low for a variety of reasons. Soil fertility is usually not a serious problem for the first and only rice crop following bush fallow. However, there is little data on the nutrient requirement of the major soils used for rice production. Zinc and sulphur are minor elements found to be deficient on certain soil types in the savanna region. The problem of iron and aluminium toxicity in hydromorphic and swamp rice cultivation also requires closer assessment, and solutions need to be found. Nitrogen applications should be split between different crop stages (sowing, tillering and flowering) as much is lost in irrigation water. Phosphorus deficiency is widespread in rice in Africa; potassium may be limiting on coarse sandy soils while zinc may be deficient in some African soils.

(g)   Water management

Moisture conservation in upland soils, drainage in valley bottoms and swamps, and controlling the water supply in irrigated rice are important factors in water management for African rice culture. Many African countries are expanding their irrigation projects where land consolidation, levelling and layout for maximum efficiency of water use is imperative. One of the factors in the decline of irrigation efficiency is the frequent breakdown of machinery, improper maintenance and lack of spare parts. Rice is most sensitive to water stress at the reproductive stage, causing increases in spikelet sterility and a decrease in grain weight. In the African rice growing systems, drainage in inland swamps and impounding water by bunding in rainfed rice are the primary needs for increased production.

(h)   Harvesting and postharvest management

Timely harvesting and appropriate postharvest management are needed to minimise losses especially with traditional varieties, which are susceptible to lodging or grain shattering. In most countries of Asia and Africa, harvesting is carried out by manual labour. Use of appropriate hand tools, is needed to increase labour efficiency. Combined or power harvesters are suitable for lodging-resistant varieties planted in rows and minimise losses.

Socio-economic constraints

The two options for increased rice production in Africa are to: (1) expand the area grown to rice, and (2) increase the yield per unit area. Africa is land rich, and rice production is labour-intensive. Therefore both options should be properly balanced in order to achieve increasing production.

(a)   Investment

This is the primary need for both options. Increasing the irrigated rice area would be the most assured way of increasing rice production in Africa, but several thousand hectares of inland swamps or flooded plains could be developed and grown to rice by application of proper drainage systems. All these development projects need capital investment and yearly expenditure for maintenance. In Africa, trained manpower for research and extension services is particularly lacking and requires long-term investment. Likewise, development of infrastructure, such as storage facilities and transport systems for input and produce, should not be ignored.

(b)   Price policies, input supply and marketing

Effective marketing, timely input supply, credit management, and price policies are essential in order to provide the farmers with assured incentives and the means of using improved technology. The price fixed by the Government should protect local rice producers against competition from imported rice which has discouraged increased production in certain African countries. Governments could aim at a balance between the guaranteed price of paddy and subsidised input prices. In Africa, rice has been promoted in many countries by permitting farmers to purchase essential services and inputs on credit and to repay the debt after harvest.

(c)   Improved seeds

A co-ordinated seed programme covering production, inspection, certification, collection and distribution is essential. In some African countries, such programmes have been intiated by the Government, and seed distribution is handled by marketing boards. Several countries are yet to strengthen this essential aspect of their rice improvement system.

(d)   Research and Extension

There is ample evidence that increased rice production in Asian countries is achieved by improving the research and extension services. In Africa, improvement is still needed in research and extension services. It is imperative that high priority be given to research areas and disciplines according to the need of each country. The basic need for strengthening research and extension services must be met through training within the country as well as at advanced institutions abroad.

(e)   The role of women

In peasant families, women are involved in most of the agricultural operations, particularly during and after harvest, but they are invariably denied access to technological advances and scientific innovations. However, greater food self-sufficiency and better nutrition will depend on removal of barriers that impede the contribution of women to agricultural development.

Areas of Future Emphasis

Genetic improvement of rice in Africa has mainly focussed on the upland ecology, but the production potential for upland rice in Africa is much below that of the other ecologies, such as hydromorphic, inland swamps and irrigated paddy. This clearly indicates the need for a change in the emphasis of genetic improvement to more productive ecosystems.

Differences in species of insect pest and disease patho-systems occur between Asia and Africa and genetic improvement must be geared toward breeding resistance to those in Africa. More information on the prevalence, incidence and extent of damage by insect pests and diseases is still needed, especially for central and east African countries. WARDA has been monitoring the pests and diseases in its member states in west Africa. However, in-depth studies are still lacking on the biology, ecology and epidemiology of pathogens, and important insect pests which are essential for a good resistance breeding programme and integrated control approach. Also needed is close monitoring of potentially destructive diseases such as bacterial leaf blight and virus diseases, especially as virus vectors such as brown planthoppers and green leafhoppers are present in Africa.

Among the climatic constraints, cool temperature and drought, when compounded with diseases like blast, become serious problems. Much systematic research needs to be done on climate – disease complexes.

Nutritional imbalances, such as Fe toxicity, salinity and alkalinity, and acid sulphate soils are localized, yet they call for more research and screening for resistance sources under specific conditions.

Agronomic research in Africa can benefit from the Asian experience if Asian cultural practices are adapted to suit African conditions. Independent research has been focused on upland rice, yet more work is needed on hydromorphic valley bottoms and mangrove swamps, where the growing conditions are somewhat different from those in Asia.

Rice has a high and rising demand in Africa as a staple food, but production levels at present are still low. However, if research is addressed to the major problems, with emphasis on genetic improvement, the potential for significant increase in production appears very good.

One can safely conclude that the need for more in-depth research on rice in Africa is imperative. Thus, the international and regional research centres such as IITA, IRRI, IRAT and WARDA need to help African national programmes, independently and in collaboration with one another, in their respective areas of competency and mandate.

The experience of many production projects and parastastals, has proved to be practical and economically viable. Many Governments are keenly interested in rice production projects as a means of reaching self-sufficiency in rice and and as an alternative to dependence on small peasant farmer production. Resolving the problem will require a two-pronged attack – millions of small farmers as well as parastastals.

Finally, the overriding factor is the policy of the national Governments; their economic goals and priorities will ultimately influence rice production in Africa.

A fuller version of this contribution is to be published by the International Rice Research Institute (IRRI).

References

Akinsola, E.A. (1982) Insect pests of upland rice in Africa. *Upland Rice Research Seminar, Bouake, Ivory Coast, October 4-8, 1982*

Barr, B.A.; Koechlert, C.S.; Smith, R.F. (1975) Crop losses to insects, diseases, weeds and other pests. *UC/AID Pest Management and Related Environmental Protection Project.* Berkeley; University of California.

Cramer, H.H. (1967) Plant protection and world crop production. *Pflanzenschutz-Nachrichten Bayer* 20, 524 pp.

FAO (1983) *World Rice Situation and Outlook* 32, 51-52. Rome; FAO.

Moormann, F.R.; Veldkamp, W.J. (1977) Land and rice in Africa; constraints and potentials. In *Rice in Africa* (I.W. Buddenhagen; G.J. Persley, eds.), 29-93. New York; Academic Press.

Posner, J.L. (1978) *Solar radiation and the growth and productivity of upland rice* (Oryza sativa) *in West Africa*. PhD thesis, Cornell University, Ithaca, New York.

Yoshida, S. (1978) Tropical climate and its influence on rice. *IRRI Research Paper Series* 22, 23 pp.

Chapter 10

Maize in Africa – Planning its Protection against Pests

D.J.W. ROSE

c/o Kenya Agricultural Research Institute, P.O. Box 74, Kikuyu, Kenya.

Maize in Africa is attacked by a wide range of pests at all stages of its growth. For Zimbabwe alone, Rose (1963) lists 85 maize pests, of which 34 may sometimes be the cause of serious losses. A large amount of knowledge has now been collected about the most important of these, especially those which are major pests in many African countries. Studies of their biology, ecology and chemical control have led to the development of pest management programmes and direct control practices which seem acceptable and worthwhile, at least to the scientists who develop them. Yet, in spite of all the progress made, the application of this knowledge gained to the improvement of crop yields is disappointing. The methods recommended have been shown to work on research stations but often they have not been adopted by farmers. This contribution enquires into the reasons for this and suggests ways in which more acceptable programmes may be planned.

The problem is best illustrated by a caricature of an unlucky maize farmer in Africa, "Jones". Jones waits for the first planting rains and then plants three maize seeds at each planting station, one for the pest, one for the birds, and one for the crop. This is the traditional way, but is not ideal as pests do not attack evenly; nevertheless this is done and Jones goes home. Out comes the sun, up come the surface beetles (*Zophosis*, *Emyon* and *Gonocephalum* species), up come the eaten seeds, and down go the germinating seedlings. Jones calls his extension entomologist who explains that the trouble could have been avoided if previous crop residues had been burnt, *or* insecticide seed dressings used, *or* insecticide baits applied once the beetles were noticed in the fields. But what do I do now, Jones enquires. Nothing, it is too late says the entomologist and it is now pouring with rain so the beetles will cause no more trouble.

So home goes Jones, and it pours with rain for several days. When he goes back to his maize field the stand looks quite good, from a distance. There is an almost even stand of stunted plants with a few taller plants representative of what all the crop should have been. Again Jones calls his extension entomologist who enjoys the chance to air his knowledge and explains that the continuous rains caused the weevil grubs (*Systates exaptus* and *Mesoleurus dentipes*) to move upwards in the soil and to feed on the underground stems of the seedlings. "You are very unlucky Mr Jones" he seemed pleased to say, "I know of no insecticides which will kill these grubs. You could have killed the parent beetles before they laid eggs on your crops last year; *or* you could have sampled your soils before planting and put a resistant crop such as velvet beans into infested fields". Jones did not want to plant velvet beans; and by the time he had been chided further for an epidemic of streak disease caused by planting his maize downwind from irrigated cereals, and for crop losses through failing to heed warnings of armyworm outbreaks, his relationships with the entomologist were not at their best.

This caricature of an unlucky farmer and a rather smug extension entomologist demonstrates the importance of giving more consideration to the human factor in planning crop protection measures. Most of the measures could have been applied successfully if they had been known by and accepted by the farmer. Much time has been given by scientists to understanding and attacking the weak links in the life cycles of pests; and far too little to strengthening the links between administrators, scientists and farmers.

The closest links occur quite naturally on some large estates and co-operative organizations where the scientists are working with the administrators and labour force with one common goal. There is a flow of information between all concerned as pest control measures are developed and applied. This leads to great efficiency, with benefit if the plans are well conceived and disaster if they are not. It is not surprising that one impetus for integrated pest management programmes was the need to counter man-made disasters created on citrus and similar plantations.

Maize is widely grown on farms of all sizes, by poor and rich farmers, for mean potential yields varying from 1500 - 5000 kg ha^{-1}. Because of this, the gap between most maize farmers and the scientists and adminstrators is wide. They are not together on location or in spirit. The closest links are only maintained in some national grain boards responsible for bulk maize storage and then good pest management procedures occur.

As the gap widens the attitudes of scientists, administrators and farmers change. There becomes a natural tendency for scientists to give priority to gaining recognition in the scientific world and not to devote enough time to understanding the difficulties faced by farmers in applying their recommendations and to making suitable modifications so that they work. The adminstrators always have too much paper work to remain fully in the picture. The farmers remain independent, learn from each other, and are more cautious than estate workers in implementing new measures.

An effort has to be made to close this gap to ensure that recommended crop protection measures are acceptable to farmers; the initiative has to come from the scientists as they understand the relationships between crops and insects which cause the latter to become pests, and the alternative tactics that may be used in their management. This calls for entomologists who will reach out to study the human factors involved.

Training courses for agricultural officers and farmers provide a forum for the interaction of views on the practicality of proposed measures, and for the development of ideas on how best to overcome constraints. They should be attended by research entomologists and administrators in order to foster this community approach.

Administrators also need to draw entomologists more closely into helping with planning crop management programmes. Most pests of maize are aggravated by the farming practices adopted, and they are preventable. An evaluation of pest hazards needs to be made before changes are made in agronomic practices. Too often, entomologists are only asked for advice when pest problems have occurred. Farmers need preventative measures to be easier and simpler before they are readily accepted, except in situations where serious pest problems persist year after year. Busy farmers have to be strongly motivated before they spend much time and thought on complex pest management programmes. These may have a place if and when all relevant information and decisions become on-tap by computer; until then the greatest profit will be obtained by putting most effort into easily applied techniques.

Self help is encouraged by good prices for products, and extensions of crop protection knowledge by all appropriate methods. More needs to be invested in the distribution of advisory pamphlets and posters. Even so, the farmers footsteps over his lands are the first and best steps to be taken every day.

Community approaches are recommended to help farmers to avoid losses by giving warning of when and where pests may attack maize crops. The use of nation-wide networks of traps to monitor daily abundance of pests has proved worthwhile for migrant pests such as the African armyworm, *Spodoptera exempta* (Odiyo 1979), and this approach might be extended for other pest species by using light or pheromone traps to catch flying stages. Surveys using standard methods of sampling numbers on maize plants need to be developed for others. Also, biogeographical techniques might be applied to map localities where major pest species occur to guide farmers on local problems. For example, false wireworms (*Psammodes* species) only occur in sandy soils; and the black maize beetles (*Heteronychus* species) in poorly drained or irrigated heavy soils. These local pest problems become known to extension entomologists but are too vaguely defined to yet be of value in planning national approaches.

The research directed to develop resistant maize varieties (maize streak disease, maize stalk borers), to apply biological control, or to prevent surges in pest numbers by simple modifications of cropping systems (rotations, intercropping, destruction of crop residues) is on the right lines as the results are fairly easily and cheaply applied by the farmer.

Many years ago I attended a maize conference in Africa. First a maize breeder gave a paper describing the greatly increased yields obtained with the new hybrid varieties; secondly, an agricultural chemist showed that yields could be doubled with nitrogenous fertilisers; and finally, a biometrician demonstrated that maize yields had remained at the same low national average for the last ten years, despite the work of the maize breeders and chemists – the biometrician was speaking too soon, as yields rapidly increased in subsequent years. This contribution has also asked why it is that the improvement in maize yields remains disappointing, this time through the application of existing knowledge of maize pests and their control. Perhaps, like the biometrician, we are asking the question too soon.

References

Odiyo, P.O. (1979) Forecasting infestations of a migrant pest: the African armyworm *Spodoptera exempta* (Walk.). *Philosophical Transactions of the Royal Society of London* B, 287, 403-413.

Rose, D.J.W. (1963) Pests of maize and other cereal crops in the Rhodesias. *Federation of Rhodesia and Nyasaland Agricultural Bulletin* 2163, 1-23.

Chapter 11

Midge Problems in African Cereals

K.M. HARRIS

Commonwealth Institute of Entomology, c/o British Museum (Natural History), Cromwell Road, London SW7 5BD, UK.

Introduction

Gall midges (Diptera: Cecidomyiidae) are important pests of many cereals in temperate and tropical agriculture. In Africa there are three notable pest species: the sorghum midge, *Contarinia sorghicola* (Coquillett); the millet grain midge, *Geromyia penniseti* (Felt); and the African rice gall midge, *Orseolia oryzivora* Harris & Gagne. All three species are almost certainly indigenous to Africa and have probably been associated with the crops and with their wild precursors for many thousands of years. The African rice gall midge is still restricted to Africa, but the sorghum midge has been spread throughout the tropics and sub-tropics of the world and the millet grain midge occurs in India as well as Africa. Adequate understanding of the identity, biology and ecology of these species is essential to the long-term planning of control measures to reduce crop losses.

Sorghum midge, Contarinia sorghicola (Coquillett)

This is now one of the most widely occurring pests of any major crop. Wherever grain and forage sorghums are grown, from latitude 40°N to 40°S, the developing grain may be attacked and destroyed by larvae of this species. Adults are small, with a wing length of only about 3 mm, and are therefore relatively inconspicuous. They mostly emerge 1-2 hours after sunrise from pupae in damaged spikelets, either in seed-heads on plants or from crop debris. Adults mate within an hour of emergence and, after resting for about half an hour, the females fly in search of flowering sorghum heads. Once a suitable head has been found, eggs are deposited on the young ovaries within the glumes through the midge's long, fine, telescopic ovipositor. For some hours each female probes suitable spikelets and carefully places about fifty to a hundred eggs. Peak egg-laying usually occurs before mid-day and most females finish laying and die before sunset. Eggs hatch about four days later and larvae feed for about 1-2 weeks at the expense of the ovary, which shrivels and fails to develop. Attacked spikelets therefore remain tightly closed and a flat, empty appearance, which is sometimes wrongly attributed to poor fertilisation, genetic sterility, unfavourable weather or attack by head-bugs or other pests.

During the growing season a new generation of adult midges is produced about every 2-3 weeks but towards the end of the season larvae spin small silk cocoons inside attacked spikelets and can then survive in diapause for up to three years.

Diapause usually ends as humidity rises during subsequent rains and emergence of the first generation of adults in the new season generally coincides with the first appearance of flowering heads in cultivated and wild sorghum. Populations then build up through the season and tend to cause most damage to late-flowering crops.

Sorghum midge was first discovered in the USA in 1895 and was formally named in 1899. At first it was thought to be an American species that was later spread to other parts of the world, but recent research involving dissection of old herbarium specimens of *Sorghum* at the Royal Botanic Gardens, Kew, has shown that it was present in Africa long before it was discovered in the USA and well before the dates of first field records from Africa. It now seems much more likely that the species originated in Africa, where it breeds on both cultivated and wild species of *Sorghum*, and that it was later spread by man as diapause larvae with contaminated seed and in seed-heads. It has now been recorded from most countries where *Sorghum* grows and is a major pest of grain sorghums in the USA, Argentina, Brazil, tropical Africa, India and northern Australia (Harris 1970, 1976).

The most accurate assessments of crop loss have been made in the USA where the midge is considered to be the most damaging of all sorghum insects (Wiseman & Morrison 1981). Recurrent annual losses are estimated at 4% of the grain sorghum crop and in Texas alone estimates of losses have exceeded 10 million dollars per annum on several occasions (Wiseman *et al.* 1976). A similar level of overall loss was estimated in Nigeria in 1958 (Harris 1961) and recurrent losses of 5-10% of the crop are probably typical of most major sorghum-growing areas. Local losses in tropical Africa and Asia may exceed 50% and complete loss of some crops is not uncommon.

A bibliography of the world literature for the period 1898-1975 (Wiseman *et al.* 1976) lists 185 publications on the taxonomy, biology, ecology and control of this pest and a recent bibliography of abstracts from the *Review of Applied Entomology (Series A)* for the period 1973-1983 contains 119 references (Annotated Bibliography E.104, *Contarinia sorghicola*, CAB, 1983). Harris (1976) reviewed sources of information on its biology, crop losses and control and additional useful reviews of information are included in Young (1970), Young & Teetes (1977) and Gahukar & Jotwani (1980). Current research papers are recorded in *Sorghum and Millets Abstracts* (CAB) and in *SMIC Newsletter* (International Crops Research Institute for the Semi-Arid Tropics).

The general position on control measures indicates that chemical control is difficult and, although insecticides may be of use in protecting high-value/high-risk crops, practical control is best achieved by developing integrated pest management programmes using cultural measures to reduce carry-over populations of diapause larvae in crop residues and to restrict the flowering periods of growing crops. In addition, a number of sources of varietal resistance to midge have been identified in recent years and are being used to develop midge-resistant cultivars.

In Africa the significance of losses to sorghum midge must be considered in the context of the other factors limiting sorghum production. The most important of these are undoubtedly climatic, edaphic and sociological with, in addition, major pests, pathogens and weeds (Table 1).

Table 1   Major pests, pathogens and weeds affecting sorghum production in Africa.

Pests
   Stored products pests (*Sitophilus oryzae, S. zeamais, Sitotroga cerealella*, etc.)
   Birds (*Quelea quelea*, and other species)
   Sorghum midge (*Contarinia sorghicola*)
   Shoot fly (*Atherigona soccata*)
   Lepidopterous stem borers (*Chilo partellus, Busseola fusca, Acigona ignefusalis*, etc.)

Dieseases
   Smuts (*Sphacelotheca sorghi, S. cruenta, S. reiliana*)
   Grain moulds (*Aspergillus, Curvularia, Fusarium*, etc.)
   Downy mildew (*Peronosclerospora sorghi*)

Weeds
   Witchweed (*Striga hermonthica*)

During the past decade, published information on sorghum midge research in Africa has been almost entirely from West Africa, although there are active research programmes in other areas. In Upper Volta, Bonzi (1979) has studied the biology and ecology of the midge, which reaches peak populations in October-November. Late-flowering crops are most severely attacked and a combination of cultural measures with the use of resistant varieties is thought to offer the best prospects of control. Breniere (1981) has also reported on the situation in Upper Volta where an integrated control programme is being implemented against the midge, which is especially damaging in the Bobo-Dioulasso district where two crops are grown each year. The control measures recommended are spatial isolation of early and late flowering crops; choice of varieties with short flowering periods and the use of insecticides as a temporary measure until populations are reduced by cultural methods.

In Sénégal, Coutin (1970) reported a 50-95 % grain loss to midge in 1967-69 and concluded that cultural and chemical control measures failed to give adequate control. He also reported that natural control by parasites, especially *Eupelmus popa* Girault and *Aprostocetus diplosidis* Crawford, and by the predator *Orius punctaticollis* (Reuter) had little effect on midge populations. Gahukar (1979), also reporting on the situation in Sénégal, recorded that midge caused 0.2 to 47.6 % infestation during the rains with levels of infestation varying according to region; the presence of grasses; the varieties grown and the humidity.

Finally, Barry (1980) has reported selection in N. Nigeria of 45 lines with resistance to midge and work on midge resistance has also been reported by the Ministry of Agriculture and Natural Resources, Gambia (*Review of Applied Entomology* 68, abstract 4182).

Millet grain midge, Geromyia penniseti (Felt)

Although this pest is not closely related to the sorghum midge, there are similarities in its biology and in the damage that it causes. The species was first described from southern India but, as in the case of sorghum midge, it almost certainly orginated in Africa. It has now been recorded from Sénégal, Mali, Upper Volta, Niger, the Sudan, Ghana, Nigeria, Uganda and Madagascar. The only detailed published study of the biology of this species has been made by Coutin in Sénégal (Coutin & Harris 1969). Adults emerge after sunset and are most numerous and active from 21.00 h to 01.00 h. Numbers then decline and all activity ceases by 04.00 h. Females lay eggs on flowering heads of bulrush millet, *Pennisetum americanum*, and larvae, which hatch about 3 days later, feed on the developing ovary for 6-7 days. The life-cycle is completed in about two weeks and 4-5 generations develop during the growing season. Towards the end of the growing season larvae diapause in the spikelets of infested heads and remain dormant for 8-9 months until diapause is terminated in the following rains so that the first adult generation coincides with the flowering of early "souna" millets. Grain losses are greatest on "souna" millets and the later-planted "sanio" millets are less affected, possibly because parasites limit midge populations later in the season. In September 1967 85 % parasitisation by *Eupelmus*, *Tetrastichus*, *Platygaster* and *Aphanogmus* was recorded in a field trial at Bambey and failure of this parasite complex would probably cause greater yield losses in "sanio" millets, which are normally most productive and widely cultivated.

African rice gall midge, + Orseolia oryzivora Harris & Gagné

The Asian rice gall midge, *Orseolia oryzae* (Wood-Mason) has long been known as a major pest of rice in many Asian countries. Substantial losses are caused by larvae feeding in shoot apices which then develop into characteristic hollow leaf galls resembling onion leaves. This prevents panicle development and reduces yield. Until recently it was thought that this species also occurred in Africa but it has now been established that the African species is distinct and probably evolved in Africa on *Oryza glaberrima* (Harris & Gagné 1982). Although there has been considerable research on the Asian species, the African species is still inadequately known. It has been recorded from Sénégal, Upper Volta, Ivory Coast, Nigeria, the Sudan and Malawi and in Upper Volta it is considered to be the most important pest of the crop (Bonzi 1980).

Conclusions

Enough technical information about sorghum midge is available to support the general planning of integrated pest management systems to reduce crop losses. Chemical control seems unlikely to be generally effective since most insecticides fail to give adequate control of midge or, if midge is controlled, phytotoxicity may cancel out potential benefits. In addition, insecticides and machinery are expensive and side-effects on predators and parasites, on other pest complexes, and on the general environment may be harmful. Biological control also seems likely to be inoperable, since the main hymenopterous parasites of the midge have already been transported with the pest, but research on the parasite/predator complexes associated with midge in Ethiopia, the Sudan and East Africa, might produce useful results.

Cultural methods of control certainly work, as demonstrated in Texas (Young & Teetes 1977), but require the close co-operation of the farming community with research and extension services. In the semi-arid areas of Africa, dry-season carry-over could be reduced and flowering could be limited if farmers would take concerted action to implement the following measures:

(a) Destroy all old seed-heads and threshing trash during the dry season, either by burning or burying.

(b) Cut down self-sown or ratoon plants that come into flower early in the season. These are often left around villages and roadsides and there is a natural tendency to let them grow, which assists the establishment of midge early in the growing season.

(c) Keep farm areas clear of flowering wild *Sorghum* grasses, such as *S. arundinaceum* and *S. verticilliflorum*, which also assist early establishment of midge.

(d) Sow varieties that come into flower at the same time and cultivate the crop well so that growth is uniform and tillering is limited.

(e) If varieties with different main periods of flowering have to be grown in the same locality, sow late-flowering crops up-wind from early-flowering crops to limit the spread of adult midges.

Implementation of such measures requires labour rather than capital in-puts but is not easily achieved, and it therefore seems that long-term prospects of effective midge control must depend on the future development of resistant varieties. Considerable progress has been made in recent years in the USA (Johnson & Teetes 1980, Teetes 1980) and in India (Jotwani & Davies 1980). One advantage in doing this is that only the single species *Contarinia sorghicola* is involved, which means that results of research carried out in North or South America, India, Australia, Hawaii or other parts of the world may be directly applicable in Africa.

This is the reverse of the situation with the rice gall midges where distinct species of pest are involved in Africa and Asia and it cannot be assumed that research on the biology, ecology and control of the Asian rice gall midge will apply in Africa.

There is obviously a great need for international co-operation on a world basis to exchange technical expertise and information and there is an equal need to provide relevant information to African farmers who will carry the main burden of increasing African agricultural production.

References

Barry, B.D. (1980) Where are we, and where are we going with insect resistance in sorghum? *African Journal of Plant Protection* 2, 149-159.

Bonzi, M. (1979) La cecidomyie du sorgho, *Contarinia sorghicola* Coq., en Haute-Volta. Possibilities de lutte. *Congres sur la lutte contre les insectes en milieu tropical. Marseilles, 13-16 Mars 1979. Compte rendu des travaux.* I (2) *Sante humaine et animale*, 531-541.

Bonzi, M. (1980) Wild host plants of the rice gall midge *Orseolia oryzae* W.M. (Dipt. Cecidomyidae) in Upper Volta. *Technical Newsletter West African Rice Development Association* 2(2), 5-6.

Breniere, J. (1981) La lutte integree contre les ravageurs des cultures vivrieres tropicales. *Agronomie Tropicale* 36, 78-81.

Coutin, R. (1970) Biologie de la cecidomyie du sorgho (*Contarinia sorghicola* Coq.) et lutte chimique. *Phytiatrie-Phytopharmacie* 19, 65-83.

Coutin, R.; Harris, K.M. (1969) The taxonomy, distribution, biology and economic importance of the millet grain midge, *Geromyia penniseti* (Felt), gen.n., comb.n. (Dipt., Cecidomyiidae). *Bulletin of Entomological Research* 59, 259-273.

Gahukar, R.T. (1979) Etat actuel de l'incidence et de la lutte des insectes ravageurs du sorgho au Sénégal. *Congres sur la lutte contre les insectes en milieu tropical. Marseilles, 13-16 Mars 1979. Compte rendu des travaux.* I(2) *Sante humaine et animale*, 543-548.

Gahukar, R.T.; Jotwani, M.G. (1980) Present status of field pests of sorghum and millets in India. *Tropical Pest Management* 26, 138-151.

Harris, K.M. (1961) The sorghum midge, *Contarinia sorghicola* (Coq.), in Nigeria. *Bulletin of Entomological Research* 52, 129-146.

Harris, K.M. (1970) The sorghum midge. *Pest Articles and News Summary* 16, 36-42.

Harris, K.M. (1976) The sorghum midge. *Annals of Applied Biology* 84, 114-118.

Harris, K.M.; Gagné, R.J. (1982) Description of the African rice gall midge, *Orseolia oryzivora* sp.n., with comparative notes on the Asian rice gall midge, *O. oryzae* (Wood-Mason) (Diptera: Cecidomyiidae). *Bulletin of Entomological Research* 72, 467-472.

Johnson, J.W.; Teetes, G.L. (1980) Breeding for arthropod resistance in sorghum. In *Biology and Breeding for Resistance to Arthropods and Pathogens in Agricultural Plants* (M.K. Harris, ed.), 168-179. Texas A&M University.

Jotwani, M.G.; Davies, J.C. (1980) Insect resistance studies on sorghum at International Institutes and National Programs with special reference to India. In *Biology and Breeding for Resistance to Arthropods and Pathogens in Agricultural Plants* (M.K. Harris, ed.), 224-236. Texas A&M University.

Teetes, G.L. (1980) Overview of pest management and host plant resistance in U.S. sorghum. In *Biology and Breeding for Resistance to Arthropods and Pathogens in Agricultural Plants* (K.M. Harris, ed.), 181-223. Texas A&M University.

Wiseman, B.R.; Morrison, W.P. (1981) Components for management of field corn and grain sorghum insects and mites in the United States. *USDA Agricultural Research Service, Agricultural Reviews and Manuals, Southern Series* 18, i-iv, 1-18.

Wiseman, B.R.; McMillian, W.W.; Widstrom, N.W. (1976) The sorghum midge: a bibliography, 1898-1975. *USDA Agricultural Research Service* ARS-S-139, 1-8.

Young, W.R. (1970) Sorghum insects. In *Sorghum Production and Utilization* (J.S. Wall; W.M. Ross, eds), 235-287. Westport, Conn.; Avis Publishing.

Young, W.R.; Teetes, G.L. (1977) Sorghum entomology. *Annual Review of Entomology* 22, 193-218.

Chapter 12

Prostephanus truncatus, a New Storage Pest in Africa

P. GOLOB

Tropical Research and Development Institute, London Road, Slough, Berks., UK.

Introduction

Much of Africa south of the Sahara is threatened by an insect pest which has recently been introduced into the continent from Central America. *Prostephanus truncatus* (Horn), the Larger Grain Borer, is a pest of stored maize and cassava. Its presence was first noted in western Tanzania in 1981 when it was found to be causing considerable losses to farm-stored maize cobs.

The problem

That year a survey was undertaken to determine the extent of the damage caused by the insect and its distribution. Maize cobs collected from 56 farms in the Tabora region exhibited a mean weight loss of 8.7 % after 3.5–6 months storage, some cobs sustained more than 30 % loss and were unfit for human consumption. Such losses are uncharacteristic of farm-stored maize in this area of Africa, where levels of loss of 1–3 % would be expected in the dry part of the storage season. As well as maize, this beetle has been found to breed successfully in dried cassava and to damage groundnuts, beans and household articles such as wooden cooking utensils.

The beetle was found on farms throughout the Tabora region and in markets of nearby trading towns of Shinyanga and Mwanza. Since then records have shown that the beetle has spread by movement of grain through trade, to twelve regions in Tanzania, to the Taveta district of Kenya and is also now in Togo and possibly Uganda.

The biology and ecology of P. truncatus in Tanzania

P. truncatus is very tolerant of hot, dry conditions; it is able to develop over a wider climatic range than the other common storage pests found on farms in the region, in particular *Sitophilus* species. *P. truncatus* was found to develop in maize with a moisture content of 10 per cent and less, levels which occur during the early part of the storage season when *Sitophilus* numbers remain low. Part of the success of this beetle can be attributed to the lack of competition during the early part of the storage season when the insect population is beginning to expand. Lack of predation is another factor for its success. In Central America *Teretriosoma nigrisans*, a Histeridae beetle, predates on *P. truncatus*. This insect has not been recorded in Tanzania. Perhaps the main reason for the success of *P. truncatus* in western Tanzania is as a result of the traditional manner in which the farmer stores maize. Most farmers store maize as grain on-the-cob without removing the sheathing leaves. It has been clearly demonstrated that *P. truncatus* develops far more successfully on cob maize than it does on loose grain.

Control

There are two ways to control this beetle. Firstly, changes can be made to the traditional storage system so that it is ecologically less favourable for beetle development. Secondly, synthetic insecticides can be applied to the produce to kill insects directly. An obvious change to the normal practice that can be recommended is to store loose grain rather than cobs. However, implementing such a change in procedure is not an easy task. Farmers are naturally cautious and will only respond to a recommendation if a direct benefit can be clearly and obviously demonstrated. The onus for success of such an operation clearly falls on the local extension officer who needs to be completely convinced of the truth of the argument himself. Even if farmers are willing to store grain they have neither the containers in which to store loose grain such as sacks and oil drums, nor in many areas the skills to construct alternative containers such as mudded baskets. To overcome these problems, appropriate materials and skills will have to be introduced, either from other areas of the country or from abroad. As *P. truncatus* has not, until the last four years, been regarded as a pest of major economic significance, very little work has been undertaken to investigate appropriate synthetic insecticides with which to control the beetle. Recently laboratory and field trials in Tanzania have, however, demonstrated that the new synthetic pyrethroid compounds provide very effective control of this pest. Organophosphorus compounds, however, are not particularly effective. The responses of *P. truncatus* to these compounds are similar to the response of *Rhyzopertha dominica*, the lesser grain borer, a closely related, cosmopolitan pest of stored cereals.

Table 1 Percentage number of damaged grains in maize samples collected after different storage periods in Tabora.

Insecticide dust	Dosage in ppm active ingredient	Grain experiment			Dosage in ppm active ingredient	Cob experiment		
		Duration of storage in months				Duration of storage in months		
		6	8	10		6	8	10
Control	0	29.4	59.8	100.0	0	57.7	98.0	100.0
Permethrin (0.5%)	2.5	0.8	1.3	4.0	5	0.5	3.0	59.8
Pirimiphos methyl (2.0%)	10	13.5	20.9	81.6	20	7.4	80.1	99.7
Malathion (1%)	12	16.0	45.8	98.4	24	13.6	68.1	99.8
Chlorpyrifos methyl (2%)	10	0.5	6.2	33.2	20	4.3	74.4	100.0
Methacrifos (2%)	10	9.1	39.3	90.4	20	20.2	77.3	99.7
Carbaryl (2%)	10	3.8	49.7	89.8	20	–	–	–
Pirimiphos methyl + Carbaryl	4 + 8	6.2	46.9	78.2	8 + 6	1.3	38.8	100.0
Permethrin + Carbaryl	2 + 8	0.8	1.3	12.5	4 + 16	0.3	5.3	28.7

Note: Each datum is the mean of 4 replicates. The initial damage in both trials was less than 1%.
All the formulations were obtained in UK except for malathion which was manufactured in Tanzania.

Some of the results obtained in trials conducted in Tabora are shown in Table 1. It is clear that permethrin provided the best protection to maize grain and also that loose grain was protected more effectively than cobs. The cobs in this trial were treated by sprinkling dilute insecticide dusts over cobs after their sheathing leaves had been removed. Removal of the sheath allows the insecticide to come into intimate contact with the insect but it is not typical of farm practice. Current trials are assessing the effect of spraying insecticide emulsions to cobs with their husks intact. In this trial the pyrethroid deltamethrin, which has recently been found to be very effective in laboratory trials, permethrin and pirimiphos methyl are being tested. It is thought that pirimiphos methyl will be effective because its relatively high vapour pressure will allow active ingredients to penetrate between the sheathing leaves and come into direct contact with the insects which will not happen with the more stable pyrethroids. Preliminary indications are that these treatments would provide effective control.

Another interesting observation is that *P. truncatus* tends to bore into the base of the cob and then along the core. Dipping the cob into either an emulsion of permethrin or a dilute dust of the compound can provide excellent protection against *P. truncatus* (Hodges, pers. communication). This effect has only recently been observed in very small scale laboratory trials and it is our intention to repeat the work and confirm the response on a larger scale in Tanzania. Although synthetic compounds offer the best short-term remedy for controlling the beetle, the large sums of foreign exchange required to purchase these chemicals make their use on a longer term prohibitively expensive for a country such as Tanzania. Experiments have been conducted to investigate alternative means to protect produce. Grain has been mixed with tobacco dust, cotton seed oil, ashes from burning maize cores and powdered goat dung. Some methods have shown promise but in order to fully utilise the potential of these readily available, cheap materials, much more investigation is required.

Monitoring the spread of the beetle

In order restrict the spread of the beetle, efficient monitoring methods are necessary so that the beetle can be intercepted when only a small population is present. The identification and isolation of a male aggregation pheromone of *R. dominica* which has been found to be effective in trapping *P. truncatus* has subsequently led to the isolation of a specific pheromone for *P. truncatus*. This compound "Truncal", identified and isolated by my colleagues at TDRI, is currently being tested in Tabora. Without an efficient monitoring service the pest will undoubtedly spread through the rest of Tanzania and into surrounding countries. It is imperative that the neighbouring countries undertake rigorous inspection of all imported foodstuffs to prevent entry of the pest.

The Current Situation

The climatic tolerance of the beetle is such that it would be able to survive and develop in most countries in sub-Saharan Africa. We need to be extremely vigilant and decisive in action if we are to prevent spread to other countries.

The Government of Tanzania has initiated a campaign firstly to contain the beetle within its existing boundaries (northern Tanzania) and secondly to control the beetle and reduce the degree of infestation and, therefore, food losses within these boundaries.

A four-man team funded by the UK of which I am a member is at present in Tanzania to assist the Government in establishing a control and containment programme. The programme will include provision of insecticides and sacks for farmers, training of agricultural extension workers and farmers and the development of alternative storage structures and strategies of the kind I have previously mentioned. A system for monitoring and inspecting produce moving through trade will be established. Specialist teams to treat infested bulk commodities in villages will be constituted. The programme will be coordinated in the long term by FAO. Farmers are extremely concerned about *P. truncatus* and are very willing to participate in any problem to control this pest. With similar enthusiasm from all those engaged in the agricultural sector it will be possible to restrict the damage to a level whereby it is no longer of economic importance and is not the subject of the front page headlines of national daily newspapers.

Chapter 13

Striga – Advances in Control

H. DOGGETT
15 Bandon Road, Girton, Cambridge CB3 0LU, UK.

Is Striga control possible?

Striga can be controlled. Given good farming with ample inputs, this weed is only a minor nuisance. *Striga* is widespread in Southern Africa, yet it is kept well under control by increasing soil fertility (especially through the addition of nitrogenous fertiliser), herbicides, crop rotation, and trap cropping. Combinations of these methods work well.

The eradication of *Striga* would be an impossible task in Africa or India, since the witchweed is a parasite of many wild grasses. Eradication is being attempted in the USA where *S. asiatica* was accidentally introduced to the Carolinas, probably towards the end of the war. Approximately 200 000 ha are affected, and *Striga* has been eradicated on about 90 000 of these, but the process is very costly (Eplee 1983).

The Problem for Small Farmers

Striga does much damage to cereals grown by the small farmer.

As a generalization, the poorer the land, the poorer the management, the fewer the inputs, then the greater the amount of *Striga* damage. The parasite can be devastating. Yields of only 70-340 kg ha^{-1} have been recorded in Tanzania. In the Sudan, grain yield losses were quoted as 70% under severe infestation, 60% in Nigeria, 25 thousand tons of sorghum grain annually are lost in Andhra Pradesh (Doggett 1965, Hamdoun & El Tigani 1977, Obilana 1979, Gebisa 1980, Rana *et al.* 1980, Hunamantha Rao *et al.* 1981, Ramaiah & Parker 1982).

The same trend is now becoming apparent in India. Local farmers had selected *Striga* resistant types: the new hybrids are largely based on exotic germplasm, and are very susceptible. On the red soils of Maharashtra, the farmers have had to replace the sorghum hybrids with pearl (bulrush) millet, and problems are reported on other soils also (Sanghi & Vishnu Murthy 1982, Tarhalkar 1982). The damage done by *Striga* is seldom realized. The losses are borne by the small subsistence farmer, and do not get into the statistics. Traditionally, he may just abandon his land, or grow an alternative crop. I have no doubt that the light soils of the Sukumaland catena were sown to millet (mawele) because of the burden of the seed in the soil of *S. asiatica* which attacks sorghum (and soon builds up on maize). With increasing population pressures on the land, *Striga* is steadily becoming more serious.

Advances in control

(1) *Resistance* is the primary control measure for small farmers, although it must be supported by good cultural practices. Good resistance sources are being identified in Africa, in the ICRISAT project in Upper Volta, at Samaru in Northern Nigeria, and in the *Striga* project in the Sudan. Framida, N-13, Najjad, IS9830, SRN 6838A, SPV 103, and the derivative of a Framida cross, "entry 39" (Ramaiah 1983).

Table 1   Agronomic characteristics of some *Striga*-resistant sorghum varieties (results from different trials), (from Ramaiah 1983).

Variety	Origin	Days to flower	Plant ht(cm)	Grain yield (kg/ha)	Grain colour	Mechanism of resistance
N-13	India	89	306	760	Yellow	Anti-haustorial factors
Framida	Africa	80	253	1930	Brown	Low stimulant + anti-haustorial factors
IS 9830	Sudan	61	281	920	White	Low stimulant
SPV 103	India	65	153	910	White	Anti-haustorial or antibiosis factors
Najjad	Sudan	78	216	1010	White	Anti-haustorial factors
Ent. 39 (148 x Framida)	India	70	150	1500	White	Low stimulant

Framida has shown convincing evidence of good resistance to *S. hermonthica* in farmers' trials in West Africa, as well as in formal trials. Seed has been sent for trial in eastern Africa, the Ethiopian lowlands and the Sudan. Some of the cv's in the table, especially Framida, have been used as parents in crosses. Framida has a brown grain, unacceptable to many people. "Entry 39" in the table has an improved grain type, *Striga*-resistant derivatives with much better grain quality are showing promise in trials in West Africa. Lines 82-S-47, 50, 52, 59 and 79 all look hopeful, especially 59 and 79 which are white-grained versions of Framida (Ramaiah 1983). Two lines with resistance combined with good grain quality, SAR-1 and SAR-2 have been released in India. (Vasudeva Rao 1983).

(2) *Selection Methodologies*. Selection for *Striga* resistance presents many problems. *Striga* seed distribution in the soil is very uneven. Good progress has been made at ICRISAT in improved methodologies: (a) Comparing two varieties in farmers' fields by planting in strips, and sampling (Fig. 1). (b) Screening sorghum seedlings for low stimulant production. This helps to reduce numbers, as more low-stimulant varieties show *Striga* resistance than do high stimulant cultivars. (c) The systematic use of check plots, and relating *Striga* incidence to the adjoining checks, is working well (Fig. 2).

(3) *Nitrogenous Fertiliser*. There are indications that nitrogenous fertiliser may enhance the degree of resistance shown by resistant sorghum cultivars (Ramaiah & Parker 1982). Very heavy dressings of ammonium nitrate (over 1100 kg ha^{-1} of N) completely suppressed *Striga* in the USA (Shaw *et al.* 1962). Stewart *et al.* (1983) have shown that the nitrate ion is toxic to *Striga* except at very low concentrations. Reports of the effects of rather low levels of nitrogen application vary: sometimes more *Striga* plants occur on the treated plots: in other places, their numbers may be fewer. The probable explanation lies in the fact that there are sometimes many *Striga* below ground which do not get enough resources from the host plants to emerge. Nitrogen improves the growth of the host, so more of these *Striga* are able to emerge. As the burden of *Striga* seed in the soil diminishes, so the benefit of nitrogen becomes more apparent, perhaps especially so on resistant varieties. There can be no doubt that steadily increasing the fertility of the land, which includes providing a good level of nitrogen in the fertiliser used, is an essential component of any *Striga* control system designed to produce better cereal yields.

(4) *Herbicides*. Dinitroaniline herbicides such as "Treflan" used pre-emergence act as a barrier to *Striga* emergence, the parasite grows below ground but does not emerge. Some diphenyl ethers, such as "Goal" and "Flex" are proving to be excellent post-emergence herbicides against *Striga*. 2, 4-D is the most commonly used herbicide in the USA, but it can be very damaging to some crop

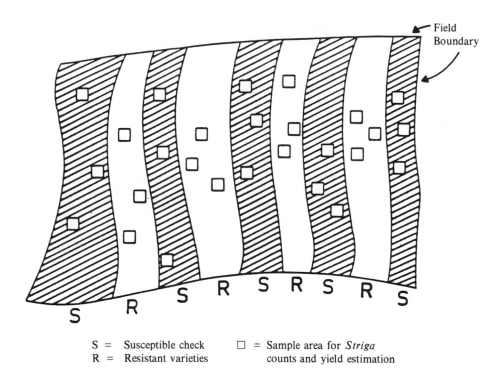

S = Susceptible check   □ = Sample area for *Striga*
R = Resistant varieties   counts and yield estimation

Fig. 1.   Farmer's field testing of *Striga*-resistant varieties in alternate strips.

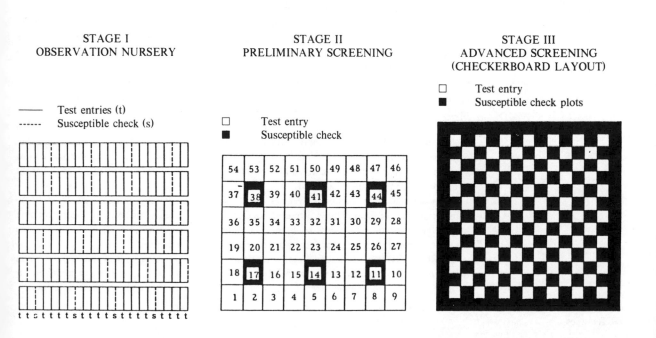

STAGE I
OBSERVATION NURSERY

—— Test entries (t)
------ Susceptible check (s)

t t s t t t t s t t t t s t t t t s t t t t

STAGE II
PRELIMINARY SCREENING

□   Test entry
■   Susceptible check

54	53	52	51	50	49	48	47	46
37	38	39	40	41	42	43	44	45
36	35	34	33	32	31	30	29	28
19	20	21	22	23	24	25	26	27
18	17	16	15	14	13	12	11	10
1	2	3	4	5	6	7	8	9

STAGE III
ADVANCED SCREENING
(CHECKERBOARD LAYOUT)

□   Test entry
■   Susceptible check plots

Fig. 2.   A three-stage screening methodology for *Striga* resistance breeding in sorghum (Vasudeva Rao 1983).

plants, such as cotton. Paraquat is also useful, so long as the cereal plants themselves are not touched by the herbicide (Eplee 1983). The work of Stewart *et al.* (1983) has shown that mannitol is very important in the carbohydrate metabolism of *Striga*, and this is true also of some other parasites. This holds out the possibility of identifying a herbicide which would be specific to a number of parasitic species, such as *Orobanche*. This wider market could make it worthwhile manufacturing such a herbicide. At present, the costs of launching new chemical compounds for agricultural use are very high because of the health safety testing programme which must be done to satisfy regulations.

(5) *Germination stimulants.* Ethylene is an effective agent for germinating *Striga* seed in the soil when the necessary pre-treatment moisture and temperature conditions have been satisfied. It is easily applied by tractor-mounted equipment on a field scale, and a knapsack injector has been devised for small, difficult patches. This treats about 1 m³ at a time. Recently, a probe injector system has been devised, which has two wheels, and can be towed along. It injects every 90 cm, and a 30 lb cylinder of ethylene can treat 2-4 ha^{-1} in a day (Eplee 1983). Again, the practicality of this for small subsistence farmers remains to be demonstrated.

Analogues of one of the natural *Striga* seed stimulants ("Strigol") which are not very difficult to produce, show promise. These are known in the USA as "Strigalogs". GR24, developed at Sussex University (Johnson *et al.* 1976) is quite effective at germinating the *Striga* seed, and could be particularly useful applied to a rotation crop, such as cotton, which itself produces *Strigol*. The combination of GR24 and "Goal" herbicide is particularly effective (Norris & Eplee 1983). Since these can be applied with a knapsack sprayer, they should be within the reach of the small farmer to use. Unfortunately, the costs of launching GR24 appear to be much greater than the profits to be expected from a market consisting largely of subsistance farmers.

Control of Striga by small farmers

The basic requirement of adequate, stable prices was mentioned in Chapter 7. In any area where shifting cultivation is no longer economic, inputs will be needed: The first requirement for small farmers is the *Striga* resistant cultivar; this needs an effective seed production and distribution service, handling certified seed. The second requirement is fertiliser, which needs a relatively large nitrogen component; farmyard manure is valuable, but some added nitrogen will prove worth while. A third requirement is weed control; *Striga* is extremely difficult to control by weeding out the parasite and an appropriate herbicide must be used in most situations. And a fourth requirement is crop rotation, using non-cereal crops which germinate some *Striga* seed; cotton in India and groundnuts in West Africa have been used in this way by farmers for many years.

This appears to be the minimum farming level needed to contain *Striga*. Inputs are low: the cost of seed, the cost of some nitrogenous fertiliser (and complete fertiliser if farmyard manure is not available), and the cost of herbicide. Clearly, there is much scope here for cropping/farming systems work, a practicable system really controlling *Striga* would be of such great evident benefit to the farmer that it would be adopted.

From the standpoint of donor agencies and Governments, it may be necessary to clean up substantial areas of land. This will require operations such as ethylene injection, trap cropping (ploughing under susceptible cereal cultivars before the *Striga* germinated by them can flower) and the use of some of the new herbicides on appropriate, well-farmed crops.

The development of commercial scale production of the "Strigalog" GR24, and research and development for a specific *Striga* herbicide based on blocking the synthesis of mannitol in the parasite, both need to be undertaken. Both could be useful tools in the hands of the small farmer.

References

Doggett, H. (1965) *Striga hermonthica* on sorghum in East Africa. *Journal of Agricultural Science* 65, 183.
Eplee, R.E. (1983) Chemical control of *Striga*. In *ICSU Workshop on* Striga, *Dakar, Senegal, November 1983*, in press. ICSU/IDRC.

Gebisa Ejeta (1980) Status of sorghum improvement in the Sudan. *Report of the ICRISAT-West African Co-operative. Programme, Gezira Research Station.* Wad Medani.

Hamdoun, A.M.; El Tigani, K.B. (1977) Weed control problems in the Sudan. *PANS* 23, 190.

Hanumantha Rao, C.; Sanghi, N.K.; Rao, U.M.B. (1981) Screening technique and evaluation of sorghum for *Striga* tolerance. A.I.C. R.P.D.A., Hyderabad. [Paper presented at the annual sorghum workshop, Navasari, May 11-14, 1981.]

Johnson, A.W.; Roseberry, G.; Parker, C. (1976) A novel approach to *Striga* and *Orobanche* control using synthetic germination stimulants. *Weed Research* 16, 223.

Norris, R.S.; Eplee, R.E. (1983) Effect of stimulants plus herbicide on *Striga* germination. *Proceedings of the 2nd International Striga Workshop, ICRISAT, Patancheru, India*, 43.

Obilana, A.; Tunde (1979) Search for *Striga* control through resistance in guinea-corn in Nigeria. *3rd NAFPP Workshop Sorghum/Millet/Wheat.* Zaria, Nigeria; National Centre.

Ramaiah, K.V. (1983) Patterns of *Striga* resistance in sorghum and millets with special emphasis on Africa. In *ICSU Workshop on* Striga, *Dakar, Senegal, November 1983*, in press. ICDU/IDRC.

Ramaiah, K.V.; Parker, C. (1982) *Striga* and other weeds in sorghum. In *Sorghum in the Eighties*, 291. Patancheru, India; ICRISAT.

Rana, B.S.; Enserink, H.J.; Rutto, J.K.; Ochanda, N. (1980) Progress Report 1980 L.R., Busia. *Field Document 6, UNDP/FAO Ministry of Agriculture, Nairobi*, 42.

Sanghi, N.K.; Vishnu Murthy, T. (1982) *Striga* — an experience in the farmers' fields. *Proceedings of the ICRISAT-ICAR Working Group Meeting on* Striga *Control, October 1982* [Sorghum Breeding Department Report SB4.], 68. Patancheru, India; ICRISAT.

Shaw, W.C.; Shepherd, D.R.; Robinson, E.L.; Sand, P.F. (1962). Advances in witchweed control. *Weeds* 10, 182.

Stewart, G.R.; Nour, J.; MacQueen M.; Shah, N. (1983) Aspects of the biochemistry of *Striga*. In *ICSU Workshop on* Striga, *Dakar, Senegal, November 1983*, in press. ICSU/IDRC.

Tarhalkar, P.P. (1982) Some considerations for *Striga* management in sorghum. *Proceedings of the ICRISAT-ICAR Working Group Meeting in* Striga *Control, October 1982* [Sorghum Breeding Deptartment Report SB4.], 226. Patancheru, India; ICRISAT.

Vasudeva Rao, M.J. (1983) Patterns of resistance to *Striga asiatica* in sorghum and millets, with specific reference to Asia. In *ICSU Workshop on* Striga, *Dakar, Senegal, November 1983*, in press. ICSU/IDRC.

Chapter 14

Sorghum and Pearl Millet Production in Africa: Problems and Prospects with New Varieties

D.J. ANDREWS[1], L.K. MUGHOGHO[1] and S.L. BALL[2]

[1]ICRISAT, Pantacheru P.O., Andhra Pradesh 502 324, India; [2]Department of Agriculture and Horticulture, University of Reading, Reading RG6 2AS, UK.

Introduction

Sorghum and millet, the indigenous cereals of Africa, are important food crops which accounted for 43 per cent of cereals and 23 per cent of all major food staples produced in sub-Saharan Africa from 1973 to 1977 (IFPRI 1983). They are grown primarily as subsistence crops on small farms and grain yields are miserably low: 400–600 kg ha^{-1} compared with 2000–4000 kg ha^{-1} in more developed regions where they are produced primarily for animal feed (FAO 1981). Although total production from 1969/71 to 1981 increased by 22.7 per cent and 6.7 per cent for sorghum and millet, respectively, sorghum yield (kg ha^{-1}) increased by only 4.7 per cent and millet yield declined by 2.3 per cent (Table 1). During the same period, population increased by 30 per cent; thus per capita production of the two crops declined by 15 per cent. This decline must contribute to the steadily deteriorating food situation that affects most African countries (World Bank 1979, IADS 1981). There is thus an urgent need to increase grain production from these two basic food crops in order to alleviate hunger and malnutrition and consequent human suffering, sub-optimal activity, and socio-political unrest.

Table 1   Change in sorghum and millet production in Africa from 1969/71 to 1981*.

		1969/71	1981	Change (%)
Sorghum:	Area (1000ha)	13,073	15,312	+ 17.1
	Production (1000Mt)	9,107	11,174	+ 22.7
	Yield (kg ha^{-1})	697	730	+ 4.7
Millet:	Area	15,359	16,691	+ 8.7
	Production	9,680	10,282	+ 6.7
	Yield	630	616	− 2.3

* Source: FAO (1981).

We recognize that a number of factors (biological, environmental, management and socio-economic) are responsible for the low grain yields of these two basic crops, and that there are no simple, easily implementable solutions. In this paper our primary concern is with the development of improved varieties and production technology which in the appropriate social-political-economic climate will result in increased production on a sustained basis.

The Place of Sorghum and Millet in African Agriculture

Sorghum and millet were the predominant cereals over most of Africa before the introduction and widespread cultivation of maize. In the more fertile and better-watered regions maize is now the preferred crop because it outyields sorghum, stores better, is hardly damaged by birds and is easier to process for food. However, there are large areas where rainfall is low and uncertain and soils are of marginal fertility. In these areas sorghum and millet are the more reliable crops because of their adaptability to a wide range of ecological conditions (Doggett *et al.* 1970). Sorghum and millet will continue to have an important place in African agriculture as population pressure requires the cultivation of marginal lands unsuitable for maize or other cereals.

Crop Improvement

Traditional Landrace varieties

The existing landrace varieties of sorghum and millet which contribute the major part of cereal food production in Africa show very high levels of genetic adaption to numerous severe biological and physical stresses. In consequence many possess location specific adaptation, which accounts for the wide range of diversity to be found in the African germplasm of these two crops.

The modes of adaptation, which have been developed by generations of exposure in historical farming environments, (in which individual stresses may not occur each season), involve resistance, tolerance, and avoidance. Resistance is shown for instance to downy mildew in many West African pearl millet populations where most plants remain free of the disease, but not all, because the resistance is partly dominant and pearl millet is a cross-pollinating crop. Adaptation to parasitism by the witchweed *Striga* appears to be largely by tolerance; and grain moulds and head bugs in sorghum are avoided because flowering has been adjusted so that grain ripening occurs after the rains finish in conditions of declining humidity unfavourable for mould and bugs (Curtis 1968).

Landrace varieties are also well adapted to the length of season, particularly the more photoperiod-sensitive cultivars used in traditional farming methods. In these methods, the length of time in which there is moisture is a vital asset which has been maximised and must not be discarded. Intercropping is a system which well exploits time (Andrews & Kassam 1975). In response possibly to weed competition and damage from pests and disease, landrace varieties tend to be "vegetative" and perform best at low plant densities. While the biomass production of these varieties under the prevailing circumstances, is very high, the harvest index is often below 20 % as compared to over 40 % of improved high yield potential varieties. There are other adaptation features which are much less well known, such as the microbial activity associated with roots of these two cereals . The bacteria and fungi found in close proximity to, and even in roots are certainly involved in nitrogen economy, and mineral uptake from soils of limited fertility (Dart & Wani 1982, Krishna & Dart 1983). The total measure of adaptation of landrace varieties is well known to those who work with these crops in Africa.

Table 2   Technologies available to overcome major limitations to production of sorghum and millet.

Limitation	Technology
Crop establishment	Planting technique, timely planting, superior varieities
Nutrient stress	Timely planting, fertilizers, rotations, weed control, efficient varieties.
Moisture stress	Timely planting, correct plant densities, mixed cropping, weed control, efficient varieties.
Pests/diseases	Seed treatment, insecticides/pesticides, good husbandry, genetic resistance.
Witchweed (*Striga*)	Genetic resistance, rotation, fertilizer, or herbicides.
Birds	Uniform maturity, short plant height, tannins/pearling (sorghum).
Post-harvest losses	Good storage techniques, new storage containers, rodent/insect control.
Market stability	Government policies, support prices.

Varietal Improvement

Varietal improvement must be considered against the background of the many dimensions of adaptation to biological and physical stresses, and against probable changes in other factors controlling production, e.g. fertilizer use, mechanization, and pesticide use.

Varietal improvement is going on in national, regional and international programmes and generally proceeds in two phases.

(1) *Improvement in landrace varieties.* Essentially this has, by reselection, reduced inherent defects in landrace populations, and exploited what positive variation exists. Occasionally similar

landrace varieties have been pooled to advantage. However, the phenotype of these improved cultivars is not greatly different and while they fit well into traditional systems their yield advantages rarely exceed 15%. Examples are Souna III millet from Senegal, and Farafara BL-3-1-6 and SK 5912 sorghum in Nigeria.

(2) *Development of new phenotypes* (varieties* and hybrids) from hybridisation and selection. This step represents a radical change in phenotype and permits potentially large increases in yield, 50% or more than landrace varieties are capable of. Initially these new cultivars have shown yield *instability* because of inadequate incorporation of adaptive traits. It is here that it must be realized that plant breeding is a time-consuming process and will involve several cycles of crossing, backcrossing and selection, each taking 4–6 generations (or the equivalent in population breeding) before high yield is adequately combined with adaptation. Examples are sorghums CE90 (Niger), 189 (Nigeria), and millets 3/4 HK (Niger), IBV 8001 (Senegal). Adapted long season sorghum hybrids which have been tested experimentally in Nigeria have shown up to 40% benefit from heterosis compared to new varieties (Andrews 1975).

In this context it is important to note that improved varieties are but *one* of the factors which contribute to increased production. There are other factors which are only partly, even minimally dependent, on new varieties, which can be employed *today*, to increase production, such as fertilizers, pesticides, and cultivation equipment, as well as institutionally controlled supply demand factors (Table 2). *Production in African countries can be increased with existing technology and does not have to wait for new varieties.* However, when new varieties are available, they will increase the effectiveness of other production factors, and will require the development of quality seed production and distribution services. The present low average yield levels of both sorghum and millet could be raised 3 to 4 fold with improved technology (FAO 1978).

Problems: Biotic and Abiotic Stresses

There are extensive lists of pests, diseases, moisture and nutritional constraints to sorghum and millet production and their importance varies from region to region. Since food grain quality is required in both sorghum and millet, birds are a universal problem to which there is no adequate genetic resistance. In sorghum, *Striga*, anthracnose, sooty stripe, grey and zonate leaf spots, downy mildew, grain and long smut, grain moulds, stemborers and midge are major problems. In millet, downy mildew, rust, blast, ergot and smut, *Striga*, stemborers and head-worms are important. Pearl millet downy mildew and sorghum grain moulds will be described as examples of the disease problems with new varieties and what is being done to deal with them.

(a) Pearl Millet Downy Mildew

Downy mildew caused by *Sclerospora graminicola* is endemic in many areas where the crop is grown. The disease is systemic and can result in total loss of grain. In traditional farming systems where seed from the previous crop is used in the following season, an equilibrium between host and pathogen has been established over the years and disease incidence remains constant around 10–15% within the crop in "normal" seasons. However, pearl millet is an outcrossing crop, and *S. graminicola* is heterothallic which means that the pathogen can also outbreed. Therefore a great deal of heterogeneity exists within the pathosystem. The most effective control strategy should involve presenting the pathogen with a heterogenous crop in terms of resistance factors where total crop uniformity is not essential. The composite breeding project at ICRISAT in India, which incorporates and preserves heterogeneity for resistance to diseases while raising yields is producing high yielding pearl millet varieties which have remained resistant in downy mildew epidemic conditions.

Ideally resistance in a host should be durable for the term of use of a particular cultivar and much research effort is directed towards achieving this. Quantitative resistance characters which may confer partial resistance are more likely to be due to accumulation of minor genes and therefore are more of a problem for the pathogen to overcome than single genes.

*The term varieties has been used in its widest sense in the paper except in this section.

Recently an acylalanine fungicide, metalaxyl, has been found effective as a seed dressing to control downy mildew in pearl millet (Williams & Singh 1981, Singh 1983). Metalaxyl also possesses symptom remissive properties when applied as a foliar spray (1–2 kg a.i/ha) (Singh et al. 1984). However, in common with most of the newer systemic fungicides, there is the risk of increasing the frequency of tolerant phenotypes within the fungus population through adaptation. Consequently, it is advisable to use metalaxyl in moderation as part of a multiple component strategy based on host resistance.

(b) Sorghum Grain Moulds

Traditional, photoperiod-sensitive sorghums mature grain under dry conditions after the rains and consequently produce clean, bright non-weathered grain. However, because of the declining soil moisture situation during grain filling, the low plant population levels typical of these late maturing varieties cannot be increased. In order to obtain higher yields new improved varieties and hybrids have been developed which are early maturing, photoperiod-insensitive and fill their grains before the end of the rains. Such cultivars usually mature grain under wet, humid conditions, and as a result can be severely infected by moulds. Grain moulds affect not only the appearance and germination of the grain, but also make it unsuitable for food.

The only practical control measure for grain moulds is host resistance. In the ICRISAT sorghum improvement program, resistance screening techniques have been developed and high levels of resistance identified in coloured-grain sorghums. These sources of resistance are being used to breed grain mould resistant white-grain varieties.

Breeding Procedures to Overcome Stress

Almost invariably new genotypes need to be selected in situ so that the necessary level of adaptation can be identified and retained. Adaptation demands require that some local parentage is involved in the crossing program. Yield potential requires that parents with high yield potential and generally of a shorter statured phenotype (with increased tillering in millet) are also involved. When individual factors controlling adaptation do not appear at consistantly high levels in the selection environments in which the progeny of these crosses are grown e.g. downy mildew, arrangements must be made to install the facilities for reliably inducing the disease, or, the test material must be sent to reliable hot spots (which are not as freely available as popularly thought).

Because of the multidimensional nature of adaptation, population breeding (recurrent selection) offers the best potential for the simultaneous incorporation and improvement of a number of traits, in both sorghum and millet. Methods for recurrent selection vary from simple to complex; however, in recent years mass selection, or simple variations of it (Andrews et al. 1982) has been shown to be quite effective. Many breeders have however, used more conventional pedigree methods and there are a number of sorghum varieties which testify to its effectiveness. Whatever the method of breeding employed the end products need to be new phenotypes with better harvest indexes, which, if they produce biomass at the same rate of production as existing varieties, will result in increased grain yield.

ICRISAT is very much involved with the development of new phenotypes of sorghum and pearl millet for Africa, but has also demonstrated with pearl millet that with 1 or 2 cycles of S1/S2 selection, local varieties can be radically improved for downy mildew resistance, and their "shibra" content (shattering weedy types) greatly reduced.

Methodologies have been produced to screen for many of the major diseases (Williams & Andrews 1982) and some pests. These need to be intensively used in Africa to screen more rapidly for resistance. In both pearl millet and sorghum new varieties are beginning to emerge from this work, with partly changed phenotypes. To mention some – pearl millets IBV 8001 and 8004 from the ICRISAT/Senegal Cooperative Program, INMB 12 and INMB 70 from the ICRISAT/ Nigerian Program; and ICRISAT sorghum varieties SPV 386 released in Zambia as ZSV 1 and Melkamesh and Kebomesh released in Ethiopia, and the hybrid Hageen Durra No. 1 in the Sudan. It appears possible in Zambia, that pearl millet varieties and hybrids from India may be sufficiently adapted for direct use. In general the use of hybrids of both sorghum and millet in the normal season will depend on the breeding of seed parents adapted to the agroclimatic zone involved; this is still at an early stage. The list of available new varieties however, of both sorghum and millet is likely to become steadily longer over the next few years.

Conclusion

The key to the welfare of mankind in Africa (or indeed elsewhere in the world), is increased agricultural production to meet the ever-growing food, feed and fibre requirements. Governments should recognise that sufficient technology is known in most countries to increase production *now*. While short-term solutions may help alleviate human suffering, it is the long-term production strategies that will sustain development. To this end, a high level of Government investment is essential in agricultural research and extension coupled with appropriate incentives for farmers to increase production. More agricultural scientists must be trained and accorded adequate status and remuneration within the community.

References

Andrews, D.J. (1975) Sorghum grain hybrids in Nigeria. *Experimental Agriculture* 11, 119-127.

Andrews, D.J.; Rai, K.N.; Pheru Singh (1982) Recurrent selection in pearl millet populations. *AICMIP Workshop, Coimbatore, India. April 1982.*

Andrews, D.J.; Kassam, A.H. (1975) The importance of multiple cropping in increasing world food supplies. In *Multiple Cropping* [American Society of Agronomy Special Publication No. 27].

Curtis, D.L. (1968) The relation between the date of heading of Nigerian sorghums and the duration of the growing season. *Journal of Applied Ecology* 5, 215-226.

Dart, P.J.; Wani, S.P. (1982) Non-symbiotic nitrogen fixation and soil fertility. In *Non-symbiotic Nitrogen Fixation and Organic matter in the Tropics. Symposia Papers. I* [Transactions of the 12th International Congress of Soil Science, New Delhi, India. February 1982.], 3-26.

Doggett, H.; Curtis, D.L.; Lauschner, F.X.; Webster, O.J. (1970) Sorghum in Africa. In *Sorghum Production and Utilization* (J.S. Wall; W.M. Ross, eds), 288-327. Westport, Conn.; Avi Publishing, 702 pp.

FAO (1978) Agro-ecological zones project. In Methodology and Results for Africa. *World Soil Resources Report* 48(1), Rome; FAO.

FAO (1981) *FAO Production Yearbook*, vol. 35. Rome; FAO.

IADS (1981) *Agricultural Development Indicators, Statistical Comparison of 139 Developing Countries.* 17 pp. New York; International Agricultural Development Services.

IFPRI (1983) Research perspectives: demand considerations for policies affecting traditional food grain production in semi arid West Africa. *IFPRI Report* 5(1) 1 and 4. Washington, DC; International Food Policy Research Institute.

Krishna, K.R.; Dart, P.J. (1983) Variation in growth and phosphorus uptake responses of pearl millet due to vesicular-arbuscular mycorrhizal symbiosis. *Proceedings of the Conference on the Impact of Microbiology on Tropical Agriculture.* University Pertanian Malaysia, Serdang. October 1983.

Singh, S.D. (1983) Variable cultivar response to metalaxyl treatment in pearl millet. *Plant Disease* 67, 1013-1015.

Singh, S.D.; Gopinath, R.; Luther, K.D.M.; Reddy, P.M.; Panwar, M.N. (1984) Systemic remissive property of Metalaxyl against downy mildew in pearl millet. *Plant Disease* (in press).

Williams, R.J.; Andrews, D.J. (1982) Breeding for disease and pest resistance in pearl millet. Paper presented at the *Conference on Breeding for Durable Resistance in Tropical West Africa, October 1982.* Ibadan; International Institute of Tropical Agriculture.

Williams, R.J.; Singh, S.D. (1981). Control of pearl millet downy mildew by seed treatment with metalaxyl. *Annals of Applied Biology* 97, 263-268.

World Bank (1979) *World Development Report, 1979.* 188 pp. Washington, DC, USA; World Bank.

Chapter 15

Utilization, Production Constraints and Improvement Potential of Tropical Root Crops

S.K. HAHN

International Institute of Tropical Agriculture (IITA), P.M.B. 5320, Ibadan, Nigeria.

Introduction

Importance

The major tropical root crops such as cassava, yams, sweet potato and cocoyams, are widely grown and mostly used as subsistence staples in many parts of the tropics and subtropics in Africa. They are the major source of dietary energy for well over 200M people in the continent, and are also grown, to some extent, as industrial raw material and for livestock feeds. They account for 31% of the major staples produced in sub-Saharan Africa (Paulino & Yeung 1981), and the trend in root crops production is a steady increase of 2.7% y^{-1} (FAO 1982).

Advantages

In Africa cassava is grown from 0-1800 m altitude and from the sub-Sahelian semi-arid region to 25°S. Sweet potato is grown from 0-2300 m and between 30°N and 30°S. Yams are grown from 0-800 m and between 25°N and 15°S. Cocoyams are grown mostly in the lowland humid tropics. Root crops, except yams, are grown on a wide range of soils and can give satisfactory yields on poor acid soils. Cassava and sweet potato are grown from high rainfall humid to semi-arid regions due to their ability to tolerate drought and their wide adaptability in terms of soil requirements. They play vital roles in alleviating famine conditions by providing a sustained food supply when other crops fail. Cocoyams are best adapted to wet and inundated areas, and are also shade tolerant in forests. Root crops are, in general, well-adapted to diverse traditional farming systems under the wide range of environmental conditions in Africa.

Root crops are biologically more highly efficient producers of calories than any other food crops, particularly in the tropics. Their efficiency as food producers arises partly from the structure of the plants; strength in other parts of the plant is not needed to support the bulky and heavy roots and tubers so that increase in size of the edible parts need not be associated with increased production of non-edible tissue. Not only are root crops capable of relatively high efficiency in production of edible carbohydrate, but their efficiency for protein production is also higher than commonly realized (Table 1).

Despite their advantages as staple food crops research and development activities have not been commensurate with their importance as major staples and increasing importance as industrial crops and animal feed. They are usually considered inferior crops because they are cheap in relation to cereals, and believed to be very low in protein content. An inverse relationship between root crop consumption and the standard of living is often also assumed, but more important is that large quantities of imported grains have discouraged farmers from producing more locally adapted root crops. However, there is now increasing political and scientific awareness in many countries of Africa of the importance of improving production of locally well adapted root crops.

The potential of root crops in the continent will be realized once modern technology is applied to their production and utilization. This potential is particularly relevant in view of the diminishing resources, and increasing population, and in sites where ecological conditions make them a logical source of carbohydrate (Haynes 1974). Techniques of production and utilization developed need to be evaluated and made available to farmers.

Table 1   Calorie and protein productivity of various food crops in West Africa (after Coursey & Booth 1977).

	Calorific production ($cal \times 10^6$ ha^{-1})	Protein production ($kg\ ha^{-1}$)
Cassava	8.2	37
Yams	5.7	107
Sweet potato	7.4	96
Cocoyams	4.5	80
Irish potato	4.7	128
Maize	3.2	82
Rice	3.2	72
Sorghum	2.4	70
Soybean	0.8	78

Cassava

Constraints to production

In Africa, the major biological constraints to stable production of cassava are diseases discussed in Chapters 18 and 19. In addition, traditional unimproved agronomic practices and processing methods have seriously limited production.

Overcoming the constraints

(a)   Varietal improvement

Cassava breeding with particular emphasis on cassava mosaic virus (CMD) resistance was initiated in Tanzania by Storey in 1937; this was a turning point in cassava improvement that merits special attention. Storey made an interspecific cross between cultivated cassava (*Manihot esculenta*) and a related species (*M. glaziovii*) and obtained 20 F1 plants which were backcrossed thrice onto cultivated cassava. Some progenies of the backcrosses showed promise of good CMD resistance. In 1956, Beck (1980) introduced to Nigeria from Tanzania seeds resulting from Storey's interspecific crosses and selected a clone ("58308") that still remains resistant to CMD. Hahn *et al.* (1980) used this clone as a resistance source for both CMD and cassava bacterial blight (CBB), and also as a source of low cyanide in early breeding programmes around 1971-73. Using the clone as a parent, improved cassava breeding populations were produced from which many good selections have been made under local conditions in many countries in Africa.

Such rapid progress was facilitated by IITA being located at site with environmental conditions representing the major cassava growing areas in tropical Africa. Many improved disease resistant and high yielding varieties have been established in Gabon, Liberia, Nigeria, Seychelles, Sierra Leone, and Tanzania (Zanzibar). In Nigeria, these varieties are grown on over 200 000 ha. The varieties are also sent out to 31 countries in Africa as tissue cultures. Dry matter content is also tested for each clone, and consumer acceptance assessed at the advanced stage of breeding.

Low cyanide clones and populations with high yield potential and disease resistance have been developed. Sources of resistance to CMB and green spider mite (CGM) have been identified and incorporated into susceptible but high yielding and disease resistant clones and populations.

(b)   Biological control

The biocontrol of cassava pests and diseases is discussed in Chapter 19.

(c)   Agronomic practices

Cassava, as other root crops such as yams and cocoyams, is a long-duration crop (10-24 months), a characteristic which is sometimes advantageous and sometimes disadvantageous. This does not permit a rapid turnover in a rotation scheme so that land can be released for short duration crops (2-4 months). However, cassava is traditionally grown in a mixture with many short duration crops and is ideal for intercropping (see below) since they attain their full development and maturity when cassava is just attaining its maximum leaf area development and is initiating and increased rate of bulking of tubers.

(1) *Land preparation:* Under good soil conditions, cassava is normally grown on flat land, while when soils are poor it is grown on mounds or on ridges. Mounds made by earthing up the soil, from 75 cm diam × 1.5 m tall to 4 m × 1.5 m, may have as many as 12 crop species.

(2) *Shade:* Cassava is often grown under the shade of oil palm, coconut, or cocoa in high rainfall forest regions, when yield is serverely reduced.

(3) *Soils:* As land is continuously used, soils become poorer, and productivity decreased. However, on poor soils, there seems to be a response to N application, liming, and with potassium application for certain varieties.

(4) *Planting material:* The quality of planting material affects stands, vigour, tuberization and final yield; stem cuttings from basal portions are most successful.

(5) *Intercropping:* Crop associations with cassava may include maize, sugarcane, beans, peanuts, melons, bananas, and assorted vegetables (Okigbo 1977). The advantages include (1) better protection of the soil against erosion, (2) insurance against crop failure, (3) more efficient use of labour, and (4) more or less continuous harvesting of crops, providing balanced food continuously and minimizing storage problems. Disadvantages include the (1) difficulty of mechanical harvesting, (2) difficulty of applying improved inputs such as fertilizers and herbicides, etc., and (3) difficulty of large scale production.

(6) *Leaf harvest:* Cassava leaves are often used as a vegetable, particularly in Central Africa, and have a potential use as animal feed. When leaves are harvested at intervals of one, two or three months, however, tuber yields are reduced by 62, 52 and 25 % respectively.

(7) *Weed control:* Weeds reduce cassava yield by 59 % on average and complete yield losses have been reported, particularly due to speargrass (*Imperata cylindrica*). Minimum weeding two or three times is required before cassava fully develops its canopy. Surface mulching can also have a significant effect on weed control and cassava yield.

(8) *Continuous production:* When root crops are grown continuously, yield reduction normally occurs. Growing root crops continuously by incorporating organic matter was tried with cassava, yams, and maize at Umudike, Nigeria, for 20 years (1940-60) and showed that continuous cropping of cassava, yams, and maize was possible without yield decline provided that organic matter is continuously incorporated.

(d)   Processing and hydrogen cyanide

Cassava contains hydrogen cyanide (HCN) in both edible tubers and leaves in the form of cyanogenic glucosides, which release HCN on hydrolysis when tissues are destroyed. No acyanogenic variety has been reported and low cyanide varieties are generally low yielding compared to high cyanide varieties. To eliminate cyanide and improve palatability, processing is necessary (e.g. peeling, grating, fermenting, dehydrating, sun drying, frying, boiling).

Yams

Yams are of great ethno-agricultural importance in West Africa, where a considerable amount of ritualism surrounds their production and utilization (Onwueme 1978). Most important are white yam (*Dioscorea rotundata*), yellow yam (*D. cayenensis*), trifoliate yam (*D. dumetorum*), and water yam (*D. alata*); all are indigenous to West Africa except water yam, which originated in Asia. In Africa, white yam is the most important in terms of production and preference, followed by water yam and yellow yam.

Yams are adapted to fairly high rainfall areas with a distinct dry season of not more than five months. The species have an annually repeated cycle of growth for 6-10 months, and a dormancy of 2-4 months, the phases corresponding approximately to the wet and dry seasons (Coursey 1967). Yams are propagated by tubers and produce vines to 2-7 m long, usually supported on poles. Tuber maturity is generally reached 6-8 months after planting. Male and female plants are separate, and the plants are relatively tolerant to shade in the high rainfall forest regions.

Constraints to production

Yam production requires high inputs, is labour intensive (1185 man-hours ha^{-1}) and difficult because (1) seed yams are expensive and often unavailable, (2) seed-bed preparation is laborious and tedious, (3) staking is necessary but staking material is expensive and often not available, (4) weeding is required for a long period of time (4-5 months), and (5) harvesting needs special care and is laborious. For these reasons there is a declining trend in yam production. Research needs to be directed towards reducing inputs and labour requirements. Consumers prefer larger tubers, which implies (a) larger seed-beds, (2) larger planting setts, (3) small plant population per unit area, (4) staking necessary for leaf display, (5) the number of tubers per plant should be small, and (6) there is more chance of tuber damage at harvest. These factors make production laborious, expensive, difficult, and inefficient. Nematodes cause severe pre- and post-harvest damage to tubers reducing not only yield but also market and keeping quality.

Overcoming the constraints

(a)   Germplasm evaluation and genetic improvement

600 accessions of white yam and 40 of water yam germplasm were collected from West Africa in the early 1970s and screened for desirable agronomic characteristics. Several promising races were selected and improved white yam clones have been produced. Several water yam clones have also been selected from the collection which show promise in terms of resistance to necrosis and yield potential, and with thick periderms so that they are seldom bruised during harvesting and handling; several are poundable in food preparation, important for consumer preference.

(b)   Agronomic practices

The yield and the plant characteristics that affect yield are influenced by the size of mound or ridge, the size and quality of planting sett, time of planting, weeding, and staking (Enyi 1973).

(1) *Land preparation:* As for cassava, yams are normally grown on mounds or ridges when the soil water table is high and soils are poor, or on flat land after loosening the soil. Little scientific information is available on the rationale of growing yams on mounds, however, recent results indicate that tuber yield increases with mound size. Although there seem to be advantages in growing yams the traditional way in large mounds, making the mound with traditional hand tools is very laborious; ridges made by machines or animal traction can, however, be used (Onwueme 1980).

(2) *Sett size:* The yield of both white and water yams increases with sett size at two plant population levels (8000 and 10 000 plants ha^{-1}), but then brings about a smaller multiplication rate or input-output ratio, thus becoming uneconomic. To be profitable, sett size should not exceed 1 kg.

(3) *Mulching:* Mulching with crop residues gave a 20% higher yam yield than no mulching, and tuber diameter was also larger. Mulching helps reduce soil temperature, resulting in good sprouting and higher yields (Lal & Hahn 1973), and also improves soil moisture-temperature-fertility interactions. Plastic mulch can significantly reduce necrosis of white yam, eliminate weeding, prolong growth duration, and conserve soil moisture; labour inputs can be reduced 23%.

(4) *Staking:* Normally unstaked white yam leaves become senescent much earlier than staked ones; the leaf area duration is shorter resulting in a lower tuber yield. Staked white yam can give a

48-62% higher yield than the unstaked. However, staking is costly, laborious, difficult to mechanize (Onwueme 1980), and staking material is not always available. Several water yam clones can be grown without staking and give reasonable yields, but white yam clones suitable to production without staking in high rainfall areas are yet to be found. Yams are slow in developing their full canopy so that staked vines leave the ground surface exposed to rain and render the soil vulnerable to erosion.

(5) *Weed control:* The growth habit of yams and their inability to completely shade the ground makes them very susceptible to weed competition. Weeds can reduce the yield of white yam by 43% and of water yam by 61%.

(6) *Intercropping:* Yams are often grown in a mixture with other crops, as is cassava. When white yam was intercropped with maize or cowpea, its yield was reduced by 33% and seems to be more affected by maize or cowpea than cassava.

(7) *Nematode control:* The yam nematode and root-knot nematode cause great problems to yams, reducing the yield of white yam by 72%. Rotation with cover crops (e.g. *Centrosema pubescens, Cynodon nlemfuensis, Stylosanthes gracilis*) would be helpful in controlling yam nematodes. The white yam is the more susceptible to nematodes, but a land race ("Abi") shows only light root-knot galling on the tubers with both *Meloidogyne incognita* race 2 and *M. javanica*.

(8) *Harvesting:* Yam harvesting is carried out almost entirely with hand tools. Special care when lifting and handling tubers is necessary to avoid damage which shortens shelf-life. Consumers prefer the longer and larger tubers, which implies a greater mean depth for extraction and increased harvesting time. Alternatives to traditional harvesting should be sought by growing varieties of a round and uniform tuber shape and a thick periderm.

(9) *Storage:* During the 4-5 month storage period tubers can lose 40% of their weight due to sprouting, moisture evaporation, respiration activities, and tuber rots (Waitt 1963). The principal factor responsible for poor storage is pre-harvest nematode attacks that open avenues for storage rots, although mechanical tuber damage is also important. Farm storage systems should be examined and improvement sought to determine (1) a storage structure that can provide adequate shade, light and ventilation, (2) varieties than can be stored longer, and (3) possibilities for processing tubers into products such as yam flour that can be stored for a long period and are easier to transport.

(10) "Seed" yam production: The cost of seed yam is approximately one-third of the total outlay of production. Seed yams are expensive and often unavailable to poor farmers, and those that are available are often of an inferior quality. A system is needed by which superior seed yams can be produced in quantity and are made available at a lower cost. A mini-sett method with which yam can be rapidly multiplied has been applied to seed yam production systems using plastic mulch; by eliminating weeding and staking the multiplication ratio was increased.

Sweet Potato

Among the root crops grown in the tropics, sweet potato has the shortest growth cycle, 4-5 months. Its yield potential is illustrated by an improved clone with 40 t ha + in four months without fertilizers and with good management and soil conditions. Sweet potato can adapt to a wide range of environmental conditions.

Constraints to production

Weevils, nematodes and viruses are prevalent as pests and diseases, often causing severe damage. Storage has also been a problem. Where it is difficult to preserve planting material during the long dry season, this becomes a factor limiting yields, particularly where the rainy season is short. In such areas, even though fields are wet enough for planting after the first good rain, planting has to be delayed by 2-3 months until planting material is available from the vines of plants grown in the field. Sweet potato is therefore usually planted towards the end of the rainy season and permitted to mature during the dry season when the weevil population is largest; damage to tubers is therefore higher. Another limiting factor is the use of old vines as planting material instead of young apical shoots which establish better and result in more rapid growth and higher yields.

Overcoming the constraints

(a)   Varietal improvement

Germplasm (over 600 accessions) has been assembled in seed form from many countries in Africa, Asia and the Americas; this has been evaluated for resistance to weevils, the sweet potato virus complex, and other important agronomic characteristics. Sources of resistance to the insect and virus complex have been identified and incorporated into breeding populations which have been further improved in terms of yield potential, virus and weevil resistance and quality through recombination and selection. Seeds from these improved populations have been distributed to many national programs for selection and evaluation under local conditions. After processing through tissue culture followed by virus indexing, improved clones have been sent to 30 countries in Africa. Local researchers select promising clones and release them as recommended varieties.

(b)   Agronomic practices

(1) *Less erosion in the sweet potato field:* Sweet potato provides early effective ground cover and there is less soil loss due to erosion compared to other root crops.

(2) *Control of weevils in the field and storage.* (i) *Cultural control in the field:* Increase in plant population results in a decrease in root size, which in turn decreases root exposure to weevil infestation at the soil surface. Earthing-up also helps to reduce weevil damage, as does planting soon after the first good rain and harvesting before or immediately after the end of the rainy season. (ii) *Control in the storage:* When weevils (adults) are buried in the soil, most die after 8 days; underground storage for 2-3 months can be effective in their control. Hot water treatment can control not only weevils but also nematodes and storage rot associated pathogens, while lower temperatures extend the development period and drastically affect survival. Sweet potatoes grown in highland areas where the temperature is lower consequently have fewer weevil problems.

Cocoyams

Colocasia and *Xanthosoma* species are important staple food crops in many parts of Africa. *Colocasia* is indigenous to south-east Asia, and *Xanthosoma* to tropical America. *Colocasia* reached Africa first is often called "old cocoyam" and *Xanthosoma* "new cocoyam". Cocoyams are essentially warm weather lowland crops, with the best yields occurring at mean temperatures above 21°C with a rainfall exceeding 2000 mm y^{-1}. *Xanthosoma* is preferred to *Colocasia* in West Africa due to its superiority in terms of yield, taste and adaptability.

Within *Colocasia* there are two major types, eddoe and dasheen, the former performing better under drier conditions and the latter under flooded conditions. *Xanthosoma* does better on deep, well-drained soil than in inundated areas. The edible parts of *Colocasia* and *Xanthosoma* are the corms and cormels. Young leaves, petioles and flowers of *Colocasia* are often used as a vegetable. Propagation is vegetative using small corms, cormels, stem cuttings or pieces of these.

Constraints to production

A root rot complex, principally caused by *Pythium myriotylum,* is the most limiting factor of *Xanthosoma* production in Africa. The plants develop chlorotic leaves, brown blight developing from the periphery and extending towards the petioles which drop, the leaves drying up and being shed prematurely. Infested plants remain stunted with necrotic shrivelled leaves. The root system is greatly reduced from decay. In advanced stages the cortical tissue turns brown and disintegrates leaving a nonfunctional vascular skeleton. Because of this disease *Xanthosoma* is being entirely replaced with root rot complex resistant *Colocasia* in the high rainfall areas of Cameroon, although *Xanthosoma* is the preferred crop.

Overcoming the constraints

(a)   Genetic improvement

GA at the concentration of 1500 ppm gives the best results for flower induction for both *Colocasia* and *Xanthosoma*. Viable hybrid seeds have been produced, seedlings established, and selections made. Hybrids have been evaluated for resistance to cocoyam root rot complex and promising results are being obtained. Over 100 germplasm accessions of *Xanthosoma* have been introduced and evaluated in Cameroon. Cytogenetics and crossability of cocoyams have been studied with encouraging results for developing future improvement strategies.

(b) Cultural control of cocoyam root rot complex

Foliar symptoms of the cocoyam root rot complex are reduced by wider spacing (1.5 m) and early planting gives higher yields.

References

Beck, B.D.A. (1980) Historical perspectives of cassava breeding in Africa. In *Root Crops in Eastern Africa: Proceedings of a Workshop held in Kigali, Rwanda, 23-27 November 1980,* 13-18.

Coursey, D.G.; Booth, R.H. (1977) Root and tuber crops. In *Food Crops of the Lowland Tropics* (C.L.A. Leaky, ed.), 75-86. Oxford; Oxford University Press.

Enyi, B.A.C. (1973) Growth, development and yield of some tropical root crops. In *Proceedings of the Third Symposium of International Society for Tropical Root Crops held on 2-9 December 1973.* Ibadan; IITA.

FAO (1982) *1982 FAO Production Yearbook.* Rome; FAO.

Hahn, S.K.; Terry, E.R.; Leuschner, K. (1980) Breeding cassava for resistance to cassava mosaic disease. *Euphytica* 29, 677-683.

Haynes, P.H. (1974) Tropical root crops: a modern perspective. *Span* 17 (3).

Lal, R.; Hahn, S.K. (1973) Effect of method of seedbed preparation, mulching and time of planting on yam in Western Nigeria. In *Proceedings of the Third Symposium of International Society for Tropical Root Crops held on 2-9 December 1973,* 293-306. Ibadan; IITA.

Okigbo, B.N. (1977) Preliminary cassava intercropping trials. In The First National Accelerated Food Production Programme (NAFPP) National Cassava Workshop held at Umudike, Nigeria, Jan. 1977.

Onwueme, I.C. (1978) *The Tropical Tuber Crops.* New York; J. Wiley and Sons, 234 pp.

Onwueme, I.C. (1980) Strategies for progress in yam research in Africa. In *Tropical Root Crops Research Strategies for the 1980s,* 173-176. Ibadan; IITA.

Paulino, L.A.; Yeung, P. (1981) The food situation in sub-Saharan Africa: A preliminary assessment. In *Food Policy Issues and Concerns in sub-Saharan Africa.* [Discussion Paper.]

Waitt, A.W. (1963) Yam: *Dioscorea* species. *Field Crop Abstracts* 16, 145-157.

Chapter 16

Nematodes and their Control in Root and Vegetable Crops

F.E. CAVENESS[1] and J. BRIDGE[2]

[1]International Institute of Tropical Agriculture PMB 5320 Ibadan Nigeria. [2]Commonwealth Institute of Parasitology, 395A Hatfield Road, St Albans, AL4 0XU, UK.

Introduction

Nematodes are probably the most numerous amongst the multicellular animals. In a rich agricultural soil, many tens of millions could be living in a single hectare. Nematodes are cosmopolitan, occurring wherever there is food to support life and particularly in agricultural soils and plant roots. Some of the nematodes inhabiting soils feed on bacteria, fungi, algae and other small animals (including other nematodes), while others function as plant parasites.

Parasitic nematodes are recognized as a major pest of foodstuffs, especially of root and vegetable crops in the tropics. Many different nematode species are known to cause damage; some have a limited distribution whereas others are widespread. These nematodes infect and damage all root and vegetable crops, often at considerable economic cost.

Importance of root and vegetable crops

Vegetable crops are of prime importance as the major part of the food intake of the African nations is from this source. The vegetable root crops, cassava, potato, sweet potato, yam and cocoyam, provide most of the starch staple food in the human diet in Africa south of the Sahara. Vegetables are also of great importance in the human diet, supplying components needed for health which are deficient in other food sources. They are important in providing bulk and roughage, carbohydrate, minerals, vitamins, and much-needed protein.

Many vegetables are highly tolerant of environments and crop well over an extended range of edaphic and climatic conditions; many are also interchangeable on the land enabling farming systems to remain sensitive to market conditions or other situations. Some, for example cassava, show considerable drought tolerance and recovery ability after devastation by insects, hail, or fire. Root and vegetable crops are also important as feed for livestock and as industrial components.

As most of the African production of vegetables is consumed within the country of origin, with little entering international trade, this is of major domestic economic importance. Other considerations include the generally perishable nature of vegetable crops, and in West Africa yams are of great ethno-agricultural importance as traditional customs have developed around their production and utilization.

A classification based on the plant part used as a food would group crops grown for their leaves or stems (cole, salad crops, potherbs, or greens), crops grown for their fruits (melons, tomatoes, garden egg, beans, peas, cowpea, soybean), and crops grown for their underground roots, tubers, bulbs or corms (potato, sweet potato, taro, tannia, carrot, onion, garlic).

Nematode damage

Virtually every plant and crop has nematode parasites, and some nematodes are parasitic on a variety of crops. Crop yields are affected by nematode species and numbers, with the interplay of environmental factors such as soil texture, fertility, moisture, and temperature. Many nematodes cause plant disease directly, or function as an active agent in a disease complex with bacteria or fungi. Others harbour and transmit plant virus diseases.

Plant injury arises from the feeding and presence of nematodes. Injury extends from simple mechanical damage to complex nematode-plant chemical interactions. Endoparasitic nematodes cause injury or destruction to individual cells either by feeding directly on the protoplast or by mechanical pressure when moving through plant tissues. Chemical interactions due to compounds introduced by the nematode result either in a physiological change in the host, or chemical substances being produced by the plant in response to the attack.

Crop damage from the same species and numbers of nematodes may vary from year to year, depending on the plant stress. Under high stress, such as high temperature, low soil moisture, low soil fertility, other pathogens, or as is often the case a combination of these factors, growth may be severely affected or the plant may die, while under more favourable conditions the plant might still produce an economically acceptable harvest.

Nematodes are frequently subtle and insidious crop pests. Yield reductions of less than 10 % can pass undetected unless carefully managed control plots are established to monitor changes. In certain root and tuber crops, nematode damage and disfiguration may lead to serious losses due to consumer rejection.

Nematodes are increasing in importance as human population growth places further demands on arable land. Shortened crop rotation, coupled with the more frequent planting of food crops, particularly near large centres of population, favours the build-up and maintenance of high levels of nematode soil populations. Greater nematode problems are therefore to be expected in future.

Field symptoms are generally nonspecific and other possible causes for poor plant growth need to be considered in addition to nematode attack. An area of reduced plant growth, usually circular, in a field of a well-growing crop is often a sympton of parasitic nematode presence. The outer circumference of stunted plants generally enlarges with each new crop. On warm days the affected plants tend to wilt a few hours before moisture stress affects healthy plants. In addition, affected plants may be chlorotic, showing signs of a nutritient deficiency caused by a nematode-reduced root system. A general reduction in plant growth, with a corresponding reduction in canopy, leaves more ground unshaded and so weed growth increases. Affected plants are sometimes forced into early senescence, reducing the yield-producing time and so lowering the final yield. The reduced root system and canopy also impair the normal processes of water and nutrient availability during the periods of flush growth for leafy vegetables, and pod, root or tuber-filling times of other vegetables, again resulting in reduced yields. One easily recognized symptom of nematode parasitism on plants with a reduced root system is that the plants are easy to pull up.

Gross symptoms of root tissue damage can appear as knots or galls (these also disrupting vascular elements), the killing of individual cells which coalesce to form lesions (causing root pruning in severe cases), the cessation of root growth giving a stubby root appearance, and especially among fleshy rooted plants, a general necrosis due to invasions by secondary pathogens. The more heavily attacked root systems, especially during the seedling stage of growth, are greatly reduced with a lack of fine feeder roots. Sometimes damage is so severe that the plant cannot survive a period of stress.

Nematode control

All control principles are based on the imposition of certain stress factors on nematode populations. These imposed conditions affect the nematode's ability to feed, reproduce, and survive as a population. Some cultural practices limit or reduce nematode populations over various periods of time, in contrast to quick population kills by the use of heat or chemicals.

Control relates to the crop and type of nematode involved, and economically satisfactory control may vary with both the market price and consumer sophistication. The means of nematode control are many and diverse. Feasibility and application depend on individual farm and farmer situations and the socio-economic conditions prevailing.

Control of nematodes is also achieved where cultural practices prevent their initial increase to damage-causing levels.

Control methods can be placed in categories such as cultural control, physical control, chemical control, biological control and legal restrictions.

Then with control methods, environments, farmers and markets — all in the plural; a workable concept for control of plant-parasitic nematodes must be broad and flexible and based on local knowledge of crop cultivars, soil, climate, nematode species and races and the farmer. Nematode control then must be based on trained extension staff, a solid research base and an enlightened farmer clientele. The goal is to increase and stabilize crop yields using minimum farmer input methods and to obtain maximum outputs.

Control strategies

International and national agricultural research institutes, universities and agricultural extension services must meet the needs of all commercial enterprises, smallholder market farmers and subsistence farmers.

The commercial enterprise farmer, and possibly a few smallholder farmers, could use control methods necessitating monetary inputs and technical knowledge of handling and the application of nematicides. However, for most farmers in Africa the purchase of nematicides is beyond their means, the technical knowledge for their judicial use is not known, and most nematicides are lethal or too dangerous to health to place in the hands of farmers lacking the knowledge or equipment to take appropriate safety precautions.

A similar situation exists in the application of cultural methods for the control of nematodes. The more advanced farmers have the tools and expertise to implement control recommendations, while the subsistence farmer, who is mainly concerned with growing food crops, is limited to fewer crops and traditional growing methods.

Control methods

The great number of species and races of plant-parasitic nematodes, the diverse environments, and multitude of crop varieties, has made it necessary to develop a range of control methods to suit particular situations. Control measures include most of the methods used in general plant protection practices, exclusion by quarantine, sanitation (fields and equipment), crop rotation (including mixed cropping and careful and knowledgeable selection of cultivars), organic and non-organic soil amendments, time of planting, breeding and selecting resistant plants with tolerance, trap crops, physical methods of heat or hot water on soil and plants, flooding, clean seed and planting stock, the use of chemicals on growing and dormant plants, and disinfestation of the soil. These control measures are modified to fit the problem, crop involved, facilities available, and inputs of labour, equipment and funds deemed appropriate.

Role of Governments, research establishments, and extension services.

Effective farmer control of nematodes is the end product of a system of specialized service institutions whose many functions are related and reinforcing. Governments must provide policy, decisions and regulation of activity within and between the other segments of the system. The universities and agricultural research establishments must supply the in-depth knowledge on nematode presence and pest value for crop damage and the improved cultivars, meshing these with cultural management practices for nematode control. It falls upon the extension service to expedite the flow of new information to the farm level. At the subsistence level of agriculture these services are provided by the family itself, or the immediate community, with resulting lack of change and improvement.

Control methods for Africa

Recommendations for the control of nematodes, no matter how experimentally valid, will not be effective if they are unrealistic, requiring implementation beyond the means and capabilities of the farmer. The appropriate input and technology level depend on the farmer audience. Nations would benefit whether there was a large yield increase from a few advanced farmers, or a small yield increase from a great number of smallholder farmers. It must also be recognized that crop response to nematode control measures does not occur in isolation from all other environmental influences.

(1) *Crop rotation* is most effective where the nematode species involved is relatively host specific. However, many tropical soils contain several plant-parasitic nematodes and development of a successful crop rotation scheme is consquently complicated. This also emphasizes the need for

accurate information on cultivar resistance and nematode biology. Where control inputs are limited or crop value is low, crop rotation may be the only method that can be employed profitably.

(2) *Nematode resistant crops*, where available, are complementary to a successful crop rotation scheme. The cost of development of a resistant cultivar is borne by Governments or other agencies and the benefit is passed on to the farmer, who need not even be aware of this kind of crop improvement.

(3) *Organic materials and mulching* has often resulted in improved plant growth, though probably as much due to improved plant growing conditions as biotic competition with the plant-parasitic nematodes. The development of alley cropping methods could help make soil amendments and mulching more attractive as hauling distance is consequently drastically reduced.

(4) *Fallow periods* of weeds, grass or bush tend to reduce plant-parasitic nematode populations as the preferred food source is reduced on a unit area basis. The eradication of volunteer food plants is important for the success of this practice.

(5) *Sanitation* and the use of *clean planting* stock are useful adjuncts to other control methods, especially in situations where the transplants of a communal seedbed are moved to a number of fields.

(6) *Heat treatment* of soil is effective for small quantities of soil. The burning of wood fires over seedbeds will kill most nematodes to a depth of 30-40 cm. Recent research utilizing solar heat trapped under black plastic sheets indicates that this may render control over larger areas increasingly feasible.

(7) *Post-harvest control* of nematodes in tubers, corms and tuberous roots and plant roots improves their storability and value as planting stock for the next season. Hot water steeps, chemical dips, and pruning of infected tissue and roots, are beneficial.

(8) *Nematicides* are effective and can be used where economic and technologically safe. In Africa their use is generally restricted by the lack of sufficient national infrastructure to develop and sustain an agricultural chemical control industry.

(9) *Plant quarantine measures* are instituted by Government to restrict the international movement of plant pests and diseases. These measures are effective when respected by the populace.

Chapter 17

Recent Advances in Control of Legume Diseases

D.J. ALLEN[1] and B.J. NDUNGURU[2]

[1]Centro Internacional de Agricultura Tropical, Cali, Colombia; [2]Department of Crop Science, University of Dar es Salaam, Morogoro, Tanzania.

Introduction

Groundnuts, beans (*Phaseolus vulgaris*) and cowpeas are the principal food legumes of tropical Africa where they provide a cheap source of dietary protein. Beans and cowpeas are important chiefly as dry seed although the leaves of each species are locally used fresh as a spinach or dried for use in soups; the use of immature seed and green pods are relatively unimportant in Africa. Groundnuts are grown both for their oil and as a pulse. Estimates of the annual production of groundnuts "in shell" from Africa are of the order of 5.4 M tonnes, Nigeria, Senegal and the Sudan being the largest producers (Gibbons 1980 a). Bean production in Africa stands at about 1.6 M tonnes of dry seed. Rwanda, Burundi, Uganda, Kenya, Tanzania and Malawi, being among the principal producing countries (Londono *et al.* 1980). Estimates of the production of cowpeas suggest that somewhat more than one million tonnes are produced annually in Africa; Nigeria is the major producer. Significant amounts of cowpea grain are produced also in Niger and Upper Volta.

In Africa, each of these three legume crops is grown principally as a subordinate component of the complex crop mixtures that typify traditional agriculture, and as such they receive few inputs. It is estimated that some 95 per cent of Nigerian groundnuts are intercropped especially with cereals (Gibbons 1980a). In these savannahs of West Africa cowpeas are also components of cereal farming systems dominated by bulrush millet and sorghum, a system where traditional cultivars of these crops are adapted with surprising precision both to their social uses and to the environment (Bunting 1975). The bean crop, which is the most important of the legumes at altitudes above 1000m, is most often intercropped with maize, and in Uganda, it has been estimated that about 75-90% of all beans are grown in mixtures with coffee, bananas, or cereals on small farms. Essentially similar bean farming systems are found elsewhere in Africa (CIAT 1981).

In each of these crops, yields are notoriously low (range 400-700 kg ha^{-1}). The extent to which diseases are constraints to increased productivity, and strategies to manage disease without recourse to heavy inputs, are the subjects of this paper which relies substantially on a recent review (Allen 1983).

Disease as a production constraint

Although it is firmly established that diseases constitute major constraints to production, there are few reliable data quantifying losses in food legumes in Africa. An exception is the study of the relationship between disease progress and crop loss in *Cercospora* leaf spot of cowpea (Schneider *et al.* 1976). Fewer still have attempted to analyse the relative importance of production constraints, such as was revealed in a crop loss profile of bean production in the Cauca Valley of Colombia (Pinstrup-Andersen *et al.* 1976). In certain instances, however, there can be no doubt about the economic importance of an outbreak of a disease; an epidemic of groundnut rosette in West Africa in 1975 led to almost total loss of the groundnut crop, with an estimated loss to Nigeria alone of about US$300 M (Yayock 1977). Nigerian production has remained low in subsequent years, for which this unprecedented epidemic of rosette, as well as the recent spread of rust, are held partly responsible (Misari *et al.* 1980). Furthermore, it has been estimated that the annual loss of groundnut kernels caused by the *Cercospora* leaf fungi alone is about 3 M tonnes (Gibbons 1980b), which further emphasizes the need in groundnut improvement to give priority to disease management.

Disease control has been at the centre of CIAT's bean improvement strategy since the programme's inception. High priority has been assigned to bean common mosaic, anthracnose, angular leaf spot, rust and common blight which are seen as the principal diseases of the crop throughout its range, recognizing that their relative importance varies markedly with ecological zone. *Ascochyta* blight and halo blight are more important under cooler conditions at high altitudes. Similarly, recent work at IITA on cowpea improvement has given prominence to the control of diseases and, especially, pests; without some protection, grain yields are negligible (Singh & Allen 1980). As with beans, the economic importance of cowpea pathogens varies considerably with ecological zone. Thus, web blight, *Cercospora* leaf spot, anthracnose, rust, cowpea (yellow) mosaic and bacterial pustule are considered the major diseases of the West African forest belt, while scab (*Sphaceloma* sp.), *Septoria* leaf spot, brown blotch (*Colletotrichum* spp.), cowpea aphid-borne mosaic and bacterial blight are the principal diseases of the African savannas. *Ascochyta* blight is important in Africa at altitudes above 1000 m.

Recent advances in disease management

We have emphasized that the extent of crop loss incurred from disease has seldom been adequately quantified. Similarly, very little is known of the cost-effectiveness of chemical control measures (Mukiibi 1969, McDonald & Fowler 1976). Much depends on the value of the crop, as well as on such factors as the persistence and spectrum of activity of the chemical. In general, it seems that foliar spraying to control disease on food legumes in Africa is seldom a viable practice except for seed multiplication. In dry areas, the scarcity of water adds to the unacceptability of conventional techniques of application. However, there is potential for the application of fungicides as ultra low-volume sprays, as dusts and, perhaps particularly, as seed dressings. There is also potential for the wider use of insecticides in the control of virus diseases, especially those transmitted in a persistent manner by their vectors.

Among the control strategies available, host plant resistance is now widely recognized as the pivot of integrated disease management, to which not only chemical control measures but also cultural practices may contribute. The potential for the biological control of legume disease remains virtually unexplored. Resistant cultivars cost the farmer nothing, nor does their adoption necessarily disrupt his farming system.

The potential for controlling disease of legume crops by host plant resistance was first realized more than 80 years ago when Orton (1902) reported differences in reaction to *Fusarium* wilt among cowpea cultivars. Anthracnose resistant cultivars of bean were to follow (Barrus 1911) but despite these early efforts, relatively little progress was made in developing satisfactory resistance in agronomically superior genotypes of these legumes for tropical areas until the late 1960s and early 1970s, when an increased awareness of the need to improve this group of food crops led to the establishment of various national and international programmes.

As a consequence of intensive work in the last 10-15 years, it is now no longer true to say, as Borlaug (1973) stated, that "neither new high yielding varieties (of grain legumes) nor improved technology have been developed". Genetically superior cultivars of the food legumes *are* now available, and the incorporation of disease resistance into a range of plant types possessing heavy-yield potential is an important part of that improvement. Certain genotypes of each of the three legumes with which this paper is concerned have been shown to possess combined resistance to two or more diseases. For instance, some Peruvian accessions of groundnut from Tarapoto have resistance to rust and to both "early" and "late" leaf spots. A few rosette-resistant cultivars also possess some resistance to late leaf spot. Similarly in beans, resistance to bean common mosaic virus has been successfully combined with resistance to anthracnose and, in Kenya, also with resistance to halo blight (van Rheenan, pers. comm.). Combined resistance appears to be relatively common in cowpea. On screening a germplasm collection of 5000 accessions in Nigeria, Williams (1977) found that 208 (4.1 %) were resistant to four diseases, and now several cultivars (e.g. cv. Iron) of cowpea are now known to combine resistance to as many as ten diseases.

The bases for recent advances in disease resistance breeding appear to have been: (1) The very extensive collections of germplasm held at the international centres (more than 10 000 of groundnuts and cowpeas; more than 30 000 of beans). (2) Improved methods of evaluation of the available genetic diversity, including more reliable methods for the identification of disease resistance. (3) The use of off-season nurseries to accelerate breeding (3 or even 4 generations per year). And (4) the establishment of interdisciplinary research teams working on a single commodity.

Future challenges

New methods are now available for monitoring the distribution and importance of plant pathogens, including the viruses which are the most difficult group with which to work. Serological indexing procedures are starting to demonstrate their potential for the more reliable identification and mapping of legume viruses many of which remain incompletely characterized. For instance, recent work in Germany suggests that polyviruses other than BCMV are present in beans in Eastern Africa, so having implications for future breeding programmes.

There are opportunities for the greater exploitation of genetic resources in resistance breeding through "wide crossing" perhaps with the long-term objective of producing a permanent immunity to disease. Interspecific hybridization in *Arachis* is given prominence by ICRISAT, and recent work at CIAT suggests that further interspecific crosses in *Phaseolus* are likely to raise the level of resistance to *Ascochyta* blight and common blight (*Xanthomonas campestris* pv. *phaseoli*) in beans.

Growing disenchantment with the objective of wide adaptability in crop improvement is leading at the international centres, to increased emphasis on "decentralization" of selection exemplified by efforts at defining agroecological zones which underlie specific adaptation, and thus also the need for a "character deployment" in breeding. The ecological deployment of combined resistance to disease is a part of this strategy.

The progress which has been made with disease resistance in the legumes has generally depended on relatively conventional methods of in-breeding, and the extent it affords only transient protection in agriculture remains uncertain. While there are strategies by which transient resistance might be stabilized, perhaps including the exploitation of cultivar mixtures, one of the greatest challenges now facing plant pathologists is to provide guidance to breeders on the means of selecting durable resistance.

We have emphasized that genetically superior cultivars of the major legumes are now available. But it is true to say that such genetic advance as has been made is yet to be translated into agricultural progress in Africa: no spectacular yield increments have occurred outside research stations, though they have been demonstrated in farmers' fields without dependence on inputs. Food legume yields are, on average, in the range of 400-700 kg ha^{-1} yet, in all cases, their potential productivity is from three to six times larger than that average.

The development of heavy-yielding cultivars capable of responding spectacularly to inputs, principally fertilizer and water, was the basis for the large cereal yield increases which occurred in Asia during the late 1960s. The challenge confronting us now is to devise means whereby the Green Revolution can be brought to the subsistence farmer, without recourse to heavy input, through the development of sustainable cropping systems of which improved cultivars of legumes are likely to be an integral part (Greenland 1975). Experience gained from the cereals boom seems unlikely to be relevant because, in that case, much emphasis was given to the transfer of technology from temperate agriculture. We suggest that the places to look for guidance are the existing systems of traditional agriculture themselves, for it is here that stability between co-evolved hosts and parasites has often been attained. An increased awareness of the need to understand farmers' practices, including the possibly protective effects of intercropping and of growing mixtures of genotypes, is a challenge to future collaboration between pathologists and agronomists to devise more productive cropping systems without losing their intrinsic balance.

We suggest, too, that these challenges are best met by increased international co-operation, stimulated by the establishment of regional programmes on legume improvement such as the IITA/SAFGRAD project on cowpeas based in Upper Volta, the ICRISAT project on groundnuts in Malawi, and most recently the CIAT/SDC project on beans based in Rwanda. The success of each depends heavily on the development of networks for the effective exchange of germplasm, information and ideas.

If the materials and methods now emerging from the international centres are used adequately by national research teams to adapt them to local requirements, and to farming systems which themselves afford some protection from disease, then we may expect that greater and more enduring productivity will follow.

References

Allen, D.J. (1983) *The Pathology of Tropical Food Legumes: Disease Resistance in Crop Improvement.* Chichester; John Wiley, 413 pp.

Barrus, M.F. (1911) Variation of varieties of beans in their susceptibility to anthracnose. *Phytopathology* 1, 190-195.

Borlaug, N.E. (1973) Building a protein revolution on grain legumes. In *Nutritional Improvement of Food Legumes by Breeding* (M. Milner ed.), 7-11. New York; John Wiley.

Bunting, A.H. (1975) Time, phenology and the yields of crops. *Weather* 30, 312-325.

CIAT (1981) *Potential for Field Beans in Eastern Africa. Proceedings of a Regional Workshop held in Lilongwe, Malawi, 9-14 March 1980.* Cali; International Center of Tropical Agriculture, 226 pp.

Gibbons, R.W. (1980a) Adaptation and utilization of groundnuts in different environments and farming systems. In *Advances in Legume Science* (R.J. Summerfield; A.H. Bunting, eds), 483-493. London; HMSO.

Gibbons, R.W. (1980b) The ICRISAT groundnut program. In *Proceedings of the International Workshop Groundnuts*, 12-16. Hyderabad; International Crops Research Institute for the Semi-Arid Tropics.

Greenland, D.J. (1975) Bringing the green revolution to the shifting cultivator. *Science, New York* 190, 841-844.

Leakey, C.L.A. (1970) The improvement of beans (*Phaseolus vulgaris*) in East Africa. In *Crop Improvement in East Africa* (C.L.A. Leakey, ed.), 99-128. Farnham Royal; Commonwealth Agricultural Bureaux.

Londono, N.R.; Gathee, J.W.; Sanders, J.H. (1980) Bean production trends in Africa, 1966-1979. In *Potential for Field Beans in Eastern Africa*, 19-29. Cali; International Center of Tropical Agriculture.

McDonald, D; Fowler, A.M. (1976) Control of *Cercospora* leaf-spot disease of groundnuts (*Arachis hypogaea*) in Nigeria. *Nigerian Journal of Plant Protection* 2, 43-59.

Misari, S.M.; Harkeness, C.; Fowler, A.M. (1980) Groundnut production, utilization, research problems and further research needs in Nigeria. In *Proceedings of the International Workshop Groundnuts*, 264-273. Hyderabad; International Crops Research Institute for the Semi-Arid Tropics.

Mukiibi, J. (1969) Control of leaf diseases of cowpeas caused by fungal pathogens. *Contributions from the 12th Meeting of the Special Committee for Agricultural Botany, Kampala, September 1969.*

Orton, W.A. (1902) The wilt disease of the cowpea and its control. *United States Department of Agriculture and Bureau of Plant Industries Bulletin.* 17, 9-20.

Pinstrup-Andersen, P.; Londono, N.de; Infante, M. (1976) A suggested procedure for estimating yield and production losses in crops. *PANS* 22, 359-365.

Schneider, R.W.; Williams, R.J.; Sinclair, J.B. (1976) *Cercospora* leaf spot of cowpea: models for estimating yield loss. *Phytopathology* 66, 384-388.

Singh, S.R.; Allen, D.J. (1980) Pests, diseases, resistance and protection in cowpeas. In *Advances in Legume Science* (R.J. Summerfield; A.H. Bunting, eds), 419-443. London; HMSO.

Williams, R.J. (1977) The identification of multiple disease resistance in cowpeas. *Tropical Agriculture (Trinidad)* 54, 53-60.

Yayock, N.Y. (1977) An epidemic of rosette disease and its effect on growth characteristics and yield of groundnuts in Nigeria. *Oleagineux* 32, 113-115.

Chapter 18

Cassava Diseases, their Spread and Control

E.R. TERRY and D. PERREAUX

International Institute of Tropical Agriculture (IITA), PMB 5320, Ibadan, Nigeria.

Introduction

The annual world production of cassava now exceeds 100M tons of fresh roots, and its significance in global agricultural production is considerable. The annual African production of cassava was estimated at about 42M tons from 55% of the total area under cultivation in 1977 and it provided 38% of the calorie needs of the continent (Phillips 1974). More than 30 diseases induced by viruses, mycoplasmas, virus-like agents, bacteria and fungi have been reported (Lozano 1978). These diseases can cause heavy losses of more than 50% of the crop or sometimes complete crop failure in certain areas (Lozano 1978). Cassava is, however, more tolerant to pests because it does not have critical periods that affect yield-forming organs. Yields are reduced when: (a) leaf-life is reduced, (b) photosynthetic rate is reduced, (c) stems are severely damaged, or (d) there is a high percentage of early plant death.

Cassava diseases

Systemic diseases

A systemic disease is defined as an infection in which the pathogen usually invades all the tissues of a host individual.

(a) *Cassava Mosaic Disease* (CMD): A foliar mosaic with alternating green and chlorotic areas. Leaf deformation is also common, and in severe cases, leaflets can be virtually reduced to the midrib. Since its first report in East Africa in 1894, CMD has been reported from all cassava growing areas in Africa and India, but not from the Americas. Overall yield losses due to CMD in Africa have been estimated at 11% (Padwick 1956), but for individual cultivars can range from 29% to 95%. Losses ranging from 24% to 69% reduction in fresh weight have also been reported (Terry & Hahn 1980). A gemini virus mechanically transmissible to *Nicotiana benthamiana* is thought to be the causal agent of CMD.

(b) *Cassava Bacterial Blight* (CBB): Watersoaked angular spots and extensive blight of leaflets are the typical foliar symptoms. When the infection develops systemically, yellow exudates appear on the stem concomitantly with a progressive wilting of the leaves, leading to the die-back of the shoots in susceptible lines. Stem die-back often induces sprouting at the base of the infected parts, with the new shoots themselves becoming systemically infected if conditions are favourable to progression. The disease was first recorded in Brazil in 1912 and is now widespread in most cassava growing areas. It attracted the attention of African scientists in the early 1970s when it reached epidemic proportions in Nigeria (Williams *et al.* 1973) and in Zaïre (Maraite & Meyer 1975). It can cause complete yield loss, and yield reductions of 50% to 90% have been recorded for susceptible lines in Colombia and up to 58% in Nigeria. Infection often results in severe defoliation leading to heavy losses of foliage which is an important source of protein in the diet of a large number of inhabitants in the cassava belt (Terry 1978). The CBB causal organism, *Xanthomonas campestris* pv. *manihotis*, is an aerobic, gram-negative, non-sporeforming straight rod, 0.6 x 1.2μm, motile with a single polar flagellum (Bradbury 1978).

(c) *Cassava Bacterial Necrosis* (CBN): Unlike the causal organism of CBB, the CBN pathogen (*Xanthomonas campestris* pv. *cassavae*) does not become truly systemic. The leaf spot stage, however, is similar to that of CBB, and stem lesions can occur in severe attacks. Wilt and dieback occur only in the distal parts after stem girdling, not by systemic infection. Spread of the disease in the plant is therefore slower and more limited. The CBN pathogen develops as single colonies which are yellow, smooth and glistening on glucose-yeast extract-carbonate agar (J.F. Bradbury pers. comm.).

Non-systemic diseases

An infection is non-systemic or localized when pathogens are restricted to a rather sharply defined area of tissue which may range from small lesions a few mm diam to large blotches several cm across.

(a) *Cassava Anthracnose Disease* (CAD): This is characterized by the presence of stem cankers, leaf-spots and tip die-back of young shoots (Lozano & Booth 1974). Stem symptoms are the most easily identified, oval necrotic cankers typically being produced, preferentially at petiole insertion points; younger stems appear more susceptible. The disease is widespread in all cassava growing regions. Accurate data on yield losses due to CAD are not available. Leaf symptoms are generally considered to be of minor importance but the incidence and severity of stem symptoms can be high, as in Zaïre where more than 90% of the local cultivars have been reported to be severely affected.

With damage to the woody stem, the quality of planting material is greatly reduced. *Colletotrichum gloeosporioides* f.sp. *manihotis* is the anamorph of the causal pathogen, producing cylindrical conidia with obtuse ends, on subepidermal, saucer-shaped acervuli; the teleomorph is *Glomerella cingulata* f.sp. *manihotis*, the perithecia of which are often found on dead twigs.

(b) *Cercospora Leaf Spots.* Brown leaf spots caused by *Cercospora henningsii* are probably the most important of all cassava fungal leaf diseases (Lozano & Booth 1974). Brown, roughly circular spots with a darker margin form; depending on environmental conditions and cultivar susceptibility, a yellow halo is sometimes found around the lesions. Brown lesions induced by *Cercospora vicosae* are more expanded, without definite borders, and cover larger areas of the leaflets which appear blighted. White leaf spots, generally 2-7 mm diam and surrounded by a yellow halo, are induced by *Cercospora caribae*. These diseases appear to be widespread in most cassava growing areas, although in Africa *C. vicosae* has been reported only from Nigeria, Benin and Cameroon. Yield reduction due to brown leaf spots and leaf blight ranged from 10–30 % (Teri *et al.* 1977), when environmental factors are favourable, and depending on cultivar susceptibility.

Endemic persistence and epidemic cycles

When a pathogen's rate of increase in a crop is zero, so that the disease is more or less pegged to a certain level, such a disease is said to be in its endemic phase. Periodic changes, however, in intensity of the disease are not excluded, therefore, endemic diseases should not necessarily be considered low level diseases. Endemicity is a common and important feature of a number of tropical pathosystems and is exhibited by certain cassava diseases. The positive growth of a parasitic population from minimum to maximum followed by a negative growth to minimum again, the epidemic cycle, is normally seasonal except in the wet tropics which are permanently warm and humid. A particular characteristic of epidemics is explosive infection rates and this feature is also exhibited by certain cassava diseases. Pathogen biology, climatic factors, cropping patterns and cultural practises interact to produce either endemic persistence or an epidemic outbreak of a disease.

Endemic persistence

(a) *Effect of climatic factors.* Cassava is a short-day plant and its distribution in Africa is therefore largely confined to areas between latitudes 15°N and 15°S (Jennings 1970). While the geographical extent of cassava cultivation appears to be limited by rainfall, i.e. it corresponds to those areas where mean annual rainfall exceeds 750 mm, temperature limits probably determine the cassava disease pathogen spectrum.

The average monthly minimum and maximum temperatures in the wet zones (rainfall 2000 – 3000 mm y^{-1}) and wet and dry zones (rainfall 1000 – 1500 mm y^{-1}) in the cassava belt are 25° and 31°C. Therefore, temperatures in the cassava belt do not reach the extremes which could eliminate or seriously reduce pathogen populations. However, the severity of symptom expression can be affected by temperature extremes; CMD symptoms are suppressed at high temperatures, 35°C; for CBB infected plants including susceptible cultivars kept under high temperature conditions viz night/day temperatures above 20° – 30°C respectively are reported to usually recover from infection. It has been suggested that the absence of any significant manifestation of CBB in the whole of the Amazon region, although most of the cultivars grown are susceptible, can be attributed to this temperature effect.

The most important cassava pathogens appear to be restricted to *Manihot* species under natural environmental conditions (Teri *et al.* 1977). These pathogens are therefore quite dependent on the availability of host tissue for their survival and endemic persistence. The long growth cycle of cassava, which is facilitated by the favourable climatic belt, ensures this spatial and temporal continuity essential for pathogen survival.

It has also been demonstrated that the epiphytic survival of *Xanthomonas campestris* pv. *manihotis* on cassava leaf surfaces throughout the long and short dry seasons within the cassava belt have contributed significantly to the persistence of this pathogen (Persley 1979).

(b) *Effect of cultural practices.* Of paramount importance for the endemic persistence of cassava diseases is the vegetative propagation of this crop. Infected cuttings greatly contribute to the inoculum persistence and are largely responsible for the continuity and dissemination of pathogens through successive growing seasons. Fungi such as *Colletotrichum manihotis* also persist as localised infection in stem cuttings and may sometimes affect sprouting.

Disease epidemics

(a) *Climatic factors.* Climatic factors certainly play an important role in the build up of inoculum, in its dispersal and in the success fo the infection process. The epidemiology of CBB in relation to climatic factors has been studied in some detail. Rainfall provides the conditions necessary for mobilization, distribution and penetration of inoculum, and it is probably the most important environmental factor affecting the disease; wind driven droplets appear to determine the direction of spread of pathogen during the growing season (Lozano & Sequeira 1973). Important attacks of CAD have been reported in the Congo at the beginning of the rainy season, whereas high incidence and severity in the later part of the rainy season was reported from Nigeria. Rain-splash significantly affects the dissemination of the anthracnose fungus, and the infection index and severity of CAD may be positively associated with monthly relative humidity and rainfall in the dry season crop and negatively correlated in the wet season. The *Cercospora* cassava pathogens are more prevalent during the rainy seasons, and their relative distribution has been found to depend on moisture and relative humidity as well as on temperature (Lozano & Booth 1974).

(b) *Biotic factors.* Epidemics of insect transmitted diseases depend indirectly on environmental factors through their action on insect populations. CMD incidence depends on the availability of inoculum, which in turns depends on the density and activity of the white fly vector (*Bemisia tabaci*); its incidence is therefore closely related to the population of whiteflies (Hahn *et al.* 1980), which are apparently favoured by (a) rainfall (150–200 mm month^{-1}), (b) temperature within the range of 27–30°C, and (c) solar radiation of 400 g cal cm^{-2}. In Kenya the slow rate of CMD spread has been attributed mainly to low transmission efficiencies and seasonally low population densities of the vector. The sap-feeding coreid *Pseudotheraptus devastans*, may have an important role in CAD etiology but the relationship under field conditions has not been reported. Insect transmission of the CBB pathogen appears to be low (Lozano & Sequeira 1973), although the bacteria have been detected in various leaf-eating species.

(c) *Effect of cultural practices.* The importance of infected planting material as a primary source of inoculum has already been emphasized. Dissemination of pathogens by means of infected tools has been reported for CBB. This method of dissemination is most important during harvesting and propagation, since these operations require extensive cutting (Lozano & Sequeira 1973).

(d) *The role of germplasm movement.* Cassava is relatively new to Africa; it was introduced at the end of the 16th century (Jennings 1970) from South America, where its pathosystem is more diversified than in Africa. In the African environment cassava encountered new pathogens, as is probably the case for CMD causal agent. Although 400 years of natural selection have produced some more tolerant lines, an adequate level of resistance to CMD has never been found within *Manihot esculenta* (Hahn *et al.* 1980). Seedlings from seeds introduced from Latin America for breeding work showed a high susceptibility to CMD, due to the absence of selection pressures in this area of origin. The reverse is also true for pathogens introduced from the area of cassava origin to Africa, where the crop has evolved without selection pressure for resistance to those pathogens. Cassava bacterial blight was probably present in Uganda in the 1930s, but no evidence of severe damage was noted up to the 1970s when epidemics developed in Zaïre and Nigeria, where most of the local cultivars appeared to be highly susceptible. This susceptibility lead to the spread of the disease over most of the cassava growing regions in less than 10 years.

Strategies for control

An integrated approach that will combine the essential elements of host plant resistance, cultural practises and crop protection through exclusion and eradication of the disease pathogens is recommended for the control of cassava diseases.

Host resistance

Breeding for resistance has been considered the most effective means of controlling cassava diseases, with major emphasis on CMD and CBB, the most devastating diseases on the African continent. IITA has succeeded in developing high-yielding tolerant cultivars to CMD, from sources of resitance originating from earlier breeding work in East Africa involving crosses between *Manihot esculenta* and *Manihot glaziovii* (Hahn *et al.* 1980). The fresh root yields of two of the improved cultivars are reduced by 24–39% respectively due to CMD, compared to a 69% reduction for the local cultivar (Terry & Hahn 1980). Resistance to CBB appeared to be positively correlated with resistance to CMD (Hahn *et al.* 1980). Improved cassava clones under severe disease stress may out-yield local cultivars by 2 to 18 times, primarily due to their resistance to diseases. Improved IITA breeding materials have been distributed to many national programmes in Africa for re-selection under local environmental conditions. It took about 7 years to incorporate CMD resistance into high-yielding, well-adapted cultivars, utilizing breeding materials developed in a breeding programme which was initiated about 30 years ago. The improved, tolerant cultivars have since been seriously challenged by two newly introduced pests, the cassava green spider mite and the cassava mealybug. This highlights the necessity to develop parallel strategies for disease control based on other considerations.

Cultural practices

Sanitation, to reduce disease incidence at the subsistence farmer level, especially considering the endemic nature of tropical pathosystems, merits serious consideration. In Kenya cassava mosaic disease is reported to be controlled by the use of mosaic-free planting material, due to the comparatively low transmission efficiencies of the whitefly vector in some parts of Kenya. CBB control through sanitation by severe pruning of most above-ground parts of infected cassava plants is also possible. About half of the cassava in Africa is grown in mixed cropping systems, but there is limited data on the epidemics of cassava disease in such systems. Results from experiments in Costa Rica indicate that cropping systems significantly affect some fungal diseases of cassava and intercropping appears to reduce CBB incidence under certain conditions. The CMD agent can be eliminated by thermotherapy and meristem culture, but the full potential of this technique has been only minimally exploited due to logistic problems encountered in the multiplication and distribution of the disease-free material produced. A rapid multiplication method which assures the elimination of CBB contaminated plants and does not require sophisticated equipment has been developed in Colombia (Cook *et al.* 1976), and a similar method has been successfully adapted at IITA for use in Africa.

Regulatory control

The enforcement of quarantine regulations which restrict the movement of cassava vegetative material across boundaries within Africa and prohibit such movement from outside the continent should theoretically prevent the accidental introduction and eventual spread of the major cassava disease pathogens in areas free from vegetatively-borne pathogens. These regulations are very difficult to enforce, but true seeds reduce the risk of disease introduction. *Xanthomonas campestris* pv. *manihotis* can be seed-transmitted, but it can be eliminated by either dry-heat treatment (55–60°C) for two weeks or hot water treatment (60°C) for 20 min. Transmission of CMD by seeds has never been demonstrated. At IITA, improved cassava varieties regenerated from meristem culture after the elimination of CMD and CBB have been indexed and certified by plant quarantine authorities to 18 African countries for multiplication and evaluation. It is envisaged that these materials will form the nucleus of national cassava programmes' disease-free stock.

References

Bradbury, J.F. (1979) Identification and characteristics of *Xanthomonas manihotis*. In *Cassava Bacterial Blight in Africa* (E.R. Terry; G.J. Persley; S.A.S. Cook, eds), 1-4. Ibaden; International Institute of Tropical Agriculture.

Cook, J.H.; Wholey, D.W.; Lozano, J.C. (1976) *A Rapid Propagation System for Cassava*. [Publication Serie EE-20.] Cali; Centro International de Agricultura Tropical, 10 pp.

Hahn, S.K.; Terry, E.R.; Leuschner, K. (1980) Breeding cassava for resistance to cassava mosaic disease. *Euphytica* 29, 673-683.

Hahn, S.K.; Howland, A.K.; Terry, E.R. (1980) Correlated resistance of cassava to mosaic and bacterial blight diseases. *Euphytica* 29, 305-311.

Jennings, D.L. (1970) Cassava in Africa. *Field Crop Abstracts* 23, 271-278.

Lozano, J.C. (1978) General considerations on cassava pathology. In *Proceedings: Cassava Protection Workshop* (T. Brekelbaum; A. Belloti; J.C. Lozano, eds). Cali; Centro International de Agricultura Tropical.

Lozano, J.C.; Booth, R.H. (1974) Diseases of cassava (*Manihot esculenta* Crantz). *PANS* 20, 30-54.

Maraite, H.; Meyer, J.A. (1975) *Xanthomonas manihotis* (Arthaud-Berther) Starr, causal agent of bacterial wilt, blight and leafspots of cassava in Zaïre. *PANS* 21, 27-37.

Padwick, G.W. (1956) Losses caused by plant disesease in the colonies. *Phytopathological Papers* 1, 1-60.

Phillips, T.P. (1974) *Cassava Utilization and Potential Markets.* [IDRC No. 0202.] Canada; IDRC, 182 pp.

Teri, J.M.; Thurston, H.D.; Lozano, J.C. (1978) The *Cercospora* leaf diseases of cassava. In *Proceedings: Cassava Protection Workshop* (T. Brekelbaum; A. Belloti; J.C. Lozano, eds). Cali; Centro International de Agricultura Tropical.

Terry, E.R. (1978) Cassava bacterial diseases. In *Proceedings: Cassava Protection Workshop* (T. Brekelbaum; A. Belloti; J.C. Lozano, eds). Cali; Centro International de Agricultura Tropical.

Terry, E.R.; Hahn, S.K. (1980) The effect of cassava mosaic disease on growth and yield of a local and an improved variety of cassava. *Tropical Pest Management* 26, 34-47.

Williams, R.J.; Agboola, S.D.; Schenider, R.W. (1973) Bacterial wilt of cassava in Nigeria. *Plant Disease Reporter* 57, 824-827.

Chapter 19

Cassava Pests, their Spread and Control

H.R. HERREN[1] and F.D. BENNETT[2]

[1]International Institute of Tropical Agriculture (IITA), PMB. 5320, Oyo Road, Ibadan, Nigeria; [2]Commonwealth Institute of Biological Control (CIBC), Gordon Street, Curepe, Trinidad and Tobago.

Introduction

Cassava (*Manihot esculenta*) is the staple food of 200M Africans living in the humid and subhumid tropics and also an emergency food reserve in many arid zones. In the cassava belt, more than 50% of the calories consumed come from cassava roots, and in many countries the leaves also are eaten as high protein diet complement; about 10M ha of cassava are planted annually throughout the cassava belt. Africa has 53% of the world cassava acreage but its production is only 38% (FAO 1981); production ha^{-1} in Africa is low in comparison to South America and Asia.

Since its introduction into Africa over 500 years ago, cassava has remained relatively pest-free. Except for regional and seasonal attacks by the variegated grasshoppers (*Zonocerus* species) and white fly (*Bemisia tabaci*) the vector of the cassava mosaic virus, no major pests were known. This changed dramatically with the accidental introduction of two pests from the area of origin of the crop in the late 1960s or early 1970s.

The cassava green spider mite complex (*Mononychellus* species) and the cassava mealybug (*Phenacoccus manihoti*) are now threatening the existence of cassava in many areas. The dramatic effect can be expressed by the estimated loss of about $2.0Bn y^{-1} or 30% of the total root production (Herren *et al.* 1983); these losses do not include loss of planting material and leaves or additional costs for weeding and clearing more land to compensate for part of the losses. The pest problem is contributing to by: (1) an exodus from rural areas into cities; (2) an extension of the cropping areas without a corresponding increase in production leading to accelerated soil erosion; and (3) an increasing need for food imports; drought has worsened the situation.

Pseudotheraptus devastans should also be mentioned as this is involved in the transmission of anthracnose disease caused by the fungus *Colletotrichum manihotis*.

Origin and Spread of Cassava Pests

Cassava mealybug (CM)

CM was first described from the Republic of Congo and Zaïre in 1977; a few specimens found on cassava in Brazil were also described as this species which led to the assumption that *P. manihoti* originated in the neotropics (Bennett & Greathead 1978). CIBC and IITA started exploration trips in 1977 and 1980 respectively, to locate the area of origin of the pest and search for natural enemies; many species of the Pseudococcidae were found but these explorations failed to discover *P. manihoti*. Instead mealybug specimens collected from northern South America were described as a new species, *Phenacoccus herreni* (Cox & Williams 1981). Early South American reports of *P. manihoti* are considered to be misidentifications, for *P. herreni*, but in 1981 *P. manihoti* was discovered for the first time in the neotropics in Paraguay. The positive identification of *P. manihoti* from the Americas validates the theory of its neotropical origin and accidental introduction into Africa. Since it was first spotted in Africa in 1973, CM has spread into 18 countries throughout the cassava belt; about 55% of the total cassava growing area is now infested, and the rest will soon be infested also if the spread continues at its present speed. CM is dispersed passively by wind (first instar CM), and by man on planting materials carried from one area to another.

Green Spider Mite Complex (GSM)

The GSM complex originated in the Americas, where the three species of the complex, *M. tanajoa*, *M. caribbeanae* and *M. progressivus*, occur on cassava. *M. tanajoa* was first discovered in Uganda in 1972 and since then green spider mites have been reported in 20 countries. Studies of specimens found in several African countries have shown that other species than *M. tanajoa* are involved. *M. progressivus* has so far been found in Nigeria, Kenya, Gabon and Zaire; specimens from additional countries are still being examined which will increase our understanding of the different species, dispersal pattern, and perhaps their different areas of introduction.

Variegated Grasshoppers

There are two species of endemic variegated grasshoppers known in Africa: *Zonocerus variegatus* and *Z. elegans*. *Z. variegatus* occurs throughout the cassava belt except areas below 5°S, whereas *Z. elegans* occurs in the eastern cassava belt, from Kenya to South Africa.

White flies

Bemisia tabaci is a cosmopolitan Aleyrodidae found in all cassava-growing areas of Africa. There is still confusion on the identity of several *Bemisia* species found in different ecological zones of Africa, and more collection and identification work is required to set this problem straight.

Economic losses

Cassava mealybug and green spider mite

Cassava mealybug and green spider mite yield losses are analysed together, since in most countries the pests occur together and the losses cannot be proportionally attributed to one. Of the approximately 10M ha of cassava, 4.5M are already affected by one or both of the pests (Hennessey & Herren 1984). Root losses average 30%, with ranges from 10 to 80%, and reach a value of US$2Bn. This dollar loss estimate is conservative since it takes into consideration only the loss of roots; losses from damaged leaves (an important source of protein in some areas), the reduced amount of cuttings available for the next crop, and the increased weeding and erosion resulting from the open canopy are more difficult to assess and few statistics have so far been collected on these aspects. Most important is probably the loss of good planting material; when poor cuttings are used, the new crop has a bad start, exposing it even more to the pests and the loss of leaves. In many Sahelian countries cassava is grown as a vegetable, or kept in backyards as a food reserve in case of drought or other crop failure. With the arrival of CM and GSM, even this hardy and often last-resort food is threatened.

In the past two years, a severe drought has extended from the Sahel as far south as the Gulf of Guinea coast, causing failure in maize and grain legume crops. With the pests striking cassava, which is the staple in most countries around the Gulf, the food supply in Ghana, Togo, Benin and Nigeria is becoming critical. Hunger has been kept at bay only by massive imports of grains.

Grasshoppers

There are no good statistics on yield loss to be attributed to *Zonocerus* species, but observations in Nigeria have shown that cassava can be completely destroyed by an attack. Usually, they attack cassava in pockets and fields close to the forest or bush where grasshoppers breed. Ghana, Togo, Benin and Nigeria are hit most severely, especially in the southern part of these countries. Grasshoppers are also vectors of bacterial blight, also a serious disease.

White flies

White flies, *Bemisia* species, may damage cassava plants by sucking the sap, mainly when heavy infestations occur on young plants, but the principle damage occurs through the transmission of mosaic disease. Yield losses due to this disease of 25 – 30% and even up to 90% have been reported (Hahn *et al.* 1980).

Control of Cassava Pests

Cassava mealybug and green spider mites

At a workshop on CM in Zaïre (Leuschner & Nwanze 1978) it was decided that the best approaches to CM would be biological control, host-plant resistance, and cultural practices, leaving chemical control for cutting treatment. Since we were dealing with two exotic pests that are not problems in their area of origin, the biological control approach looked both suitable and promising to solve the problems quickly and permanently with the minimum involvement of the farmer and no detrimental effect on the environment (Bennett & Greathead 1978, Herren 1980). Since the Zaïre workshop, IITA in collaboration with CIBC and CIAT has been actively exploring the Americas for natural enemies of the two pests (Herren 1982). Since areas of occurrence of the CM were identified in Paraguay, Brazil and Bolivia, 27 species of natural enemies have been found, 11 in numbers large enough to be shipped to CIBC's quarantine laboratory in the UK. At present, eight species are being multiplied for detailed bionomic studies and experimental releases in different ecological zones. Two species, *Apoanagyrus lopezi* and a *Diomus* species, have become established and shown great potential in experimental releases in Nigeria (Herren *et al.* 1983), and are now being released in the Central African Republic, Guinea Bissau, Congo and Senegal.

For the GSM, CIBC and CIAT have identified 15 species of predatory mites (Bellotti *et al.* 1984) on the *Mononychellus* species complex; this indicates the great potential for biocontrol using phytoseiid mite predators. So far, IITA has introduced one phytoseiid species from Colombia via the CIBC quarantine laboratory in the UK; experimental releases will be made as soon as preliminary laboratory studies on biology and behaviour are completed and rearing methods developed. Efforts are also underway to introduce the other known phytoseiid species from South America. CIBC has made releases of the staphylinid predator *Oligota minuta* in Kenya in 1977 and 1982–83 but no recoveries have been made yet.

Local natural enemies have been found on both pests, but no evidence of effective control has been shown. This is to be expected, since the local natural enemies are not ecologically adapted to their new hosts.

Breeding for resistance or tolerance to the two pests is in progress at IITA and CIAT. There are good chances for the development of resistant cultivars as several clones showing resistance to CM and GSM as well as antibiosis against CM have been identified. The recommended cultural practice to reduce yield losses due to CM and GSM is early planting at the onset of the rainy season. This will allow the cassava plant to grow to a respectable size before the dry season, and therefore be more tolerant of the defoliation.

Grasshoppers

Little research has been done on control possibilities against *Zonocerus* species. Chemical control is not recommended, except for the use of bait or for local treatment of nymphs, taking advantage of their gregarious behaviour. Biological and microbial control needs to be looked into more seriously, since natural enemies are known to exist as well as fungal diseases; the potential surely exists and may provide the best solution.

Host-plant resistance is given high hopes by some authors providing that more research is carried out. Observations by ourselves and Leuschner show that although some cassava clones may show repellent characteristics, they will be consumed by hungry grasshoppers after the more preferred (less repellent clones) have been destroyed. The fact that *Zonocerus* species are polyphagus makes host-plant resistance an unlikely solution. Furthermore, cassava is usually one of the few host plants remaining green during the dry season, making it very attractive to the grasshoppers.

White flies

Since the white fly problem is mainly due to the mosaic virus disease they transmit, its solution has been sought through resistance breeding against the disease. Work in the 1930s in East Africa forms the basis for the resistant cassava clones known today. IITA has been active in this field and has selected many different clones with high resistance to near immunity (Hahn *et al.* 1980). Resistant clones have been made available through IITA's Tissue Culture Laboratory to many countries, a great step forward in the dissemination of improved and resistant characters. Control of mosaic by roguing infected plants appears feasible and effective under Kenyan, but not Nigerian, conditions.

Conclusion

The major cassava pests, CM and GSM, have been studied in detail and control measures, including biological control and host plant resistance are in good progress. For the biological control part, IITA and the OAU/STRC are now launching an Africa-wide programme, with additional exploration in the Americas (Herren & Lawson 1983), a mass rearing laboratory in Ibadan, and an aerial release scheme to cover any African country requesting help. The programme also includes the intensive training for 56 biocontrol specialists. It is expected that within a five year period, half the cassava growing areas could be supplied with natural enemies of both pests; natural spread would progressively take care of the remaining areas. Control is effected when the released natural enemies establish an equilibrium situation between their hosts and themselves.

With respect to host-plant resistance, further testing and selection of clones continues, hand in hand with the rapid multiplication of those already selected.

In view of the synergistic effect of disease and insect pests on cassava yield losses, it is of foremost importance to multiply and distribute as many disease resistant cassava clones as possible. Disease free plants will tolerate a much higher insect population, and also recover much faster after the attack. An increased number of national breeding programmes will facilitate this approach.

Partial solutions are therefore already available and should be given high priority. Host plant resistance and biological control will eventually be integrated and provide a solid basis towards an integrated approach to the control of the major cassava pests.

References

Bellotti, A.C.; Byrne, D.H.; Hershy, C.H.; Vargus, O.; Varela, A.M. (1984) The potential of host plant resistance in cassava for the control of mites and mealybugs. In *Proceedings, International Workshop on Biological Control and Host Plant Resistance against the Cassava Mealybug and Green Spider Mite*, in press, Ibadan; IITA.

Bennett, F.D.; Greathead, D.J. (1978) Biological control of the mealybug *Phenacoccus manihoti;* prospects and necessity. In *Proceedings: Cassava Protection Workshop, CIAT, Cali, Colombia, 7–12 November, 1977* (Brekelbaum, T.; Bellotti, A.; Lozano, J.C. eds), 181-194. Cali; CIAT.

Cox, J.M.; Williams, D.J. (1981) An account of cassava mealybug (Hemiptera: Pseudococcidae) with a description of a new species. *Bulletin of Entomological Research* 71, 247-258.

FAO (1981) *Production Yearbook.* Rome; FAO.

Hahn, S.K.; Terry, E.R.; Leuschner, K. (1980) Breeding cassava for resistance to cassava mosaic disease. *Euphytica* 29, 673-683.

Hennessey, R.D.; Herren, H.R. (1984) *Proceedings, International Workshop on Biological Control and Host Plant Resistance against the Cassava Mealybug and Green Spider Mite.* Ibadan; IITA, in press.

Herren, H.R. (1980) Biological control of the cassava mealybug. In *Proceedings of the First Triennial Root Crops Symposium of the International Society for Tropical Root Crops – African Branch, 8–12 Sept. 1980, Ibadan, Nigeria,* 79-80. Ibadan; IITA.

Herren, H.R. (1982) Cassava mealybug: an example of international collaboration. *Biocontrol News and Information* 3(1), 1.

Herren, H.R.; Lawson, T.L. (1983) The Africa-wide programme for biological control of the cassava mealybug and green spider mites: the importance of thorough foreign exploration for beneficial insects and mites. In *3rd Brazilian Cassava Congress,* 7-11. Brazilia; in press.

Herren, H.R.; Lema, K.M.; Neuenschwander, P. (1983) Biological control of the mealybug, *Phenacoccus manihoti,* and the green spider mite complex, *Mononychellus* spp. on cassava, *Manihot esculenta,* in Africa. In *Proceedings of the Plant Protection Congress, Brighton,* 20-25. Brighton; International Congress on Plant Protection.

Leuschner, K.; Nwanze, K.F. (1978) Preliminary observations of the mealybug (Hemiptera: Pseudococcidae) in Zaïre. In *Proceedings: Cassava Protection Workshop, CIAT, Cali, Colombia, 7–12 November, 1977,* (Brekelbaum, T.; Bellotti, A.; Lozano, J.C. eds). 195-198. Cali; CIAT.

Chapter 20

Potatoes:
I. New Approaches to Disease Control

O.T. PAGE
Director of Research, International Potato Center (CIP), Apartado 5969, Lima, Peru.

Introduction

Worldwide, the potato is mankind's fourth most important food crop after wheat, rice and maize. It is cultivated in 33 African countries as well as in 96 other countries. Potatoes are a highly adaptable crop being grown through more than 50 degrees of latitude on each side of the equator and beyond a height of 4 kilometres in their native Andean home in Peru. No other major food crop has this adaptive capacity.

Whereas the average European eats about 73 kg (160 lb) of potatoes each year, the average African enjoys only about 9 kg (20 lb) of this nutritious food (Table 1).

The International Potato Center (CIP) is devoted to the development and dissemination of knowledge for the greater use of the potato as a basic food. CIP has seven Regional Research Stations where technologies are evaluated in accordance with the needs of a country or region. This includes evaluation of the adaptability of advanced potato clones to climatic and edaphic factors as

Table 1   Daily energy and protein production of potato and other selected food crops (Horton 1982).

Crop	Production (ha⁻¹ day⁻¹)	
	Energy (10^3 kcal)	Protein (kg)
Potato	55	1.4
Sweet potato	52	0.7
Cereals*	37	1.0
Cassava	34	0.2
Dry bean	10	0.7

* Average for wheat, rice and maize.

well as resistance to selected pathogens and pests of which a total of 268 are known to attack the crop (Hooker 1981, Jensen *et al.* 1979, Shands & Landis, 1964).

During the past 11 years, through direct association with over 40 developing country potato programs, CIP has developed a comprehensive view of the pathogens and pests which are annually most destructive. Losses may be very substantial, contributing to the low yields in Africa (Table 2).

In contrast to disease and pest control measures used in developed countries, which depend very heavily on chemical inputs, small poor farmers in developing countries usually cannot afford costly chemicals. As a consequence, the development of practical, low cost methods of disease and pest control often requires different approaches than those used in industrialized countries.

Table 2   Potato production, area and yield in 1980 (from Horton 1982).

Region	Area (1000 ha)	Production (000t)	Yield (t ha⁻¹)
World	18 026	225 676	12.5
North & Central America	675	17 447	25.8
Europe	12 590	158 388	12.6
Africa	595	4 918	8.3
South America	995	9 332	9.4
Asia	3 124	34 392	11.0

Important Diseases and Pests

Priority concern is with those particular diseases and pests that are generally epidemic in nature, causing significant yield losses season after season. The late blight fungus is a good example of a common pathogen which causes serious losses in many African countries. The Potato Leafroll Virus (PLRV), which can causes losses up to 90 %, is transmitted by aphids and through infected tubers (Burton 1966). Bacterial Wilt, sometimes called "brown rot" often limits potato production in warm regions. Root-knot nematodes, found worldwide in the tropics, in addition to direct damage to the potato plant contribute to root invasion by wilt bacteria. The potato moth, another warm climate pest, is destructive of foliage, stems and tubers. All six of these economically important pathogens and pests are controllable. The problem is, at what cost to a small farmer with few resources?

Potato varieties developed in the northern industrialized countries are not bred for disease and pest resistance. While these varieties may produce excellent yields of attractive tubers under high input conditions, disease and/or pest resistance is usually incidental, indeed accidental. That is, the varieties have not been purposefully bred with disease or pest resistance as a priority objective. Because of reliance on chemical control measures and tuber seed certification programmes and the need to meet special consumer preferences, breeding for disease and pest resistance is relegated to a low priority. However, the relatively high cost of oil used to produce agrochemicals may be changing priority breeding objectives in developed countries.

Control of Important Diseases and Pests

The first research priority in the control of diseases and pests is breeding for resistance. Breeding for disease and pest resistance to reduce small farmer risk requires a philosophy and methodology different from developed country breeding strategies. For example, breeding and selection in the developed countries has depended almost exclusively on *Solanum tuberosum*, which is only one of more than 150 wild and cultivated tuber-bearing species of potato. CIP's World Potato Collection contains many thousands of primitive varieties and wild species. Probably less than one percent of the available genetic material has been utilized in breeding for resistance and other desirable characteristics in the modern potato.

To adequately utilize the potential of the germplasm pool requires a population breeding strategy (Mendoza & Rowe 1977, Mendoza & Estrada 1979). This involves the application of continuous cycles of recurrent phenotypic selection in order to maintain wide genetic variability. This heterozygosity provides the basis for good yields and stability of performance over a range of tropical environments.

As an example, in breeding for resistance to the root-knot nematode (*Meloidogyne incognita*) from a sample of 1020 cultivated diploid clones only 2.7% resistant plants were identified. Crosses were made to concentrate genes for resistance. Through one generation of selection 3.6% representing 52 out of 1540 plants were found to be resistant to the nematode. These resistant plants in turn were crossed with two other species of *Solanum*, *S. sparsipilum* and *S. chacoense*. The interspecific hybrids had 655, or 62% of 1058 plants which were resistant to root-knot nematodes. In another example involving crosses between diploid and tetraploid plants 715 or 57% of 1254 genotypes were resistant. These high levels of resistance resulted from increasing the frequency of resistant genes inherent in the populations of diploids through cycles of recurrent selection.

Under tropical conditions resistance to bacterial wilt can be broken by root-knot nematodes penetrating into the roots of the wilt-resistant plants. Fortuitously, dual resistance has been found in a number of intra- and interspecific hybrids involving diploid *Solanum* species such as *S. sparsipilum* and *S. chacoense* (Goméz et al. 1983).

Non-reliance on chemical controls leads to greater reliance on alternate methods of control involving various integrated management practices. A practical example is the control of root-knot nematodes by means of the fungus *Paecilomyces lilacinus* which parasitizes the eggs of this nematode (Anon 1982). The technology is relatively simple, involving growing the fungus on cooked rice which is placed in a plastic bag and inoculated with the fungus. After growth of the fungus for several days, the infected rice is spread in the furrow at time of planting. The fungus has been found effective in controlling root-knot nematodes not only on potatoes but it also shows promise in controlling similar nematodes on citrus and sugar cane. On potatoes the fungus is as effective as nematicides due in part to the continuing activity of the fungus during the growth of the crop outlasting the effective period of nematicide control.

During the hilling of potatoes roots are frequently cut providing an entrance for pathogens such as the wilt bacteria. Effective management of bacterial wilt and the root-knot nematode complex may require an integrated programme. This programme would include not only resistant potato clones but also the biological control of nematodes as well as appropriate agronomic practices. Crop rotation and minimal cultivation are helpful.

The two most troublesome potato viruses, the leafroll virus (PLRV) and potato "Y" virus (PVY), are spread by aphids. Control measures could focus on breeding for resistance to either virus and/or the aphid vector. In practice, it is relatively easy to develop resistance to PVY since resistance is simply inherited. That is resistance appears to be monogenic. Because potato virus "X" (PVX) is also controlled by a single resistance gene breeding for dual resistance to PVX and PVY is not difficult to achieve. It should be noted that chemical control of aphid vectors of PVY is generally not effective because of the short virus acquisition and infection feeding periods by the aphids.

Although breeding for PLRV resistance is well advanced, it is more difficult because of the multigenic nature of resistance. Presently available forms of resistance derived from such sources such as *S. etuberosum* crossed with *S. pennatisectum* and also from crosses involving three other *Solanum* species appear to have resistance to infection. The basic objective in developing potato clones with combined virus resistance is to permit a farmer to save his own seed tubers for planting the next crop. In the absence of seed certification programs, which rely on an infrastructure of trained field staff coupled with legal standards for certification, virus resistant potato varieties offer a viable alternative.

Caution must be exercised in the storage of seed to prevent transmission of PLRV and PVY during storage. In an experiment in Peru in 1983, PLRV incidence increased by 86% and PVY by 20% in stores. The application of insecticides did not significantly reduce the spread of these two viruses.

The potato tuber moth is a very serious pest of potatoes both in the field and in stores. An effective field control is by means of field traps to which male tuber moths are attracted by synthetic female pheromones. These sex attractants are both inexpensive and last for many months in the field. By trapping the males the female moths remain unfertilized (Raman 1982). In stores, layering seed tubers with repellant plant material such as *Lantana* leaves is very effective in reducing both tuber moth and aphid infestations.

In research conducted at Cornell University (Mehlenbacher *et al.* 1983), Rothamsted (Gibson 1979) and CIP (Anon 1982), glandular leaf hairs have been found effective in trapping insects. These glandular trichomes which are of two types are selectively effective in trapping different size insects and also mites. The wild potato species *S. berthaultii* from CIP's germ plasm bank has been the most proliferate source of leaf trichomes.

While not a significant pest in Africa, the leaf miner fly, *Liriomyze huidobrensis*, is readily attracted to greenish yellow placards. A coloured placard with a sticky surface is an efficient trap which might be useful in trapping other insects.

A survey of disease control in potatoes would be incomplete without mention of late blight. It is obvious that the testing protocol is complex. Due to the history of failure of so-called major gene resistance CIP, and also research contracted by CIP, has concentrated on developing generalized or field resistance to the late blight fungus (*Phytophthora infestans*). Advanced clones with good, practical levels of field resistance to late blight have developed. However, since fungicidal control of late blight is often practised in developing countries, it will remain a useful method of control. This is particularly true where there is consumer demand for a particular potato variety which may be susceptible to late blight. Also, there are many clones with intermediate levels of resistance which may be high yielding but require perhaps only two or three applications of fungicide during a growing season. CIP is carrying out very extensive field trials in several countries adjacent to Tanzania to select high yielding late blight-resistant clones for specific environments. Present research emphasis is on breeding for late blight resistance combined with earliness.

The organization of a certified seed tuber programme in developing countries to minimize virus transfer is analogous to putting a roof on a house. It should be the last part built. For most developing countries a basic seed certification scheme should probably also be the last programme to be structured.

There are a number of alternatives to a seed tuber certification programme to ensure availability of low virus seed tubers. The use of true potato seed (TPS) is one of the more innovative approaches. The use of TPS is a disease and pest management strategy which offers the following advantages (Page 1980): (1) Practical freedom of TPS from transmissable pathogens and pests. (2) The one or two tons of seed tubers required to plant 1 ha (about 0.75 tons acre^{-1}) can be consumed as food; about 100 g TPS will plant 1 ha (about 1.5 oz acre^{-1}). (3) The use of TPS eliminates the need for storing and transporting seed tubers. And (4) TPS may be stored for six to seven years under dry conditions without significant loss of germination.

True potato seed has been used in China for a number of years for large scale production of potatoes. Excellent progress in the use of TPS is being made in Bangladesh, India, the Philippines, Rwanda, South Korea and Sri Lanka. Experience has shown that either of two methods offer the best way to use TPS. Both start by growing plants in a seed bed. Seedlings may then be transplanted to the field or tubers produced in a seed bed for subsequent field planting. Up to 10 kg of tubers m^{-3} of seed bed have been produced. In the field excellent yields of uniform tubers have been grown from TPS seedling transplants (Macaso & Peloquin 1983, Martin 1983).

In addition to the practical aspects of TPS technology in producing food, I would like to emphasize the disease control value of growing a crop from TPS: (1) Viruses such as PLRV, PVY and PVX do not pass from the parental plant to the seed. (2) Bacterial and fungal pathogens do not pass from parent to seed. And (3) because the storage of seed tubers is eliminated, storage losses due to bacterial and fungal rots are eliminated.

No other major crop produces more energy and protein per unit area per day than the potato. Africa needs to increase its yield of this nutritious crop. Disease and pest control *is* available to reduce losses and improve yield.

References

Anonymous (1982) Glandular trichomes. *Annual Report, International Potato Center, Lima, Peru,* 1982, 62.

Burton, W.G. (1966) *The Potato.* 2nd edn. Wageningen; Veenman & Zonen.

Gibson, R.W. (1979) The geographical distribution, inheritance, and pest-resisting properties of sticky-tipped foliar hairs on potato species. *Potato Research* 22, 223-236.

Goméz, P.L.; Plaistd, R.L.; Thurston, H.D. (1983) Combining resistance to *Meloidogyne incognita, M. javanica, M. arenaria* and *Pseudomonas solanacearum* in potatoes. *American Potato Journal* 60, 353-360.

Hooker, W.J. (ed.) (1981) *Compendium of Potato Diseases.* American Phytopathological Society, 125 pp.

Horton, D. (1982) *World Food Facts.* Lima; The International Potato Center, 54 pp.

Jatala, P.; Salas, R.; Kaltenbach, R.; Bocangel, M. (1979) Multiple application and long term effect of *Paecilomyces lilacinus* in controlling *Meloidogyne incognita* under field conditions. *Journal of Nematology* 13, abstract.

Jensen, H.J.; Armstrong, J.; Jatala, P. (1979) *Annotated Bibliography of Nematode Pests of Potato.* International Potato Center & Oregon State University Agricultural Experimental Station, 315 pp.

Macaso-Khwaja, A.C.; Peloquin, S.J. (1983) Tuber yields of families from open pollinated and hybrid true potato seed. *American Potato Journal* 60, 645-651.

Martin, M.W. (1983) Techniques for successful field seeding of true potato seed. *American Potato Journal* 60, 245-259.

Mehlenbacher, S.A.; Plaisted, R.L.; Tingey, W.M. (1983). Inheritance of glandular trichomes in crosses with *Solanum berthaultii. American Potato Journal* 60, 699-708.

Mendoza, H.A.; Rowe, P.R. (1977) Strategy for population breeding for adaptation to the lowland tropics. *American Potato Journal Abstracts* 54, 488.

Mendoza, H.A.; Estrada, N. (1979) Breeding potatoes for tolerance to stress: heat and frost. In *Stress Physiology in Crop Plants* (Mussel & Staple, eds.). New York; John Wiley.

Page, O.T. (1980) The use of botanical seed: A practical approach to potato production. In *The Potato in Developing Countries,* 53-55. Indian Potato Association.

Raman, K.V. (1982) Field trials with the sex pheromone of the potato tuberosum moth. *Environmental Entomology* 11, 367-370.

Shands, W.A.; Landis, B.J. (eds) (1964). *Potato Insects. Their biology and biological control.* [Handbook No. 264.] United States Department of Agriculture.

II. Removing Constraints which Limit Cultivation and Utilization

S. NGANGA[1], G.L.T. HUNT[1] and A. RAMOS[2]

[1]International Potato Centre Regional Office, Nairobi, Kenya; [2]National Agricultural Laboratories, Nairobi, Kenya.

Introduction

Some data on potato production and consumption and the constraints imposed by pests and diseases have been discussed in the first part of this Chapter. Other problems barring the way to increasing the utilization of potatoes concern seed production, agronomic factors and post-harvest constraints including marketing and storage. Methods of overcoming these problems are discussed with special reference to Africa.

Seed constraints

The cost of clean seed is one of the major constraints in potato production. In Africa seed has often to be imported from Europe. Problems including lack of foreign exchange and unco-ordinated shipping logistics result in poor quality of the shipped seed and resultant low crop yield. These seed problems can be reduced by countries adopting appropriate seed production and multiplication techniques. These include the one time import of a small quantity of seed which can then be multiplied up in isolated areas that are disease free. Several multiplications can be carried out particularly in the highland tropics where virus spread is minimal. Germplasm with resistance to seed-borne viruses would eliminate the need for strict choices of sites for seed multiplications.

Field crop control techniques can also be used for seed selection; the farmer can select some suitably yielding plants for seed usage from among his own crop. The selection procedure involves positive and negative selection over several generations. Seed from such a crop results in yield increases of 50 − 80 % compared to a non-selected crop under similar cultural practices. This system is very simple and can be directly practiced by a "low resource" farmer within his own seasonal cropping pattern.

The development of sophisticated seed production could result in a complete Certified Seed Scheme. Clonal selection by seed plot techniques and multiplication of clones by a Research Station with careful disease checks and indexing work produce foundation seed stocks. These are made available to registered seed growers who are controlled by an independent seed quality service.

Techniques to aid seed scheme multiplications

Tissue culture has become a useful tool for multiplication and storage of potato material. Rapid multiplication of *in vitro* plantlets and tuberlets by induction has been achieved through manipulation of the growth media (Roca *et al.* 1979). Work at CIP has shown that starting with a clean plantlet over 1000 fold multiplication can be achieved in just under six months of *in vitro* tuberlet production in a restricted space. A room 3 x 3 m will be sufficient to hold a multiplication scheme of over 5000 plantlets per clone each season. This technique is now available for national seed programmes.

Tuber sprout multiplication utilises well sprouted tubers. Multiple tuber sprouts of 1 − 2 cm are removed, and tubers subjected to alternate light and dark and a growth hormone applied to induce new elongating sprouts. These are removed, divided aseptically into small portions and planted immediately in a suitable rooting medium. The subsequent plantlets are planted in field plots after 2 − 3 weeks. Each mother tuber can produce over 200 sprouts in three to four desproutings over one to two weeks, and can then be planted to produce a mother plant with more tubers.

Single node cuttings can be taken from side shoots produced from healthy vigorous young plantlets which have had their apices removed at the 4 − 6 leaf stage to encourage shoot production from leaf axils. These are transferred to a clean rooting/growth media (coarse sand) and subsequently planted out. At least ten harvests can be made and multiplication is very rapid. Further cuttings can be taken from the nodal cuttings to produce multiple increases of the crop (Bryan *et al.* 1981).

Stem cuttings utilize older but vegetatively active plants. The parent plant, usually one month old, is cut off at the apical shoot. Side shoots develop to 10 cm in length and are then harvested and planted in a suitable rooting/growth medium. The cuttings can be dipped in hormonal preparations to encourage rapid rooting. Each mother plant produces 100 – 300 stem cuttings depending on the variety. Usually most indeterminate varieties produce many side shoots and stolons all of which grow to new plantlets after cutting.

The use of leaf bud tuberlets has been described by Nganga *et al.* (1980). Old mother plants that are close to tuber maturity are cut out and all the green mature but senescing leaves are removed. Portions of stem, each with a leaf and its subtending bud, are planted in sterilised rooting/growth medium. After 5 – 14 days these buds differentiate to tuberlets, are then harvested and kept under cold storage for 6 months or treated with hormone to break dormancy if immediate use is necessary. The multiplication rate of leaf bud tuberlets is 100 to 200 tuberlets per mother plant plus the tubers produced originally. The technique is efficient in utilising healthy plants that would otherwise be thrown away at harvest.

The use of true potato seed has already been discussed in the earlier part of this Chapter.

Crop Production

Ngugi (1982) has reviewed the agronomic constraints to potato production by small farmers in Kenya. Small scale farmers try to intensify production on their small plots (0.2 – 2 ha) through mixed cropping, family labour and their own, often degenerated seed. Inputs such as fertilisers, pest and disease control chemicals constitute a financial burden in the face of scarce income. Problems of plant spacings, tuber seed sizes and costs, the use of adapted disease resistant varieties, and the timing of cultural operations are all locality specific problems. They need to be addressed by national researchers.

Post-Harvest Constraints

The constraints which exist at harvest time are connected with the sale or future use of the tubers and imposed by the nature of the potato itself.

Marketing and future use

The market for table potatoes is relatively steady and the periodic surplus and shortage of supply caused by unavoidable seasonal factors results in a cycle of low and high prices, a disadvantage to both producer and consumer.

The reaction of the farmer is to grow only enough potatoes to satisfy his home requirement and a surplus which he can be reasonably sure of selling at or soon after harvest. Thus any real surplus to help satisfy the urban longer term demand is not available (Durr & Lorenzl 1980).

The solution to variations in production, is to store the surplus at harvest time for later sale thus maintaining a steady supply and an improved financial return to the farmer. In many potato producing areas there are no effective storage methods and little attention is being given to this problem.

Seed potatoes may have to be stored for a long period from harvest to planting. Where two crops are grown per year, potato dormancy may not break in time for the next planting and seed must be held for many months to plant a later crop. Highly effective seed storage methods are available and with selection of varieties having appropriate dormancy, seed may be maintained in a healthy condition up to planting.

The potato itself is a constraint because as a relatively fragile living organism, it survives normally by successive periods of growth and dormancy. However, when grown and intended for use as food or seed at some distant time, the interference of man causes problems. Also in much of the tropics, the potato is out of its usual environment and this adds to the problems of storage compared to the cool temperate zones.

Solutions to Post-harvest Constraints

In the tropical potato growing areas, there is a diurnal temperature range of 10 − 15°C; this range and particularly the mean minimum temperature makes on-farm low cost storage possible. Potatoes tend to maintain an ambient temperature so that if they are protected from the daytime heat and ventilated at night with cool moist air, the storage temperature will remain at an acceptable point. Thus at a typical 15 − 17°C, a storage time of three months with a good storing potato cultivar is possible and as long as five months if a sprout supressant chemical is used (Hunt et al. 1982).

For seed storage, low cost structures which admit free ventilation and shaded daylight enable farmers to store purchased or home grown seed over several months (Booth & Shaw 1981). Systemic insecticides and dusts can control pests which transmit virus diseases and damage tubers and sprouts. Promising techniques involving the integrated control of tuber moth attack in the field and the elimination of potato storage infestation with aromatic and repellant plant material are being developed and offer very low cost methods for the small scale farmer.

The development of potato storage in tropical Africa has been undertaken extensively since 1977 (Homann et al. 1979, Hunt et al. 1982). Design and management methods are proven and extension is well underway especially for seed storage. Farmers have been shown to prefer in-house storage (Durr & Lorenzl 1980). Traditional containers may often be adapted very easily to enable the safe keeping of potatoes to be undertaken. Perhaps the most difficult task is to train extension workers to understand thoroughly the principles involved so that they may be able to help adapt existing structures.

Potato processing can allow long life storage and ultimately, market presentation in perhaps a more attractive form than tubers. Whether a particular country is ready for processing is quite a different matter. In Kenya for example some 10 tonnes per week were processed into crisps in 1979 (Durr & Lorenzl 1980) out of a commercial production of 400 000 to 500 000 tonnes per year. In much of Africa, the potato is a seconday starch food treated as a side vegetable and usually eaten as a boiled tuber in a stew. Simple solar drying processing methods for potatoes exist (Booth & Shaw 1979) and in areas with one potato crop per year, would be highly effective for long-term low volume storage in hot conditions. Other processes and products which maintain the high quality of the potato are being developed, so that appropriate technology can be introduced when demand for processed potato foods materialises.

Both potato production and storage should be associated. There is little point in growing potatoes if many are wasted due to lack of appropriate post-harvest technology. This realisation is not universal and is recommended for the attention of planners and officials. The partnership of a well grown crop, free of serious disease and insect pests, followed by safe storage, is the only way to utilise fully the input of human effort and resources to provide the maximum possible nourishment and future crop potential.

Potato Research and Technology Transfer

A proper link between research, development extension and farmer information services is needed in national programmes. Some important aspects of this are as follows. Research should be based upon real awareness of farmers' problems and requires "hands on" involvement by the researchers. Research results should be evaluated at station level but often unfortunately the process stops at this point. Research should not become an end in itself; a correct adaptive evaluation leads to field testing in representative locations and ultimately to the design of the extension package with the involvement of "farmer-back-to-farmer" research approach. This approach requires relevant training of manpower and direction of resources to bring about the required result. The "train and visit" system could be adopted for example.

Conclusions

Potato production has gradually increased while grain production has declined in recent years (Weerasinghe 1982); 40 % of the food grain deficit is provided by root and tuber crops of which the potato plays a major part. In Africa the place of the potato in basic food is complementary to other crops. The potato is an efficient crop, contributes to income and can alleviate Africa's food deficit. Governments could maintain policies on prices and income that would encourage farmers to grow potatoes to meet demand from low salaried populations, particularly in cities.

Reduction of problems associated with seed import is possible through applying technology accessible to small farmers. The potato technology available at CIP shows that this commodity has adequate flexibility to fit into the basic farming systems of Africa. There is an urgency to develop national capability to adapt research efficiently and to transfer potato technology with the farmers involvement at the adaptation stages. Future potato development should be judged not only by increases in production, but also by the quantities and frequencies with which this commodity will complement Africa's diets.

References

Booth, R.H.; Shaw, R.L. (1979) *Simple Processing of Dehydrated Potatoes and Potato Starch.* Lima; International Potato Center.

Booth, R.H.; Shaw, R.L. (1981) *Principles of Potato Storage.* Lima; International Potato Center.

Bryan, J.E.; Jackson, M.; Melendes, N. (1981) *Rapid Multiplication Methods.* Lima; International Potato Center.

Durr, G.; Lorenzl, G. (1980) *Potato Production and Utilisation in Kenya.* Lima; International Potato Center, 133 pp.

Homann, J.; Zettelmeyer, W.J. (1979) *Potato Storage: An Example of the Small Scale Farm Level in Kenya.* Bonn; German Agency for Technical Cooperation.

Hunt, G.L.T.; Luitjens, E.J.; Wiersema, S.G. (1982) Low cost ware potato storage in Kenya. In *Research for Potatoes in Year 2000,* 99-100. Lima; International Potato Center.

Nganga, S.; Quevedo, M.; Melendez, N. (1980) *Leaf Bud Tuberlets: A Simple Technique for Seed Production.* Lima; International Potato Center.

Ngugi, D.W. (1982) Potato production in Kenya: potential and limitations. In *Research for Potatoes in Year 2000,* 240-242. Lima; International Potato Center.

Roca, W.M.; Espinoza, W.O.; Roca, M.; Bryan, J.E. (1979) A tissue culture method for rapid propagation of potatoes. *American Potato Journal* 55, 691-701.

Weerasinghe, D.P.R. (1982) Potato for the hot humid tropics. In *Research for Potatoes in Year 2000,* 183-184. Lima; International Potato Center.

Chapter 21

The Place of Annual Crops in Tanzanian Agriculture

H.Y. KAYUMBO

Faculty of Agriculture, Forestry and Veterinary Science, Morogoro, Tanzania.

Tanzania has a population of nearly 20M people which is increasing at about 2.3% a year. About 90% of the economically active population is engaged in agriculture including crop and livestock farming. Over much of the country rainfall is the most important climatic factor influencing agriculture, having the largest effect in determining the production potential of an area, the crops which can be grown, the farming system, the sequence and timing of farming operations, and the distribution of the human population. Agriculture contributes roughly 40% to Tanzania's GNP, half of which is accounted for by subsistence production, and generates over 70% of total export earnings.

Agriculture thus provides the largest source, not only of income but also of employment, and is the predominant source of foreign exchange earnings. Economic growth is thus dependent on a breakthrough in agriculture. Furthermore, resources for development in the rest of the economy are expected to be generated out of an increase in output and income in agriculture.

Despite the dominant role of the agricultural sector in the national economy, agricultural productivity is relatively low; the agricultural population has an output per economically active participant of only one sixth the average for other sectors. It is estimated that agriculture had a real growth rate of about 2.2% per year during the 1972-82 period. The non-agricultural sector achieved an annual average of 7.5% and the whole economy achieved 5% growth during this same period.

Tanzania is predominantly a nation of small peasant farmers who cultivate almost 90% of the total areas farmed. Large-scale agriculture consists of a limited number of estates engaged in coffee, tea and tobacco production, and state farms concentrating on wheat, rice, sugar, sisal and livestock. Estate production has declined in importance since 1961, and the state farm programme remains small. Peasant cultivation techniques are based on manual labour and the simple handhoe (*jembe*), and are inherently low in productivity. The use of manure, fertilizers and pesticides is not widespread. Since the major inputs into Tanzania's agricultural production are labour and land, the level of production and output on a farm is currently a function of the size of the land-base which the individual farmer or village holds, and, to a lesser extent, of the input of labour provided by the farmer and his family.

At present Tanzania is not self-sufficient in food, and in the cash crop subsector production performance of the main agricultural export commodities has not been satisfactory in recent years.

A feature of agriculture in Tanzania, as in other tropical African countries, is the diversity of its annual crops (Table 1). In order of importance, the cereals maize, sorghum and rice are the most important being the major staple foods of a large part of the country. Any major effort to improve food production in Tanzania, as indeed in most sub-Saharan African countries, requires a substantial involvement with these crops. In addition several of the commodities of world trade, including tobacco, cotton and oil seeds (including sunflower, sesame and groundnuts in Tanzania), all of which are suitable for growth in wide areas of Tanzania, are also important. Properly managed, they can generate substantial incomes for small and large scale growers.

Table 1   Food crops production for 15 years (1966/67 − 80/81) in Tanzania ('000 tons).

Year	Rice	Paddy	Wheat	Cassava	Sorghum Millets	Pulses (grain)	Irish Potatoes	Sweet Potatoes	Bananas	Sugar	Oil Crops
1966/67	880	140	45	3300	389	145	24	237	1345	71.8	–
1967/68	750	114	43	3500	344	174	35	354	891	–	–
1968/69	770	131	40	3600	374	172	46	253	140	–	–
1969/70	730	144	41	3500	372	159	62	238	185	–	–
1970/71	870	192	60	3444	413	180	74	248	261	–	–
1971/72	850	202	77	3209	367	183	67	229	998	–	–
1972/73	980	178	67	3189	409	224	120	234	1206	105	20
1973/74	750	193	49	3388	423	193	165	296	1400	96	22
1974/75	750	141	32	3688	280	182	101	302	1440	103	18
1975/76	625	157	46	3800	440	181	87	320	1500	112	18
1976/77	897	180	58	3900	390	210	92	330	1540	99	16
1977/78	968	203	35	4000	390	219	96	335	1580	136	18
1978/79	1000	260	38	4450	410	210	85	330	1466	129	25
1979/80	900	250	30	4550	380	213	85	330	1492	118	31
1980/81	800	180	–	4600	169	219	84	332	1500	122	26

Data from The Tanzania National Agricultural Policy (Final Report), Dar es Salaam, October 1982.

Cotton is a typical small-holder crop, mainly grown in the lake regions (Kagera, Mwanza, Mara, Tabora and Shinyanga). The number of cotton farmers is estimated at 500 000 familes, or some 3M people; roughly 15% of Tanzania's population. Cotton has been one of the most important cash crops in Tanzania, both in terms of the hectares under cultivation and its contribution to foreign exchange earnings. Between 1971/72 and 1974/75 cotton production slumped to 234 308 bales. The low 1975/76 production was attributed to a very heavy concentration on food crop production during the 1974/75 planting season in the main cotton areas, and to adverse weather conditions in early 1975. In 1976/77 production was 370 000 bales, but due to rain and pest problems in 1977 the 1977/78 production fell to 280 000 bales.

While cotton production increased by over 150% in the early 1980s, the position is now stagnating at around 360 000 bales per season. This stagnation is due to most farmers in the cotton belt already cultivating cotton so that large increases in planted area cannot be expected without technological changes, improved farm power, irrigation, and a more effective input distribution programme. Two development programmes, in which cotton is a major crop, were initiated during the 1970s, although their impact is yet to be evaluated.

(1) The Geita Cotton Project was financed by the International Development Association (IDA) and aimed at increasing yields from the present 400 kg ha^{-1} to 1000 kg ha^{-1} and expanding the area under cotton and maize grown by 2900 small-holders. The impact of the project on national cotton production was forecast to be around 27 000 tons of seed cotton by 1984/85, or an increase in national production of 1.3% yr^{-1}.

(2) The Kigoma Rural Integrated Development Programme, also financed by an IDA credit, started operating in 1975 with the target of 14 300 tons of seed cotton by 1981/82.

These efforts should enable the country to achieve the target of 500 000 bales of cotton, or 100 000 tons of seed cotton, in the mid-1980s. However, as a result of other contraints in the economy such as transportation, spareparts for ginneries, and increased costs of fertilizers and pesticides, this target may not be realized.

Tobacco is produced by both estates and small-holders, small-holder production being the larger. Tanzania produces two major kinds of tobacco; flue-cured and fire-cured, with the flue-cured accounting for around 7% of production. Tobacco production increased rapidly in the 1960s from about 2200 tons in 1963/64 to about 11 000 tons in 1969/70, and continued its upward trend in the 1970s, reaching an estimated 20 000 tons in 1977/78. Flue-cured tobacco is an ideal small-growers crops with good economic potential. With investment it seems destined to provide a high return for many small growers. Arrangements for credits, inputs and sales have been well organized through cooperatives and the Rural Development Bank.

An IDA-financed tobacco development project was set up in the late 1970s to establish new tobacco complexes to boost small-holder tobacco production by 9000 tons and to settle 15 000 families. Under this programme a three year tobacco handling project is being executed which aims to reduce losses in curing, grading and bulking. While the average yield potential of one hectare is 12 000 kg of wet leaf, the farmer loses 10% of his crop due to lack of barn capacity, 10% due to loss in the barn through poor curing techniques, and 25% in inefficient baling and grading operations. The implementation of this project, together with efforts of the Tobacco Authority to recruit farmers and to supply them with all the necessary inputs (fertilizers, seed-bed packets, barns, flues, etc.), is expected to maintain the growth rate in tobacco production.

Tobacco and cotton demand much more attention to detail than the cereal, legume and root crops commonly grown in the farming/cropping systems generally practised in various areas. Other papers being presented here deal with these crops in detail, presenting production constraints and problems and suggesting appropriate remedies for their solution.

Oil seeds are important crops in Tanzania and, as in other African countries, their production has continued to decline over the past 10 years. Consequently, the edible oil mills built by the Government during the 1970s are operating below capacity largely because of the lack of adequate raw materials.

Table 2   Imports of cereals 1975-82 ('000 tons).

	Maize	Rice	Wheat
1975/76	213.0	64.0	101.4
1976/77	73.8	8.8	33.6
1977/78	34.3	48.1	40.5
1978/79	–	41.2	61.3
1979/80	32.5	54.7	32.5
1980/81	274.6	65.2	48.7
1981/82	231.6	66.5	73.0
1982/83	122.7	29.4	9.4

Data from the Ministry of Agriculture, Marketing Development Bureau Statistics on Food Imports, Dar es Salaam, 1983.

Any discussion of annual cash crop production in Tanzania, as in other tropical African countries, must be viewed in relation to food crop production. The importance of annual cereal crops in Tanzania's agriculture has been emphasized in recent years through Tanzania being forced to import large quantities of cereals to off-set shortages brought about by successive crop failures due to drought. The high level of imports (Table 2) is indicative of the precarious food balance which has been developing over several years. In view of this situation, plans intended to bring the country to self-sufficiency in food during the 1980s have been initiated. These include the development of small-holder food grain production as well as large scale mechanized cereal production.

The place of annual cereal crops is best illustrated by maize, by far the most important staple food crop in the country. It is produced by more than half of the Tanzanian farmers who, in many districts, are also involved in the production of annual cash crops. Concentration on tobacco or cotton in some years has resulted in low production of maize, making it necessary to import grain. Estimated total maize production in the last 15 years has increased only slightly from about 880 000 tons to about 1 025 000 tons in 1982 (Table 1) corresponding to an annual growth rate of only

2.0%, barely equal to Tanzania's population growth of about 2.3%. In 1975 the Government started, with IDA support, the National Maize Project aimed at boosting small-holder maize production by providing farmers with broad input and extension packages designed to increase maize production to 300 000 tons ha^{-1}. The target incremental production was 195 000 tons by 1982/83, which corresponds to an increase in the growth rate from 2.5 to 5.5%. The total maize production target of 1.3M tons by 1981/82 has not been realized. In addition to small-holder development, the National Agriculture and Food Corporation (NAFCO) has also been involved in the production of maize on large scale farms. NAFCO is currently producing approximately 15 000 tons of maize. The Tanzania Sisal Authority also produces about 5000 tons yr^{-1} under its diversification programme, while a certain number of parastatal organizations, schools, prisons, etc., also produce maize for their own consumption.

Recently the Ministry of Agriculture has put forward a policy aimed at assisting farmers to increase production per unit area of both annual cash and food crops. It must be remembered however that under subsistence cropping practices it is not easy to have an accurate and continuing measure even of the areas planted to annual cash and food crops, and much less a measure of yields per unit areas. This problem is exacerbated under natural rainfall production where yield increase due to the introduction of better practices may be masked or exaggerated by yearly differences in rainfall. For example, in 1980/81 25% more fertilizers were distributed to maize farmers, but the yields were low due to a failure of the short rains in October and November and erratic distribution of the main rains in March and April.

In summary, cereals, maize, rice and sorghum, are the most important annual crops, being the major staple foods, in Tanzania. In addition, tobacco and cotton are also grown annually for cash. Yields are generally low because the farmers normally use no fertilizers, plant the local traditional varieties, provide limited weed and pest protection, and follow few improved practices. Variation in rainfall patterns and volume from season to season also have an important impact. Farmers often interplant their cereals with other crops, such as grain legumes, cassava or sweet-potatoes, so as to ensure some food in case the cereal crop fails. In contrast, yields of the annual cash crops, tobacco and cotton, are relatively high as farmers use adapted high-yielding improved varieties, good cultural practices, and some chemicals to control weeds and pests. Attempts are now being made to increase the yields of basic food crops through the transformation of areas of traditional subsistence agriculture into commercial production.

Chapter 22

The Oilseeds Crisis in Africa

B.R. TAYLOR

Agricultural Research Institute, Naliendele, P.O. Box 509, Miwara, Tanzania.

Production

Oilseeds production in sub-Saharan Africa* comes mainly from groundnut, slightly less that 5M t of in-shell nuts being produced annually (FAO 1982). Excluding cottonseed, this compares to 0.4M t sesame, 0.1M t sunflower and, 0.2M t soybean, and smaller quantities of castor, safflower, rapeseed and linseed. Part of production consumed locally, never enters official markets so that these quantities are estimates.

Although there is an expanding world market for oilseeds, oils and oilmeals, including those from plantation crops such as coconuts and oilpalm not dealt with here, supplies are adequate at present (FAO 1983). However, some African countries (for example Nigeria and Tanzania) have reduced or stopped oilseed exports and processing mills are under utilised. Increased local demands for refined vegetable oils, are often not satisfied. Tanzania's newest mill with a capacity to process 100 000 to 200 000 tonnes of oilseeds annually runs at about 30% capacity, and many Nigerian mills are under utilised or not utilised at all (Onwuka 1981).

Oilseeds production in Africa has been dominated by groundnuts in West Africa, notably Nigeria and Senegal where 1.8M and 1.0M ha were sown annually to groundnuts between 1969–71 (FAO 1982). By 1981 the area in Senegal was little changed, in Nigeria it had fallen to 0.6M ha, but Sudan is now the largest groundnut exporter in Africa with 1.0M ha (Cummins & Jackson 1982). Production in Africa in 1981 was about 10% less than 10 years before following large fluctuations in total production.

Fluctuations in groundnut production have been caused by drought and disease in the West African crop. Drought affected Nigerian yields from 1971 to 1973, groundnut rosette virus was severe in 1975 and a general decline in production has been reinforced by low producer prices relative to other crops and lack of labour (Misari *et al.* 1980). In Senegal a general decline in rainfall since 1967 has reduced yields in the north and centre of the country by 50% (Forest 1982).

For sesame, Sudan is again a major producer, followed by Nigeria, Uganda and Ethiopia. Tanzania's production is estimated to be 15 000 t y^{-1}, although only 5000 t is sold to the official buying agency. Production of sesame has declined slightly since 1971. Whilst the area sown in the Sudan has increased, yields there are estimated to have fallen to less than 250 kg seed per ha. In Nigeria both area sown and yield have increased.

Sunflower production has increased steadily since 1970, but total production in Africa was estimated to be only 123 000 t in 1981 (FAO 1982). Tanzania is the largest producer (estimated 40 000 t).

Table 1 Seed yields, oil contents and oil yields of some oilseeds.

Crop	Seed yield[1] (kg ha^{-1})	Oil content[2] (%)	Oil yield (kg ha^{-1})
Groundnuts	542	45	244
Sesame	284	50	142
Sunflower	569	35	199
Soybean	640	20	128
Castor	565	50	283
Safflower	477	35	167
Rapeseed	406	44	179

[1]Based on different producer countries; data from FAO (1982).
[2]Various sources.
[3]Assumed 70% shelling.

*Annual oilseeds and sub-Saharan Africa are subsequently referred to as oilseeds and Africa. This paper is concerned mainly with edible oilseeds.

A more dramatic increase has taken place in soybean, which from an annual production of about 75 000 t in 1970 has increased to 178 000 t in 1981. Nigeria (75 000 t) and Zimbabwe (64 000 t) are major producers and yields range from 240 to 1800 kg ha^{-1}. Appreciable production of soybean is confined to a few countries, the crop having yet to gain the confidence of small farmers in much of Africa.

Seed yields, oil contents and oil yields of some important oil-seeds are given in Table 1. After oil extraction the oil-meal or cake is a valuable livestock feed. Where livestock industries are developed near large towns and cities oilmeals will be particularly important (Doorn 1978).

The problem of vegetable oil shortages is not new. A world shortage after 1945 prompted the British Overseas Food Corporation to attempt large scale mechanised groundnut production in parts of what was then Tanganyika, the so-called Groundnut Scheme. That the scheme failed appears to have been due more to its scale and administrative problems than to soil and climate (Wood 1950), but Nachingwea in southern Tanzania, is still an important small-holder goundnut growing area.

Ecology

Groundnut is a native of South America, sunflower of North America and soybean of Asia. Sesame and castor are thought to have originated in Africa, sesame being taken at an early date to India where a secondary centre of diversity developed.

Groundnut and sesame are adapted to areas of moderate to poor rainfall and high temperatures; they may be found on sandy, nutrient – deficient soils of the semi-arid tropics primarily because they are one of the few cashcrops that can be grown economically in these regions. Groundnut growing regions are seasonally dry and have rainy seasons subject to uncertain starts, early finishes and mid-season droughts. Perhaps it is not surprising that oilseed production fluctuates under these conditions.

Sunflower is also a crop of semi-arid regions having a degree of drought resistance and tolerance to low soil fertility (Arnon 1972). It is better adapted to the cooler conditions of medium altitudes and is not widely grown in sub-Saharan Africa, although more sunflower enters official markets in Tanzania than other annual oilseeds, production being mainly in areas having altitudes above 1000m. Sunflower was an alternative crop considered for the Groundnut Scheme, but attempts to grow it at Kongwa were thwarted by late sowing, drought and unsuitable varieties (Wood 1950).

Soybean is a crop of the humid tropics, normally requiring more rainfall during the growing season than other oilseeds. The crop responds well to fertilization, but requires a strain of *Rhizobium japonicum* for symbiotic nitrogen fixation not normally found in African soils (Rachie & Sylvestre 1977). Where the crop has not been grown before yield responses may be expected from inoculation of the seed.

Cultivation and marketing

Oilseeds are not especially difficult for small holders to grow, requiring few purchased inputs. Labour demands depend on crop and level of husbandry, groundnuts being labour intensive at harvest, and so-called improved practices often including dense sowing, thinning and additional weeding. Groundnuts and soybeans may give yield responses to phosphate fertilizers, though this could be supplied in the residues of applications given to other crops. Sesame has shown responses to N and P, but small farmers often use sesame as an opening crop on land out of fallow, applying no fertilizer (Tribe 1967). Yields of older sunflower varieties benefitted most from phosphorus (Blackman 1951), but newer varieties and hybrids are also likely to respond to nitrogen. Seed availability should not restrain the production of sesame and sunflower by small farmers, requirements being only a few kg per ha, although the viability of sunflower seed declines at high temperatures and humidities, and is damaged by storage pests. Soybean requires about 50 kg seed per ha, and groundnuts up to 100 kg in order to achieve the populations necessary for high yields, and this of a potential food at what is often a hungry time of year. The multiplication rate for sesame and sunflower is about one hundred, but for groundnut only ten, which along with the special machinery needed to handle seed of this crop, make it an unattractive proposition to seed companies.

Few, if any, small farmers practice chemical crop protection, not only because of lack of training, but also because of the cost of materials and equipment and the scarcity of water. Cramer (*in* Weiss 1983) estimates that 40% of potential groundnut production and 23% of potential sunflower production in Africa is lost to pests, diseases and weeds. Many of the pests and diseases which attack sesame in Africa are of local importance; such as sesame flea beetle, *Alocypha* (*Aphthona*) *bimaculata*, in south-east Tanzania.

Annual oilseeds are frequently intercropped by small farmers. Combinations found in Tanzania include groundnut and cassava, sesame and late sorghum, and sunflower and maize. The yield and stability advantages of intercropping are given by Willey (1979). Oilseeds are often found secondary to a staple crop in the first intercropping situation recognised by Willey "Where intercropping must give full yield of a 'main' crop and some yield of a second crop", a situation which, he points out, has not been the subject of much research in the past. Although intercropping places limits on output, precluding the use of optimum practices it may always be an important method by which African farmers produce oilseeds.

Groundnut differs from other oilseeds in being both a food and a cash crop. Sesame and sunflower are used to some extent by producers (in Tanzania at least), but much is sold often to a marketing board or parastatal buying organisation at a fixed price. Groundnuts, on the other hand, can easily enter local markets when the offiical price is unattractive to growers. This is reflected in marketing board purchases. For example, in 1981 when the official purchase price of groundnuts in Tanzania was TzShs. 4.80 kg^{-1} and the local market price TzShs. 15.00 to TzShs. 20.00 kg^{-1}, official purchases were 227 of an estimated production of about 39 000 t of kernels.

Prospects

Present small holder husbandry methods for oilseeds require no sophisticated technology nor large inputs of fertilizer or plant protection materials. Advances in food grain production in some parts of the developing world, have not been matched in oilseeds. Even where improved crop production practices are available to African farmers, they may have little impact on yields (Eicher, 1982).

The following are some important examples of current developments likely to benefit oilseeds production in Africa. Groundnut cultivars with resistance to leaf spots and rust have been developed in the USA (Norden 1980) and in India (Subrahmanyam *et al.* 1980), and are being tested in Africa. Work on resistance to groundnut rosette virus has been done in Senegal and Upper Volta and resistant cultivars are being tried in parts of Nigeria (Gillier 1980, Misari *et al.* 1980).

Drought resistant cultivars adapted to the length of the rainfall cycle have been developed in Senegal (Gautreau 1982). However, many local problems, including those of soil pH and organic matter have still to be identified (see Cummins & Jackson 1982). Small-scale equipment like fertilizer-spreaders, threshers, shellers, and groundnut washers have been produced in some countries to reduce labour requirements (Gillier 1980).

Tribe (1967) suggested that a yield of 1100 kg ha^{-1} was necessary to make sesame worthwhile, but this is rarely achieved by African farmers who regard it a low input crop, so that yields may be restricted by poor farming practice. The diversity of genetic material available to plant breeders has made it possible to increase potential yields and to match varieties to different environments (Weiss 1983). In Tanzania yield increases have been achieved with crosses of local and Venezuelan cultivars (Auckland 1981). The local nature of many pests and diseases, including variations in the virulence of pathogens and epidemiological stress between different areas hamper the development of reliably resistant varieties. Although work in some parts of the world includes the development of non-shattering (indehiscent) cultivars of sesame for mechanised production (Yermanos 1981), this character is of less interest to the small farmer who can harvest at the optimum time by hand.

Sunflower, though important on a world scale, is not widely grown in sub-Saharan Africa. High yields of the mechanised crop in the USA has been achieved by synthetic hybrids but on small farms in Africa their uniformity and self-compatability are not important and seed prices might deter farmers from using them. The control of many diseases may not be practical or economic in sunflower, except in so far as field hygiene allows (Weiss 1983) and less insect pests are reported on sunflower in Africa than in other parts of the world (Rajamohan 1976) though this may be a reflection of amount of research effort.

There appears to be a greater potential for soybean in Africa. Yields of 2000 kg per ha have been reported from Tanzania (Auckland 1970) and of 3000 kg per ha from Nigeria (Rachie & Sylvestre 1977), but average yields in Africa are more likely to be $500 - 1000$ kg ha^{-1} (Weiss 1983). Other oilseeds which are important in a few African countries include niger seed, notably in Ethiopia, rapeseed which can be grown at higher elevations, and safflower, again found in Ethiopia and tolerant to dry and saline conditions (Weiss 1983). Other possible oil seed crops include the winged bean (*Psophocarpus tetragonolobus*), some cucurbits and *Hibiscus* spp. though perhaps more promise is shown by some tree crops such as the African mango, the shea butter tree and the African oil bean (Onwuka 1981).

Although plant breeders are developing cultivars with high yields, high oil contents and pest and disease resistance, accompanied by packages of recommended practices, it is ultimately the farmer who decides whether or not the new cultivars will be grown. Where there are risks of drought farmers may incline towards late or phased sowing, the use of drought-resistant low yielding cultivars, and minimum labour and fertilizer inputs especially for oilseeds, which are often regarded as secondary to food crops. The failure of soybean to become more important in Africa is likely to be related to its unfamiliarity as a food crop (Rachie & Sylvestre 1977) and therefore farmers require incentives to grow it for processing or export.

There is evidence that farmers respond to price incentives, especially those offered for specific crops (Lipton 1977). The prices offered by marketing boards will help determine production levels, although low official prices will divert output especially of groundnuts to domestic consumption and unofficial markets (the so-called parallel market in Tanzania). Marketing boards, which are intended to stabilise producer prices, tend in practice to force prices down, farmers being made to pay the administrative costs of price stabilisation (Lipton 1977). Moreover, price incentives to production, may be eroded by inflation (FAO 1983). Other incentives which can be offered to farmers may include free seed and the provision of some subsidised inputs (including credit).

The outlook for increased African oilseed exports is not especially promising. The world oilseed, oil and oilmeal economy is expanding, but production is meeting or even oversupplying this demand (FAO 1983) particularly by increased production of American soya bean, Russian sunflower and European oilseed rape. Prices for most oilseeds, oils and oilmeals reached their lowest level for several years in 1982, and in real terms were 37% lower for oils and 23% lower for oilmeals than in the mid-1970s (FAO 1983). Nevertheless, the relative prices of the most important oils have been quite stable over the last two decades, with groundnut commanding a higher price than sunflower or soya oil.

Conclusions

Oilseed production in Africa has been affected by drought and disease and by social and economic problems. There have been large fluctuations in groundnut production which has declined by about 10% since 1971.

A number of steps are being taken to help overcome the technical problems of producing oilseed crops in the semi-arid areas where many of them are found. However, domestic supplies are inadequate in many countries and the satisfying of these requirements and the reducing of imports should be a first priority. For this improved crop packages will have to be backed up by Government encouragement.

References

Arnon, I. (1972) *Crop Production in dry regions, vol. 2. Systematic treatment of the principal crops*. London; Leonard Hill.

Auckland, A.K. (1970) Soya bean improvement in East Africa. In *Crop improvement in East Africa* (Leakey, C.L.A. ed.), 129-156. Farnham Royal; Commonwealth Agricultural Bureaux.

Auckland, A.K. (1981) Sesame breeding and selection in East Africa. In *Sesame: Status and Improvement*. [FAO Plant Production and Protection Paper No. 29.] 129-131.

Blackman, G.E. (1951) The sunflower. *World Crops* 3, 51-53.

Cummins, D.G.; Jackson, C.R. (1982) *World Peanut Production, Untilization and Research*. [Special Publication No. 16.] University of Georgia, College of Agriculture Experiment Station.

Doorn, J.J.L. van (1978) Sunflower role in developing countries. In *Proceedings 8th International Sunflower Conference, Minneapolis, July 1978*, 1-6.

Eicher, C.K. (1982) Facing up to Africa's food crisis, *Foreign Affairs* 61, 151-174.

FAO (1982) *Production Yearbooks 1973-1981*. Rome; FAO.

FAO (1983) Committee on commodity problems. *Report of 17th Session Inter-governmental Group on Oilseed, Oils, and Fats*. Rome; FAO.

Forest, F. (1982) Evolution de la pluviometrie en zone Soudano − Sahelienne au cours de la periode 1940-1979. *Agronomie Tropicale* 37, 17-23.

Gautreau, J. (1982) Amerliorations agronomiques par le developpement de varietes d'arachides adaptees aux contraints pluviometriques. *Oleagineux* 37, 469-475.

Gillier, P. (1980) Role and function of the IRHO in groundnut research and development. In *Proceedings of the International Workshop on Groundnuts, October, 1980, ICRISAT, Patancheru, A.P.*, 25-28.

Lipton, M. (1977) *Why Poor People Stay Poor*. London; Temple Smith.

Misari, S.M.; Harkness, C.; Fowler, A.M. (1980) Groundnut production, utilization, research problems and further research needs in Nigeria. In *Proceedings of the International Workshop on Groundnuts. October 1980, ICRISAT Patancheru, A.P.*, 264-273.

Onwuka, N.D. (1981) The vegetable oils processing industries in Nigeria's economy. *Rivista Italiana delle Sostanze Grasse* 58, 630-636.

Rachie, K.O.; Sylvestre, P. (1977) Grain legumes. In *Food crops of the lowland tropics* (Leaky, C.L.A.; Wills, J.B. eds), 41-74. Oxford; Oxford University Press.

Rajamohan, N. (1976) Pest complex of sunflower − a bibliography. *PANS* 22, 546-563.

Subramanyam, P.; Mehan, V.K.; Nevill, D.J.; McDonald, D. (1980) Reseach on fungal diseases of groundnut at ICRISAT. In *Proceedings of the International Workshop on Groundnuts October 1980, ICRISAT Patancheru, A.P.*, 193-198.

Tribe, A.J. (1967) Sesame. *Field Crop Abstracts* 20, 189-194.

Weiss, E.A. (1983) *Oilseed Crops*. London; Longman, 660 pp.

Willey, R.W. (1979) Intercropping − its importance and research needs. Part 1. Competition and yield advantages. *Field Crop Abstracts* 32, 1-10.

Wood, A. (1950) *The Groundnut Affair*. London; Bodley Head.

Yermanos, D.M. (1981) Sesame breeding objectives in California. In *Sesame: status and improvement*. [FAO Plant Production and Protection Paper No. 29.] Rome; FAO.

Chapter 23

The Need of Integrated Pest Control (IPC) in Cotton Growing

A. WODAGENEH

Programme Leader, African Inter-Country Integrated Pest Control Development Programme, FAO/UN, P.O. Box 913, Khartoum, Sudan.

Introduction

The significance of cotton, the magnitude of pest problems, the short- and long-term hazards associated with widespread of pesticides urgently call for implementation of Integrated Pest Control (IPC) in cotton growing.

Significance

Cotton is an important cash crop and grown both under irrigation and rainfed conditions. It has much significance in the social, economic and political life of those associated with it. Increased production of artificial fibre did not remove it from its position as number one crop amongst all fibrous plants. It also plays a major role in households and has countless other widespread uses. It is a source of both edible oil and animal feeds, and, therefore, cannot be separated from other foodcrops.

Drawbacks

There are major drawbacks in cotton production. It is a crop subject to attack by a multitude of insect pests and diseases. Its abundant green leaves, succulent stems, and young branches, buds, flowers and bolls all offer a haven and nutrition, optimal living conditions for pests. It provides the best microclimatic conditions conducive to attacking pests both underneath the leaves and within the canopy of the whole plant population. Added to this, almost invariably, the intensity and gravity of cotton pest problems continues to be rampant and may become irreversible when man interferes with its delicately balanced agro-ecosystem. The usual means of interference is the unrestrained and wide use of chemical pesticides. Incidences of chemically induced complexities in cotton pest problems have been well-documented from early experiences in Latin American countries and the USA. The severity of pest problems in cotton production almost always passes through distinct phases of development, (1) the subsistence phase, (2) the exploitation phase, (3) the crisis phase, (4) the disaster phase, (5) the integrated phase and (6) the deterioration phase. Once the severity reaches the last phase, the situation is almost irreversible.

In the *subsistence phase* yields are low and organized forms of crop protection usually absent.

The *exploitation phase* takes place mainly under irrigation. Complete dependency on chemicals becomes more and more common and intensive. Application is usually on fixed schedules irrespective of pest presence or absence. Initially the outcome is rewarding and high yields of both seed and lint are achievable.

Next follows the *crisis phase*. After a number of years of intensive use of chemicals and exploitation, the effect of diminished returns becomes obvious. Therefore, in an attempt to reverse the situation, applications of pesticides are usually stepped up and yet effective control becomes increasingly difficult or impossible. Pest populations escalate, becoming rampant after every additional treatment with pesticides having little or no effect. Either substitution or the mixing of pesticides becomes ineffective. Instead, resistance and cross resistance to pesticides, including resurgence and secondary outbreaks of pests, become prominent and the cost of production becomes increasingly prohibitive.

Soon the *disaster phase* follows. As a result of the increased costs of pesticides the cotton is no longer profitable and therefore marginal land and marginal farmers concerned with cotton disappear. The only alternative to this worsening condition is an *integrated phase*, the development and combination and utilization of various compatible methods to obtain the best control possible with the least disruption of the environment. This involves natural mortality factors, biological, cultural and crop manipulation, the use of resistant varieties, pest prediction, and use of suitable and/or selective pesticides. Integrated control is a recovery stage, however, and a failure of its strict and continued use triggers the *deterioration phase*, returning events to the disaster phase.

Integrated pest control is not new and has been gradually acquiring a firmer basis from the 1860s to the late 1920s. However, the advent and introduction of synthetic insecticides, for example wide spectrum pesticides such as DDT in the 1930s, and organophosphorus pesticides in the 1940s-1950s, retarded the continued development and application of IPC.

A Case Study in the Sudan

In Africa, particularly amongst the north-east African countries, Sudan cotton production has passed through the various phases of pest problems described above. The Gezira Scheme in the Sudan is under single management and the largest cotton production scheme in the world.

From about 1940 to the late 1950s, Sudan was in an exploitation phase. Pests such as whitefly and jassids posed only minor problems, apart from leaf curl virus disease transmitted by whitefly. However, that disease was controlled by cultural and sanitary measures.

Although organophosphorus pesticides were widely available between 1945 and 1960, application was limited only to early infestations and yields increased. After 1960, jassids were no longer effectively controlled by pesticides such as DDT and endrin, and soon whitefly started to return, this time as a feeder and serious source of honey dews rather than as a vector of leaf curl virus disease.

All alternative pesticides and mixtures tried were only short-lived as control agents. Whitefly became increasingly serious. In 1963 a serious outbreak of American bollworm forced the introduction of mixtures of DDT and dimethoate to control three insects simultaneously. Later on the effect became nil and the cost of application was soaring.

Between 1960 and 1970 the Sudan was in the crisis phase. The cost of pesticide sprays, aggravated by the oil crisis of the early 1970s, rendered the production of cotton unprofitable. The increase in the cost of protection between 1972 and the early 1980s exceeded 600%.

By 1975 the area under cotton had started to decline, yields decreased, and marginal lands were taken out of production. Immediately after, in 1975/76 the fourth phase the *disaster phase* was becoming evident. The various phases of pest development problems became distinctly obvious as more and more liberal use of pesticides was preferred. Earlier, as the conventional approach of pest control had started with DDT, a misleading justification of pesticide use arose as yields increased. The wide-spectrum effect of DDT, although known to be a precursor and increasingly damaging to beneficial insects, unfortunately continued in use into the early 1980s. In addition the widespread and unrestricted use of pesticides and "package deal" agreements encouraged unlimited numbers of applications by insecticide-selling companies. Although the object was to attain a guaranteed level of yields, it led to the whole cotton production system entering a disastrous phase. Consequently, the number and frequency of sprays increased from one in 1961 to eight or more in 1980/81, depending on location and the severity of pest attacks.

The above negative phases and the urgency of the situation demanded a restructuring of the pest control programme, providing an opportunity for the introduction of IPC. As similar problems and incidences exist in other regions connected with the increasing use of pesticides, FAO has now made it a policy to introduce IPC on a global basis not only in cotton production but in other crops as well.

In 1976 the Government of the Sudan requested that activities under the FAO/UNEP Global Programme be initiated in the Sudan. In 1978 FAO/UNEP started an African inter-country programme for the development and application of integrated pest control in cotton growing to cover the Sudan and five neighbouring countries. However, owing to the seriousness and complex of cotton pests in the Sudan at the time, the first two years were entirely devoted to identifying the causes of the problem and in drawing up a national Sudanese Project Document. By 1979, the Sudanese national IPC programme began to operate under the title "Development and Application of Integrated Pest Control in Cotton and Rotational Food Crops" with contributions from the Dutch Government with FAO acting as executing agency.

The project programme of work was mainly in two directions: increasing host plant resistance, and identifying natural enemies of whitefly and the American bollworm. The work on host plant resistance showed that okra and super okra leaf varieties offered considerable potential as the microclimatic conditions with low humidity were less favourable to whitefly. Work on cotton characteristics conferring less susceptibility to bollworm infestation is underway. Work on natural enemies is also in progress but requires increased attention. Studies on the dynamics of whitefly and bollworm populations is also in progress. This has shown the importance of cultural practices, rotational changes, varieties, sowing dates and distance, spacing, thining, and plant population, fertilization, weeding, water use, and time of picking in affecting individual pests and pest complexes.

The role of FAO/UNEP as a catalytic agent has assisted in the formulation of an integrated pest approach in the Sudan designed to solve the cotton pest problem. The establishement of the FAO/UNEP IPC programme also helped the abolition of "package deals" and the use of DDT and its mixtures. Ineffective insecticides were withdrawn, the scheduling of pesticide applications on the basis of pest surveillance has become more prevalent, and the refinement of spray techonology and the use of pre-emergency herbicides are encouraged.

Other Countries

The problems of depending on pesticides alone for the control of cotton pests also exist in Ethiopia, Uganda, Kenya and Tanzania. In all these countries, at least one third of the cost of cotton production is pesticide purchase despite the fact that yields remain low. Of course, low yields are not only due to pest problems. Socio-economic conditions related to prices, lack of adequate transport, delayed payments, lack of consumer items, competition from other crops, and population pressure all contribute significantly by either decreasing yields or reducing the areas cultivated.

In addition, persistent drought contributes to decreasing yields. Therefore, in view of the widespread nature of problems in cotton production, FAO/UNEP are extending their catalytical role by assisting further countries to start IPC programmes. In October 1983 the first African Inter-Country IPC Steering Committee Session was held in Khartoum. Amongst the resolutions passed was one recognizing the need to establish a national IPC commission in each country to both initiate and activate IPC programmes and serve as a link with a FAO/UNEP IPC base office. An IPC Project Document has been drafted in the light of existing cotton pest problems in each country and at present is pending discussions with the Government authorities concerned. A second IPC Steering Committee Session will be held in Egypt in May 1984 to discuss the future development and strengthening of the regional IPC programme. The African inter-country IPC programme has both short- and long-term objectives. The main objective is to improve cotton growing and cotton production, relying mainly on the transfer of existing knowledge through demonstration and training, by IPC implementation at the farmer level through extension services, and by the improvement of local knowledge and by adaptive research. The main drawback in the promotion and development of IPC programmes is finance, and reliance has to be placed on donor contributions, for example in the form of expertise, equipment, and training facilities.

Conclusion

The farmer, in his eagerness to increase yields, will adopt any means of pest control and agricultural inputs available to him. He is at the mercy of those around him offering advice and guidance on pesticides; the pesticide dealer has one aim, to sell his chemicals, and may not take note of their short- and long-term effects. The promotion and the effort of introducing alternative methods of pest control such as IPC has to compete with short-sighted interests. Pesticides are an effective means of pest control, and will remain in agriculture for many years to come, but a rational use of them is essential.

Chapter 24

The Development of Integrated Pest Management of *Heliothis armigera* – Current Status and Future Possibilities in East and Central Africa

B. T. NYAMBO

Agricultural Research Institute, Ukiriguru, P.O. Box 1433, Mwanza, Tanzania.

Introduction

The genus *Heliothis* contains several well known pests of agricultural crops and *H. armigera* is the most damaging species in the Old World (Hardwick 1965) especially in the tropical areas of many developing countries. It is an important pest of cotton in East and Central Africa and has been recorded from a wide range of food crops. Although well documented as a pest of cotton in Africa (Reed 1965, de Lima 1977, Balla 1981) its insidious effect on food and minor crops which form an integral part of less intensive agriculture and intercropping systems practiced by small farmers is less appreciated. *H. armigera* is noted for its ability to thrive on a wide range of host plants and its success in tolerating many insecticides and developing resistance to several. The pest appears to become relatively more important as standards of agriculture improve. Accurate data on actual losses caused by *H. armigera* in the developing countries is meagre; it was responsible for the abandonment of cotton cultivation in the Ord River Scheme in Australia and continues to attack intended replacement crops including sorghum. Reed *et al.* (1981) indicated that losses to cotton alone in Tanzania must be more than USA $20M in most years, which is massive in relation to the country's economy. In Sudan, *Heliothis* became established as a major annual pest of cotton in the

Gezira scheme after the 1960s causing severe losses in most seasons and attacking sorghum and groundnut which serves as an alternative host.

Previous Research on Biology and Control of H. armigera in East and Central Africa

Studies in the biology and ecology of *H. armigera* in East and Central Africa were carried out for many years on the cotton crop (Coaker 1959, Davies & Kasule 1964, Reed 1965). These studies showed that maize is a potential host of the pest. Using maize as a trap crop, or adjusting sowing dates for cotton and/or maize to minimize damage through exposure to successive generations of *Heliothis* from maize were found to be largely impractical in the small farmer context. Delayed sowing of maize was unacceptable because of its importance as a food crop and repeated sowings necessary to provide flowering maize to act as a trap crop during the cotton fruiting season was also not feasible because of labour constraints.

Subsequent studies (de Lima 1977, Topper 1978) have stressed the importance of suppressing *H. armigera* on other crops to reduce the pressure of the pest on cotton. Although there have been regular attempts to assess the effects of parasites and predators on *H. armigera* (Coaker 1959, Reed 1965, Balla 1981), generally little is known about their effectiveness. A large number of parasites have been identified but no effort has been made to augment their activity because little is known about their biology or about the identity or prevalence of most of the predators.

Despite some earlier success with local strains of nuclear polyhedrosis virus (NPV) and *Baccillus thuringensis* in biocontrol (McKinley 1971, Roome 1975) the work has not been developed. Little success has been achieved in identifying cultivars which either discourage breeding by *H. armigera* or have resistance to the pest (Nyambo 1981).

Most research was directed to control *H. armigera* with insecticides as these provided quantifiable yield increments and clear visual evidence of insect mortality. Before the mid 1960's, insecticide use on cotton, the largest insecticide consumer, was based on calendar spraying regimes but this ignored differences in pest infestation levels in time and space and led to indiscriminate use of insecticides. Furthermore, heavy reliance on insecticides can have a number of economic and environmental limitations. Resistance of *Heliothis* to a wide range of insecticides has been reported from several countries including Zimbabwe where Gledhill (1981) reported resistance to both DDT and endosulfan and South Africa where *Heliothis* showed resistance to endosulfan and parathion (Whitlock 1973). Widespread insecticide resistance in the African *H. armigera* population will further complicate contol of the pest. Most insecticides are nonselective and kill both the pest and its natural enemies thus removing natural checks to population increase. In the Sudan the level of larval parasitism has declined in recent years to almost nil in the 1970s after intensive use of insecticides (Balla 1981). Similarly, at Ukiriguru Reed (1965) observed a decline in the level of larval parasitism from 27% in 1962 to only 6.4% by 1964 after a short period of intensive use of insecticides at the station. To optimize the use of insecticides on cotton, significant progress was made in Central Africa (Matthews & Tunstall 1968) where spraying regimes based on infestation threshold levels were developed, but this method has not gained general acceptance despite using less insecticides.

Current Research in Tanzania with Emphasis on Integrated Pest Management

Crop interactions

During the 1980–81 season, a start was made to identify the cropping sequence of alternative hosts of *H. armigera* in the vicinity of Ukiriguru and the bionomics of the pest with emphasis on control strategies. The study was carried out in a relay intercropping system used by small farmers around the station and included maize, sorghum, cotton, chickpea, tomato and *Cleome* sp. a wild host plant of *H. armigera*, and common weed in the area which flowers throughout the year. The flowering pattern is an indication of the attractive stage of the host plants to *H. armigera*. These overlap in such a way as to give continuous moth activity throughout the year by influencing the population dynamics of *Heliothis armigera*. Early in the season, a high population of *Heliothis* builds up on early sown maize and sorghum and the adults emerging from this migrate to early sown cotton towards the end of January and early February and re-infest sorghum and maize in the flowering stage. Infestation of cotton early in the season thus depends on early crops of sorghum and maize and the availability of their flowers in January and February. The sequential sowings of maize and sorghum between October and December provide alternative hosts and with favourable weather conditions allow early sown cotton to escape severe bollworm infestation.

Later in the season, timely sown cotton would have set a good crop and the long maturity maize would deviate the infestation from cotton since maize is a prefered host. Factors affecting *Heliothis* population changes on each host were also examined. *Cleome* generates *Heliothis* moths throughout the year except between August and mid-October, although the plants flower throughout the year. Natural larval mortality caused by parasitism was dominant between October and December on *Cleome* and diseases accounted for a very low adult moth emergence in January to May. However the intensity of the diseases and parasitism levels fluctuated between and within the seasons on this host plant. The level of larval parasitism and the range of parasitic groups involved also varied between host plants, as did the pattern and intensity of diseases. NPV was more prevalent on larvae collected on maize and *Cleome* than on any other host; and very rare on sorghum and cotton. Reed (1965) observed a similar distribution. Augmentation or introduction of the NPV disease may thus be more successful on certain host plants than on others, an area which needs further investigation. On the other hand, bacterial diseases were more widespread among the host plants but more common on chickpea, sorghum and tomatoes.

Monitoring and forecasting

Field monitoring or scouting as practiced in central Africa has not so far been successful in Tanzania because the *H. armigera* infestation level on cotton remained consistently low during trials and the thresholds used were unable to give clear cut results. However, the need to minimize insecticide use to reduce costs made it necessary to review monitoring possibilities. Source monitoring together with light trap data was used successfully by Reed (1965) to warn the extension service of a heavy infestation in 1964. It is possible to monitor *Heliothis* on maize and sorghum early in the season in order to predict early infestation in cotton and to apply control measures on sorghum to reduce the number of survivors which will infest cotton.

Light trap data has been used successfully in East Africa to monitor and forecast armyworm outbreaks, but although the same traps collect *H. armigera* moths, the data has not been used to facilitate *H. armigera* control. Reed (1965) was able to find some relationship between moth catches and cotton infestation between 1961–62 and 1963–64 seasons and more recent light trap data at Ukiriguru and at Ilonga show that peak moth flights lag behind the main rainfall peaks. Similar observations were made by Robertson (1977). At Ukiriguru during 1975–76 to 1981–82 (except 1978) *Heliothis* infestation on cotton was very low as peak moth activity occurred towards the end of the season. However, a grid of traps will be needed to monitor and forecast *H. armigera* accurately. Lack of electricity and skilled labour to sort the catch may not permit the expansion of the existing light trap network so the use of pheromone traps in the western cotton growing area is being studied. Pheromone traps are simple to run, do not require skilled labour because they are very selective, and do not use electricity, so they can easily be set up even in very remote areas. The pheromone capsules were obtained from the UK Tropical Products Institute, and one trap design was chosen after a preliminary evaluation of several.

At Ukiriguru, four traps were set up to study the activity of the pest within the station; others were set up at Mabuki and at Lubaga to monitor the seasonal activity over a more varied area. Peak activity between May and June was followed by a period of low activity up to the beginning of October when the population began to build up at Ukiriguru and Mabuki. The pattern of activity at Ukiriguru in the pheromone trap was almost identical to that in the light trap during 1982 providing local evidence on the reliability of the pheromone trap for monitoring the activity of *H. armigera*.

Breeding for insect resistance in host plants

In Tanzania, the commercial cotton varieties are already resistant to jassid damage and have the ability to compensate for loss caused by early bollworm attack provided soil moisture and nutrients are adequante. An attempt has also been made to incorporate *Heliothis* resistance characters into the commercial cotton varieties with emphasis on frego bract, nectariless, high gossypol and glandless characters. The frego bract selections have consistently given promising results and therefore continue to be evaluated.

Constraints to the Adoption of Integrated Control and Research Required

Biological control

Natural enemies are an important element in the control of *H. armigera* populations in Africa and elsewhere, however the species is far from an ideal target for biological control by arthropod parasites and predators (Greathead & Girling 1981). If there is much population movement both locally (Haggis 1981) and on a migratory scale (Bowden 1973, Rainey 1975), enhancement of parasites or predators by artificial augmentation is likely to be a dubious proposition. However, in the small farmer situation much remains to be discovered of the role of different elements of mixed cropping which may affect parasitoid increase. The differential attractiveness of different crops (or cultivars) to parasites as opposed to the pests themselves, is little researched. Natural augmentation particularly early on in mixed cropping situations may be very important and needs investigating.

In central and eastern Africa the main egg parasites are in the Trichogrammatidae and Scelionidae, whereas the larvae and pupae stages are attacked by the Ichneumonidae, Braconidae and Tachnidae. Some of the common species, e.g. *Apanteles* spp., appear to be widespread, but the records of all parasite and predatory species are fragmentary and research has been limited to species identification and percentage parasitism over restricted periods.

The main hope of success with biological control of *H. armigera* appears to lie in the use of microbial pesticides. The current status of these for *Heliothis* is well summarized by Bell (1981) and McKinley (1981), both of whom stress their potential, particularly within integrated pest management schemes, as supplements to chemical control. There is good information based on the use of pathogens in the developed world, but a great deal remains unresearched in Africa. A particular area which requires increased attention is formulation and application methodology.

Given the life cycle of *Heliothis* and its host range, there is a real need for further investigation of the use of pathogens in Africa and in the tropics generally. Cultivars which have antibiotic characteristics may extend the life cycles of pest insects thereby allowing more time to control them using pathogens, a useful feature for these generally slow acting agents. In cotton in the USA, Fernandez *et al.* (1969) tested various combinations of resistant cottons with NPV against two species of *Heliothis* and found that control was comparable to that obtained with organophosphate insecticides.

Crop interactions

Crop − insect interactions occur at many different levels in *Heliothis armigera*, and are greatly complicated by the large host range including many cultivated species. A wide range of cultivars of different susceptibilities and agronomic characteristics is grown over large areas, more or less simultaneously but with considerable spread of sowing dates. The interactions resulting from this have been little studied in depth and in Africa are very complex as farming systems may include a range of plant species, several which are at one stage or another susceptible to attack by *H. armigera*. Interactions at the three recognized categories − tolerance, non-preference and antibiosis, begin as early as egg laying. It is well known (and has been utilized in control) for instance, that, *H. armigera* displays clear preference for maize for oviposition. An ovipositional preference of maize> tobacco> soybeans> cotton, demonstrated in the USA, is related to plant pubescence.

Resistance to Heliothis armigera in host plants

Examples of resistant cultivars giving complete control of an insect pest do exist e.g. jassid resistance in cotton, but host resistance has made little impact on *Heliothis* control although plant characteristics which restrict population increase or delay the necessity of insecticide use have been determined (Lukefahr *et al.* 1971, Wiseman *et al.* 1978).

Refai *et al.* (1979) found that *H. armigera* appears to have different levels of susceptibility to insecticides as a result of feeding on different host plants. Larvae fed on tomatoes were most sensitive to insecticides, those fed on castor and cotton only moderately susceptible and those on maize least susceptible. Thus, the use of several resistant crop species could result in a more efficient suppression of the pest by insecticides especially on crops grown seasonally in the small farmer situations. This might be achieved either chemically or by the manipulation of cultivars.

The physical characteristics of cultivars can also effect the development and survival of *Heliothis armigera* populations; tight panicled sorghums were shown by Doggett (1964) to harbour many more *H. armigera* larvae than the more open headed types which may allow better access by birds and other predators. Lukefahr (1981) in his review of the prospects for use of host plant resistance notes that, although the principles were first applied to *Heliothis* spp. the convenience of insecticides resulted in a reduced emphasis on this aspect. He considers that the prospect for the use of resistance is not good in crops with high production inputs and high fixed costs particularly with the discovery of the synthetic pyrethroids which are both highly effective and give stable yields.

Although these conditions do not apply to much of Africa, the necessary inputs in terms of funding or manpower seem unlikely to be channelled to host plant resistance breeding. In cash crops such as cotton and groundnuts, which are hosts of *H. armigera*, there is likely to be a quicker and easier return from experimentation on dosages and application techniques with the new synthetic pyrethroids than from breeding and selecting for resistant cultivars. However, although insecticides gave consistently stable yields over a number of years in developed countries other means of controlling *Heliothis*, including host plant resistance, were investigated after the pest developed resistance to chlorinated hydrocarbons and carbamates.

Possible Areas for Co-operation within Africa and with developed countries

There is scope for collaboration in work on *Heliothis armigera* in several main areas of work as follows.

Monitoring and forecasting Long range migrations of *H. armigera* in Africa and localized movements jeopardize the effective use of arthropod parasites and predators by creating population/parasite-predator imbalances, and facilitating the spread of insecticide resistant strains. To enable practical control strategies and tactics to be formulated, intensive studies of local movements using data from light traps correlated with meteorological data is necessary as already used successfully to monitor and predict *Spodoptera exempta* Walk movements. Existing data could be used to detect inter-regional movement of *H. armigera* with possibilities of setting up an early warning system. Pheromone traps could be used to increase information on local movements since these can be located in remote areas. Hopefully, a collaborative project will soon be established to make use of past records from agricultural research stations in East and Central Africa.

Biological Control The most common parasitic and predatory species of *H. armigera* present in east and central Africa have been identified but there are large gaps in the knowledge of their biology and ecology. Such information is vital if augmentation and artificial introductions are to be contemplated. Research on *H. armigera* pathogens needs increased funding particularly to develop good formulation and application techniques and assess local strains; techniques perfected in developed countries can be adapted to local situations by transfer of knowledge and expertise.

Host plant resistance To date little success has been achieved in using resistant plant cultivars to control *H. armigera*. However, evidence from developed countries shows that there is scope to use plant species diversity in the small farmer situation and to breed and select for *Heliothis* resistance. Exchange of promising material and information between countries could hasten the development of *Heliothis* resistant varieties. The long-term use of resistant cultivars can change overall populations of pest species and reduce their numbers if grown on large enough areas.

Intercrop relationship The small farmers of Africa often use a mixed cropping system and sowing date sequence which enables the pest to thrive. The interactions of this system with the predator-parasite complex could be exploited to the benefit of the farmer if work in this most difficult research area was increased particularly if carried out over a range of ecosystems to allow comparative evaluation.

Insecticides   Insecticides will continue to be an essential component in the pest control package. There is a need to identify selective insecticides with minimum hazard to the non-target organisms to maintain a balance between the pest and its natural enemies. Brettell (1982) in Zimbabwe has made a good start in identifying such insecticides which minimize damage to Chrysopids and other natural enemies of *H. armigera* in cotton fields. More of such work and exchange of information should be encouraged within Africa.

References

Balla, A.N. (1981) Progress in research and development for *Heliothis* management in the Sudan. In *Proceedings of the International Workshop on* Heliothis *Management, November 1981, Patancheru, India.*

Bell, M.R. (1981) The potential use of microbials in *Heliothis* management. In *Proceedings of the International Workshop on* Heliothis *Management, ICRISAT Center*, November 1981, *Patancheru, India.*

Bowden, J. (1973) Migration of pests in the tropics. *Mededelingen van de Faculteit Landbouwwetenschappen Rijksuniversiteit Gent* 38(1), 785-796.

Brettell, J.H. (1982) Green Lacewings (Neuroptera: Chrysopidae) of cotton fields in Central Zimbabwe. 2. Biology of *Chrysopa congrua* Walker and *C. pudica* Navas and toxicity of certain insecticides to their larvae. *Zimbabwe. Journal of Agricultural Research*, 20, 77-84.

Coaker, T.H. (1959) Investigations on *Heliothis armigera* (Hb.) in Uganda. *Bulletin of Entomological Research* 50, 487-506.

Davis, J.C.; Kasula, F.K. (1964) A note on the relative importance of Heteroptera and bollworms as pests of cotton in Eastern Uganda. *East African Agriculture and Forestry Journal* 30, 69-73.

Doggett, H. (1964) A note on the incidence of American bollworm, *Heliothis armigera* (Hüb) (Noctuidae) in Sorghum. *East African Agriculture and Forestry Journal* 24, 348-349.

de Lima, C.P.F. (1977) Interrelations and control of insects, attacking cotton and food crops, with particular reference to *Heliothis armigera*. In *Advances in Medical, Veterinary and Agricultural Entomology in Eastern Africa. Proceedings of the first East African Conference on Entomology and Pest Control, December 1976, Nairobi*, (C.P.E. de Lima, ed.).

Fernandez, A.T.; Graham, H.M.; Lukefahr, M.J.; Bullock, H.R.; Hernandez, N.S. (1969) A field test comparing resistant varieties plus applications of polyhedral virus with insecticides for control of *Heliothis* spp. and other pests. *Journal of Economic Entomology* 62, 173-177.

Gledhil, J.A. (1981) Progress and problems in *Heliothis* management in tropical southern Africa. In *Proceedings of the International Workshop on Heliothis Management, ICRISAT Center, November 1981, Patancheru, India.*

Greathead, D.J.; Girling, D.J. (1981) Possibilities for natural enemies in *Heliothis* management and the contribution of the Commonwealth Institute of Biological Control. In: *Proceedings of the International Workshop on Heliothis Management, ICRISAT Center, November 1981, Patancheru, India.*

Haggis, M.J. (1981) Distribution of *Heliothis armigera* eggs on cotton in the Sudan Gezira. Spatial and temporal changes in their possible relation to weather. In *Proceedings of the International Workshop on* Heliothis Management, ICRISAT Center, November 1981, Patancheru, India.

Hardwick, D.F. (1965) The corn earworm complex. *Memoirs of the Entomology Society of Canada* 40, 247 pp.

Lukefahr, M.J.; Houghtaling, J.E.; Graham, H.M. (1971) Supression of *Heliothis* populations with glabrous cotton strains. *Journal of Economic Entomology* 64, 486-488.

Lukefahr, M.J. (1981) A review of the problems, progress and prospects for host plant resistance to *Heliothis* species. In *Proceedings of the International Workshop on Heliothis. Management, ICRISAT Center, November, 1981, Patancheru, India.*

Matthews, G.A.; Tunstall, P.J. (1968) Scouting for pests and the timing of spray application. *Cotton Growing Review* 48, 115-127.

McKinley, D.J. (1971) Nuclear polyhedrosis virus of the cotton bollworm in Central Africa. *Cotton Growing Review* 48, 297-303.

McKinley, D.J. (1981) The prospects for the use of nuclear polyhedrosis virus in *Heliothis* Management. In *Proceedings of the International Workshop on Heliothis Management, ICRISAT Center, November, 1981, Patancheru, India.*

Nyambo, B.T. (1981) Problems and progress in *Heliothis* management in Tanzania, with special reference to cotton. In *Proceedings of the International Workshop on Heliothis Management ICRISAT Center, November, 1981. Patancheru, India.*

Rainey, R.C. (1975) *Heliothis* outbreaks in the Sudan Gezira as indicated by egg infestations and their relationship to transport and concentration of airborne moths in wind fields. In *Ciba Geigy Seminar on the Strategy for Cotton Pest Control in the Sudan Gezira, February, 1975, Wad Medani, Sudan.*

Reed, W. (1965) *Heliothis armigera* in Western Tanganyika. *Bulletin of Entomological Research* 56, 117-140.

Reed, W.; Pawar, C.S. (1981) *Heliothis*: a global problem. In *Proceedings of the International Workshop on Heliothis Management ICRISAT Center, November, 1981, Patancheru, India.*

Refai, A.; El-Guindy, M.A.; Abdel-Sattar, M.M. (1979) Variation in sensitivity to insecticides of *Heliothis armigera* Hb fed on different host plants. *Zeitschrift für Angewante Entomologie* 88, 107-111.

Robertson, I.A.D. (1977) *Records of insects taken at light traps in Tanzania. VII – Seasonal changes in catches and effect of the lunar cycle on the adults of several pest species of Lepidoptera (Lepidoptera: Noctuidae).* [Miscellaneous Report no. 38.] London; Centre for Overseas Pest Research.

Roome, R.E. (1975) Field trials with a nuclear polyhedrosis virus and *Bacillus thuringiensis* against larvae of *H. armigera* Hb (Lepidoptera: Noctuidae) on sorghum and cotton in Botswana. *Bulletin of Entomological Research* 65, 507-514.

Topper, C. (1978) The incidence of *Heliothis armigera* larvae and adults on groundnuts (and sorghum) and the prediction of oviposition on cotton. In *Third Ciba-Geigy Seminar on the Strategy for Cotton Pest Control in the Sudan, May, 1978, Basle, Switzerland.*

Whitlock, V.H. (1973) Studies on insecticidal resistance in the bollworm, *H. armigera. Phytophylactica* 5, 71-74.

Wiseman, B.R.; McMillian, W.W.; Widstrom, N.W. (1978) Potential of resistant corn to reduce corn earworm production. *Florida Entomologist* 61, 92.

Chapter 25

Problems of Perennial Crop Production by Smallholders

J.N.R. KASEMBE

Tanzania Agricultural Research Organisation, P.O. Box 9761, Dar es Salaam, Tanzania.

Introduction

As an introduction to the series of papers on perennial crops (Chapters 26-30), I shall outline the many problems of perennial crop production faced by small holders in Africa. These problems are not all common to all the perrenial crops, coffee, tea, sugarcane, coconuts, cashew, sisal, bananas, cloves, cocoa, palm oil, forest trees, rubber, wattle, citrus, and even jojoba, and the problems can be divided into four groups treated separately below.

Human factors

Many perennial crops are in fields far away from homesteads, and are therefore more likely to be destroyed by bush fires caused by either irresponsible people or accidents. Crops such as coconuts, cashew, palm oil and forest trees are all vulnerable to bush fires.

Sometimes attitudes towards certain crops tend to affect production efforts. For instance, the cultivation of fuel-woodlots in hedges of fields is not yet widespread even in areas where fuelwood is scarce. *Lucaena* species are a good example of a fuelwood which is useful for that purpose, but not yet fully utilized.

Natural factors

Natural factors also contribute to the low production of some crops, particularly the life-cycle of most perennial crops. Coconuts and cashew take three to sometimes seven years to bear fruit, thereby delaying production from the crop. Sometimes small holders are discouraged from planting these crops and instead sow annual crops generating quicker returns from their efforts.

Government policies

I shall discuss only two of the many different policies which may affect production.

The low availability of foreign exchange for the purchase of inputs such as pesticides, machinery, and farm equipment including transport and processing equipment. This is serious, more so when most of the foreign exchange earned in African countries is from perennial export crops. A lack of priority for export crops in foreign exchange allocations heavily affects not only those exchange earner crops, but also food crops, which are mostly annual. The latter may contribute to the foreign exchange position by substitution for imported food, but if Africa is to be self-sufficient in food, we must look at this question very seriously.

Producer prices for the perennial crops are generally low. This is also partly dependent on our policies and partly due to lack of institutional arrangements to allow producers to have a say in the determination of prices.

Research problems

(1) Many crops suffer from pests including weeds, insects and diseases. The plant protection technology developed for some crops such as cloves, cashew and coconuts are inadequate or non existent. Further research is therefore required.

(2) Fertilizer placement and its effects are not available to the small holder either because the technology is not fully developed, as in cashew, or lacking, as in the case with tree crops (except for boron and nitrogen).

(3) Information on the ecology, management, and cultural systems sometimes does not reach the small holder even if this information is already available to scientists. There is a major difficulty in information dissemination in most of rural Africa. Joint efforts between research and extension institutions are required in most countries for most perennial crops.

These problems can be considered more critically in reference to the following papers concerned with sugarcane, arabica coffee, tea, coconuts and cashew, but our attention should also bear in mind other crops in order to cover all aspects of perennial crop production and hence identify ways and means of enhancing agricultural production in Africa.

Chapter 26

A Study of Self-Sufficient Sugarcane Production: The Mumias Outgrower Scheme

A. BEEVERS and R. M. D. GLASFORD

Mumias Sugar Co Ltd, Mumias, Western Province, Kenya

Introduction

The Mumias sugar scheme in western Kenya attracts interest because of its dependence on sugarcane grown by small scale farmers. The factory expects to grind 1.55M tonnes of cane this year of which 1.37M tonnes will be supplied by farmers whose average cane holding is 1.34 ha.

The scheme is located at Mumias in the Kakamega District of Western Province about 400 km north west of Nairobi. A pre-feasibility study was carried out in 1964-65. In 1967, a detailed feasibility study and pilot project was begun by Booker Agricultural and Technical Services (now Booker Agriculture International) under contract to the Kenya Government. The objectives of the project included the creation of wage earning employment; creation of an import-saving industry; achievement of self-sufficiency in sugar production, and the provision of a further source of cash income for farmers.

Before the start of the scheme, the Mumias area was seriously under developed. Land utilization was poor. Farmers grew small areas of subsistence food crops but bush and rough grazing predominated. The relative remoteness of the area and poor communications had prevented the development of an active market economy. However, one factor favoured the proposed development. Land adjudication had been carried out and the farmers had freehold title to their land. The average size of a holding is 4 ha, which is relatively large for western Kenya and reflects the historically unprofitable nature of the farming activity in the Mumias area. The combination of rainfall, altitude and soil is not particularly favourable for traditional cash crops but is reasonably suitable for sugarcane. The average annual rainfall of about 2000 mm is uniformly distributed. The mean maximum temperature is 29.4°C and the mean minimum temperature is 14.1°C. The low average temperatures discourage rapid cane growth but sunshine receipts are high which favour the production of high purity juices throughout the year. The soils have poor reserves of plant nutrients and cane responds well to fertilizer application.

Outgrower Farming Systems

Under these circumstances, the farmer could offer his experience as a cultivator, his labour and that of his family, and his land. He had no machinery nor cash resources with which to purchase agricultural inputs. The company resolved these problems by providing the machinery for land preparation, supplying fresh, disease free planting material and supplying the farmer with the necessary fertilizers.

In turn the farmer and his family would plant and cover the cane setts, apply fertilizer and maintain the crop in a weed-free condition.

Any grower wishing to become a cane farmer is required to register with the company's outgrower services section. The land is inspected for suitability for cane growing. In addition to a visual assessment of the land, it is neccessary to ensure that if the plot is not of sufficient size to constitute a field, then other adjacent farmers must be willing to be associated with the plot. The minimum size is 6 hectares and few single owners can meet the requirement. A minimum plot size of 1 hectare is the objective, but frequently less is accepted. The necessity for these minima is that for a field of less than 6 hectares, the operations of land preparation equipment is uneconomical. It should be noted that although a field may be comprised of the plots of several owners, the plots are distinct, and the growing of the cane remains the job of the individual plot owner. It has been considered important to ensure that a farm is not completely contracted for cane and that the farmer retains an area large enough to produce sufficient food crops to feed his family.

At the outset of the scheme the credit was supplied by the company, but in 1975 the Mumias Outgrowers Company (MOCO) was formed whose principal objectives were to represent the interests of the outgrower with Mumias Sugar Company, Government and other outside bodies; to negotiate on behalf of grower members the terms for the supply of sugar cane and services from the company; and to finance outgrower credit.

When a field has been accepted for cane growing, the individual plot owners become members of MOCO, sign cane farming contracts and become contracted outgrowers. The contract sets out the obligations and rights of the farmer on one hand and of the sugar cane company on the other.

Land Preparation and Cultivation

The owner of a plot is obliged to remove bushes and stumps. The company undertakes the ploughing, harrowing and furrowing necessary to create a proper seed bed for the cane. Having prepared the land, the company delivers seed cane and fertilizers to the field. The farmer organizes, under the guidance of a company-employed headman, the planting and fertilizing. Later work consists of a top-dressing of fertilizer and manual weeding, and is the responsibility of the farmer. The company outgrower's staff in the field keep a very close eye on the planting operation and subsequent cultivations to ensure that they are performed in a timely and thorough manner.

Harvesting and Transport

The climatic conditions at Mumias are such that there is a long planting season extending from early March to the end of October. Planting in November-February is usually avoided because one or more months in this period may be dry.

The harvesting of the cane is performed by the company, as is the cane transport. All cane cutting is manual and outgrower cane is cut green. The cutters stack the cane in heaps of a size calculated to make 6 tonnes of cane. Cane bundles are tagged at cutting to identify the owner. These tags are attached to weighbridge tickets as the trailers enter the factory. Payment is made only on the basis of weight of cane delivered.

Ratoon Cultivation

For the ratoon crops, cultivations are fairly simple. The trash has to be placed in the interrow to allow the cane shoots to develop. Fertilizers are applied. The trash blanket greatly limits weed growth, but some manual weeding is neccessary. The farmer is responsible for trash lining, fertilizer application and any necessary manual weeding. The company supplies the fertilizer. All services and goods supplied by the company are charged to the farmers account, and the cost plus interest is recovered at the time payment for the cane is made.

It will be seen that the system is simple. It is deliberately kept this way, and part of the success of the scheme is due to its simplicity. The system is also deliberately labour intensive and it is necessary to ensure that the farmers return represents as far as possible a return for his own labour, and is not reduced by purchase of inputs which would substitute for his own labour.

The normal crop cycle is plant plus two ratoons which may be spread over a five or six years period. The harvest age of plants is generally in the range of 19 − 26 months, and of ratoon cane in the range of 17 − 23 months.

Financial Returns

Cane productivity has been good and even in bad years there has been a positive income for the farmer. Net payments to farmers for 1983 totalled Ksh 128M.

Table 1   Comparative yield and income data are given.

	1979	1980	1981	1982	1983
Area under cane (ha)	18 500	21 300	24 300	27 400	30 300
Cane yield (tc ha^{-1})	150.0	154.0	104.3	72.0	75.5
Cane productivity (tc ha − month $^{-1}$)	5.50	4.95	3.98	4.06	4.40
Net income Ksh ha^{-1} month^{-1}	350	292	262	271	415

Community Benefits

The Mumias scheme has made a tremendous impact on development in western Kenya. Throughout the development period more than 3100 housing units have been constructed on five separate sites on local communities. Major roads have been improved in these areas and an integral part of the development was the construction of a network of 316 feeder roads and 200 km of access roads and tracks to facilitate cane, personnel and product transport.

MSCO currently employs 5000 permanent workers with a further 9000 employed on a contract basis to cut outgrower cane. Taken together with the over 22 000 contracted farmers and other persons engaged in spin-off activities, it is conservatively estimated that, today, there are upwards of 350 000 persons dependent on the Mumias Sugar Scheme in some way for their livelihood.

The Mumias scheme represents a massive Government investment in land, equipment and people with total assets today in excess of Ksh 1200M and is, undoubtedly, one of the most successful agro-industrial ventures in the country today making a vital contribution to the economy.

Key Factors in Successful Development

(1) *Careful Project Planning*. The feasibility study and pilot project work were conducted thoroughly over a sufficient period of time to enable a blue-print development to be prepared.

(2) *Indigenous Population*. The indigenous population are essentially farmers. Prior to the introduction of the scheme they had little awareness of science based agriculture, and no resources to practice it. But being farmers they have responded well and willingly adopted the new farming practices which have been introduced.

(3) *Relationship between Government, Company and Management*. The responsibilities of the various parties involved have been clearly defined. Policy is primarily determined by the Government as the majority shareholder. Implementation of policy is clearly the responsibility of the contracted managers. Monitoring of management performance is the responsibility of the board.

(4) *Adequate finance*. The flow of funds to finance the project development has in general been adequate and timely. There has been no interruption to the development programme because of fund shortages, although minor adaptations have been necessary from time to time.

(5) *Profit-based operation*. The organization of outgrower development relates returns to the individual farmer directly to the individual farmer's success as a cane grower.

(6) *Control of Mumias Sugar Company*. The company's interest is ensuring in both the short and the long term that cane sufficient to fill the factory capacity is available. The orderly development of outgrower farming by the contract system helps to ensure this; close control by the company ensures that a very large proportion of the farmers produce good crops which are profitable, and this in turn ensures long-term cane supply prospects.

Possible Future Developments

It is considered that the scale of present sugar production operations cannot be significantly expanded. The collection radius for cane, given the existing road infrastructure, is as large as is economically viable.

It will be necessary for the outgrowers scheme to adapt to changing conditions in the 1980's. This may require some amalgamation of holdings so that the average cane plot size is increased beyond the present 1.34 ha level.

It may also be necessary, because of a diminishing availability of labour, to introduce a greater degree of mechanization. In any event, it is believed that the scheme, having been soundly established, will be able to adapt as necessary to meet changing external conditions.

Chapter 27

New Varieties to Combat old Problems in Arabica Coffee Production in Kenya

D.J. WALYARO, H.A.M. van der VOSSEN, J.B.O. OWUOR and D.M. MASABA

Coffee Research Station, Ruiru, Kenya.

Introduction

Coffee is one of the most valuable export commodity produced by the developing world. Africa, the homeland of most *Coffea* species, produces almost 30% of all the world's exported coffee (Anon 1983) and is likely to continue to be a major producer. However, in virtually all the arabica coffee producing countries in Africa, two diseases have had devastating effects on this tree crop. The two diseases, coffee berry diseases (CBD) caused by *Colletotrichum coffeanum* Noak (*sensu* Hindorf) and coffee rust *Hemileia vastatrix* B. & Br. are still regarded as the most serious constraint to arabica coffee production. This paper gives an account of the efforts in Kenya to combat these disease problems through breeding for new varieties.

Arabica coffee is the leading foreign exchange earner in Kenya accounting for between 25 – 30% of the total domestic exports. Coffee berry disease in particular has had the most traumatic effect on the coffee industry in Kenya. During the very wet years 1961–62 and 1967–68, it was estimated that Kenya lost between 40 – 50% of the total crop due to CBD (Wallis & Firman 1967, Griffiths & Gibbs 1969). The Coffee Research Foundation (CRF) however, has continued to work towards providing the coffee growers with more effective and economic chemical disease control measures enabling arabica coffee still to be grown in Kenya today. Nevertheless, growing coffee is becoming less profitable even to the large scale growers who spend 30 – 35% of their total production cost on CBD control. For the small-holders who produce more than 60% of the exportable coffee, the recommended disease control measures are simply beyond their capability financially and technically. On the other hand, the control of CBD and rust provides a major market for pesticides in Kenya (Njagi 1982); coffee is the leading consumer of pesticides, most of these being fungicides. The importation of such quantities of fungicides can clearly be viewed at the national level as a major drain on the country's foreign exchange.

Because of these disease problems, the CRF and the Ministry of Agriculture decided that as an alternative to the chemical control measures, efforts should be made to develop new arabica varieties that are inherently resistant to both diseases. The breeding programme was initiated in 1971 under a bilateral technical aid programme between the Government of the Netherlands and the Kenya Government. The aim of the programme was to breed new varieties of arabica coffee that combine resistance to CBD and rust with improved yield and quality. This programme also provided possibilities of breeding for more compact (dwarf) plant types that are better adapted to intensive coffee planting.

The Breeding Programme

Germplasm screening

The starting point in the improvement programme was the screening of germplasm available on the Coffee Research Station based primarily on the following characters: resistance to CBD and rust, productivity, bean size and cup quality, and compact growth characters. Although a number of accessions were found that possessed a high level of resistance to CBD and rust, no single accession was found that possessed all the desired characters at an acceptable level. Table 1 gives a summary of the characteristics of a few of the varieties that were eventually selected as progenitors in the improvement programme.

Table 1   A summary of the performance on basis of yield quality and disease resistance for a number of important varieties.

Parent varieties	Annual yield first cycle clean coffee (t ha⁻¹)	Coffee Quality		Resistance in the field[3]	
		Bean size AA[1] (%)	Liquor St[2] (0–7)	CBD (%)	Rust (0–10)
SL28	3.6	41	2.0	S	S
Caturra	2.7	15	3.8	S	S
Pandang	3.9	37	3.6	MR	S
Rume Sudan (RS)	0.9	3	3.3	R	MR
Hibrido de Timor (HT)	1.7	39	4.0	R	–

(1) Bean grading: AA = fraction of heavy beans retained on No. 18 (1.15 mm) screen.
(2) Liquor quality St = Overall standard (0 = fine; 7 = very poor).
(3) Disease Resistance (average per progeny). CBD scored as % infection. Rust scored from 0 (no infection) – 10 (maximum infection). S = Susceptible, R = Resistant.

Rume Sudan and Hibrido de Timor proved to be the best sources of CBD resistance; Hibrido de Timor was also a good source of rust resistance. These two varieties have since been extensively used in the breeding programme. On screening the Ethiopian collection that became available later on the Station, again no genotypes were found that could be used directly as new varieties under the Kenyan situation although some possessed resistance to CBD and rust. These have already been crossed with local cultivars to broaden the genetic base of our new population especially with regard to disease resistance.

Development of elite breeding populations

In order to combine disease resistance, compact growth, improved yield and quality in individual plant genotypes, advanced breeding procedures had to be applied (van der Vossen & Walyaro 1981, Walyaro *et al.* 1982) to produce a plant type suitable for the exacting requirements of the Kenyan coffee industry. The first phase of the breeding programme led to development of two distinct breeding materials which now form the parent populations for the new coffee variety that will be produced for commercial growing.

The first type of material consists of outstanding selections that were developed from the initial main breeding programme. These are tall types combining resistance to CBD and rust with improved yield and quality and were obtained using such methods as single crosses, multiple crosses and eventually backcrossing. Plants produced by these methods are heterozygous especially at gene loci conditioning disease resistance. Because arabica coffee is largely self-pollinating, the only rapid way of large scale propagation of such types would have been through some form of vegetative propagation.

The second type of breeding material known as Catimor had been derived through pedigree breeding and was obtained through an exchange programme with Colombia. Catimor types have a compact growth habit, good yield and broadly based disease resistance. This was further improved, mainly through progeny testing to select genotypes that are homozygous both for compact growth and for resistance to CBD and coffee rust.

On examining the results of the entire breeding populations, it became obvious that large scale multiplication of the new coffee varieties would be done best by artificially cross pollinating the two populations to produce hybrid seeds rather than by vegetative propagation. The hybrid progenies would provide a population of uniformly compact trees which would combine not only improved productivity and quality but also a broad spectrum of genes conditioning resistance to CBD and rust.

Fundamental Research as an Aid to the Breeding Strategies

Selection methods and the nature of disease resistance

Selection for CBD resistance in the breeding programme was greatly improved only after the development of a reliable preselection test for resistance performed on 6 week old seedlings (van der Vossen *et al.* 1976, van der Vossen & Waweru 1976). Furthermore, the identification of genes conferring resistance to CBD in the important progenitors, Rume Sudan (2 genes on the R– and K– loci) and Hibrido de Timor (one gene on the T- locus) (van der Vossen & Walyaro 1980) have led to further refinement of the breeding programme to ensure that all the 3 different genes are represented in the new improved material.

No differential pathogenicity has been found among hundreds of isolates of the CBD pathogen tested to date, including those from variety plots but differences in aggresiveness have been noticed. Van der Graaff (1981) also found the same situation with isolates collected in Ethiopia. This on its own however does not necessarily imply that this is a durable type of resistance, but the recent histological studies carried out at Ruiru (Masaba 1982) have indicated that the defence mechanism in the host which involves the formation of cork barriers soon after infection, is a type that may not be easily overcome by changes in the pathogen. Thus, although resistance is likely to be of a stable nature, pyramiding of all the available resistance genes as is being done in our breeding programme, may be one important way of enhancing this stability (Parlevliet & Zadoks 1977, Nelson 1973).

In the course of the improvement programme, it has also been possible to screen for rust resistance more effectively by use of the leaf disk inoculation test (Owuor 1980, 1983). Using this procedure, it is also possible to differentiate reaction due to incomplete resistance from that of the hypersensitive type. If for example a true source of horizontal resistance is identified, it would be possible using such a procedure to combine both horizontal and race specific types of resistance within a single population (Walyaro 1983).

Quantitative characters

Yield, bean size, and cup quality in coffee are all characters of this nature. Biometrical genetic studies (Walyaro & van der Vossen 1979, Walyaro 1983) have provided valuable information on how certain characters can be used to improve the efficiency of selection for yield and quality and have helped to elucidate the effects of genotype-environment interactions. For instance it has been demonstrated that in arabica coffee heterozygosity imparts improved stability in differing environments. It was also concluded from such studies that improvement of bean size and cup quality characteristics is an objective that can be easily attained through proper breeding techniques.

Hybrid varieties in arabica coffee

It has been pointed out (Walyaro 1983) that there are considerable benefits to be gained from use of hybrid varieties as compared to inbred line varieties. In this breeding programme, the first type of selected breeding material has already been progeny tested for CBD resistance. Most has 2 or even 3 CBD resistance genes, though in a heterozygous form. The liquor quality is also generally superior to the Catimor progenies. In the Catimor material the CBD resistance is derived from Hibrido de Timor and is known to be governed by only a single gene (van der Vossen & Walyaro 1980). However if the selected genotypes from the first population with for example RRKktt or RRKkTt gene combinations, are crossed to the Catimor material homozygous for the T gene it can be shown in principle that 75% of the progeny population would carry CBD resistance genes on 2 or 3 loci. In addition, there would be no susceptible plants in such F1 progenies. Such a mixed population should probably give adequate protection against CBD epidemics. The results of CBD preselection seedling tests performed on progenies of the hybrids show a very low percentage of susceptible seedlings with a mean CBD score well within the resistant category. This confirms that hybrids derived according to our crossing scheme can withstand even the most severe CBD epidemics and can therefore be safely used to establish new coffee plantations. It is also possible to derive a population with a higher level of CBD resistance if all the genotypes in the population were to carry at least 3 CBD resistance genes as in a scheme such as that proposed by Walyaro (1983).

Resistance to coffee rust in Catimor progenies is governed by the S_H6 gene derived from Hibrido de Timor which is known to confer resistance to most of the known rust races (Bettencourt 1981). Among the other breeding populations, a number of genotypes have been found to exhibit an apparent horizontal type of resistance to rust i.e. a susceptible reaction but a long latent period and reduced sporulation (Owuor 1983). Hybrids obtained from crossing two such populations could therefore have more durable resistance to coffee rust.

Finally, there is a more fundamental argument in favour of hybrid seed production. Although the hybrids will be all resistant to CBD and rust and have the required phenotypic uniformity in plant type because the compact character is dominant, they will represent a fairly heterogenous populations on basis of the genotypes. Such populations have considerable advantages over homozygously uniform populations as they tend to be well buffered against environmental variation and may exhibit considerable heterosis if the two parental populations happen to represent genetically divergent populations (Walyaro 1983).

Multiplication of the New Disease Resistant and Compact Varieties

At present, a number of investigations are being conducted on various aspects related to large scale multiplication of the new plant material through production of seeds for hybrid varieties.

The aim is that by 1986 the first batch of seeds should be available and the seedlings ready for field planting would be sold to the growers in 1987. Since 1981 research in the Breeding and other Sections, has been reoriented to meet these targets.

The establishment of a seed garden where all the hybrid seeds will be produced is now complete. According to our projections, the garden should be able to cope with the national demand for seedlings of the new variety. This is estimated at between 10–18M seedlings per annum, assuming a replacement period of between 15 – 20 years.

A number of trials with disease resistant and compact materials have been established, and others are planned. Information from such trials will form the basis for tentative recommendations on spacing, fertilizer requirements, pruning systems and insect control measures appropriate for the new variety.

At the same time, the first series of hybrids referred to as the 'CBU hybrid' are undergoing thorough evaluation in adaptation trials aimed at verifying the performance of these hybrids in all the arabica growing zones in Kenya. These trials are situated on the CRF substations and a number of large estates situated in widely differing ecological zones. Trials with the same hybrids are also to be established on small-holders' plots during the course of 1984 in practically all the coffee growing districts in Kenya in conjunction with the Ministry of Agriculture extension staff.

Finally, owing to the complexity of the operations needed to ensure production of a large enough quantity of hybrid seed, a number of investigations on practical aspects associated with such operations are being carried out. These include isolation methods, bulk methods of pollen handling, and the labour requirements and rate of performance for emasculation and pollination on a fairly large scale.

Conclusion

The coffee breeding programme in Kenya is now set to achieve its goal of developing material that is resistant both to CBD and rust and of a compact growth type within a remarkably short period of time. It has been stated (Njoroge *et al.* 1981) that the gradual replanting of Kenya coffee orchards with the new varieties is likely to have considerable impact on the coffee industry and the national economy as a whole. Not only will the downward trend in profitability of coffee growing be reversed but the competitiveness and viability of the Kenya coffee industry will be assured. In addition it will save the country millions of pounds in foreign exchange spent on importation of fungicides, spray machinery and fuel. Furthermore there is evidence to suggest that the much closer spacing of compact varieties could easily result in doubling the yield per hectare. However, since the amount of Kenya coffee that can be sold is restricted under the present ICO quota system, it would be unrealistic to aim at production targets of double the present amount. However, producing an equivalent of the present Kenya export quota on half the area presently under coffee would release the rest of the fertile land under coffee for alternative farm enterprises, especially production of food crops. This aspect is of crucial importance both at the farm level, at a time when there is growing demand for more fertile land to grow food crops, and to the nation as a whole in striving to restore self sufficiency in food production.

References

Anonymous (1983) Africa coffee. *Bulletin of the Interafrican Coffee Organization* 1 (September 1983).

Bettencourt, A.J. (1981) *Melhoramento genetico do cafeeiro transferencia de factores de resistencia a* Hemileia vastatrix B. & Br. *para as principais cultivares de* Coffea arabica *L.* Lisboa; Junta de Investigacoes Cientificas do ultramar, 93 pp.

Griffiths E.; Gibbs, J.N. (1969) Early season sprays for control of coffee berry disease. *Annals of Applied Biology* 67, 45-74.

Masaba, D.M.; van der Vossen, H.A.M. (1982) Evidence of cork barrier formation as a resistance mechanisms to coffee berry disease(*Colletotrichum coffeanum*) in arabica coffee. *Netherlands Journal of Plant Pathology* 88, 19-32.

Nelson, R.R. (1973) *Breeding plants for Disease Resistance. Concepts of applications.* Pennsylvania State University Press.

Njagi, S.B.C. (1982) Economics aspects of CBD control. In *First AAASA/EEC Regional Workshop on Coffee Berry Disease.* Addis-Ababa.

Njoroge, I.N.; Njuguna, S.K.; Sparnaaij, L.D.; van Santen, C.E. (1981) *Final evaluation of the Coffee Breeding Project 1971 – 1981.* 37 pp.

Owuor, J.B.O. (1980) Selections for coffee rust using the leaf disk inoculation test. *Annual Report for 1979/80, Coffee Research Foundation, Kenya* 6, 5-66.

Owuor, J.B.O. (1983) Selection for resistance to coffee rust *Hemileia vastatrix* B. et Br. in the breeding programme for resistance to coffee berry disease in Kenya. In *Symposium on Coffee Rusts (CIFC), Oeiras, Portugal, 17 – 20 October 1983.*

Perlevliet, J.E.; Zadocks, J.C. (1977) The integrated concept of disease resistance, a new view including horizontal and vertical resistance in plants. *Euphytica* 26, 5-12.

van der Graaff, N.A. (1981) Selection of arabica coffee types resistant to coffee berry disease in Ethiopia. *Mededelingen Landbourwhogeschol Wageningen* 81-11.

van der Vossen, H.A.M.; Cook, R.T.A.; Murakaru, G.N.M. (1976) Breeding for resistance to coffee berry disease caused by *Colletotrichum coffeanum* Noak (*sensu* Hindorf) in *Coffea arabica* L. I. Methods of preselection for resistance, *Euphytica* 25, 733-745.

van der Vossen, H.A.M.; Waweru, J.W. (1976) A temperature controlled room to increase the efficiency of preselection for resistance to coffee berry disease. *Kenya Coffee* 41, 164-167.

van der Vossen, H.A.M.; Walyaro, D.J. (1980) Breeding for resistance to coffee berry disease in *Coffea arabica* L. II. Inheritance of the resistance. *Euphytica* 29, 777–791.

van der Vossen, H.A.M.; Walyaro, D.J. (1981) The coffee breeding programme in Kenya. A review of the progress made since 1977 and plan of action for the coming years. *Kenya Coffee* 46, 113-130.

Wallis, J.A.N.; Firman, I.D. (1967) A comparison of fungicide spray volumes for the control of coffee berry disease. *Annals of Applied Biology* 59, 111-122.

Walyaro, D.J. (1983) *Considerations in breeding for improved yield and quality in arabica coffee* (Coffea arabica *L.*). Thesis, Agricultural University of Wageningen, 118 pp.

Walyaro, D.J.; van der Vossen, H.A.M. (1979) Early determination of yield potential in arabica coffee by applying index selection. *Euphytica* 28, 465-475.

Walyaro, D.J.; van der Vossen, H.A.M.; Pwuor, J.B.O. (1982) Breeding arabica coffee in Kenya for resistance to coffee berry disease. In *First Regional Workshop on Coffee Berry Disease. AAASA/EEC.* Addis Ababa.

Chapter 28

Tea: Quality, Processing and Smallholder Production

R. LEWIS

Commonwealth Development Corporation, P.O. Box 43233, Nairobi, Kenya.

Introduction

This is a large subject, so I shall confine myself strictly to one very successful smallholder tea development project and the apparent reasons for its success, the smallholder tea industry in Kenya under the Kenya Tea Development Authority (KTDA).

Kenya Tea Development Authority

This began with a pilot scheme on about 600 ha in the early 1950s. The responsibility for the fledgling smallholder tea industry was assumed by KTDA on its incorporation in 1964 under a Board whose Chairman is appointed by the Minister of Agriculture and members represent the Tea Board of Kenya (the controlling body responsible for promoting Kenyan teas), Ministry of Agriculture, external financiers, and elected smallholder representatives.

Planning and achievements

Four carefully designed 6-7 year plans have been implemented and the development phase is now drawing to a close, with the last six of KTDA's 39 factories to be commissioned in the first half of 1984. To-date 145 000 growers have planted almost 55 000 ha of tea, an average plot of 0.38 ha yielding 946 kg of made tea per mature ha in 1982/83. KTDA is now the largest single tea growing organization in the world, and will soon produce more than half of all Kenyan teas.

Planting programmes

KTDA prepared planting programmes for each of 12 tea growing districts based on the area of suitable land on which growers wished to plant and the available supervisory staff, in units to provide sufficient leaf for factories requiring $5.2 - 6.5M$ kg green leaf each year. Estimates were originally based on an expected average yield of 1120 kg made tea ha^{-1} in year 6, a target yet to be achieved.

Selection of growers

A licence is obtained from the Tea Board to develop a stated number of hectares in a scheduled area and planting licences are then issued to individual growers with land titles. Growers are selected only within a line demarcating potential tea areas according to climate, topography and soils described in a Department of Agriculture Survey of 1964 ("Brown's Tea Line"). A KTDA team carries out soil tests in areas where soils are variable. Other criteria to be satisfied are accessibility for leaf collection and proximity to a factory site.

Statutory control

The Tea Cultivation Order 1964 banned planting by non-registered growers; originally, offenders were prosecuted and even jailed, but a more conciliatory approach was adopted later. The same Order prescribed land preparation, planting, pruning and plucking practices to be adopted according to the instructions of KTDA field staff.

The extensionist and tea plot size

Field supervision and the maintenance of grower records is provided by Tea Officers (one per District) and their Agricultural Assistants (roughly one per hundred growers), seconded at KTDA's cost from the Ministry of Agriculture. The Agricultural Assistant advises the new grower where to plant, how to prepare the land, and what soil conservation measures to take. As very small plantings are uneconomic to supervise and give a poor return, there is a minimum of 0.1 ha and growers are encouraged to plant 0.4 ha over one or several seasons. There is no maximum provided that the grower has sufficient labour to maintain the tea alongside his subsistence crops.

Planting materials

Initially, seedling stumps were issued to growers from KTDA nurseries, but costs were reduced by 75 % after 1967/68 with the introduction of vegetatively propagated material obtained from clonal bushes selected for high quality as well as yield. Growers are now issued with a kit consisting of 1200 polythene sleeves and the same number of single node cuttings, $\frac{1}{2}$ kg bags of SSP and NPK fertilizer, and a large polythene sheet under which the cuttings are rooted. Overhead and side shade is provided by the grower. The grower is expected to obtain at least 800 plants net of deaths and infills. Four such kits will plant 0.4 ha at the recommended spacing (7173 to 8611 ha^{-1} depending on location). These materials are supplied to the grower on credit along with windbreak plants (*Hakea saligna*) if considered necessary. The credit, along with a development charge to cover supervision costs, was to be recovered through a Capital Cess; KTDA does not give cash advances.

Cultural operations

Extension staff demonstrate planting to new growers at the start of the main rains, and later also the other operations necessary to bring the tea to bearing: artificial shade for the first six months, manual weeding and mulching, and initial prune at 15 cm, frame formation by pegging, tipping at 60 cm and then pruning every third year (commonly carried out by skilled pruners on contract). No fertilizer is given until after the tea is mature when a Fertilizer Credit Scheme and other measures ensure the grower obtains suitable fertilizer on time at the lowest possible price. In some years KTDA has made arrangements with local dealers, but when KTDA has access to the necessary funds it imports and distributes directly, for example 10 000 t of 25:5:5:5 S in 1983. A low application rate (90 kg N ha^{-1} +) is recommended to ensure good returns and high quality leaf.

Extension and training

Extension staff visit growers at least quarterly and keep records of recommendations given. The quality of plucking and maintenance is judged and an aggregate score calculated which assists both the extensionist/grower relationship and KTDA's supervision of the extensionist. The best growers are rewarded with a cash prize and a cup. Training is provided through one – week courses at KTDA's training farm school. Transport, board and lodging are provided at nominal charge and courses are popular. To overcome communication problems emphasis is placed on the practical side, especially plucking. Junior extension staff are trained on one month courses and their confirmation of appointment depends on performance.

Leaf inspection

In order to maintain quality, strict standards of plucking are required and the inspection and delivery of leaf is the responsibility of the Leaf Officer (one per factory) and Leaf Collectors who are direct employees of KTDA. Growers carry their leaf up to 4 km to Buying Centres, built by growers on a self-help basis with a small KTDA grant for specified materials. A 0.5 kg sample of leaf is sorted by the Leaf Collector into grades according to its quality and condition. Two leaves and a bud are the quality standard. If more than 25 % of the leaf is coarse, loose, bruised, or stalky, the grower is required to resort it. Badly fermented samples are also rejected.

Packing of leaf

The weight of accepted leaf is recorded on the grower's card and a receipt given. At this point the grower is guaranteed payment and the risk passes to KTDA during transport to the factory. Leaf is repacked in sacks of 12 kg which on the arrival of the leaf collection vehicle, are hung from racks to ensure good air circulation, and avoid bruising and crushing which cause premature fermentation. Other methods of packing, for example in wire cages, were found more costly and less effective.

Delivery to the factory

Based on the quantity of leaf available, Buying Centres are opened according to a schedule worked out by the Leaf Officer and communicated to all growers. Leaf collection is administered by the Leaf Officer from a Leaf Base adjacent to the factory compound. In addition to his office there are parking and basic maintenance facilities for vehicles and housing for a mechanic, drivers and clerks. Lorries designed to carry 200 sacks make a round of the Buying Centres dropping off Leaf Collectors and empty sacks on the way out, and bringing back staff and filled sacks on the return journey. Careful planning is required to fill the lorries as fully as possible. High vehicle running costs have been a cause of concern to KTDA; tachographs enable a 30 mph speed limit to be enforced and a driver bonus scheme is operated to encourage good maintenance. Several types of vehicles were tried, including landrovers and tractors with trailers, but two-wheel drive diesel lorries (fitted with chains in wet weather) were found most cost effective. In 1982/83 KTDA operated a fleet of 285 lorries over 3.9M km to collect leaf from 870 Buying Centres and deliver it to 35 factories.

Factory operations and quality control

On arrival at the factory weighbridge the leaf is inspected. If accepted it becomes the property of the factory and is weighed, unloaded and transferred to the withering troughs. Green leaf is wilted in troughs under forced air draft for 10-12 h, then delivered to rotorvane and CTC (Cut-Tear-Curl) machines which cut the leaf into uniformly sized particles, and by breaking the cell walls allow the process of aerobic fermentation by leaf enzymes to begin. During fermentation polyphenols in the leaf are oxidised and the leaf darkens. Temperature (25-30°C), airflow and the period of fermentation (approximately 1 ½ h) are all critical to the production of high quality black teas. Continuous monitoring with airflow meters and thermometers is practised and the leaf turned several times.

Fermentation is brought to an end by "firing" the tea; oven drying at 49-52°C for about 16 min to about 3 % moisture content. Again temperature is critical to quality.

The remaining operations are designed to remove lumps, fluff, fibre and other impurities (by the use of polythene wrapped rollers and a winnower); to sort it into grades (by the use of vibrating screens), and finally to pack it into moisture-tight tea chests. Samples of the teas are examined, brewed, and tasted to detect any faults arising in the various stages of manufacture.

Results

KTDA factories achieve some 87-90 % of production of first grade teas (Broken Pekoe 10 %, Pekoe Fannings 50 %, Pekoe Dust 25 %, Dust 12 % approximately) which command premium prices in the world market. The average conversion ratio of green leaf to made tea is about 22 %. Because of substantial savings in processing costs (to 50 %) and foreign exchange, KTDA is converting its factories from oil- to wood-firing as quickly as it can locate suitable sources of firewood; seven have now been converted.

Marketing

KTDA's record for quality is good, and last year KTDA teas fetched a 3 % premium over the average of all teas on the world market. KTDA's teas are auctioned regularly in Mombasa and London. 15 % is, by Government order, marketed for local consumption. The total value of KTDA teas sold in 1982/83 was K£ 48M.

Grower payments

The weight of leaf delivered to the factory is reconciled with the weights purchased from individual growers. At KTDA Head Office the data from growers' cards is transferred to punched computer cards and the amount owing to the grower calculated after deductions. At the end of the month the Leaf Officer draws cash or cheques to pay his growers their First Payments along with a payment advice slip.

Factory management

Initially smallholder leaf was sold to commercial tea companies for processing. In the second phase, alongside commercial partners, KTDA built its own factories with management provided by commercial agents on a commission basis. Latterly, KTDA has designed, constructed and managed factories in which it is sole shareholder. KTDA now manages and acts as selling agent for all its factories.

Ownership of factory companies

Each factory is a separate financial entity, with initial share and loan capital subscribed by KTDA often alongside the Commonwealth Development Corporation and commerical partners. As long-term loans are repaid, KTDA-registered growers are invited to buy into the companies. Take up by growers has been fairly slow, but at June 1983 had reached 2.2M shares (K£ 11.1M) by 20 468 grower-shareholders in 23 factory companies.

First payment to growers

The factory company enters into a "Green Leaf Agreement" with KTDA to buy a specified quantity of leaf and to make a down payment to KTDA on receipt of leaf, at a fixed rate per kg for all factories. After deductions of cesses, KTDA pays the grower a First Payment, originally quarterly but monthly since 1965.

Cesses

In addition to recovering debts under the Fertilizer Credit Scheme, two types of deductions were previously made at a fixed rate per pound to cover KTDA's operating costs. A Revenue Cess, designed to meet KTDA's recurrent costs, was paid by all growers whilst a Capital Cess was paid only until the grower had refunded KTDA the cost of planting materials, supervision and its own capital development, normally about ten years. The Capital Cess was poorly understood and unpopular with growers so the two were combined into a consolidated cess, which also prevented growers avoiding payment of the Capital Cess by selling on the card of a neighbour who had completed repayment. It had the disadvantage of not being entirely equitable amongst all growers. The rate was to have been reduced but inflation has necessitated increases. KTDA's total income from cesses in 1982/83 was K£ 3.15M from 206M kg green leaf.

Second payments to growers

At the end of each year the tea company Boards authorise a final payment to KTDA which is the distribution of profits after appropriation to reserves, loan repayments, and an 8 % dividend on share capital. The sum available depends on efficiency, throughput, and prices realised so the rate per kg will vary from one factory to another. At one time these were averaged over a group of factories before distribution to growers as a Second Payment to avoid problems when growers were required to switch delivery of their leaf from one factory to another. KTDA has now reverted to payments on an individual factory basis with some compensation for newly opened factories. In 1982/83 Second Payments ranged from KShs 1.64 to 3.21 kg^{-1} green leaf compared to a First

Payment of KShs 1.11 kg^{-1} green leaf. Until recently KTDA maintained a Green Leaf Price Reserve but this price stabilisation mechanism has also been dropped in favour of exposing the grower to the full effects of fluctuations in the world market price, a policy for and against which strong cases can be argued.

Problems faced by KTDA and the Industry

Irregular and underplucking by growers

When growers pluck infrequently quality is reduced by coarse, stalky leaf and more frequent pruning is necessary. The commonest cause is a labour shortage due to children being at school, and others are commitments such as coffee harvesting, and religious and other festivals. A lack of supervision by absentee growers and the use of tea as a "bank" can also cause underplucking. Further problems arise at the factory as daily deliveries vary by a factor of three with surges at month-ends and when school fees are payable. This results in poor utilization of factory capacity and increased costs. Growers have resisted encouragements to even-out deliveries. Congestion also occurs daily as deliveries are concentrated into the early evening, risking overheating and uneven withering. Early delivery of part of the crop increases costs due to inefficiency in collection.

Tea roads

Construction, upgrading and maintenance of tea roads is a costly obligation of Government and has long been a source of concern to the industry. The recommendation by consultants, financiers, growers and KTDA that all roads be improved to an all-weather standard was accepted by the Government but progress in implementation has been slow.

Shortage of revenues

Sharp rises in the costs of fuel, spares and exchange losses on foreign exchange loans have increased KTDA's costs. To improve KTDA's revenue base substantial increases were allowed by the Government in 1983 in KTDA's factory management commission (3 to 5% of sales value), in cess income (31 to 38 cents kg^{-1} green leaf), and interest rate on its loans to factories (8 to 12.5% y^{-1}). There is also a need for KTDA to reduce costs by a rationalisation of its technical staff now that the development phase is ending.

Yeilds

Sustained efforts are being made to improve yields, an area in which research plays a vital role.

Apparent Reasons for the success of the KTDA Model

Suitable crop

Tea is an excellent smallholder crop as it can provide year-round employment and income, its cultivation requires few expensive inputs (it is remarkably free from pests and diseases in Kenya) and it has a relatively stable world market price in comparison to many tropical commodities.

Careful planning

As KTDA started the industry almost from nothing, it developed free from old inefficient, badly designed institutions and was designed to have the sole direct relationship with the grower and to control almost all the stages. In those few areas which were outside KTDA's direct control, for example research (conducted by the Tea Research Institute) and tea roads, the KTDA has generally received good support.

Control and accountability

KTDA benefitted from strong Government support and avoided political pressure to allow tea to be planted in unsuitable areas, to build factories where insufficient tea was available, and to over-expose itself financially. Decisions were made on a sound commercial basis and a strict administration has ensured careful control of finance and field staff, aspects in which many parastatals have failed.

Estate sector

KTDA derived considerable benefit from the transfer of technology and experience from the estate sector.

Restriction of credit	KTDA considered but rejected the offering of cash credit for tea planting because of the high cost and poor repayment record of similar schemes, because it attracts an undesirable category of farmers preoccupied with credit so that field staff would become swamped and diverted from their key supervising role.
Method of payment	Assured monthly first payments not subject to delay are very popular with growers; second payments are too as the lump-sum can be invested, for example in the purchase of a cow. The KTDA grower is confident that he will be paid a fair price for his crop.
Competent, well-trained field staff	Note that the extension service is not provided free as in many similar schemes.
Participation of Growers	Growers are represented by elected members on the District Tea Committees, Provincial Tea Boards, and KTDA Board. KTDA management and Ministry of Agriculture staff are represented on each of these Committees and Boards which facilitates two-way communication between growers and policy makers and provides a forum for the discussion of problems. The participation of growers as factory shareholders, and their elected representatives in factory Boards, completes the representation of growers at all levels of the industry.
Efficiency	The efficiency of KTDA and the tea factory companies can be measured by the fact that the smallholder receives 72 % of the world market price fetched for high quality teas, despite the fact that a complex processing operation is involved. (The KTDA itself receives another 6 % in cesses; the balance covers manufacturing, transport and marketing cost.) Very few similar industries anywhere in the world could match that record.

Replicability of the KTDA Model

It is unlikely that an identical system could be implemented elsewhere for tea or any other smallholder crop as there are always factors which necessitate modification. Ultimately, KTDA's success is the result of a well-conceived development implemented under good management to produce a high quality product ensuring consistent marketability.

Chapter 29

Coconuts

D.H. ROMNEY

National Coconut Development Programme, P.O. Box 6226, Dar es Salaam, Tanzania.

Introduction

Coconuts have long been regarded as a crop planted by small farmers as a legacy to their grand-children, or grown extensively on large tracts of land without infrastructure. Over recent decades the crop has enjoyed close attention from estate managers and systematic study from scientists. Their experience and results can be used to advance coconut production in Africa.

The coconut crop has certain advantages: (1) Some of the environments where it grows successfully will hardly support any other crop. (2) Its management and labour demands are flexible, and can be satisfied when the demands of other crop and livestock enterprises allow. (3) Edible oil/fat is important in diets to raise palatability, and nutritionally to provide the essential fatty acids, i.e. those not manufactured by the liver. (4) Surplus coconuts can be made by a farm operation into copra, a relatively non-perishable item capable of earning foreign exchange. (5) Where distances to market are long or transport constrained by shortage of foreign exchange, coconut has the advantage that properly made copra is not perishable and has a low weight/value ratio.

Environment

The sunlight, high temperatures and intense rainfall of the tropics cause arable cropping to destroy soil organic matter, and this is accentuated by fire, leading to deterioration in soil structure, permeability and fertility. Properly managed tree-crops are one solution, including coconuts where suitable.

Because coconuts sometimes grow well where many other crops do not, and as small farmers seldom pass judgment on the productivity of their crops (coconuts may yield under suboptimal conditions, but not at an economic level), hence coconuts are sometimes grown in places where returns are poor and where national investment is ill-advised. A coconut development programme should therefore be based upon a land suitability survey such as that carried out by NCDP (1980) in Tanzania.

Development of a coconut industry with techniques which require foreign exchange to provide inputs, either to farmers or to the agency coordinating the development, is likely to fail in a country such as Tanzania. Farming inputs should be of local origin where possible, production targets set realistically low, and development techniques designed with this in mind (see below). The development agency should perhaps concentrate its efforts on target areas likely to yield maximum response; pressures to spread and so dilute efforts should be resisted.

Management at Farm Level

Experience from other countries, for example Malaysia, Sri Lanka, the Phillippines, the Ivory Coast and Jamaica, shows that coconut trees respond to management as any other crop. Where coconuts are grown on poor soils, or under inadequate or poorly distributed rainfall (e.g. Tanzania), agriculturalists have to adapt management methods accordingly. With very small farms, methods of management must be within the capabilities of the operators of such farms.

Fires must be avoided to prevent damage to both soil and trees. Coconut trees seriously affected by fire are frequently seen in Tanzania. Wide spacing may be advantageous under conditions of moisture deficit, but encourages weeds and bush which in turn encourage fires.

Weed control in dry weather to reduce competition for moisture is very effective. Romney (1982) reports yield increases of up to 100% by weed control in coastal Tanzania. Due to the extreme competitiveness of permanent weeds and shrubs in the tropics, inter-cropping is often beneficial to coconuts because weeds are removed to establish the intercrop. In the seasonally dry conditions of coastal Tanzania, short-term crops grown during the brief rainy seasons improve coconut yields by 12–20% (Romney 1982).

Fertilizers and herbicides are used to great economic advantage in coconut territories with no foreign exchange problem or with local resources. However, fair coconut production can be achieved with manual weeding and no fertilizer. Agronomists can play an important part in determining the *limiting* nutrient(s) for each main coconut soil, with the aim of removing a constraint at minimum expense. For instance, although response to phosphate by coconuts is uncommon, Romney (1983) showed that young coconuts on sandy soils formed from Neogene sandstone in coastal Tanzania grow almost 20 % faster with phosphate alone. Incidence of leaf-spot also declined significantly when phosphate was applied.

The coconut varieties grown traditionally are virtually wild. Farmers and Governments in many countries have attempted to improve productivity by selecting mother-palms, but progress has been limited, partly because tall varieties are obligatorily cross-pollinated most of the time, and the pollen source is uncontrolled. F1 hybrids in many countries (e.g. the Ivory Coast, Jamaica, Malaysia, Indonesia, the Phillippines) outyield local talls by 50–100 % (Ollagnier 1983). They are valuable because they come to bear some two years earlier than talls (which facilitates repayment of loans), with smaller crowns (enabling more trees to grow per hectare). De Nuce *et al.* (1983) report that, even in the absence of fertilizer, hybrids are superior to talls. Field trials are in progress in Tanzania (Romney 1983).

In some situations certain F1 hybrids have better disease resistance than talls. In Jamaica, a hybrid which was sufficiently resistant to lethal yellowing disease for commercial planting was found in 1981. Field tests are proceeding in West Africa and Tanzania (Schuiling 1982) with the object of finding hybrids or varieties resistant to lethal disease there.

As coconut cultivation is likely to be relatively extensive in countries such as Tanzania, a lower level of resistance to diseases or pests may be acceptable to the farmer than in countries where land is valuable and coconut farming needs to be efficient to compete.

There is a risk that uninformed farmers will take seeds from F1 hybrids and plant the F2 with disappointing results. It is probable with coconuts, provided that the F1 does not carry some vital disease resistance factor, that the F2 would be superior on average to the tall. However, the risk of farmers planting F2 can be combatted by having F1 plants always available. Owing to the cost of and demand for F1 hybrid seedlings, emasculation and pollination usually proceed year-round. Seedlings are therefore grown in large polybags to enable them to be kept at the nursery until a rainy season.

Coconuts are very light-demanding, and the habit in many countries, including Tanzania, of planting replacements in the shade of existing trees must be actively discouraged. Wild pigs dig round coconut plants to eat the endosperm, usually destroying the plant; damage may occur up to 18 months after planting. Romney (1983) found that hybrids now on test are damaged more than the local tall variety. Electric fencing has been found effective but is beyond small farmers' resources. Perimeter trenches and brush-wood fences are effective if properly made and maintained. High weeds act as hiding-places for pigs and should be removed. On a long-term level, the pigs need to be killed by traps or shooting, although ammunition is scarce.

Organisational Criteria for Development

Small farmers may collectively comprise the major fraction of development; in Tanzania over 90 % of coconut trees belong to small farmers. However, most small farmers have a limited ability to understand, to look after the crop, and to manage its finances. For rapid national development, agricultural companies and large farms, including communal farms, should be included in the plans; in such large-scale entities, management training is often needed.

In developed countries, the rental or mortgage cost of land is one factor determining whether a crop is viable but, in many peasant-farming areas of Africa, land is not tradeable. Land which is owned, tradeable and subject to tax is more appreciated and better used. Farm boundary demarcation and issue of ownership papers are capital inputs, needed only once, after which the land constitutes collateral for loans for coconut development. A necessary incentive is a realistic price for the product. A farmer who cannot produce at that price might be advised to try something else.

The responsibility of Government is seen as providing umbrella items which an individual farmer cannot. The most important are:

(1) *Legislation* relating to coconut production including (a) *Quarantine laws*. In Tanzania the movement of live coconut plant material from diseased to disease-free areas should be restricted.

Seeds can then be produced in disease-free areas and, as far as possible, nurseries also sited in these areas. (b) *Larceny laws*. The stealing of coconuts is a disincentive to farms and in Tanzania leads to the harvesting of immature nuts with consequent production of copra with poor storage qualities and low oil content.

(2) *Vermin control*. The damage from wild pigs and uncontrolled farm stock is a serious disincentive in Tanzania. Government support is needed in their control.

(3) *Pest and disease control*. Where large scale spraying or release of parasites are involved (e.g. if viral or fungal control of Rhinoceros beetle, now under research by NCDP in Tanzania, becomes a practical proposition) Government or its contractee must supervise this work.

(4) *Plants*. The production of adequate numbers of good quality plants is a large-scale long-term operation requiring appropriate technical and financial controls. Government could ensure that plants are available throughout the planting seasons at sites accessible to farmers; this could involve a delivery system. For F1 hybrids, polybags are needed for much of the year, but plants becoming a plantable size in a reliable rainy season (March–April in Coast and Tanga Regions; November–December south of the Rufiji River) would be more economically produced and distributed bare-root as polybags require foreign exchange and a lorry can carry 3000 plants bare-root compared with 300 polybagged. Plants sold at cost would encourage careful handling both before and after planting.

(5) *Essential inputs*. Farming items such as tools, insecticides against termites, vermin traps, etc. need to be made available in the rural areas to save farmers wasting time searching in the towns.

(6) *Research*, and extension of research findings, should be crop-orientated and on-going. This not only gets relevant information to the growers, but give growers confidence to invest effort and funds. Although the publication of scientific papers is necessary, as a stimulus to professional research workers and as a vehicle for interchange of information, the release of findings to advisory staff should be the first priority.

(7) *Copra grades* should be defined and price differentials set large enough to dissuade producers from making bad copra.

Governments have achieved the above umbrella activities in other successful coconut territories (the Phillippines, India, Sri Lanka, Jamaica) by placing responsibility with an organization which grades and purchases the copra. Such organizations have their own Board of Management; in Jamaica, a majority of Board members are copra producers elected by other producers, but with Government representatives also on the Board. An organization may be financed by a cess on copra or coconut products; it provides services, and advisory staff drive load-carrying vehicles to deliver plants and agricultural chemicals in the course of normal advisory visits.

Farm workers also need incentives. The availability of goods in rural shops acts as an incentive to workers to earn money.

Permission to submit this paper was obtained from GTZ (Deutsche Gesellschaft für Technische Zusammenarbeit; GmbH). Constructive comments from Mr E.A. Momber are also gratefully acknowledged.

References

NCDP (1980) *Land Suitability Survey, Phase* I. Dar es Salaam; National Coconut Development Programme.

de Nucé de Lamothe, M.; Pomier, M.; de Taffin, G. (1983) Local coconut or hybrid coconut in the village environment. *Oleaginéux* 38, 183.

Ollagnier, M. (1983) Oil palms and coconuts. *The Courier* 82, 72.

Romney, D.H. (1982) *Use of old coconut trials in planning new trials.* Dar es Salaam; National Coconut Development Programme.

Romney, D.H. (1983) *Annual Technical Report, Agronomy,* 1982-83. Dar es Salaam; National Coconut Development Programme.

Schuiling, M. (1982) *Annual Technical Report, Disease Control,* 1981-82. Dar es Salaam; National Coconut Development Programme.

Chapter 30

Cashew Production in East Africa

L.C. BROWN[1], E. MINJA[2] and A.S. HAMAD[3]

[1]Mtwara/Lindi RIDEP Project, Mtwara, Tanzania; [2]Agricultural Research Institute, Naliendele, Tanzania; [3]UTAFITI, Dar es Salaam, Tanzania.

Introduction

World production of raw cashew nuts reached a peak in 1973/74 at 480 000 t, of which three-quarters came from East Africa. Within East Africa, Mozambique contributes the largest share, followed by Tanzania with Kenya a relatively small producer. In all three countries cashew is predominantly a smallholders crop. Until the early 1970s most of the raw nuts were exported to India for processing, but in recent years the East African countries have developed their own processing industries.

From the end of the second world war until 1973/74 production in Mozambique and Tanzania rose rapidly but since then it has declined steeply in both countries. By 1981 output from Mozambique and Tanzania was less than one third the peak level.

The result on world trade of the declining production in East Africa has been a rising trend in international prices. Also, because of the development of processing industries in all of the major producing countries, the price of the raw nut has increased faster than that of the processed kernel.

In Tanzania and Mozambique the decline in production has led to substantial losses of potential foreign exchange earnings. In Tanzania, the other immediate and more serious consequence relates to the underutilisation of processing capacity. An ambitious programme of expansion of the existing processing industry was initiated in 1974. This involved the construction of ten new factories financed by international loans. The country now has a total rated processing capacity of 113 000 t against a production in 1982/83 of between 32 and 33 000 t.

Reasons for the Decline in Tanzania

Many reasons for the fall in production since 1973/74 have been suggested, but there has been a tendency to interpret symptoms of the decline as underlying causes in their own right. However, it is important to distinguish between cause and effect in order to formulate suitable policies for the future. The most frequently cited explanations are discussed below.

Unfavourable weather conditions

Drought, particularly, is often blamed for crop failures including cashew. It seems unlikely, however, that the persistence of the decline can be attributed simply to unfavourable weather conditions since the period of expansion in production up to the mid 1970s contained climatic variations similar to those recorded over the last ten years. Nevertheless weather conditions do cause fluctuations in production. In very wet years for instance, disease problems may adversely affect yield.

Poor standards of husbandry

There is little if any evidence in the literature to suggest that the growth in production up to 1973/74 was associated with good husbandry techniques. There is, however, much evidence of neglect and lack of interest in cashew over the last decade which is attributable to low profitability relative to other crops.

Bush fires

The apparently increasing incidence of burning in and around plantations also seems to be a symptom rather than a cause of the decline.

Insect attack

The most damaging pests are *Helopeltis* bugs and thrips. There is some evidence of a build up of pests in recent years which has been linked with close planting, the age of trees (in the case of thrips the older trees regularly turn out to be the ones most seriously attacked) and the increase of neglected and abandoned plantations (de Stefano 1983).

Disease problems

The most destructive diseases of cashew in Tanzania are powdery mildew (*Oidium anacardii* Noack) and anthracnose (*Colletotrichum gloeosporioides* Penz.) It has been suggested that powdery mildew in cashew is a relatively recent introduction, i.e. probably reaching Tanzania around 1974 or shortly before and that its spread is one of the main causes of the decline (Intini 1983). However, the failure to identify *O. anacardii* in Tanzania before 1978 may simply be a reflection of the fact that until then there had never been a plant pathologist working specifically on cashew. Both powdery mildew and anthracnose in cashew have a wide distribution and may well have been present in East Africa for some time.

According to interviews with farmers in the major district of Newala, powdery mildew has been observed mildly for a long time but attacks have intensified over the last ten years. Further evidence for a relatively early introduction is the absence of a focal centre to the production decline. In fact production started to decline in several, geographically widely separated districts, apparently simultaneously in 1974/75 and thereafter was fairly consistent everywhere.

The increasing severity of diseases is associated with neglect and abandonment of plantations combined with the ageing and high density of tree populations. Disease problems are generally less serious in vigorous, young, well-spaced plantings but as trees age and canopies close, competition leads to loss of vigour and microclimatic conditions within the canopies become more conducive to the spread of disease.

The old age of trees

The majority of trees in the main cashew growing areas of Tanzania were planted in the 1950s and 1960s, while very few trees were planted in the 1970s. Ignoring variable environmental conditions, the productivity of a cashew plantation depends on the age of trees and their spacing. Cashew is a peripheral bearer and fruit production becomes almost nil on branches of trees that intermingle with neighbouring trees, but, even before the canopies close, competition for soil moisture may limit production per tree.

The present recommended spacing for cashew is 12×12 m (69 plants ha^{-1}) but in actual fact farmers in Tanzania have never planted in rows. In most instances the number of trees per hectare is higher than the recommended 69 and irregular spacing more often results in much closer spacings than is desirable between mature trees.

Measurements have been made in Kenya of the development of canopy area and yield over time at particular spacings (van Eijnatten 1979) and the results agree with data available from experiments carried out at Mtwara, Tanzania. From this work we conclude that the majority of cashew plantings in Tanzania, owing to their close spacing, would be at peak production between seven and 16 years old. The average age of trees surveyed in Mtwara region in 1982 (Brown, in preparation) was 22 years implying peak yields between 1967 and 1976. This coincides with the national production peak between 1968 and 1973.

Abandonment of plantations as a result of villagisation

The villagisation programme, in which many people were moved into villages and away from their cashew trees, is often cited as a major cause of the cashew decline since the main thrust of the programme in the principal cashew regions occurred in 1974.

Marketing problems

The grading system introduced in 1968 proved very unpopular with farmers and surprisingly bears little relation to factory requirements. More serious, however, has been the deteriorating efficiency of the procurement system over the last decade resulting in long delays in the collection of raw nuts from villages and in making payments to farmers.

The producer price of raw nuts

Although the producer price for raw cashew nuts has steadily increased in money terms from the early 1960s, the real producer price (deflated by a cost of living index) fell continuously from 1969 to 1977. By 1977 the value of 1 kg of cashew nuts to the farmer was probably less than half its value nine years earlier and it was not until a very large price increase was introduced in 1981 that cashew recovered its 1969 value. In contrast, over the 1969-1977 period most of the other major crops of the cashew regions maintained their real values or even substantially increased in value. Only since 1981 has cashew been able to compete in profitability with the other major crops.

Conclusions

There can be little doubt that the declining producer price of cashew relative to inflation and the price levels of other crops during the 1970s is a fundamental cause of the decline in production in Tanzania. It seems likely that the problems associated with marketing and villagisation are also basic contributory causes but in themselves, without the unfavourable price situation, would probably not have had such impact. Ellis (1980) identified villagisation as the "precipitating" cause of the decline, coming as it did in 1974.

The other fundamental problem associated with the decline is the close and irregular tree spacing. Combined with an ageing tree population this appears to have led to premature decline in yield and to an intensification of disease and pest problems.

Considering the impact of declining price combined with unsatisfactory tree spacing over time, therefore, the first effect was the disincentive to plant new trees. The fall-off in new plantings probably began in the late 1960s. The next effect was increasing neglect of plantations. Until 1973 harvesting, at least, was maintained at its full potential, but villagisation in 1974 and subsequently declining yields as a result of ageing, neglect and the build up of pests and diseases led to the massive abandonment of cashew in favour of other more lucrative crops.

Reasons for the Decline in Mozambique

The striking similarity in the simultaneous declines in production in Tanzania and Mozambique leads one to question whether the underlying causes might not be related. However, it seems likely that the decline in Mozambique after 1973 was primarily the result of disruption in the marketing system during the immediate post-independence period. Before independence in 1975 trade in raw nuts was controlled by Indian and Portuguese merchants. After independence many of these merchants left the country and the successive falls in raw nut production up to 1977 may well have been due to a breakdown in the procurement system. A marketing organisation similar to that in Tanzania was established and since 1977 production seems to have stabilized.

Further evidence that the similarity with the decline in Tanzania is coincidental are reports of substantial new plantings in Mozambique in the early 1970s (Ohler 1979), indicating that the industry was in a much healthier condition than that prevailing in Tanzania at the same period.

Prospects for the Future

The prospects for cashew in East Africa ought to be good; world market prices are high and demand for processed nuts is increasing. However, the current production problems raise serious doubts for the future. The decline in Mozambique may prove to be a temporary problem but in Tanzania production constraints are serious and are not likely to be resolved quickly or easily.

The problem of low producer price was recognised some years ago and led to substantial price rises in 1980 and 1981. Production increased after the first of these but the harvest following the larger price rise in 1981 was disappointing and the 1982 harvest even more so. The absence of a quick response to price changes from a tree crop, however, is not surprising since there is frequently a time lag between cause and effect in the management of perennial crops and it may in fact take several years of favourable prices to restore farmers' confidence.

Other efforts to reverse the decline in production should be addressed to fundamental causes rather than symptoms, for instance to replanting programmes with special emphasis on sufficiently wide and even spacing and the clearing of abandoned plantations not worth rehabilitating rather than the importation of large quantities of chemicals and motorised sprayers to attack pests and diseases which could be controlled by better management.

One encouraging fact in the present situation is that the Tanzanian research programme has identified several clones which combine high yield with low susceptibility to disease and sufficient seed is available from this material to plant over 4000 ha y^{-1}. Nevertheless further research in cashew management is urgently required. Coppicing and hedgerow planting systems are currently being investigated in Kenya. Coppicing, for example, might have application in Tanzania as a method of rejuvenation of mature plantations.

Finally the importance of establishing an efficient marketing organisation in which farmers have confidence cannot be overemphasised. In view of the Tanzania Government's intention to revive the co-operative unions this would be an appropriate time for a thorough appraisal of the whole marketing structure.

References

Brown, L.C. (in preparation) Cashew production in four villages in Mtwara region, Tanzania. *Mtwara/Lindi Regional Integrated Development Programme Bulletin.*

de Stefano, M. (1983) The insects of the cashew-tree. In *Report on the activity of the Italian Technical Cooperation Team concerning cashew-tree development programme in Tanzania.* Florence; Ministero Degli Affari Esteri, Dipartimento per la Cooperazione allo Sviluppo, Instituto Agronomico per l'Oltremare.

Ellis, F. (1980) *A preliminary analysis of the decline in Tanzanian cashewnut production 1974-1979: Causes, possible remedies and lessons for rural development policy.* Dar es Salaam; Economic Research Bureau, University of Dar es Salaam.

Intino, M. (1983) The diseases of the cashew-tree. In *Report on the activity of the Italian Technical Cooperation Team concerning the cashew-tree development programme in Tanzania.* Florence; Ministero Degli Affari Esteri. Dipartimento per la Cooperazione allo Sviluppo, Instituto Agronomico per l'Oltremare.

Ohler, J.G. (1979) Cashew. Department of Agricultural Research. *Royal Tropical Institute, Amsterdam, Communication* 71.

van Eijnatten, C.L.M. (1979) Canopy development in cashew trees. *Coast Agricultural Research Station, Kikambala, Communication* 5.

Chapter 31

Fertilizers in Africa's Agricultural Future

V.L. SHELDON, D.L. McCUNE and D.H. PARISH

International Fertilizer Development Center, P.O. Box 2040, Muscle Shoals, Alabama 35662, USA.

Introduction

In Africa, there is a great potential for fertilizer inputs to rectify the food situation. Africa has become the largest importer of food, surpassing developing Asia, which has five and one-half times its population. The amount of cereal imported into Africa today is 10 times that of the 1970s and now exceeds the entire cereal exports of Canada (FAO 1981).

Given good genetic plant material and sound agronomic practices, water and soil fertility are the keys to high crop production levels. The importance of water in crop production cannot be overstressed; but generally irrigation plays only a minor role in Africa's crop production. Africa's agricultural future lies in the rainfed areas, and improvements in soil and water conservation at the farm level are key elements in maintaining and improving crop yields. Drought is a factor in the rainfed areas, but should not be overemphasized. Throughout Africa, low-fertility soils are a major constraint on crop production even in the Sahel. Much of the blame for the poor performance of African agriculture can be laid on the lack of soundly planned and executed crop production programmes in which the development of an efficient fertilizer sector has been invariably neglected.

Fertilizers are expensive ex-factory and their distribution and storage costs are very high because of their seasonal use and bulky nature. By careless management the farmer can turn a potential profit source into a net loss and waste both his restricted reserves and national foreign exchange.

Fertilizers can contribute to increased crop production only in areas with adequate physical and institutional infrastructures, such as adequate transport, storage facilities, marketing services including credit and an adequate information base. In Africa, those rather restricted areas having a sound infrastructure will enable fertilizers to provide profits for the farmers and increased economic wealth for the country.

Soil Fertility in Africa

A farmer will only show interest in the use of fertilizers when he has been shown that fertilizer use increases yields, and he has seen convincing proof that the practice is profitable. The base of the fertilizer sector pyramid therefore is a site- and situation-specific knowledge of the fertilizer requirements of the farmer.

The African farmer is aware of the low fertility levels of African soils since traditionally he has attempted to grow crops only for a short time, then abandoned the plots to regenerate through fallow. This practice of shifting cultivation, is the only crop production system that will produce good yields without the use of fertilizers. Careful management of organic residues can and should play a role in maintaining and upgrading soil fertility, but is labour intensive and does not increase the soil fertility of an area; it merely shifts nutrients from one place to another.

With the notable exception of nitrogen, only inputs of fertilizers can increase the fertility of a farm. The potential for the fixation of atmospheric nitrogen by symbiotic bacteria in theory is almost limitless but is slight in practice. Grain legumes are grown widely in Africa, often in association with cereal crops, but the contribution they make as hosts of rhizobia to the nitrogen economy of the farming system is low. Newer systems of using leguminous crops as nitrogen sources are under intensive study at the international centres for Agricultural Research and are being developed nationally. Although symbiotic nitrogen will play an increasingly important role in supplying soil nitrogen, in the major cereal-producing areas it is complementary to fertilizer nitrogen use rather than replacing it. Even with legumes, however, phosphate and sulphur and other limiting essential nutrient elements must be introduced as fertilizers.

Soil fertility research in Africa is a long-standing scientific activity, and much invaluable information on crop needs for fertilizers has been developed but it has tended to be concentrated on research stations or in areas of low population pressure. Neither of these are truly representative of the current small farmer situation.

The Food and Agriculture Organization of the United Nations (FAO) fertilizer demonstration programme has confirmed that under small farm conditions nitrogen is almost universally limiting and phosphorus is a close second (Wrigley 1981). Potash deficiencies are localized, but under a more intensive crop production system the nutrient can be expected to become increasingly important. Sulphur is now widely recognized as being as important as nitrogen and phosphate in many areas. In fact, much of the popularity of ammonium sulphate and single superphosphate in the past was because of their content of sulphur.

The generally low base levels of African soils and widespread deficiencies of trace elements indicate that the nutrients that fertilizers carry and the ratios in which they are made are also critical. These facts have an important bearing on the fertilizer products that should be made available to the farmer and on his satisfaction with their use. The work of Jones (1972) in Uganda clearly shows how cotton production is limited by fertilizer use.

Fertilizer Use

Historically, the use of fertilizer in Africa started with the plantation industries and the small-farmer cash crops such as groundnut and cotton. The steady market for these cash crops gave the farmer confidence to invest in purchased inputs not only for the cash crop but also for use on his subsistence crops. An unreliable cash crop market will kill this confidence.

Fertilizer use in Africa is very low compared with the average for developing countries. Africa uses about 18 kg ha^{-1} of arable land and land in permanent crops compared with an average of 67 kg ha^{-1} used in Asia and 218 kg ha^{-1} used in Western Europe. In 1981/82 Africa consumed 3.6M mt of fertilizer nutrients; more than half of this was used in Egypt and South Africa and the 10 largest consuming countries in Africa use 86% of the fertilizer (FAO 1981b). Thus, for many African countries fertilizer use is almost negligible. Consumption has increased at the rate of about seven percent annually during the past 10 years compared with about 10 per cent in Asia and Latin America. Current trends indicate that consumption in 1985/86 will be 4.3M mt of nutrient and 5.0M mt in 1990/91. These increases are not nearly large enough to increase yields and agricultural production to the extent needed for Africa to become food self-sufficient (IFDC 1981).

Fertilizer Production

The present situation

In 1981/82 about 1.2M mt of nitrogen and 1.4M mt of phosphate fertilizer were produced in Africa. No potash was produced. The deficit between fertilizer production and consumption has been increasing and is more significant when the 0.5M nutrient mt of phosphate fertilizer exported by North Africa is considered. Fertilizer production in Africa has been increasing about 5 per cent annually compared with increases of about 10 per cent annually achieved in Asia and Latin America. Imports in 1981/82 totalled 1.8M mt of nutrients. African fertilizer imports have increased from the previous year in 8 out of the last 10 years.

Potential Resources

All of the primary raw materials and many of the secondary raw materials needed to produce fertilizer occur on the African continent. Unfortunately, not all occur in deposits of sufficient size and quality to permit commercial exploitation at present prices.

Twelve African countries have sufficient proven reserves of oil or natural gas or both to produce ammonia for nitrogen fertilizers (IFCD 1981); some are already doing so, others plan to. Several others have identified fossil fuel resources for future use. Excess hydroelectric power development could lead to electrolytic hydrogen-based ammonia production.

Of the 36 African countries reported to have phosphate deposits, 11 have deposits that have been classified as economic reserves and nine produced more than 100 000 t of rock in 1979 (IFCD 1981). An additional seven countries have deposits classified as subeconomic resources, and 18 others have deposits that have not been classified yet. These unclassified indigenous ores often present unique processing problems; those unsuited for chemical processing might be used for direct application. Many phosphate ores contain valuable accessory components such as uranium which should be considered in any economic evaluation. Niobium and valuable rare earth minerals sometimes occur in phosphatic carbonatites.

There is no known production of potash in Africa at the present time although resources are reported in 13 countries (IFCD 1981). Between 1969 and 1977, potash was mined in the Congo Republic but the mine was flooded from overlying aquifers and has not been reopened. Solar evaporation of potash brines has been proposed in several countries.

There are no large, economically minable deposits of elemental sulphur (brimstone) in Africa (IFCD 1981). Some elemental sulphur resources are known in Angola, Egypt, and Ethiopia. The remaining requirements are met by recovering sulphur from fuel or ores. Fourteen African countries have sulphide ore resources, and seven are reported to be producing sulphidic acid from them. No doubt there are many other pyrite and sulphide ore occurrences that might be important in local situations.

Sulphur deficiencies are so serious in many agricultural areas that gypsum should be considered as a nutrient source. Deposits of gypsum and anhydrite are reported in at least 22 African countries. Finely ground pyrite (up to 50 % sulphur) may be a good direct application source of sulphur.

Increased production of fertilizers using local resources and relevant technology must be the preferred route of development for many countries. The major problem in developing nonconventional fertilizer production is the reluctance of the international and bilateral lending agencies to supply funds which are only offered for capital-intensive technology that may not necessarily be the best solution since it neglects the use of local mineral resources and labour market.

Fertilizer Marketing

To achieve a breakthrough to adequate fertilizer use by African farmers, at least five major constraints must be removed: (1) lack of fertilizer; (2) lack of knowledge about correct fertilizer use; (3) lack of money/credit; (4) unfavourable price relationships between fertilizer and farm produce; and (5) untimely distribution. Modern marketing systems operate to resolve these constraints (Mathieu de la Vega 1978).

IFDC training courses stress that "marketing" is the identification of a farmer's need and the meeting of that need in the most cost-effective manner possible rather than deciding what the farmer needs and then attempting to supply the needs by a bureaucratic allocation system; an approach used for cash crops such as cotton and groundnuts grown under a marketing board monopoly.

In its simplest terms, fertilizer marketing can be divided into four major decision areas. (1) Product – What types of fertilizers and quantities does the farmer need? (2) Place – Where is the fertilizer needed and when? (3) Promotion – What education is needed to teach the farmer proper fertilizer use? (4) Price – What prices should the farmer pay for the fertilizer when considering crop prices and credit needs? These are critical components needed to develop an effective fertilizer marketing system. Some of the issues raised are as follows:

Product

Historically, ammonium sulphate (AS) and single superphosphate (SSP) were popular fertilizers in Africa. Although these were sold as single-nutrient low-analysis materials having very high costs per unit of nutrient content, they were, in fact, multinutrient fertilizers containing sulphur and calcium. High-analysis materials (urea, triple superphosphate and ammonium phosphate) have replaced low-analysis materials that entail higher transport and handling costs in Africa.

The problem therefore in African agriculture, which requires a relatively complex supply of fertilizer materials, is determining how to reduce costs by using high-analysis materials while ensuring that this does not lead to neglect of the non-NPK components of fertilizers needed by the farmer. Supplying the farmer with the products that he needs requires considerable thought based on sound agronomic knowledge, the socioeconomic condition of the farmer, and national-level economics. There are many examples in Africa of "prestige" or short-term profit motive-based fertilizer import and production decisions that have cost national economies very dearly.

The objective must be to deliver the most cost-effective fertilizers while encouraging the use of local resources to the greatest degree possible. Until large fertilizer markets have been built up, a low level of capital investment combined with product flexibility, such as bulk handling of imports and local blending and bagging operations, is needed.

Place

Fertilizer is only of interest to the farmer if it is available to him in the form and at the place and time that he needs it. An effectively planned and executed distribution system is required to make a timely supply of fertilizer conveniently available to the farmer. Africa's current transportation infrastructure and the sheer distances involved enhance the fertilizer distribution problems. Most fertilizers are imported, so that overall movement is almost all inward from the coastal areas involving long supply lines especially to landlocked countries. These countries constitute a significant proportion of the land area and have double problems of very high costs for transportation of imports and exports.

For land transportation over long distances, the most practical and economic method is by rail. Unfortunately, the railway systems throughout Africa remain fragmented and generally in need of modernization. Practical problems such as the different principal gauges used and political difficulties hinder movement between countries. There is a lack of the skilled management necessary for both planning and operating of the existing systems. For example, Governments may hold freight rates at historic, now uneconomic, levels and fail to provide the neccessary capital or operating funds for running the system efficiently. Road transport remains the mainstay of most transport systems. Specific development funds must be devoted to expanding the range of farm-to-market transport means.

Clearly, not only is the provision of the physical facilities important but also their effective management is needed. In order to reduce costs, movement of fertilizer should be year round even though the farmer only uses it at certain defined periods. Some sort of buffer system must be provided by developing the necessary storage capacity. The design of a good storage system must be integrated with the distribution capacity of the transport and consumption pattern. The cheapest storage is on the farm or with the fertilizer retailer. A least cost system for distributing and storing fertilizer must be developed nationally if costs are to be kept at the minimum. Some of the physical distribution costs incurred by African countries are given by Trupke (1983).

Promotion

In a sound marketing system, promotion is *education* and *awareness*. National agricultural centres and extension staff in many African countries meet the needs of the farmer adequately. However, with intensive production programmes, input suppliers have a key and complementary role to play.

One of the important functions of modern fertilizer marketing is its built-in capacity to improve fertilizer use efficiency through its educational efforts. A fertilizer marketing organization requires a strong agronomic staff to assist in these efforts through their retailers and to provide technical information for sales promotion and advise farmers on *all* aspects of agricultural production. The close coordination of these activities with the sales and distribution functions in the form of special sales campaigns provides a synergistic impact on fertilizer use development. To reach the small farmer sector, a need exists to establish extensive well-informed retailer networks. For example, the National Accelerated Food Production Project (NAFPP) in Nigeria installed Agro-Service Centers through which all goods, services, and crop production techniques flow. This marketing support of the normal extension educational effort reinforces the adoption of proven food production technologies.

The transfer of skills for both production and marketing requires lengthy on the job training undertaken by experienced manager/teachers. The lessons to be learned from this approach are that (1) Government extension services should *not* be involved in sales to farmers and (2) sales agents must be involved in farmer education.

Prices

A truly free market system does not exist in any country since political pressures everywhere interfere. However, a World Bank report (1981) presents substantial support for the view that pricing policies are a root cause of the food agricultural crisis in Africa.

There are several *factors* related to price that limit fertilizer use. The farmer may not be able to justify the use of fertilizer, because prices are too high, crop prices are too low, or there is no market to sell the excess crop production. The farmer may not have the money to buy fertilizer and *credit* may not be available or its cost may be too high. The price *risk* associated with having a crop failure whereby the farmer loses his total investment in fertilizer may be a constraint. On the other hand, the price decline that can occur because every farmer has a bumper crop presents an equal risk.

The Governments of most African countries have a "cheap food policy" for urban consumers which means that farmers are paid low prices. This is generally done by administratively setting prices of major agricultural inputs, such as fertilizer. The major reasons for intervening in agricultural pricing have been to (1) stabilize prices and production, (2) foster self-sufficiency, (3) generate tax revenue, (4) curb the profit of middle men, and (5) control the cost of living for urban consumers.

However, farmers in the long run may well decide that farming is not profitable and move to the cities. Agricultural production declines and a large countrywide food deficit results in low nutritional standards and a high dependence on food imports. This dependence on imports coupled with low export earnings increases foreign exchange problems.

To partially offset the effects of a cheap food policy, many African Governments have tried to reduce farmer's production costs and encourage fertilizer use by subsidies, but increased fertilizer consumption will then cost the national economy more.

Generally the use of higher crop prices is a better way to encourage production than the use of subsidies because (1) subsidies will not prevent loss of fertilizer investment; (2) with price supports the Government is encouraged to develop a marketing system that pays the farmer a higher price than the support prices; (3) higher crop prices encourage not only the use of fertilizer but also other complementary inputs; (4) higher prices encourage additional crop acreage; and (5) out-of-country sales associated with subsidized fertilizer are eliminated.

Funding

The investment required to establish a marketing system are about the same as those needed for equivalent fertilizer production. Where volumes are large, a marketing system can operate at about US \$150 mt^{-1} or about 50% of the retail price. Funding agencies need to consider the monetary requirements for marketing along with production facilites in any fertilizer sector development project.

Conclusions

Fertilizer will play a vital role in increasing food production in Africa. An efficient fertilizer sector must integrate agricultural research, education, handling and distribution, pricing, and production or importation of fertilizer in an effective manner. This requires a multidisciplinary approach using experts with many years of practical experience under a wide range of conditions. Unless development of the fertilizer sector is based on sound planning instead of fertilizer use as a beneficial development tool, Governments will find themselves paying dearly in terms of subsidies, wasted capital investments, and large foreign exchange import bills.

References

FAO (1981*a*) *FAO Trade Yearbook.* Rome; FAO.

FAO (1981*b*) *FAO Fertilizer Yearbook.* Rome; FAO.

IFDC (1981) *Fertilizer raw materials in Africa.* [Unpublished paper.] Muscle Shoals, Alabama.

International Food Policy Research Institute (1981) *Food Policy Issues and Concerns in Sub-Saharan Africa.* Washington, DC.

Jones, E. (1972) Principles for using fertilizers to improve red ferralitic soils in Uganda. *Experimental Agriculture* 8, 315-320.

Mathieu, M.; de la Vega, J. (1978) Constraints to increased fertilizer use in developing countries and means to overcome them. *Proceedings of the Fertilizer Society* 173, 1-30.

Sanchez, P.A. (1976) *Properties and Management of Soils in the Tropics.* New York; John Wiley and Sons.

Trupke, H. (1983) *Marketing Costs and Margins of Fertilizers in selected African Countries.* Rome; FAO.

Wrigley, G. (1981) *Tropical Agriculture.* New York; Longman.

Chapter 32

Plant Breeding for Crop Improvement with Special Reference to Africa

M.H. ARNOLD[1] and N.L. INNES[2]

[1]Plant Breeding Institute, Maris Lane, Trumpington, Cambridge CB2 2LQ, UK; [2]National Vegetable Research Station, Wellesbourne CV35 9EF, UK.

Introduction

Although the application of new technology to agricultural production has forestalled some of the worst consequences of growth in human populations, increased food production in Africa has become an urgent problem. In the further development of African agriculture, plant breeding will continue to play a crucial role. New varieties provide the potential for increased productivity at minimum cost, as well as the focus for promoting successful packages of new technology.

Some of the possibilities for crop improvement through plant breeding in relation to the problems of tropical environments in Africa, priorities for current programmes, some recent developments in methodology and the need for strengthening collaboration are discussed in this paper.

Early Breeding Programmes in Africa

During the first fifty years of the present century much work was done in Africa on crop introduction, selection and improvement by Government officials, teachers and businessmen. The more enterprising individuals made crosses and developed successful breeding programmes.

From the beginning of the 1950s, greater cohesion was given to the work through the activities of regional research organisations. These were conceived because of the urgent need for work, the difficulty of funding work for all countries that needed it and consequently the desirability of centralizing the work, in order to concentrate on problems of wide applicability. Examples are the East African Agriculture and Forestry Research Organization (EAAFRO) and in West Africa, the Maize Research Unit and the Rice Research Station. These were regional in concept and were better placed than national research stations to draw on the knowledge and resources of institutions in the industrialised countries, with which they maintained close professional contacts.

Much of the work accomplished in these earlier periods is not well-known, partly because the information is difficult to find or was never published in the scientific literature. Nonetheless, the following examples show that, contrary to current opinion, work was done on important food crops in these earlier times.

The history of maize breeding in the USA and the impact that hybrids made on maize production there is well known, but similar work was undertaken in Africa. In Zimbabwe, for example, although maize varieties were improved by mass selection in the 1920s, the potential for hybrids was quickly recognized. The programme initiated by H.C. Arnold in 1932 was later to revolutionize maize production in central Africa. Indeed, hybrids such as SR52 have contributed to increased yields in many countries, either directly or as a source of germplasm.

By the 1950s, successful maize-breeding programmes had been developed in many other African countries. One of the most comprehensive was that developed in Kenya (Harrison 1970, Dowker

1971) where a range of open pollinated varieties and hybrids was produced, each adapted to a particular environment and made available to breeders in other countries.

Breeding programmes on soybeans and cowpeas were begun in Zimbabwe in the 1930s and subsequently developed in other countries. A valuable rice-breeding programme in Tanzania gave rise to some promising new varieties during the 1950s (Doggett 1965). Some of these earlier programmes such as that on cassava laid the foundations for much more extensive work that came later.

Cassava has been grown in Zanzibar since 1799 and later it spread rapidly in eastern Africa when it became recognized as a useful reserve of food when other crops failed. It could survive long periods of drought as well as attacks by locusts, but among the clones then available, those most affected by African cassava mosaic virus could lose up to 95 % of their crop (Briant & John 1940). During the 1930s H.H. Storey and his colleagues demonstrated that the virus was transmitted by *Bemisia* sp. (Storey & Nicols 1938) and used this discovery to screen a large collection of cassava clones and related species. A programme of interspecific hybridization and backcrossing was begun in 1937 (Nichols 1947) and gave rise to a succession of new clones with improved performance as well as resistance to mosaic (Jennings 1956). Much of this material was eventually incorporated into the currently successful programme at the International Institute for Tropical Agriculture (IITA) in Nigeria.

Early work in the main cash crops in Africa is generally better known than that on the food crops, partly because of its greater economic impact and partly because it was better organized institutionally. Cotton, in particular, benefitted from the activities of two well known institutions: the Cotton Research Corporation (CRC) and the Institut de Recherches du Coton et des Textiles Exotiques (IRCT). The development of American upland cotton as a major cash crop throughout Africa and India can be thought of as an early type of "green revolution".

The Challenge to Plant Breeders in Africa

The green revolution in wheat and rice was achieved on land when water availability was not a constraint. The most urgent challenge facing plant breeders in Africa is to increase the genetic potential for sustained yields in environments that are entirely dependent on rainfall of widely differing amounts and unreliable distributions. Understanding crop water requirement in relation to rainfall probability is therefore fundamental to the achievement of stable yields at higher levels of productivity. An entirely empirical approach to evaluating crop performance under rainfed conditions in Africa can be very dangerous and makes it vital for plant breeders to work in multidisciplinary teams. As far as seasonal variation in rainfall is concerned, the lessons to be learned from earlier work, such as that on cotton (e.g. Hutchinson *et al.* 1958, Hearn 1976, Arnold & Innes 1976) and maize (Dowker 1971) do not seem to have been universally heeded by plant breeders working in Africa to-day. The definition of ideotypes, based on greater understanding of the environment including water relations and all other aspects, must be developed aggressively if sustainable progress is to be made in the rainfed agricultural environments of Africa.

The severely weathered and often infertile soils of tropical environments frequently impose severe limitations both on yields and on the plant breeder's rate of progress. There is a need for greater understanding of how more nutrients can be made available to crops at low cost. Possibilities for the biological fixation of nitrogen and increased availability of phosphate through mycorrhizal associations must be kept constantly in mind. The poverty of many soils and their prevalence to drought causes problems of variability in crop response which greatly increases the difficulties of evaluating relatively small differences in crop performance on which selection demands.

Similar limitations are imposed by equally serious problems associated with the biological environment. Crops in Africa are seldom spared the ravages of pests and diseases, unless preventative measures can be introduced. Competition from weeds is also severe and is sometimes the major limitation to increased production. Chemical control of these is usually beyond the means of the small farmer. Consequently, there is an urgent need for varieties with durable resistance to pests and diseases as well as growth characteristics that limit weed development through effective competition.

Furthermore, many decisions that the plant breeder must take in defining his aims relate closely to overriding economic or sociological considerations. For example, whether crops should be grown in pure stand or in mixtures, or whether open pollinated varieties or hybrids should be released will depend on a complex of factors none of which can safely be ignored. Finally, success in achieving the aims of a breeding programme may be limited by the availability of adequate facilities for seed multiplication, distribution and marketing.

Priorities in Current Breeding Programmes

In spite of this complex background, the strategy is clear. The cost of crop improvement through plant breeding is minimal to the farmer and the benefits are assured, provided new varieties remain stable in performance. Improved average performance is also essential to provide the overall increase in production so urgently required. Moreover, the components of quality must be kept constantly in mind, especially where markets are competitive.

Nevertheless, where quality is not of overriding importance, the primary aim of plant breeders in Africa should be to increase yields through stability of performance, while simultaneously increasing the potential for higher yields in favourable environments. Pressures to incorporate too many aims into already complex breeding programmes must be resisted. All concerned in planning research must recognize that for every character added to a breeding programme the complexity increases exponentially. In consequence, either the rate of progress slows or the resources available must be increased in a corresponding manner. For example, plant breeders are often under pressure to include nutritional elements in their breeding programmes but in many circumstances in Africa resources are best used by breeders concentrating on yield; the problems of nutrition can be solved in other ways including the development of alternative crops, especially vegetables (Bressani 1983).

Unfortunately vegetables usually come into the category of minor crops and breeding effort in Africa has been minimal. At an international level the Asian Vegetable Research and Development Center (AVRDC) in Taiwan has an enlightened policy of releasing basic breeding material, which has successfully provided new improved varieties to several Asian countries (Innes 1983). In Africa, improved landraces, adapted to local conditions, and introductions from overseas, such as the Texan Early Grano onion, have made an impact but there is a pressing need for well-planned screening trials, breeding programmes and seed production schemes at a regional and national level.

Advances in Methodology

While crop improvement through plant breeding will always require a long-term approach, innovations are continually being incorporated into the methodology that help to shorten the time required. These innovations, and some future possibilities may be considered under the three main components of plant breeding systems: variation, selection and evaluation.

Variation
Free interchange of germplasm among plant breeders is the key to maintaining adequate pools of genetic variation and the International Board for Plant Genetic Resources (IBPGR) has greatly assisted this in many ways.

Schemes to foster private investment in plant breeding are bad only if they directly inhibit progress that might otherwise be made. With an enlightened approach by Governments and commercial companies it should always be possible to give reasonable safeguards to the investor while depriving no-one of the benefits of the breeders' efforts. Although the subject is controversial, it is vital that common sense should prevail. Despite the disappearance of land races and future possibilities offered by genetic engineering the most valuable material for the plant breeder is still that represented by current successful varieties throughout the world. Their free interchange across national boundaries is therefore fundamental to continued progress. For example, the current breeding programme on pearl millet at the International Crops Research Institute for the Semi-Arid Tropics (ICRISAT) is highly dependent on recombination between African and Indian cultivars, from which both regions stand to gain.

New ways of inducing genetic changes to augment naturally occurring variation arise from research in biotechnology and genetic engineering. Some possibilities include the limited (and therefore more useful) genetic change as a result of pollen irradiation (Pandey 1975, Jinks *et al.* 1981); somaclonal variation, being investigated particularly in maize (Brettell, Thomas & Ingram 1980) and potatoes (Secor & Shepherd 1981, Gunn 1983), and various techniques for incorporating DNA into plant cells (Caplan *et al.* 1983) by such methods as micro-injection into ovules or pollen, uptake by pollen tubes from a suitable medium, prior to their transfer to the stigma, and transformation with the aid of plasmix vectors. The application to plant breeding of some of these developments in genetic manipulation has recently been reviewed by Sybenga (1983).

New sources of resistance to pests and diseases may become available such as the incorporation into crop plants of the genes coding for anti-metabolites affecting a wide spectrum of insect pests, but the side-effects of the gene products on the physiology of the host plant cannot be predicted. Similar considerations apply to the possibilities for inducing virus resistance in host cells by incorporating fragments of DNA derived from viral RNA, with the aim of inhibiting virus replication in the host plant.

The development and wide application of such general techniques will not change the principles of plant breeding in any fundamental way but could shorten the time required to produce improved varities, and make available genes that would be more difficult, or impossible, to incorporate by existing methods. However, innovations in biotechnology and genetic manipulation will be exploited widely in Africa only by strengthening collaboration among all the institutions involved.

Selection

We have already mentioned the need to define ideotypes as an aid to efficient selection. In cereal crops, such as maize, sorghum and pearl millet, selection for shorter types with stems that can withstand the force of tropical convection storms must surely have high priority, but this requirement is only one aspect of defining the ideotype. Drought tolerance is usually another. This complex character may well interact with other desirable features of the plant including stem height. Selection for such interacting combinations under controlled conditions of water supply is difficult and expensive. An alternative approach is to investigate the underlying physiological processes that determine drought tolerance and to develop easily-applied techniques for indirect selection. Work on the relationship between drought tolerance and the accumulation of abscisic acid in wheat, rice and pearl millet is already suggesting ways in which this might be done (Quarrie 1983).

Efficient techniques for selection are particularly important in screening material for resistance to pests and diseases. Simple, reliable and cheap techniques are preferable, but not always available. With respect to virus diseases, developments in molecular biology have enabled the presence of a specific viral RNA in the host tissue to be detected in a single spot of sap by "probing" with the complementary DNA derived from previously-isolated viral RNA. The technique was first applied to the detection of the potato spindle tuber viroid (Owens & Diener 1981), but is already being used for the detection of other viruses and may well have much wider application to plant breeding in the future.

With very complex characters, such as adaptation to mixed cropping, efficient selection techniques are difficult to devise and only a small proportion of the breeding material can be evaluated over the range of conditions to which the final variety will be exposed. Consequently, there is need for greater understanding of the interrelationships among the crops in the mixture, and further definition of their ideotypes, as shown by work at ICRISAT (Willey 1979) and IITA (Wien & Smithson 1981).

Evaluation

Difficulties of evaluating overall performance arising from environmental heterogeneity and genotype-environment interactions are particularly severe in many African environments. The advent of computers has facilitated approaches that enable the breeder to increase the overall scale of operation which, of itself, helps with both problems. They have also encouraged the development of new statistical techniques such as the development of the lattice principle to include any number of entries (Patterson *et al.* 1978) which assists more precise discrimination among large numbers of genotypes in field experiments and the development of multivariate analyses to provide new ways of looking at the complex patterns of varietal response over a range of different environments (Blyth *et al.* 1976, Kempton 1984). All of these applications are entirely dependent, however, on access to computers and to the appropriate software and expertise that must accompany them.

Collaboration

The complexity of many of the problems facing plant breeders in Africa, and the need to share ideas and resources, call for close collaboration among all concerned. Excellent work being done, on small remote research stations can be so much more productive if linkages can be developed with individuals and institutions in other countries. The role of the International Centres sponsored by the CGIAR in fostering the development of informal networks of co-operating plant breeders contributes greatly to the overall rate of progress. Systems of international variety trials greatly assist in evaluating breeders' material and provide vital information on genotype-environment interactions. The International Centres can also assist with the provision of central data processing facilities, and can provide a focus for the collation and dissemination of ideas, methods and materials. Moreover, the training provided by Centres is fundamental to strengthening national programmes.

It is also important to strengthen linkages with advanced institutions in other continents, particularly with those working on fundamental problems relevant to African crops. Many individuals and institutions are already contributing to the formation of these linkages and to broaden knowledge of the needs and opportunities for research in a more systematic way, the International Council of Scientific Unions (ICSU) inaugurated the Commission for the Application of Science to Agriculture, Fisheries and Aquaculture (CASAFA). The Commission seeks to harness the goodwill that exists in research institutions in the wealthier countries in ways that will direct more of the relevant fundamental research towards the needs of the developing countries.

We have outlined the problems facing plant breeders in Africa and indicated some of the ways in which future progress may be accelerated. But crop improvement will always be long-term in nature and its progress will continue to depend on collaboration that is truly interdisciplinary as well as international.

References

Arnold, M.H.; Innes, N.L. (1976) Plant breeding. In *Agricultural Research for Development: the Namulonge Contribution* (M.H. Arnold, ed.), 197-246. Cambridge: Cambridge University Press.

Bressani, R. (1983) World needs for improved nutrition and the role of vegetables and legumes. *Asian Vegetable Research and Development Center, 10th Anniversary Monograph Series*, 22. Taiwan; AVRDC.

Brettell, R.I.S.; Thomas, E.; Ingram, D.S. (1980) Reversion of Texas male-sterile cytoplasm maize in culture to give fertile, T-toxin resistant plants. *Theoretical and Applied Genetics* 58, 55-58.

Briant, A.K.; Johns, R. (1940) Cassava investigations in Zanzibar. *East African Agricultural Journal* 5, 404-412.

Blyth, D.E.; Eisemann, R.L.; de Lacy, I.H. (1976) Two-way pattern analysis of a large data set to evaluate genotypic adaptation. *Heredity* 37, 215-230.

Caplan, A.; Herrera-Estrella, L.; Inze, D.; van Haute, E.; van Montagu, M.; Schell, J.; Zambryski, P. (1983) Introduction of genetic material into plant cells. *Science, New York* 222, 815-821.

Doggett, H. (1965) A history of the work of the Mwabagule Rice Station, Lake Province, Tanzania. *East African Agricultural and Foresty Journal* 31, 16-20.

Dowker, B.D. (1971) Breeding of maize for low rainfall areas of Kenya. I. Reliability of yield of early and later maturing maizes. *Journal of Agricultural Science* 76, 523-530.

Gunn, R.E. (1984) Breeding new potato varieties from protoplasts. In *Proceedings of the Eighth Long Ashton Symposium on improvement of vegetatively propagated crops, Bristol, 12-15 September 1982.* London; Academic Press, in press.

Harrison, M.N. (1970) Maize improvement in East Africa. In *Crop Improvement in East Africa*, (C.L.A. Leakey, ed.), 21-59. Farnham Royal; Commonwealth Argicultural Bureaux.

Hearn, A.B. (1976) Crop Physiology. In *Agricultural Reseach for Development: the Namulonge Contribution* (M.H. Arnold, ed.), 77-122. Cambridge; Cambridge University Press.

Hutchinson, J.B.; Manning, H.L.; Farbrother, H.G. (1958) Crop water requirement of cotton. *Journal of Agricultural Science, Cambridge* 51, 177-88.

Innes, N.L. (1983) Breeding field vegetables. *Asian Vegetable Research and Development Centre, 10th Aniversary Monograph Series*, 34. Taiwan; AVRDC.

Jennings, D.L. (1957) Further studies in breeding cassava for virus resistance. *East African Agricultural Journal* 22, 213-219.

Jinks, J.L.; Caligari, P.D.S.; Ingram, N.R. (1981) Gene transfer in *Nicotiana rustica* using irradiated pollen. *Nature, London* 291, 586-588.

Kempton, R.A. (1984) The use of biplots in interpreting variety by environment interactions. *Journal of Agricultural Science*, in press.

Nichols, R.F.W. (1947) Breeding cassava for virus resistance. *East African Agricultural Journal* 12, 184-194.

Owens, R.A.; Diener, T.D. (1981) Sensitive and rapid diagnosis of spindle tuber viroid disease by nucleic acid hybridisation. *Science, New York* 213, 670-672

Pandey, K.K. (1975) Sexual transfer of specific genes without genetic fusion. *Nature, London* 256, 310-313.

Patterson, H.D.; Williams, E.R.; Hunter, E.A. (1978) Block designs for variety trials. *Journal of Agricultural Science* 90, 395-400.

Quarrie, S.A. (1983) Genetic differences in abscisic acid physiology and their potential uses in agriculture. In *Abscisic Acid* (F.T. Addicot, ed.), 365-419. New York; Praeger.

Secor, G.A.; Shepherd, J.F. (1981) Variability of protoplast-derived potato clones. *Crop Science* 21, 102-105.

Storey, H.H. and Nicols, R.F.W. (1938) Studies of the mosaic chiase of cassava. *Annals of Applied Biology* 25, 790-806.

Sybenga, J. (1983) Genetic manipultaion in plant breeding: somatic versus generative. *Theoretical and Applied Genetics* 66, 179-201.

Wien, H.C.; Smithson, J.B. (1981) The evaluation of genotypes for intercropping. In *Proceedings of the International Workshop on Intercropping, 10th-13th January, 1979.* 105-116. Hyderabad.

Willey, R.W. (1979) Intercropping – its importance and research needs. Part 2. Agronomy and research approaches. *Field Crop Abstracts* 32, 73-85.

Chapter 33

The Quantification and Economic Assessment of Crop Losses due to Pests, Diseases and Weeds

P.T. WALKER

Tropical Development and Research Institute, Porton Down, Salisbury, SP4 0JQ, UK.

Introduction

Losses of yield caused by crop pests, diseases and weeds have commonly been described as light or severe, and control measures have been based on the numbers of pests and weeds and disease symptoms rather than on the measured effects on yield. Only in the last few decades have yield losses been studied as an essential subject in the crop yield system and as a basis for decisions by the farmer, the scientist and the administrator on pests, diseases and weeds. An increasing body of knowledge is now available on the principles, methods and interpretation of crop loss assessment (Chiarappa 1971, 1981, Govindu *et al.* 1980). Crops are now more valuable, inputs more costly, and agriculture is even seen as a threat to other interests. The need to balance the benefits of pest control against the costs becomes ever more important as a strategy for crop improvement.

Ordish (1951) described such losses as untaken harvest. The actual losses have been estimated by Cramer (1967), and the subject reviewed by Bardner & Fletcher (1974). A discussion of all aspects of losses is provided by Walker (1983), where further references to many of the examples cited below will be found.

The Importance of Crop Loss Studies

Forecasting yield

The annual crop production is of vital importance in maintaining the supply of food, the farmers revenue, and in contributing to a nation's balance of payments through export. Some countries forecast production by an early warning system based on rainfall, but this is only one of a number of factors affecting pest, disease or weed attack: biological controls, crop variety resistance, the extent of monoculture, and the availability of pesticides. These can be monitored and predicted, and if the effect on yield is available, yields can be forecast. Deterministic models of yield accumulation based on the availability of rainfall, temperature or radiation exist for maize, sugarcane, coconut and sweet potato, which together with improved monitoring in a national or international surveillance system, should greatly increase the reliability of crop forecasts. The extreme importance of bananas, for example, to St Lucia, cocoa to Ghana, or sugar to Mauritius, and their pests and diseases needs no emphasis. Effects on the world price are not unknown.

In addition, pests, diseases and weeds are not static, as Tanzania is aware from the appearance of the *Prostephanus* borer in grains (see Chapter 12), *Mononychellus* mite on cassava and *Eldana* borer in sugarcane. World crop diseases, such as sugarcane smut and coffee rust, change their distribution, and there are national changes in the demand for crops. New varieties of different resistance to pests or diseases are grown, and new farming techniques, such as no tillage, are developed. These factors make the continuous revision of crop loss data a necessity.

**Decision-making
in pest, disease
and weed management**

A second reason for crop loss studies is in the decision-making process (Mumford & Norton 1984). Should a farmer apply a pesticide, when, and how much? How many cultivations are needed to control weeds and obtain an acceptable return on costs? In a wider context, can investment on biological control, a pheromone monitoring scheme, a disease forecasting system or a new cultural system be balanced by the losses due to pests, diseases or weeds prevented? These problems can only be answered by crop loss assessment.

It would appear to be a simple analysis, comparing the benefits in increased yield, better quality, reduced cost of harvesting or labour or more timely harvest with the costs. Costs include the standing costs of pesticide, its application and labour, or buying or breeding parasitoids, or of any other non-pesticidal method: plastic bags for bananas, the destruction of crop residues or the provision of structures for post-harvest loss prevention. As the decision-making system becomes more complex, costs can include that of professional advice, information gathering, insurance, and not least, training of farmers, extension staff and scientists.

Costs increasingly include the external diseconomies of social and environmental costs, notoriously difficult to evaluate (Gunn & Stevens 1976). The prevention of health risks to man, livestock, crops and wildlife must also be considered. These can be evaluated as loss of earnings, cost of hospitalization, poisoning of drinking water, cattle or fish, the cost of a pesticide regulation scheme, or the loss of tourist revenue from effects on birds or fish. Another factor, the human element, is considered below.

Costs and benefits can be expressed as a concept, the economic or action threshold of pest, disease or weed, the density or intensity at which the farmer decides to act. It can be used in a decision system in which the farmer surveys the situation and compares it with an economic threshold before acting or not acting to control the infestation. Efficient pest, disease or weed management, the use of several methods to reduce attack and produce greater benefit at less cost, depends heavily on monitoring the pest or disease situation, on crop loss assessment, and on making decisions on this information (Youdeowei & Service 1983).

**The allocation of
resources**

A third reason for crop loss assessment is in planning or project appraisal. The allocation of funds or an application to donors for a project in a particular crop sector depends on an assessment of the pest, disease or weed attack, the preventable loss due to pests, diseases or weeds, and projections of the increased yield over the life of the project, discounted to present values. Values can be calculated ignoring the diseconomies, and an arbitrary judgement made of their effect if included.

The decision may be between the importance of different pests, diseases or weeds. If, for example, the average yield increase of maize over several years in the absence of weeds is 28%, stem borer 12%, armyworm 5% and streak disease 2%, this may be helpful, although a simple yield increase may not be as important as savings of time, labour, energy or foreign exchange, perhaps expressed as consumer-producer surplus or added utility at family, village or national level. Loss may also be expressed as the Ordish area or loss equivalent area, the area saved by growing the yield without loss. Political expediency is not unknown as a deciding factor. The entomologist or pathologist must recognize that the major constraint to high yields is usually climate, although economic and political factors are often not far behind.

Research on yield

A study of how yield is lost is a valuable guide to yield improvement: a study of the components of yield such as tillering, panicle or seed number or weight in sorghum, a study of the physiology of yield accumulation, such as the importance of the flag or leaf area in wheat, or the interaction between plant chemicals, the pest and the weed as they affect yield, may give useful guidance to the plant breeder, or to those developing novel methods of pest, disease or weed control.

The Definition, Causes and Study of Loss

Crop loss is most meaningful if defined as the reduction caused by pests, disease or weeds in the maximum potential attainable yield obtained in the absence of such causes, expressed as a percentage of the maximum. It may be called avoidable loss, or the yield gap. It may be difficult to decide whether to take as maximum the research maximum, the good farmer average or the average farmer maximum yield. It depends on the purpose.

At the plant level, loss may be caused at various stages in the yield production system. As a system it can be modelled and simulated. The seedling may be destroyed, the plant may suffer primary damage to roots, stems, interrupting uptake and flow of water, nutrients or photosynthates to sinks and to storage or reproductive parts such as tubers, stems, fruits or seeds. Damage to leaves reduces photosynthesis. Damage to yielding parts results in a direct loss. Secondary damage by disease introduced by vectors may occur, as in the cereal virus diseases or the nematodes in coconut red ring disease. Loss may be caused by damage to supports such as stem or root, by spoilage or downgrading, and by cosmetic damage and effects on harvesting or processing.

At the crop level, loss may result from competition for available water, light or nutrients, for example by weeds. Compensation by unattacked plants for those attacked often permits a certain amount of pest, disease or weed attack without loss of overall yield, depending on the availability of the limiting factor. Such compensation may be modelled mathematically, as in tobacco plants attacked by pests in Zimbabwe or potato plants by disease in Scotland. Compensation effects are closely related to plant population effects and research on thinning rates is helpful.

Crop loss may be examined as a model of the change in yield (y) with increasing infestation (i), $y = f(i)$, usually a negative regression:

$$y = m - b(i)$$

where (m) is the maximum potential yield and (b) the slope or rate of loss of yield. Where there are several causes of loss, for example insect pests and diseases of rice, a multifactorial regression of yield on the amounts of each cause present can be calculated, and the significance of each contribution to the total loss estimated (Khosla 1977).

The relation between yield and infestation is taken to be linear, with yield inversely, or loss directly proportional to the amount of infestation. Mostly because of compensation, the relation is usually sigmoid, with a threshold infestation below which control is uneconomic. The curve may be concave or convex depending on the rate at which loss changes. If the effects of infestation are multiplicative rather than additive, a logarithmic scale may be needed. An increase in yield can result from some types of infestation, in certain anatomical or physiological situations, but this cannot be relied upon to increase national production.

Information Needed

To make rational decisions, the first requirement is quantified information on pests, diseases or weeds. Cheap, simple, accurate and reproducible methods are preferred, and much effort is spent on developing and testing them. Monitoring systems, which exist in East Africa, India, Pakistan (Sind) and the UK, are valuable.

Regarding pests, many techniques of quantifying the number of insects, rodents, nematodes or birds have been developed, some more sophisticated than their accuracy deserves. Estimates are usually made on a sample, when the sampling error can be estimated. The sample can be a direct count per unit area, which is preferred, or an indirect measure, the damage to a crop, or some other indication of attack. A sample from the environment such as in a light trap, suction trap or rodent trap will require correction for the conditions at the time, and should be related to the absolute pest density, to the damage and to crop loss. Recent developments include the use of satellite pictures, radar and sex pheromone traps.

Diseases are usually estimated in incidence or severity by their symptoms, although sometimes the causal agent is sampled. As with pests, the distribution of the disease in time and in space is essential information in forecasting disease progress and subsequent loss of crop from single or several observations. As with pests, standard damage area diagrams are valuable, with attention to the crop growth stage. Degrees of infection are often given scale ratings, but these are less informative than actual measurements.

In the case of weeds, the time and duration of growth is as critical as the dry or fresh biomass, numbers, or area of ground cover, when determining final crop yields. If the distribution of pests, diseases or weeds is non-random, transformation of the data may be needed if parametric statistics are to be used to analyse the results.

The relation between yield and infestation

When the infestation has been assessed, how is the effect on yield established? As James & Teng (1979) state, "production is the result of a multifactorial equation in which all factors can be constrained and interact." Thus the relationship is a "response surface" resulting from many variables, and regression analysis can be used to select the most significant affecting yield. Trials or surveys are needed to measure the yields resulting from a range of infestations in as many climatic, agricultural or pest-disease-weed conditions as possible. Methods are fully described in Chiarappa (1971, 1981) and Walker (1983).

Methods of crop loss assessment

The simplest method is to compare the yields of naturally attacked and unattacked plants, plots, whole fields or harvests, for example sugarcane at the mill. If possible a range of infestations is selected. As the farmer is interested in crop, not plant yield, area estimates are preferable, unless compensation by unattacked plants is corrected for, and the distribution of attack considered. If single plants, grains, or fruit are counted, the number of units multiplied by the loss per unit gives the total loss. All types of plot design may be used depending on the uniformity of the crop, of pest, disease or weed attack and on the time and money available. Unfortunately the degree of natural attack cannot be controlled.

A common method is to increase or decrease infestations artificially, e.g. by placing stem borer eggs on plants, beetle grubs in the soil, spraying plants with disease spores, introducing infected plants, or releasing virus-infested aphids. The pest or disease level may be decreased by removing pests, trapping rodents, scaring away or screening from birds. Cages may be used to confine or exclude pests. Weeds are removed by hoeing.

Thirdly, chemicals are commonly used to produce a range of infestation rates, for example insecticides to measure losses due to maize stem borer in Kenya and Tanzania, fungicides to reduce coffee diseases, rodenticides to control rodents in oil palm in Malaysia, avicides to control *Quelea* in East Africa, nematicides for banana nematodes, and herbicides on cereals. Combinations of chemicals and other techniques can be used to control selectively a combined attack by several pests or diseases. Unfortunately chemicals may have unexpected effects on yield, and interplot interference may complicate the results.

Other approaches are to simulate the loss of plants, leaves or roots by artificial removal or damage. Although more controllable, it may be difficult to simulate natural attack exactly in place and time. Another method is to compare the yields of crops which are susceptible and resistant, for example to *Empoasca* leafhopper on cowpea in Nigeria, or eyespot on wheat in England.

Crop loss surveys

As much information on crop losses should be collected as resources permit. Regular surveys by the Plant Protection Department, or by a surveillance unit, in co-operation with Statistics or Economic Planning Departments, are needed. The extent and accuracy of surveys depends on whether their purpose is to measure the quantity, or the type or distribution of losses, the minimum being governed by the principle that any information is better than none.

Surveys are either based on the actual measurement of yields by crop-cutting of attacked and unattacked crops, as done by agricultural census teams, and used in crop loss studies of rice in India, Bangladesh and Sierra Leone, of maize in Kenya and Tanzania, and cotton in Malawi. Alternatively, if a sound model or relationship between yield and infestation has been found, surveys of the pest, disease or weed can be used to estimate losses. Diseases of peas have been estimated by this method in Canada and on cereals in Britain. The difference between the estimated yield and the potential pest-free yield is the loss. If pest or disease attack or type of farming is very variable, stratification into areas or types will be more accurate. Weighting of data by crop area is almost essential, and transformation, perhaps logarithmic, before averaging is advisable, depending on the frequency distribution of losses.

Problems needing attention

Research is needed on how different causes of loss, such as a pest, a disease and weeds, interact. Multifactorial and polynomial approaches may be necessary. Mixed cropping of two or more crops growing at the same time creates problems, and is probably best dealt with in terms of the equivalent area yields of one of the crops. Most of all, data on the relation between pest, disease and weed attack and associated yield are needed on as wide a range of farming systems as possible.

Collection of crop loss data

The criticism is often made that the wide limits of error of crop loss data make the data meaningless. Despite this, the information is always in demand for commercial and government purposes, and Cramer (1967), Watts-Padwick, Haimsworth, Walker and others have attempted to summarize what is available.

FAO has supported data-collection in its Action Programme for Improved Plant Protection, and with the help of a profile from the CAB database, a crop-loss database is being compiled from which print-outs of losses by country, crop, and cause, together with the reference, are possible (Table 1).

Table 1   Examples of losses in Kenya.

maize	grain weevil	30%	wheat	birds	5%
	stemborer	15%		armyworm	5%
	armyworm	7%		rodents	4%
	soil pests	3%		soil pests	3%
	rodents	2%		aphids	2%
sorghum	grain weevil	30%	barley	armyworm	5%
	shootfly	20%		bulbfly	5%
	birds	15%			
	stemborer	10%	millets	birds	15%
	armyworm	8%		armyworm	8%
	midge	3%		soil pests	2%
	soil pests	1%			
			sugarcane	brown spot	7%
coffee	berry disease	35%		mealybug	4%
				smut	3%
				stemborer	1%

Crop Losses in Farmer Decision-making: Economic Thresholds

The crop loss caused by pests, diseases or weeds expected by a farmer is incorporated with other factors into a threshold level, density or amount of pests, diseases or weeds at which he will take action to control it. This is the economic, action or damage threshold. As the infestation rises with time, the Action Threshold (AT) is the level at which control must be applied to prevent a further rise to the economically damaging Damage Threshold (DT), or Economic Injury Level. Economic damage may be defined as when the benefit/cost ratio is less than one. The difference in time between the AT and the DT will depend on the rate of pest or disease development and the speed of control.

Factors which affect the control action threshold

The AT is an expression of all the factors influencing a farmer's decision, and may be conceptualized (Chiang 1982) as:

$$AT = \frac{C \times T \times CF}{Y \times V \times \frac{y}{i} \times CE \times F \times R}$$

C: the cost of control. The higher the costs, the higher the threshold at which pests will be controlled. Large enterprises can economize in the supply and application of inputs. The smallholder is often less efficient unless he is in a cooperative and his technology is adapted to his situation.

T: a time factor. Economic loss can occur immediately the (AT) is reached, or after a delay depending on the climate, the farming system or the speed at which control works.

Y: the yield. The greater the yield, the lower the threshold at which control will be applied. The peasant farmer with low yield will apply control at a high threshold.

V: the value of the yield. Infestations on high value crops will be controlled at low thresholds. The scale of operations is important: if a few farmers increase yield, their revenue will be higher than if all farmers do and the price elasticity of demand is low. Ordish pointed out the disastrous effect of plenty on the farmer in Shakespeare's Macbeth. World shortfalls and high prices, for example of coffee or sugar, can benefit a country that can increase its own yields.

y/i: the relation between yield and infestation. The rate of yield increase with control of infestation governs the level of infestation at which control will be applied.

CE: the control efficiency. More control per unit of input is expected on sophisticated operations and control will be at a lower threshold.

SC: a survival coefficient. The response of pest, disease or weed in the time (T) between control and an effect on yield.

Three other factors are of critical importance in understanding the use of sophisticated pest, disease or weed control in developing countries, or indeed in any farming system (Chiang 1982):

CF: a critical factor which reflects the importance given by the farmer and his community to the need for yield, perhaps of food, rather than pay the opportunity costs of improving his physical and social environment.

F: the profit level or benefit ratio used by the farmer to assess his actions or his level of inputs. Is a marginal yield increase worth the marginal cost or effort of obtaining it? Is his motive to maximize his profit or his satisfaction?

R: the farmers perception of pests, diseases, weeds and pesticides: Many farmers are ignorant of their pests, and the chemicals and methods used to control them. Increasingly an understanding of farmers decisions and their attitude to change is seen as a first need in introducing new ideas, as in the work of Tait, Norton and Mumford (Austen 1982, Mumford & Norton 1984).

The farmer element

The methods of sociology and anthropology have shown valuable results. Among the Tonga of Zambia, Abrahamse found that land tenure, family responsibilities, non-farm activities, labour migration, use of non-local language, level of education, attitude to cattle, credit, and other factors influenced a farmer's attitude to pests and diseases. In Tanzania, Cox related farmers' use of pesticide on cotton to their expectation of reliable rainfall. In Malawi, Farrington also noted that farmers adapted adapted research recommendations to their expectations. In Kenya, Goldman & Omolo found similar perceptions and practices in South Nyanza, particularly on intercropping and crop and income diversification. Some losses were underestimated, others wrongly attributed, and interesting traditional practices were revealed. In a recent survey of an outbreak of banana weevils and nematodes in Bukoba, Tanzania, Hebblethwaite, Bridge and myself found that some farmers confused the cause and extent of damage and did not appreciate the value of clean planting material, points to be emphasized in an extension campaign. Among rice farmers in Asia, pest management recommendations were modified after a study of farmers attitudes to them (Smith 1982).

Training and Commonwealth Assistance

The FAO, together with other institutes, has sponsored pre- and post-harvest loss assessment courses (Govindu *et al.* 1980). Courses have been held by ICIPE, universities, and others. The Commonwealth Secretariat has supported some courses and financed the exchange of scientists. Crop loss data collection has been furthered by CAB data-base print-outs. The CAB publish the FAO manuals and a bibliography of pest control economics (Bellamy 1972). Identifications by CIE, CMI and CIP (see Chapter 79) have played a major part.

Conclusions

(1) There is an urgent need to quantify pest, disease and weed attack, the relationship to crop yield, and the crop losses caused by pests, diseases and weeds.

(2) Training in the techniques of pest, disease and weed survey and crop loss assessment is available and should be given.

(3) The farmer's actions and attitudes to pests, diseases and weeds and the losses they cause must be understood in his social and economic background if he is to use new or improved methods of pest, disease or weed control, in a technology appropriate to his circumstances, to increase his yields.

(4) Improved crop protection must have the support of an effective infrastructure of advice, inputs, research, finance and marketing.

References

Austen, R.B. (ed.) (1982) *Decision-making in the Practice of Crop Protection.* [Monograph No.25.] Croydon; British Crop Protection Council, 238 pp.

Bardner, R.; Fletcher, K.E. (1974) Insect infestations and their effects on growth and yield of field crops: a review. *Bulletin of Entomological Research* 64, 141-160.

Bellamy, M. (1972) *Economics of plant protection.* [Annotated bibliography No.15.] Oxford; Commonwealth Bureau of Agricultural Economics, 11pp.

Chiang, H.C. (1982) Factors to be considered in refining a general model of economic threshold. *Entomophaga* 27, 99-103.

Chiarappa, L. (1971, 1981) *Crop Loss Assessment methods, FAO Manual.* [with 3 *Supplements.*] Farnham Royal; FAO & Commonwealth Agricultural Bureaux.

Cramer, H.H. (1967) Plant protection and world crop production. *Pflanzenschutznachrichten Bayer* 20, 3-524.

Govindu, H.C.; Veeresh, G.K.; Walker, P.T.; Jenkyn, J.F. (1980) *Assessment of Crop Losses due to Pests and Diseases.* Bangalore; University of Agricultural Sciences, 300 pp.

Gunn, D.L.; Stevens, J.G.R. (1976) *Pesticides and Human Welfare.* Oxford; University Press, 278 pp.

James, W.C.; Teng, P.S. (1979) The quantification of production constraints associated with plant diseases. *Applied Biology* 4, 201-267.

Khosla, R.K. (1977) Techniques for assessment of losses due to pest and diseases of rice. *Indian Journal of Agricultural Science* 47, 171-174.

Mumford, J.D.; Norton, G.A. (1984) Economics of decision-making in pest management. *Annual Review of Entomology* 29, 157-174.

Ordish, G. (1951) *Untaken harvest; man's loss of crops from pest, weed and disease.* London; Constable 171 pp.

Smith, W.H. (ed.) (1982) The role of anthropologists and other social scientists in interdisciplinary teams developing improved food production techniques. 102 pp. Manila, Philippines. IRRI.

Walker, P.T. (1983) Crop losses: the need to quantify the effects of pests, diseases and weeds on agricultural production. *Agriculture, Ecosystems & Environment* 9, 119-158.

Youdeowei, A.; Service, M.W. (1983) *Pest and Vector Management in the Tropics.* London; Longman, 399 pp.

Chapter 34

Plant Diseases: Changing Status and Diagnosis in Relation to Agricultural Development

J.M. WALLER

Commonwealth Mycological Institute, Ferry Lane, Kew, Surrey TW9 3AF, UK.

Introduction

Crop diseases restrict agricultural productivity and hinder its advancement in several ways. Firstly, they have a direct effect on the plants they infect which often culminates in a noticeable or measurable reduction of the quantity and/or quality of crop yield (Walker, Chapter 33). Secondly, the effort spent in combating the effects of disease results in an economic cost either to governments, as in funding quarantine measures and agricultural research and development sectors, or to farmers in terms of extra labour and monetary resources spent on direct disease control measures; frequently these are passed down-stream to consumers in the form of increased taxes or prices. Thirdly, plant disease can prevent new crops, improved plant varieties, or better agricultural practices from being adopted, thus directly hindering the advancement of agriculture. Time and resources spent in overcoming these hurdles may result in the loss of potential markets and the disillusionment of farmers. Many promising agricultural development ventures have revealed unsuspected problems of crop health on which some have then foundered.

Clearly, both the magnitude and effects of these problems need reducing if agricultural productivity is to be advanced. Is plant pathology sufficiently effective as an applied science in doing this and moreover are agricultural development and improvements in crop health interdependent? Realistic answers to these questions require an assessment of the influences of agricultural development on crop diseases and of the problems which face the practice of plant pathology in developing countries.

Effects of Agricultural Development on Crop Diseases

Some overall changes which developing agriculture has had on agricultural productivity can be assessed from data given in the FAO Yearbooks. Total agricultural production increased by more than a third over all developing countries during the last decade; this has been achieved partly by an increase in the area of land under cultivation and partly by an increase in the yield/unit area due to improved crop production practices involving changes in husbandry techniques and crop germplasm. Food production has just kept ahead of population increase overall except in Africa where an apparent shortfall of some 10% has occured. Although the volume of agricultural exports from all developing countries increased four times during the last decade, their value has not kept up with the increased costs of energy based imports (fuel and manufactured items). The resultant increase in the value of agricultural inputs over that of agricultural production is bound to affect agricultural efficiency and investment.

The influence which the various aspects of agricultural development have had on crop diseases has recently been reviewed (Waller 1984). The main consequential effects which the three components of increased agricultural productivity have had on crop diseases are shown in Fig. 1.

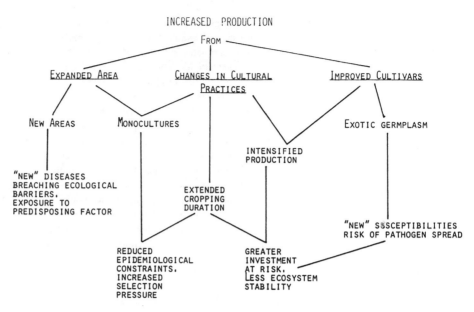

INCREASED PRODUCTION
FROM

EXPANDED AREA   CHANGES IN CULTURAL PRACTICES   IMPROVED CULTIVARS

NEW AREAS   MONOCULTURES   INTENSIFIED PRODUCTION   EXOTIC GERMPLASM

"NEW" DISEASES BREACHING ECOLOGICAL BARRIERS. EXPOSURE TO PREDISPOSING FACTOR

EXTENDED CROPPING DURATION

"NEW" SUSCEPTIBILITIES RISK OF PATHOGEN SPREAD

REDUCED EPIDEMIOLOGICAL CONSTRAINTS. INCREASED SELECTION PRESSURE

GREATER INVESTMENT AT RISK. LESS ECOSYSTEM STABILITY

Fig. 1. Some effects of increased production on crop disease

Extending the areas

Extending the areas in which crops are grown can expose them to apparently new pathogens or pathogen races previously restricted to uncultivated habitats. Much of the original expansion and development of tropical perennial crops occurred in continents remote from their centres of diversity, where the major co-evolved pathogens did not exist, but where this expansion occurs closer to their original home, they may encounter apparently new diseases harboured by related wild species. Thus a virulent form of *Colletotrichum coffeanum* Noak emerged to cause coffee berry disease when arabica coffee plantations extended into the centre of diversity of *Coffea* spp. in East and Central Africa. The vulnerability of crops to exotic pathogens has been discussed by Buddenhagen (1977).

The natural or induced spread of existing pathogens to new areas may also be facilitated by expanding agricultural areas. Ecological barriers to disease spread may be breached as happened when coffee growing areas were opened up in the Amazon basin and coffee rust (*Hemileia vastatrix*) spread to the Andean countries of South America. Urban and infrastructural development associated with human increase has resulted in crop based agriculture moving into areas progressively more marginal or more variable for crop growth. Besides the direct effect of environmental constraints on crop growth, predisposition to soil borne root diseases also occurs.

Changes in Cultural Practices

These produce variable effects on crop diseases; those practices characteristic of "developed" agriculture, tend to increase the epidemiological vulnerability of crops. Diversity is a major stabilizing factor in natural ecosystems and the spatial and temporal discontinuity of plant species and genotypes within natural ecosystems constrains the epidemiological development of crop diseases. Monoculture of crop varieties with limited genetic diversity over large areas and for long periods erodes this natural buffering capacity and enhances the selection pressure for virulent pathogen races able to exploit the monocultured genotypes. This increased vulnerability needs to be offset by special crop protection methods involving fungicide application, as now used on many perennial crop monocultures in developing countries e.g. coffee, cocoa, or by the breeding and selection of crop cultivars with enhanced resistance. Both require the use of scarce or expensive resources. By contrast, multiple cropping systems increase diversity and ecological stability, although the components may need careful matching both for agronomic and marketing reasons.

The duration of the cropping season can be extended in order to increase production. This can be achieved by using earlier maturing (short season) cultivars or by the use of irrigation to offset seasonal moisture constraints; continuous rice cultivation can now produce three harvests during one year. Such temporal monoculture can enable plant diseases to progress unchecked year after year as there is an automatic carry over of inoculum between crops. Irrigation, however, is clearly of considerable benefit to developing country agriculture in many other respects and may help in disease avoidance or control. Production of certain crops by irrigation during the dry season can avoid periods of maximum disease pressure when inoculum levels of foliar pathogens are high and supplemental irrigation can remove drought stress and predisposition to soil borne diseases.

Many other changes in cultural practices have generally improved plant health, particularly where these have produced physiologically stronger plants or have helped to reduce sources of pathogen inoculum by e.g. adequate tillage, fertilizer use and weed control. Palti (1981) has discussed how cultural practices can be used to control plant diseases; many are particularly relevant to developing tropical countries where there is a need to enhance the disruption of epidemic cycles and to destroy sources of pathogen inoculum by fairly simple techniques.

The traditional agricultural methods which have evolved over centuries in Africa have often incorporated some "unconscious" elements of disease control; these have often been lost during modernization so that cultural disease control techniques now need to be built back into the system.

Genetically improved cultivars

Genetically improved cultivars are responsible for most of the recent increase in agricultural productivity of staple crops and many plant health problems have been overcome by the breeding and selection of resistant cultivars (Chapter 22). But surmounted problems quickly become forgotten; only those which remain or appear as new are of contemporary concern as constraints to future agricultural advancement. Some disease problems have been enhanced by the use in plant breeding programmes of germplasm exotic to the traditional agroecosystem. "New diseases" are a problem particularly associated with the new improved cultivars of staple crops. Often previously minor or obscure pathogens have become important on new varieties which have less resistance to them than the older traditional local land races; the cultural regime under which new varieties are grown can also exacerbate such weaknesses. Classic examples are the previously minor foliar pathogens of rice such as bacterial blight (*Xanthomonas campestris* pv. *oryzae* (Ishiyama) Dye), sheath blight (*Corticium sasakii* (Shirai) Matsumoto) and scald (*Monographella albescens* (Thum) Parkinson), which became important on the "green revolution" semi-dwarf rices especially when grown under higher nitrogenous fertilizer regimes. The increasing genetic uniformity associated with the use of new cultivars has resulted in the loss of the resistance stability which existed in the diversity of local land race populations. Thus pearl millet downy mildew (*Sclerospora graminicola* (Sacc.) Schröter) can overcome the resistance of many of the new cultivars within a few years because the natural buffering capacity of the genetically diverse outbreeding land race population has been reduced and selection pressure on the pathogen for particular virulence patterns is more directed.

The spread of new susceptibilities allowing the emergence of previously unnoticed diseases on new cultivars has often been confused with the spread of pathogens themselves, but the more widespread transfer of crop germplasm, increased agricultural trade, and greater opportunities for rapid international travel undoubtedly facilitates the spread of pathogens and pests. The continuing geographic spread of plant diseases and the emergence of new ones is illustrated by the number of new geographic and new host records abstracted annually in *Review of Plant Pathology*. Over the last five years these have averaged between 100-120 each per year, many occurring in developing countries.

Intensified production systems

Intensified production systems are the result of using both improved cultivars and modified cultural practices and seem to be a natural progression from subsistence agriculture; but the larger inputs needed require more investment both by the state and the farmer. Greater returns are expected, particularly by the farmer, for the extra effort and money put into growing the crop. Thus the intrinsic value of the crop is increased, economic thresholds become more critical, and the identification and removal of yield constraining factors becomes more important. Yet this becomes more difficult as the cost of crop protection and other inputs increases relative to prices paid for produce. For example, control of coffee rust in terms of yield equivalent has nearly trebled since the

1977 coffee "bonanza" as the costs of agrochemicals and machinery have increased and the market price of coffee has declined. The higher yields of more intensive agriculture are subject to greater variability because the reduced ecosystem stability makes crops more sensitive to the limiting effects of external factors (Swaminathan 1983).

Crop Health and Diagnostic Capability

The foregoing shows that the changes wrought by agricultural development have frequently increased the vulnerability of crops to diseases and have often favoured the development of greater virulence in pathogens themselves. While there have been obvious advances in the control of some diseases by genetic, chemical or cultural means, this has seemingly paved the way for others to emerge. Furthermore, as science and man's expectations advance so does the concept of crop health; todays apparently healthy crop may appear diseased by tomorrows standards.

The key factor in both the achievement and maintenance of crop health is diagnosis. Successful crop protection is based on the correct diagnosis of the causes of loss against which crops require protection. Subsequent evaluation and control follow on from this initial step and the CAB identification and information services have a key role to play in this sequence of events. The problems of diagnosis are particularly acute in developing countries and stem from two main causes: problems concerned with disease causation or etiology and those concerned with pathogen identification.

Etiological problems

These are often the result of confusing disease diagnosis with pathogen identification which is only part of the process of diagnosis. The Latin name of a putative pathogen means little without the vital knowledge of the causal relationship to disease loss and the circumstances of its occurrence, but once the total disease situation is understood, it provides the key to existing knowledge on control and other attributes.

Often, pathogenic fungi isolated or otherwise identified from diseased plants may be only one of several contributing factors involved in the development of disease. Interactions with environmental factors and with man's activities need to be considered. Thus *Fusarium stilboides* is of world-wide distribution on coffee yet is only significant in certain situations where environmental stress or cultivation damage allow it to initiate *Fusarium* bark disease.

Multiple infections and interactions between several pathogenic organisms (and pests) are particularly problematic in tropical areas. Synergistic interactions can overcome host plant resistance as occurs with nematodes and *Fusarium* wilt diseases. Apportioning the blame for loss between the various components of diseases of multiple etiology, and the formulation of rational control measures often requires considerable research effort. The relevance of these problems to agricultural development has recently been discussed in more depth elsewhere (Waller & Bridge 1984).

Recognition of disease symptoms is an important initial stage of diagnosis, yet these can be an unreliable guide to the cause. Many quite unrelated pathogens (and pests) can produce similar symptoms, e.g. false rust (*Synchytrium* sp.) and true rust (*Phakopsora* sp.) on legumes, *Helopeltis* damage and anthracnose (*Colletotrichum*) on many fruits. Definitive identification of the suspected casual agent is therefore essential and only too frequently faulty diagnosis has prolonged the resolution, often for years, of relatively simple crop protection constraints. In this context it is pertinent to note that Africa's usage of the CMI identification service has declined from 16% of accessions received in 1976 to 7% in 1982; two thirds of this reduction was from non-contributing countries.

Identification problems

These mainly concern the occurrence of apparently new or unfamiliar pathogens, or those for which specialized techniques are required such as with viruses or some bacteria-like pathogens. The major fungal pathogens of the more significant agricultural crops no longer present major problems of identification, although current taxonomic usage offers little guide to pathogenicity. New techniques seem to be required to resolve this. The identification of viruses, particularly of leguminous and solanaceous crops, does present a constraint to their control. Diseases of unknown or obscure etiology thought to be caused by mycoplasma- or rickettsia-like organisms still constrain productivity of some perennial crops such as citrus, coconuts and clove.

The Knowledge Base of Plant Pathology

Following the successful diagnosis of a plant protection problem, the application of existing knowledge is required for its proper evaluation and control. A large body of information exists on the biological attributes of pathogens, necessary for an understanding of disease dynamics (source, spread and epidemic increase) and on the various methods available for control. The knowledge base comprises both published scientific knowledge, access to which can be obtained through information retrieval systems, and that local knowledge acquired through the experience of individual scientists, departments of agriculture and local farming communities. The selection of appropriate information from this knowledge base and the use to which it is put are of critical importance to disease management.

Although scientists are trained to make use of the published scientific knowledge base, it is often done incompletely or indiscriminantly. As far as agricultural development is concerned much of this conventional knowledge base needs to be treated with caution. The major criterion for publication in scientific journals is novelty and the reason is often for the advancement of individual careers; access to it is through abstracts and indices often prepared with a meagre awareness of user requirements. A vast morass of published information exists and is increasing all the while. Few scientists concerned with the immediate problems of farmers' crops have sufficient time to wade through it, sift and digest the relevant bits and distill the essence suitable for application to problems of developing agriculture. Yet an appreciation of available knowledge is essential for the economic use of scarce research and development resources.

The unpublished knowledge base of corporate experience buried in the files of departments of agriculture and gained by the experience of agricultural scientists and farmers is of special relevance to the local development scene. Much of this is untapped and is already being lost, but is needed to temper and adapt the available conventional knowledge for contemporary local use. Local scientists must be the catalyst for this, but they must have a "sincere interests in carrying laboratory results to the field..... and using their training to help their countrymen" (Baker 1980). Hence the value of and need to retain local professional manpower.

Returning to the introductory theme of this paper, it is apparent that the advancement of agricultural productivity involves the revelation and surmounting of a succession of interacting constraining factors. Plant disease is only one of these but it is inextricably linked to others which are likely to increase the relevance of crop health. Much of the basic information required for successful diagnosis and correction of crop health problems exist, although there is a need for more basic knowledge about certain groups of pathogens, the durability of genetic resistance to diseases and more efficient and reliable chemical control techniques. It is the adaptation and application of this information by combining it with local experience and synchronizing it with the many facets of agricultural development that will help to advance agricultural productivity in Africa.

References

Baker, K.F. (1980) Developments in plant pathology and mycology (1930-1980). In *Perspectives in World Agriculture*, 207-236. Farnham Royal; Commonwealth Agricultural Bureaux.

Buddenhagen, I.W. (1977) Resistance and vulnerability of tropical crops in relation to their evolution and genetics. In Day, P.R. (ed) Genetic basis of epidemics in agriculture. *Annals of the New York Academy of Science* 287, 309-326.

Palti, J. (1981) *Cultural Practices and Infectious Crop Diseases*. Berlin; Springer-Verlag, 243 pp.

Swaminathan, M.S. (1983) Plant protection for global food security. *Proceedings of the 10th International Plant Protection Congress* I, 5-12.

Waller, J.M. (1984) The influence of agricultural development on crop diseases. *Tropical Pest Management* 30(1), 86-93.

Waller, J.M.; Bridge, J. (1984) Effects of pathogen interactions on tropical crop production. In *Infection Damage and Loss* (R.K.S. Wood; G.J. Jellis, eds), 311-319. Oxford; Blackwell Scientific Publications.

Chapter 35

Role of Nematodes as Constraints to Agricultural Development in Africa

J. BRIDGE and R.L.J. MULLER

Commonwealth Institute of Parasitology, 395A Hatfield Road, St. Albans, Herts. AL4 0XU, UK.

Introduction

Africa is host to many of the world's economically important plant-parasitic nematodes; some are endemic to this continent but a high proportion have almost certainly been introduced. In order to demonstrate the many ways in which nematodes can affect agricultural production we will give examples of the different types of plant nematodes that occur on a wide range of crops in Africa (Table 1) but this is by no means an exhaustive list. A full review of most of the plant nematodes occurring in Africa is given by Taylor (1976).

Table 1   Some important plant nematodes in Africa

Nematode	Main crops affected	Nematode	Main crops affected
Aphasmatylenchus straturatus	legumes	*Meloidogyne africana*	coffee
Aphelenchoides arachidis	groundnut	*Meloidogyne decalineata*	coffee
Aphenlenchoides besseyi	rice	*Meloidogyne hapla*	pyrethrum
Ditylenchus angustus	rice	*Meloidogyne incogniata*	vegetables, cotton (and many others)
Helicotylenchus multicinctus	banana	*Meloidogyne javanica*	tobacco, vegetables
Heterodera gambiensis	millet, sorghum	*Pratylenchus brachyurus*	Maize, groundnut, pineapple, cotton, cassava
Heterodera oryzae	rice	*Pratylenchus goodeyi*	banana
Heterodera sacchari	sugarcane	*Praylenchus zeae*	rice, sugarcane
Heterodera schachtii	beet	*Radopholus similis*	banana
Hirschmanniella oryzae	rice	*Rotylenchulus reniformis*	cowpea
Hirschmanniella spinicaudata	rice	*Scutellonema bradys*	yam
Hoplolaimus seinhoristi	cowpea, cotton	*Tylenchulus semipenetrans*	citrus
Meloidogyne acronea	cotton, sorghum		

Root-knot Nematodes of Vegetables

An earlier contribution (Chapter 16) mentioned the serious losses caused by root-knot nematodes, *Meloidogyne* species, on vegetables. These nematodes are common throughout all African countries without exception and are one of the major constraints to vegetable production. They produce characteristic galling or knotting of the roots and prevent normal growth of the plants. Their very short life-cycle, 3-5 weeks with each female producing 200-500 eggs, explains their rapid population increase and their importance as pests in tropical soils. The main root-knot species are endemic to Africa and a large proportion of the indigenous weeds and plants, including the Baobab tree, are hosts for one or more of the nematodes, so that it is incorrect to assume that new or virgin land is free of this pest.

Vegetables, as a cash or food crop, tend to be grown continuously on the same land, often under irrigation. The repeated planting of susceptible crops greatly increases root-knot damage. This is further exacerbated by irrigation schemes which provide an ideal moist environment for rapid and continuous multiplication of the pests and also a means of spread of the nematodes in irrigation water. Yield losses in excess of 25%, and often of 50%, can be expected when this nematode is allowed to build up in vegetable gardens.

The main vegetable root-knot species, *Meloidogyne incognita* and *M. javanica*, are pests of many other crops in Africa and one good example is that of *M. javanica* on tobacco. Control of this nematode on tobacco has produced yield increases of over 300% in Africa and its control is an essential part of commercial tobacco production if the enterprise is to remain viable. Another example is the case of cotton, a familiar and important crop in this part of the world. Cotton can suffer from infestation by the root-knot nematode *M. incognita*. It is not a pest of the crop in all cotton growing countries but it does occur in Tanzania, Uganda and the Central African Republic where it interacts with another serious disease, *Fusarium* wilt.

Root-knot nematodes can severeley stunt cotton and reduce yields on their own. When they occur with *Fusarium* wilt, the nematodes increase the incidence of the wilt disease and, possibly more important, they also allow the wilt disease to be seed-transmitted because of the late infection of the plants (Hillocks, pers. comm.).

Banana Nematodes

Other widely distributed nematodes are some of those that occur on banana. The most well-known and widespread pest is the banana burrowing nematode, *Radopholus similis*, but another which we are now finding to be a serious pest of bananas in Africa is a root lesion nematode, *Pratylenchus goodeyi* (Table 2). In Tanzania, it is *P. goodeyi* which is the major pest of cooking and other bananas in the Kagera region, occurring in over 90% of the farms and associated with severe decline in yields.

Nematodes are the most important root parasites of banana throughout the world, causing severe root and corm necrosis which stunts plants, reduces yields and commonly leads to toppling or fall-down of bananas, due to root breakage and uprooting of the plants. *R. similis* has been introduced into Africa whereas *P. goodeyi* is possibly endemic and is one of the nematodes that has probably been overlooked as a problem in many countries.

In addition to these species, other nematodes are pests of banana in Africa, including *Helicotylenchus multicinctus* and *Meloidogyne* species.

Nematodes are disseminated in the roots and corms of infested banana planting material, the suckers, which universally are the means of propagating the crop. This untreated planting material carries the nematodes and other pests from one country or site to another, and by this means the nematode pests have rapidly spread into and throughout Africa. In Tanzania, *Pratylenchus* and, more recently, *Radopholus* have probably been introduced via Uganda. At present, infested bananas are being moved from the traditional banana growing areas in the north of Tanzania to new planting areas, spreading the banana nematodes to other parts of the country.

Table 2   Known distribution of main banana nematodes in Africa.

Banana lesion nematode (*Pratylenchus goodeyi*)		Banana burrowing nematode (*Radopholus similis*)	
Canary Islands		Cameroon	Malawi
Kenya		Central African Republic	Mauritius
Tanzania		Congo	Mozambique
Uganda		Egypt	Nigeria
		Ethiopia	Senegal
		Gabon	Somalia
		Ghana	South Africa
		Guinea	Tanzania
		Ivory Coast	Uganda
		Kenya	Zambia
		Madagascar	Zimbabwe

Rice Nematodes

Banana is a food crop in many African countries. Another important food crop is rice which has its share of the nematode burden and some of the major rice nematodes occur in Africa. Of these, the rice white tip disease nematode, *Aphelenchoides besseyi*, is one of the most well known, causing leaf chlorosis and reduced grain yield. *A. besseyi* is a foliar nematode which is seed transmitted and thus very easily disseminated.

Another rice pest is the ufra nematode, *Ditylenchus angustus*. This is an extremely important foliar pest of rice in Asia where it can cause complete yield loss but, so far, has a very limited distribution in Africa.

From Table 3, it can be seen that the seed transmitted nematode *A. besseyi* which occurs on both upland and flooded rice is very widely distributed in Africa. In contrast, *D. angustus*, which is not seed transmitted and can only survive in areas of flooding and high humidity, has only been found, so far, in Madagascar and it is hoped that this state of affairs continues.

Rice root nematodes commonly found in flooded rice growing areas of Africa are species of the genus *Hirschmanniella*. These are migratory endoparasites that produce necrotic lesions and eventual rot of roots resulting in poor growth and yield. The losses caused by *Hirschmanniella* spp. can be over 30 % but generally are less than the observable threshold level of 15 % and thus tend to go unnoticed. Root-knot nematodes, particularly *M. incognita*, are also pests of rice in African countries.

Table 3   Known distribution of main rice nematodes in Africa.

Rice white tip nematode (*Aphelenchoides besseyi*)		Rice root nematodes (*Hirschmanniella* spp.)	
Benin	Mali	Cameroon	Mauritania
Cameroon	Nigeria	Gambia	Nigeria
Central African Republic	Senegal	Ghana	Senegal
Chad	Sierra Leone	Egypt	Sierra Leone
Congo	Tanazania	Ivory Coast	Upper Volta
Gabon	Togo	Madagascar	Zaïre
Ghana	Uganda	Mali	Zambia
Ivory Coast	Upper Volta		
Kenya	Zaire	**Rice Cyst nematode** (*Heterodera oryzae*)	
Madagascar	Zambia	Ivory Coast	Senegal
Malawi	Zimbabwe		
		Ufra rice nematode (*Ditylenchus angustus*)	
		Malawi	

Citrus Nematode

An example of an important nematode parasite of a perennial tree crop is the citrus nematode, *Tylenchulus semipenetrans*, which is commonly found on all types of citrus. It is a semi-endoparasitic swollen nematode that protrudes from the roots feeding on, and eventually destroying, the cortex. There has been comparatively little study of this nematode in Africa although it is certainly causing considerable yield loss. This is unfortunate because the nematode has an easy means of spread in roots and soil of citrus rootstocks and is widely distributed in Africa (Table 4). Treatment of rootstocks is comparatively simple and the introduction of the nematode into new land could easily be prevented.

Table 4   Known distribution of citrus nematode (*Tylenchulus semipenetrans*) in Africa.

Algeria	Kenya	Morocco	Uganda
Egypt	Libya	South Africa	Zaïre
Guinea	Malawi	Tanzania	Zimbabwe
Ivory Coast			

Specialities of Africa

There are a number of important nematodes that are unique to Africa. One of these is the African cotton root nematode, *Meloidogyne acronea*, which has only been found in Malawi and South Africa. Where it occurs on cotton in Malawi, it causes considerable root damage and yield loss. It has also been shown to markedly reduce yields of sorghum and other crops. *M. acronea* is at present restricted to a particular soil type in an area of low rainfall, but it has been shown that the advent of irrigation would enable the nematode to spread and cause plant damage in all soil types (Page 1984).

Other African root-knot nematodes, *Meloidogyne africana* and *M. decalineata*, are little known but very serious pests of coffee. They have been recorded on coffee in northern Tanzania and Kenya and have also been found in Sudan. On coffee, they can replace all the feeder roots with small, rounded galls containing many nematodes and eggs.

An unusual plant nematode pest of groundnuts which to date has only been found in Nigeria is the groundnut testa nematode, *Aphelenchoides arachidis*. The nematodes feed in the roots, pods and testas of groundnuts and cause the browning and shrinkage of the confectionery nuts because of damage to tissues of the testas. It can be seed transmitted, particularly in whole pods, and so, in theory, should have a wider distribution.

Yam Nematode

The yam nematode, *Scutellonema bradys*, causes dry rot disease of yam tubers in West Africa, greatly reducing the marketable value of the crop and initiating severe rot of the tubers during storage. It has been found to occur in almost 50% of tubers harvested in parts of Nigeria. Because yams are propagated from tuber pieces, the nematode is easily disseminiated.

Cyst Nematodes

A contrasting group of nematodes are those that produce protective cysts enclosing the eggs which enable the pests to survive for very long periods in adverse conditions and in the absence of a plant host. These cyst nematodes, *Heterodera* and *Globodera* species, are economically important pests in temperate agriculture and now are becoming known in the tropics. In Africa, many cyst nematodes are found but so far with a restricted distribution: *H. gambiensis* occurs on sorghum, millet and maize in the Gambia; the rice cyst nematode, *H. oryzae*, is potentially a serious pest of rice, at present found only in Senegal and the Ivory Coast; the sugarcane cyst nematode, *H. sacchari*, is known to reduce sugarcane yields in Nigeria and also occurs in the Congo and Ivory Coast; the soybean cyst nematode, *H. glycines*, is an important pest of soybean in different parts of the world but it has only been found in Egypt on the continent of Africa; *G. rostochiensis*, the temperate potato cyst nematode, has been recorded in countries around the Mediterranean and in South Africa. *H. schachtii*, a very important pest of sugarbeet and related crops in temperate climatic zones, has been introduced into the Gambia and Senegal where it has apparently become adapted to the higher tropical temperatures. It clearly points to the very real danger of the introduction into Africa of many other nematodes, even those originally considered not able to survive there.

Nematode Interactions with Other Organisms

Plant nematodes are not only pests in their own right but can, and often do, interact with many other pathogenic organisms to cause diseases of different crops, as we have shown above with root-knot nematodes and *Fusarium* on cotton. The nematodes can break disease resistance, predispose plants to infection, act synergistically in the development of a disease and can also act as vectors of a disease. Interactions are known with both bacteria and fungi, and certain nematode genera can transmit viruses. These interactions have received little attention in Africa and are likely to prove very important.

These examples of plant nematodes in Africa illustrate a number of points in relation to their importance as agricultural pests that can be summarized as follows:

(1) The problems of plant nematodes as agricultural pests are *increasing* in Africa, in the extent of the crop losses they cause, their distribution, and in the numbers of different parasitic species present.

(2) They are very diverse pests occurring on a wide range of crops, producing many different symptoms of damage. The first and most important step towards control is their identification and recognition as pests.

(3) Many plant nematodes are very easily disseminated in corms, tubers, rootstocks and in seed. In the absence of efficient plant hygiene or a quarantine service, internally and internationally, it is inevitable that many exotic nematode pests will enter and spread throughout African countries.

(4) Nematodes are involved in interactions with other pathogens, directly or indirectly, and agricultural scientists should be aware of this fact in their work.

(5) Changes or improvements to agricultural practices can increase both the incidence and extent of nematode damage to crops in the following ways:

(a) Irrigation produces the right conditions for continued survival of nematodes in the soil, allows for a rapid build up of populations in the land and spreads nematodes in irrigation water.

(b) Mechanical cultivation compared to more traditional hand-hoeing techniques will spread soil and nematodes more rapidly throughout the cultivated areas.

(c) Continuous cropping or monocropping of a nematode-susceptible crop can produce rapid nematode population increase and more serious yield losses.

(d) New varieties or new crops introduced into an area can often prove to be highly susceptible to the endemic species or biological races of nematodes and need to be tested prior to release.

(6) The means of reducing crop losses caused by nematodes are known in many cases; in others they have to be investigated. Crop damage by nematodes is a fact of African life and for it to be reduced requires greater commitment by scientists, administrators and funding bodies in the various countries of Africa.

Medical and Veterinary Nematodes and Helminths

Plant nematodes affect man indirectly through the crop but are not parasites of man. However, there are medical and veterinary nematodes and related helminths that cause diseases in man and livestock (Table 5). These nematodes and helminths can be profoundly influenced by different agricultural practices. Undoubtedly the most notorious is the tragic spread of schistosomiasis which is following the implementation of vital irrigation schemes in so many parts of Africa. For instance, the World Bank is financing a large cotton growing scheme at Rahad, Sudan, irrigated from the Blue Nile. About 100 000 people live in the area and before irrigation in 1979 there was no schistosomiasis; now both the intestinal and the bladder forms of the disease are common with an overall prevalence of nearly 40% and an annual health bill of $48M. This is in spite of the fact that careful precautions were taken to treat all cases before resettlement and appears to have resulted from unauthorized settlement by infected West African Immigrants. Once infection has become established in irrigation canals (where the infection is transmitted by freshwater snails), control is both difficult and expensive as evidenced by the energic attempts that have been made in sugar estates in Tanzania (e.g. Arusha Chini) over the past 30 years.

In general, the morbidity and economic importance of worm infections of both man and livestock depends very much on the intensity of transmission. In many parts of Africa, prevalence is widespread but the morbidity is low because infections are light. However, irrigation of plantation crops especially can change this situation. For instance, the larvae of human hookworm develop in the soil and heavy infections occur where there are moist conditions with heavy shade such as in banana, coffee or other plantations. Larvae of nematode parasites of livestock develop under similar conditions and the severity of infection will increase under intensive grazing on well watered pasture land.

Table 5   Helminths of medical and veterinary importance associated with agriculture in Africa.

Parasite	Host	How larvae acquired
Schistosomes	Man (Cattle)	Contact in canals, lakes, streams, ponds
Liver Fluke	Cattle	Ingestion with grass around ponds
Hookworms (*Strongyloides*)	Man	Contact with shaded, moist earth and grass
Intestinal Nematodes	Cattle Sheep	Ingestion with grass – heavy infections with intensive grazing
Intestinal Nematodes	Man	Ingestion with vegetables
Filariasis (also malaria)	Man	From mosquitoes which breed in ponds

Conclusion

In conclusion, it is hoped that we have clearly shown that these mainly microscopic (not to be confused with insignificant) worms and helminths of plants, animals and man are indeed an important constraint to agricultural development in Africa and, if ignored, will continue turning against all attempts at improvement.

References

Anon. (1973) *Republic of Zambia, Department of Agriculture. Annual Report of the Research Branch 1971-72.* Zambia Ministry of Rural Development, 287 pp.

Anon. (1979) *Inter African Phytosanitary Council. Fourteenth General Meeting, Cairo, Egypt, November 1978,* Yaounde, Cameroon; IAPSC, 134 pp.

Bridge, J. (1982) Nematodes of yams. In *Yams. Ignames* (J. Miege; S.N. Lyonga, eds), 253-264. Oxford; Clarendon Press Oxford.

Bridge, J.; Bos, W.S.; Page, L.J.; Mcdonald, D. (1977) The biology and possible importance of *Aphelenchoides arachidis,* a seed-borne endoparasitic nematode of groundnuts from northern Nigeria. *Nematologica* 23, 253-259.

Bridge, J.; Manser, P.D. (1980) The beet cyst nematode *Heterodera schachtii* in tropical Africa. *Plant Disease* 64, 1036.

El Tigani; El-Amin, M.; Siddigi, M.R., (1970) Incidence of plant parasitic nematodes in the northern Fung area, the Sudan. *FAO Plant Protection Bulletin* 18, 102-106.

de Grisse, A., (1971) *Tylenchulus semipenetrans*, un nematode phytoparasitaire des agrumes au Zaïre. *Mededelingen van Fakulteit Landbouwwetenschappen, Gent* 36, 1419-1427.

Germani, G.; Luc, M., (1982) Etudes sur la "calorose voltaique" des légumineuses due au nématode *Aphasmatylenchus straturatus* Germani. II. *Revue de Nématologie* 5, 195-199.

Gichure, E.; Ondieki, J.J. (1977) A survey of banana nematodes in Kenya. *Zeitschrift fur Pflanzenkranheiten und Pflanzenschutz,* 84, 724-728.

Muller, R.L.J. (1975) *Worms and Disease. A manual of medical helminthology.* London; Heinemann, 161 pp.

Page, S.L.J., (1984) Effects of the physical properties of two tropical cotton soils on their permanent wilting point and relative humdity, in relation to survival and distribution of *Meloidogyne acronea. Revue de Nématologie* 7, in press.

Taylor, D.P. (1976) Plant nematology problems in tropical Africa. *Helminthological Abstracts,* B, 45, 1-40.

Taylor, D.P.; Netscher, C.; Germani, G. (1978) *Adansonia digitata* (Baobab), a newly discovered host for *Meloidogyne* sp. and *Rotylenchulus reniformis*: agricultural implications. *Plant Disease Reporter* 62, 276-277.

van Gundy, S.D.; Luc, M., (1979) Diseases and nematode pests in semi-arid West Africa. In *Agriculture in Semi-Arid Environments* (A.E. Hall; G.H. Cannell; H.W. Lawton, eds), 257-265. [Ecological Studies no. 34.] Berlin; Springer.

Webster J.M. (ed.) (1972) *Economic Nematology.* London & New York; Academic Press, 563 pp.

Williams, R.J., (1976) A review of the major diseases of soybean and cowpea with special reference to geographical distribution, means of dissemination and control. *African Journal of Plant Protection* 1, 83-86.

Yassin, A.M.; Zeidan, A.B., (1982) Root-knot nematodes in the Sudan, 1982 round-off report. In *Proceedings, 3rd Research and Planning Conference on Root-knot Nematodes,* Meloidogyne *spp. Region VII. September, 1982, Coimbra, Portugal.*

Chapter 36

Advancing Weed Control Strategies for Developing Countries

I.O. AKOBUNDU

International Institute of Tropical Agriculture (IITA), P.M.B. 5320, Ibadan, Nigeria.

Introduction

Adequate food and fibre for man's needs can only be supplied by a stable crop production system. Good weed control is one of the corner stones of such a crop production system. Weeds are so universally associated with reductions in crop yield that their removal should become an integral part of any farming systems programme and of every farm practice.

Most developing countries of the world fall within the tropics which has about 50 % of the world population and less than 40 % of the earth's land surface (FAO 1977). The per capita food production in most of the developing countries has declined over the past two decades while human population growth rate has increased steadily (Christensen *et al.* 1981). The challenge of providing enough food for people in the developing countries is therefore one of the world's most pressing and urgent problems. While bringing more virgin land into cultivation, increasing land use intensity, and introducing genetically improved crops that give yield stability are among the ways to increase food production and sufficiency; these must be linked with improvements in weed control practices. Practices that will minimize yield losses caused by weeds, remove the drudgery associated with traditional crop production practices in the developing countries, and improve the productivity of tropical soils are considered in this paper.

Weed Problems in Developing Countries

Crop losses caused by weeds are higher in the developing countries of the world than in the developed countries where improved technology is used in crop production (Parker & Fryer 1975). Weeds cause crop losses by interferring with normal crop growth and development through competition for nutrients, water and space, by lowering the quality of harvest through contamination with foreign matter, by serving as hosts of insects and diseases, by increasing the cost of crop production, and by creating a need for control measures which may themselves injure the crops. Weeds are also known to poison food and forage (Akobundu 1981a, Dawson & Holstrun 1971, Koch *et al.* 1983).

Weed problems are so inextricably interwoven with problems of soil and climate that the very land use systems that call for a long fallow period to "rest" the soil perpetuates the problem of perennial weeds. The long tradition of plantation agriculture for export crops has also encouraged the establishment of shade-tolerant perennials such as *Chromolaena odorata, Melanthera scandens;* climbers such as *Geophila repens, Dioscorea hirtiflora;* and many parasites and epiphytes. The most serious weeds of food crops in the semi arid tropics are the parasitic weeds of which *Striga* species (Chapter 13) are the most common. *Cuscuta australis, Cassytha filiformis* and *Buchnera hispida* also effect a wide range of crops throughout the developing countries. Advancement in weed control technology is normally related to the technological development of a country and this has complicated the solving of weed problems in the developing countries.

Advances in Weed Control

Most developing countries have little or no improved weed management to support their crop production efforts. International concern on the status of weed management in the developing countries led to a recent Expert Consultation sponsored by FAO and the International Weed Science Society (IWSS). This consultation was aimed at identifying ways to improve weed management in these countries and the recommendations have since been made widely available (FAO 1983).

Weed control in food crops in the developing countries is carried out in either monoculture or intercropping systems. Irrespective of the cropping system, weed management falls into broad categories of cultural, chemical, biocontrol (biological) and integrated weed control systems.

Cultural weed control

This relates to all methods employed by the farmer to physically remove growing weeds from the crop environment and the energy required may be provided by man, his animals or machinery. The extent to which the energy derives from man depends largely on the level of technological advancement. Hand pulling of weeds represents the most primitive form of weed control in field crops and is commonly used in the developing countries along with hand hoeing and inter-row weeding with animal- or tractor-drawn implements. These weed control methods expose the crop to risks of accidental mechanical damage to its roots or above-ground parts.

The fact that crops and weeds co-exist is well-known and each crop has a critical period in its growth cycle when it is most sensitive to weed interference. The primary purpose of cultural weed control study is to identify this critical period for each crop and the most economical method to remove weeds at this period. For most food crops with growth periods of less than 20 weeks, two timely weedings at three and six to eight weeks after planting are adequate to minimize yield reductions especially if initial seeding is in a weedfree seedbed (Akobundu 1977, 1981a, Gill 1983, Moody & Ezumah 1974). Cultural weed control methods are most effective if done at the early stages of crop establishment and subsequently to keep it free of weed competition throughout its critical growth period. The role of these cultural weeding methods on weed management in the developing countries was recently reviewed by Gill (1983).

The farmer's weeding practice accounts, in part, for the low yields he gets even when he grows genetically improved crops. This is because labour is in short supply during the time of first weeding. Hence handweeding frequency should be minimized by using a low dosage of herbicide at planting or early post-emergence to be followed later in the crop cycle with one handweeding. There is also need for better understanding of the weeding needs of crops in specific countries and ecologies. Finally there is need to increase farmer's awareness of the opportunity cost of timely first weeding. Already, weeding is a farm activity that takes up more of the farmer's time than any other crop production input in the developing countries.

Biocontrol of weeds

This method refers to weed control by the action of one or more organisms, accomplished either naturally or by the manipulation of the control organism or environment. Major developments in this area include classical biological control of weeds with animals (vertebrate and invertebrate), use of plant pathogens for weed control (myco-herbicides) and use of non-crop plants (live and dead in-situ mulch).

Although examples of classical biological weed control have been achieved in several developed countries (Delfoss & Cullen 1981, Wilson 1960), few have been reported in the developing countries. A world catalogue of agents and their target weeds was recently edited by Julien (1982). Several attempts to control siam weed (*Chromolaena odorata*) in both Asia and Africa have largely failed (Julian 1982).

Progress in use of plant pathogens for weed control (myco-herbicides) has been reported in pasture, rangeland, citrus groves and rice fields in various parts of the USA (Hay 1980, Templeton 1983, Templeton & Smith 1983), but these biocontrol tools are still a long way from field testing and use in the developing countries.

Recent studies with living and dead *in situ* mulches show that weeds can be effectively controlled with non-crop plants in crop production systems of the developing countries. This can enable the low activity clay soils of the tropics to be used intensively without the yield decline observed when conventional tillage practices are used (Akobundu 1980a, Akobundu & Okigbo 1984). Live mulch is a promising alternative to cultural weed control in the developing countries. Current research on dead mulches for weed control centres on manipulating herbaceous tropical legumes such as *Mucuna pruriens* var. *utilis* which can either be interplanted in maize or used as a fallow crop in whose dead mulch a range of other food crops can be grown (IITA 1982).

Chemical weed control

The high pest pressure in the developing countries coupled with increasing costs of labour has already made chemical weed control an attractive option. In terms of energy use, chemical weed control represents a major technological advancement from a dependence on human energy for hand weeding to the use of chemical energy. The application of pre-emergence herbicides makes it easy to cope with problems of early weed interference during crop establishment. The ability of chemical weed control to increase crop yield and reduce labour cost in the tropics has been well documented (Akobundu 1980a, Parker 1972, Li 1983). Major constraints limiting herbicide use in developing countries had been previously discussed by Akobundu (1980b). These include: (1) inadequate knowledge of herbicide recommendations and usage; (2) poor timing of herbicide application; (3) scarcity of consumer-usable small-packs of herbicides; (4) lack of reliable extension services; and (5) an acute shortage of professionally trained weed scientists.

It is 40 years since the introduction of organic herbicides and many plants previously susceptible to them have now either become resistant or show increased tolerance to various types of herbicides (LeBaron & Gressel 1982). Weed resistance to herbicides has been most associated with families or orders that are wholly or partially self-sterile and found in the monoculture type of agriculture with limited crop rotation, and where specific herbicides have been used repeatedly (Hill 1982). By contrast, fewer cases of weed resistance to herbicides have been noted in countries where crop and herbicide rotation are widely practised.

Integrated weed control

Integrated weed control involves combining two or more practices to obtain better weed control effectiveness and cost-benefit ratios than obtained by any of the component practices used singly. Problems caused by climatic extremes, poor soils, cropping variability, manpower shortages and an abundance of hard-to-kill weeds make absolute reliance on any one method of weed control difficult in the developing countries. Handweeding is laborious, unattractive and can be ineffective during the rains in the humid tropics and herbicides used at rates that are safe for the crop, fail to provide adequate weed control in long season crops. An integrated approach to weed management offers a real solution to these problems.

(a) Cultural and chemical weed control

The time of the first weeding is often delayed because it competes for labour with other pressing activities. A low rate of a pre-emergence herbicide used to provide a weedfree environment for early crop establishment will give the farmer time to carry out subsequent weedings during the lag period when labour is more readily available. In relay and other cropping systems where earthing-up is a standard practice, the first crop (e.g. maize) could be treated with a low rate of herbicide with will keep the field clean until such time when the follow-up crop can be planted.

(b) Cultural and biocontrol system

Use of low growing crops such as Egusi melon (Akobundu 1980c) combines cultural weed control with the use of melon to provide ground cover and smother weeds in such cropping patterns as sole maize, maize/cassava and maize/yam/cassava. The use of *in situ* mulch also reduces the volume of herbicide needed to provide lasting weed control in maize (IITA 1982). The use of cassava cultivars that develop good canopy to shade weeds out is an excellent example where breeding can play a role in reducing the cost and frequency of weed control.

Logistics of Weed Science Technology Transfer

Research station results from many of the developing countries show that crop yield, even with existing husbandry practices can be increased if weeds are better controlled and that more land can be cultivated if the farmer does not have to depend on hired labour for weed control. Ultimately it is the farmers in the developing countries that will have to use the new developments in weed control to produce their crops more efficiently. An efficient technology transfer system is necessary to permit the incorporation of these new developments into the farmer's cropping systems.

Weeds and their control are human affairs. Repeated, arduous hand weeding becomes both degrading and dehumanizing when society makes no effort to improve the practice. Those who do the hand weeding have little choice – the alternative is starvation; they do not have the education to secure alternative employment in the cities. They invest all they have in their children's education in the hope that these children will never return to till and weed as these courageous but often neglected farmers have had to. Unless we find a way to transfer the improved weed control methods to these farmers, the yields of our improved crops will continue to decline not just from weed competition but because no young man wants to bend or stoop to keep up the fight against weeds. Some of the ways out are as follows.

Manpower development in weed science

Very few developing countries have a citizen who is a professionally trained weed scientist. In Nigeria for example, there is at present one weed scientist to every 900 000 farmers. There is a quantitative as well as qualitative need in manpower development in the developing countries. How this deficiency can be met has recently been discussed (Akobundu 1983, FAO 1983, Matthews & Burrill 1983). There is no lasting alternative to sound national training to combat these problems.

Government Intervention

Weed problems are location-specific and no country can rely on imported manpower to solve its weed problems. The Governments of the developing countries must now take positive steps to improve manpower development in weed science, for example by offering postgraduate scholarship awards in weed science. A team of professional capable weed scientists will build a sound foundation in research and teaching to solve each country's problems. Since facilities for postgraduate training are lacking in most developing countries, weed scientists need to be trained abroad in the short-run, until local capabilities to provide such training in the country improve. The manpower needs in weed science cannot always be met by sending people for short-term refresher courses overseas as part of bilateral aid.

Role of International Research Institutions and donor-Governments

Many developed countries continue to provide aid money both to prevent starvation and improve food production capabilities in the developing countries. Many of these projects include training components but few include weed science as an area worthy of staff development. It is easier to breed for disease and pest tolerance than for better weed competition but the subsidized fertilizer programmes and the introduced high yielding crops are rejected by farmers because they fail to overcome problems of weed control. Nearly all the International Agricultural Research Institutes (IARCs) are either located in or have contact with the developing countries and advantage should taken of the training facilities and opportunities they offer to strengthen national manpower in weed science.

Role of the Agrochemical Industry

Agrochemical industries play an active part in personnel development in weed science in developed countries. Many of them operate in the developing countries where they should be seen to behave as partners in progress rather than unscrupulous traders. Agrochemical companies can support manpower development efforts in the developing countries by providing research grants for weed science programmes in local research institutions; many weed scientists in developed countries received training through such research grants.

Role of national research institutions and universities

Many research institutions and universities actively engaged in agricultural research do not have a professionally trained weed scientist. Some have attempted to solve the problem by exploring short-term solutions. Sometimes by giving staff without professional training in weed science responsibility for weed research and training. Weed science is a specialized discipline and cannot be practised as a science by a layman and specialists will train other people better than the general-purpose scientist-turned "weed control expert". Both the universities and research institutions should develop full teaching and research capabilities through balanced manpower development.

Development of research capabilities within the developing countries

Inspite of the work done by the handful of weed scientists in the developing countries within the last decade, more is needed to evaluate fully the weed problems in agriculture. Areas in which research capabilities are most needed are:

(a) Weed/crop associations

The nature of weed/crop associations, weed interference in various crops and losses caused by weeds can be adequately studied only when the developing countries attain full research capabilities in weed science. Crop and land management systems affect weed flora, intensity and distribution. Attempts at initiating sound weed control practices require an understanding of weed biology and ecology.

(b) Methods of weed control

More information is needed on methods of weed control appropriate to the various cropping systems commonly seen in developing countries. Future research should focus on how to make farming more attractive through improvements in weed control practices.

(c) Safe use of pesticides

Herbicides and other pesticides will continue to play a role in third world agriculture and research is needed on safe methods of using them. Since chemical control of crop pests involves the use of many different pesticides research capabilities should not be limited to herbicides.

Transfer of weed science technology

A first step in technology transfer is to intensify on-farm research in the developing countries. This should examine the small farmer's practices to identify areas where new technologies will improve the farmer's welfare and productivity. Infrastructural modifications may be necessary to ensure that (1) farmers have been properly prepared to receive the package, (2) the new package is economical and will not drastically change their way of life, and (3) the new technology will improve their productivity and well-being.

Conclusion

Farmers in the developing countries have for years been asked to produce their crops using traditional weed control practices. Uneven development in vital areas of crop production hampers efforts to improve crop yield in these countries. Subsidized fertilizer has tended to benefit the weeds as well as the crops because Governments have no lasting solutions to weed problems. Improved crop varieties that give excellent yields in research and demonstration plots have fallen prey to the ubiquitous weeds on the farmers' fields. There is an urgent need for a shift in emphasis.

An agricultural system that condemns a people to the drudgery of hand weeding, and ties up a majority of any country's economically active population to the primary function of food production, is certainly perpetuating mediocrity and preventing the country from advancing its economy and well-being. A better understanding of all the components required for the production of a given crop, and their interrelationships necessary to sow it, protect it until it is harvested and safely delivered to the consumers, requires a balanced development in our agricultural industry. A holistic view of crop production systems will allow weed control strategies to contribute to food sufficiency in the developing countries.

References

Akobundu, I.O. (1977) Weed control in root and tuber crops. *Proceedings of the Annual Conference of the Weed Science Society, Nigeria* 7, 19-28.

Akobundu, I.O. (1980a) Economics of weed control in African tropics and subtropics. *Proceedings of the British Crop Protection Conference 1980 – Weeds* 3, 911-919.

Akobundu, I.O. (1980b) Weed science research at the International Institute of Tropical Agriculture and research needs in Africa. *Weed Science* 28, 439-445.

Akobundu, I.O. (1980c) Weed control strategies for multiple cropping systems of the humid and subhumid tropics. In *Weeds and their Control in the Humid and Subhumid Tropics* (I.O. Akobundu, ed.), 80-100. [IITA Proceedings Series no. 3.] Ibadan; International Institute of Tropical Agriculture.

Akobundu, I.O. (1981a) Weed interference and control in white yam (*Dioscorea rotundata* Poir). *Weed Research* 21, 267-272.

Akobundu, I.O. (1981b) Weed control and food production in the humid tropics. *Proceedings of the Association for the Advancement of Agricultural Science in Africa, 3rd General Conference* 3, 225-241. Ibadan.

Akobundu, I.O. (1983) The role of conservation tillage in weed management in the advancing countries. In *Improving Weed Management* [FAO Plant Production and Protection Paper no. 44.], 23-39. Rome; FAO.

Akobundu, I.O.; Okigbo, B.N. (1984) Preliminary evaluation of ground covers for use as live mulch in maize production. *Field Crops Research* 8, in press.

Christensen, C.; A. Dommen; N. Horenstein; S. Pryor; P. Riley; S. Shapouri; Steiner, H. (1981) *Food Problems and Prospects in Sub-Saharan Africa; The Decade of the 1980s.* [Foreign Agricultural Economics Report no 166.] International Economics Division, ERS, USDA, 293 pp.

Dawson, J.H.; Holstun, J.T. (1971) Estimating losses from weeds in crops. In *Crop Loss Assessment Methods* (L. Chiarapa, ed.). Farnham Royal; Commonwealth Agricultural Bureaux.

Delfosse, E.S.; Cullen, J.M. (1981) New activities in biological control of weeds in Australia. III. St. John's wort: *Hypericum perforatum. Proceedings of the 5th International Symposium on Biological Control of Weeds, Brisbane 1980,* 575-81.

FAO (1977) *State of World Agriculture.* Rome; FAO.

FAO (1983) *Improving Weed Management.* [Plant Production and Protection Paper no. 44.] Rome; FAO, 185 pp.

Gill, H.S. (1983) The role of hand and mechanical weeding in weed management in the advancing countries. In *Improving Weed Management.* [Plant Production and Protection Paper no. 44.], 17-22. Rome; FAO.

Hay, J.R. (1980) Weed science – a changing technology. *Weed Science* 28, 617-620.

Hill, R.J. (1982) Taxonomy and biological considerations of herbicide-resistant and herbicide-tolerant biotypes. In *Herbicide Resistance in Plants* (H.M. LeBaron; J. Gressel, eds), 81-98. New York; Wiley-Interscience.

Holm, L.G.; Plucknett D.L.; Pancho, J.V.; Herberger, J.P. (1977) *The World's Worst Weeds: Distribution and ecology.* Hawaii; Honolulu East-West Centre, 609 pp.

IITA (1982) *1982 Annual Report.* Ibadan; International Institute of Tropical Agriculture.

Julien, M.H. (1982) *Biological Control of Weeds: A World Catalogue of Agents and their Target Weeds.* Farnham Royal; Commonwealth Agricultural Bureaux.

Koch, W.; Beshir, M.E.; Unterladstatter, R. (1983) Crop losses due to weeds. In *Improving Weed Management.* [FAO Plant Production and Protection Paper no. 44.], 153-163. Rome; FAO.

LeBaron, H.M.; Gressel, J. (1982) *Herbicide Resistance in Plants.* New York; Wiley-Interscience, 401 pp.

Li, Y.H. (1983) Farmers' weed control technology in rice in mainland East Asia. In *Weed Control in Rice,* 147-152. Los Banos, Philippines; International Rice Research Institute.

Matthews, L.J.; Burrill, L.C. (1983) Status of training opportunities and training needs. In *Weed Control in Rice,* 407-408. Los Banos, Philippines; International Rice Research Institute.

Moody, K.; Ezumah, H.C. (1974) Weed control in major tropical root and tuber crops – A review. *PANS* 20, 292-299.

Parker, C. (1972) The role of weed science in developing countries. *Weed Science* 20, 408-413.

Parker, C.; Fryer, J.D. (1975) Weed control problems causing major reductions in world food supplies. *FAO Plant Protection Bulletin* 23, 83-95.

Templeton, G.E. (1983) Integrating biological control of weeds in rice into a weed control program. In *Weed Control in Rice,* 219-225. Los Banos, Philippines; International Rice Research Institute.

Templeton, G.E.; Smith, Jr. R.J. (1983) The role of plant pathogenes in weed management in advancing countries. In *Improving Weed Management.* [Plant Production and Protection Paper no. 44.], 85-90. Rome; FAO.

Wilson, F. (1960) A review of the biological control of insects and weeds in Australia and Australian New Guinea. *Commonwealth Institute for Biological Control, Technical Communication* 1, 51-68.

Chapter 37

Biological Control Constraints to Agricultural Production

D.J. GREATHEAD

Commonwealth Institute of Biological Control. Silwood Park. Ascot. Berks. SL5 7PY. UK.

Introduction

The term biological control was orginally applied to the use of parasites, predators and pathogens as pest control agents. However, there have been attempts to broaden this definition to include use of behaviour-modifying chemicals, toxins, cultural controls, sterilization and genetic controls. These other biologically based control measures use different techniques and are not included in this appraisal.

The employment of predators as control agents is as old as agriculture and the domestication of cats to suppress rodents in food stores is probably the first example. Most other early attempts also employed mammals and sometimes birds. Many were unsuccessful or had unwanted results as when mongooses released on islands in the West Indies and Indian Ocean for rat control attacked chickens instead. These unfortunate results gave many people the impression that biological control is dangerous or ineffective. However, these predators are not host-specific and have adaptable behaviour which allows them to adopt new prey.

Many less highly evolved organisms have specialised anatomy and stereotyped behaviour which ensures that they cannot easily change their food preferences and so can be relied upon to attack the target pest and nothing else. The first spectacular success using such a host-specific predator, the introduction of an Australian ladybird (*Rodolia cardinalis*) against the cottony cushion scale (*Icerya purchasi*) into California in 1886, achieved permanent control and saved the Californian citrus industry from disaster. The excitement generated by this well-publicised success, achieved before effective safe pesticides were available, stimulated entomologists in many parts of the world to try biological control. Since then, there have been more than 330 successes against over 140 insect pests (e.g. Clausen 1978, Wilson 1960, McLeod *et al.* 1962, Canada 1971, Greathead 1971, Rao *et al.* 1971, Greathead 1976), and more than 50 against over 20 species of weeds using insect natural enemies on all continents (Julien 1982).

In these instances, the pest or weed was an introduced species and the control agent introduced from its area of origin. This use of biological agents – the *introduction* method often referred to as "classical" biological control – is so far the most widely used method and is the most attractive since long-term control is achieved from a single introduction or series of introductions, and it involves no further effort or expense. There are other ways of using biological control agents: *inoculation*, where an agent cannot survive indefinitely and has to be recolonised periodically; *augmentation*, when naturally occurring agents are insufficiently abundant to control a pest and so are augmented by carefully timed releases of laboratory-bred individuals; *inundation*, when an agent, native or imported, is mass-bred or cultured and used to give quick short-term control as a biological pesticide; *conservation*, when naturally occurring natural enemies are made more effective by indirect measures to increase their numbers.

These other methods are not as spectacular and are more difficult to apply. Usually, recurrent cost is involved and trained personnel are required to implement them. However, introductions are not always possible and may not be effective, then these other methods can be used. Frequently, a crop has a complex of pest diseases and weed problems and biological control cannot solve them all. It is then combined with other compatible control measures in integrated pest control (IPC).

Until synthetic pesticides became readily available in quantity after World War II, biological control and common sense combinations of other methods had to be used. There followed a period of reliance on chemical pesticides and biological control declined in popularity but now that pesticide resistance is widespread, the deleterious effects of chemicals on non-target organisms are realised and the cost of chemicals is increasing, biological control is again more widely used.

So far, arthropod (insect and mite) natural enemies have been most frequently employed as control agents but now as the biology, handling methods and production techniques for other kinds of agents are better known, their use is increasing and the range of agents now in use includes nematodes, fungi, bacteria and viruses. Usually, these agents are most effective when used as biological pesticides, and thus provide useful substitutes for broad spectrum pesticides in IPC.

The range of pests being suppressed by biological control has widened also to include some plant diseases and recent research on nematode natural enemies shows promise that there will eventually be biological controls for at least some plant pathogenic species (Papavizas 1981).

Biological control in Africa

In Africa, as elsewhere, enthusiasm for biological control has varied from country to country. Usually once a major success has demonstrated the benefits, it becomes a technique which is considered when a new problem develops. Thus, the chief practitioners on the African mainland are South Africa (since 1892) and Kenya (since 1911), and among the islands, Mauritius (since 1762). Elsewhere, few successes have been reported and many countries are yet to attempt biological control.

Biological control results throughout the Afro-tropical region up to 1969 were reviewed by Greathead (1971); these and new successes are summarized in Table 1. Successful biological control by introduction has been achieved in at least one country against 30 insect pests and nine species of weeds. In nine instances against insects and five against weeds this has been achieved by repeating introductions which had already been successful on other continents, for example against cottony cushion scale, apple woolly aphis, citrus scale insects, prickly pear cactus, and water hyacinth. On these occasions, no new research is required. Once the opportunity is known, it is a matter of locating the source of the agent, importing, breeding and releasing it. This is not costly but requires plant protection personnel who are aware of work outside their countries and have the enthusiasm to persevere in obtaining material, getting it to breed under local laboratory conditions and conducting the release programme in such a way as to give the control agents the best chance of success. There are many more opportunities to capitalise on results elsewhere. Thus, so far, only Zambia, Zimbabwe and South Africa have taken biological control of the potato tuber moth seriously. In each of these countries, the savings have soon repaid the outlay. A costing by an economist of the programme in Zambia forecast benefits of 346 000 Kw discounted over 20 years to 1980 for an outlay of 22 580 Kw (Cruickshank & Ahmed 1973).

In other instances, when a new pest has appeared, research has been required before candidate control agents could be introduced. Control of the coffee mealybug in Kenya is an instructive example. This appeared in eastern Kenya in 1922-23 and was soon jeopardizing the coffee industry. At first, it was incorrectly identified as an Indonesian species and an expedition was made to Java and sent back natural enemies which did not attack the mealybug. Once it was realized that the pest was a hitherto unknown species, a search was made for its origin, which turned out to be in Uganda. There it was under natural control, and parasites subsequently introduced from Uganda brought the mealybug under complete control by 1940. Except for some outbreaks during the 1950s, when coffee was sprayed with persistent organo-chlorine insecticides, this mealybug has ceased to be of economic importance. Melville (1959) estimated that £10M had been saved up to that time for an expenditure not exceeding £30 000.

Another important example is the eucalyptus weevil, a native of Australia which was found in South Africa in 1916; by 1925, it had become a serious pest throughout the country. After

Table 1   Targets of successful biological control in tropical Africa.

INSECTS		Country (year of introduction)*
Collembola		
Lucerne earth flea	*Sminthurus viridis*	SA (1963-9)
Orthoptera		
Red locust	*Nomadacris septemfasciata*	MRU (1962)
Hemiptera: Homoptera		
Sugarcane Planthopper	*Perkinsiella saccharicida*	MRU (1956-7)
Citrus psyllids	*Trioza erytreae*	R (1975)
	Diaphorina citri	R (1978)
Citrus black fly	*Aleurocanthus woglumi*	K (1960,66) SEY (1955-6) SA (1959-60)
Woolly apple aphis	*Eriosoma lanigerum*	K (1927-8) SA (1920-1) ZIM (1961)
Common white mealybug	*Icerya seychellarum*	MRU (1952) SEY (1930)
Cottony cushion scale	*I. purchasi*	K (1917) SEN (1954) St.H (1898) SA (1892) T (1955) ZAM (spread) ZIM (spread)
Citrus mealybug	*Planococcus citri*	SA (1900-30)
Kenya mealybug	*P. kenyae*	K (1938)
Jacaranda blight	*Orthezia insignis*	K (1953) T (1953) U (1953)
Soft green scale	*Coccus viridis*	SEY (1953)
Mussel scale	*Cornuaspis beckii*	SA (1966) ZIM (?)
Cirular purple scale	*Chrysomphalus aonidum*	SA (1962)
Citrus red scale	*Aonidiella autantii*	SA (1962-6)
Date Palm scale	*Parlaetoria blancnardii*	MAU (1967-9)
Coconut scale	*Aspidiotus destructor*	MRU (1937-9) P (1955)
Coconut scale complex	—	SEY (1936-40)
White sugarcane scale	*Aulacaspis tegalensis*	MRU (1969, 75) T (1971)
Lepidoptera		
Potato tuber moth	*Phthorimaea operculella*	MRU (1964) SA (1965-9) ZAM (1968) ZIM (1967)
Spotted stem borer	*Chilo sacchariphagus*	MAD (1960)
Pigeon pea pod borers	*Maruca testulalis*	MRU (1953-9)
	Etiella zinkenella	
Pink stem borer	*Sesamia calamistis*	MAD (1968) MRU (1951-2) R (1953-5)
Orange dogs	*Papilio demodocus*	R (1965)
Diptera		
Stable flies	*Stomoxys calcitrans*	MRU (1965-9)
	S. nigra	MRU (1972-3)
Coleoptera		
Eucalyptus weevil	*Gonipterus scutellatus*	K (1945) MAD (1948) MAL (spread) MRU (1946) StH (1958) SA (1926) U (spread) ZIM (spread)
White grubs	*Clemora smithii*	MRU (1913-39)
	Oryctes tarandus	MRU (1917)
Rhinoceros beetle	*Oryctes rhinoceros*	MRU (1970)
Acarina		
Red spider mites	*Tetranychus* spp.	SA (1979)

WEEDS

Boraginaceae
Black sage — *Cordia curassavica* — MRU (1948-50)

Cactaceae
Prickly pear — *Opuntia ficus-indica* — K (1957) MAD (1923) MRU (1928)
O. megacantha — R (189?) SA (1913-41) T (1958)
Imbricate cactus — *O. imbricata* — SA (1969)
Jointed cactus — *O. aurantiaca* — SA (1932, 79-80)

Clusiaceae
St John's wort — *Hypericum perforatum* — SA (1960-2, 71)

Pontederiaceae
Water hyacinth — *Eichhornia crassipes* — SUD (1979-81)

Proteaceae
Hakea — *Hakea sericea* — SA (1969)

Salviniaceae
Water fern — *Salvinia molesta* — ZIM (1970) ZAM (1970)

Verbenaceae
Lantana — *Lantana camara* — SA (1962) U (1960-3)

---

* K, Kenya; MAD, Madagascar; MAL, Malawi, MAU, Mauritania; MRU, Mauritius; P, Principe; R, Réunion; SA, South Africa; StH, Saint Helena; SEN, Senegal; SEY, Seychelles; SUD, Sudan; T, Tanzania; U, Uganda; ZAM, Zambia; ZIM, Zimbabwe.

investigations in Australia, an egg parasite was introduced in 1926. By 1935, it had achieved satisfactory control except at high altitude. The weevil later spread northwards into Zimbabwe and Malawai but was accompanied by its parasite and so did not cause serious damage. It appeared at Kisumu in 1944, probably arriving on flying boats carrying war supplies, and was controlled by introduction of the parasite in 1945. Similarly, when it reached Mauritius (in 1940), Madagascar (in 1948) and St Helena (before 1958), it was soon brought under control by introduction of the parasite (Greathead 1971).

Biological control results against tropical weeds have been fewer. The most widely used agents have been those for control of the rangeland weeds prickly pear cactus and lantana, following successful introductions elsewhere. Against prickly pear, introductions of cochineal insects began at the end of the 19th century in South Africa and Réunion and at later dates other species were liberated in South Africa, East Africa, Mauritius and Madagascar, as well as the moth borer (*Cactoblastis cactorum*). Results were variable with failure to achieve complete control due to poor matching of agents to weed species and predation of cochineals by ladybirds. Predation was overcome in South Africa by sub-lethal application of insecticide which killed the predators but not the control agents. Similarly, introductions of lantana insects have been made in southern, eastern and western Africa and on many of the islands. Except in central Uganda where control was spectacular, little has been achieved except dry season defoliation. In both instances, further more carefully planned introductions are required in most areas to further stress the weeds and achieve satisfactory control.

Research, pioneered by the CIBC, on the floating water weeds, water hyacinth and water fern, both from South America, has been applied only to a limited extent so far. In Lake Kariba, water fern has been reduced by introduction of an aquatic grasshopper and reports are now coming in from the Sudan that two weevils and a moth liberated during 1979-81 are materially reducing the amount of weed in the Nile.

In all except two instances, control of insects and weeds has been achieved by insect natural enemies, but a fungus is credited with partial suppression of the soft green scale in the Seychelles and a virus has been very effective against the rhinoceros beetle in Mauritius.

Now, two recently introduced cassava pests are the subject of a major biological control programme, as discussed by Herren & Bennett (this volume, Chapter 19).

Another introduced pest, the American leafminer (*Liriomyza trifolii*), is now the subject of a biological control programme in Senegal. This insect is also a damaging pest in East Africa, Mauritius and Réunion, and has also appeared in South Africa.

As well as these introduced pests, which are obvious targets for classical biological control, native pests may also be good targets but there are no outstanding successful examples from the African mainland. However, in Mauritius, a white grub of sugarcane fields has been controlled by a wasp from Madagascar, and one of the stemborer moths by a parasitic wasp from Kenya. A degree of success has also been achieved against borers in Madagascar and Réunion and against orange dog caterpillars on citrus in Reunion (Greathead 1971).

However, biological control of native pests is more likely to be achieved by using other approaches and, in most instances, integrated programmes will be required. This approach was developed on coffee in Kenya to safeguard biological control of mealybug and to combat outbreaks of other pests caused by the use of persistent chemicals against antestia bugs (Evans 1968, Bardner 1978). In Zimbabwe and South Africa, integrated control relying on natural enemies to control scale insects is employed on citrus, of necessity since the cost of routine spraying had become unbearable and was inducing outbreaks of secondary pests (Bedford 1978).

Choice of target and method

Opportunities for employing biological control vary greatly from pest to pest, crop to crop, and the conditions under which the crop is grown. These factors also determine which approach to biological control is most likely to succeed.

Some groups of pests have many natural enemies, and others few. Their number and effectiveness is affected by the lifestyle of the pest. For example, scale insects and mealybugs which are sedentary and mostly live exposed to natural enemies, have many predators and parasites and have been the subject of some 40 % of successful biological control attempts against insects. By contrast, migratory pests such as locusts and pests protected from natural enemies by boring into plants or the soil, make poor targets for long-term suppression by predators and parasites but may be good targets for biopesticides.

In practice, the nature of the damage done is also important as pests causing direct damage to marketable produce must be subjected to a much higher degree of control than indirect pests (such as leaf feeders and root feeders) and this may be difficult to achieve using biological control. Disease vectors which have to be held at extremely low density to prevent significant disease transmission are also poor targets – most aphids, some leaf hoppers and antestia bugs on coffee, fall into this category.

Similarly, with weeds, opportunities vary and, on the whole, the best targets are perennial or woody species which develop major infestations in pasture (e.g. *Lantana*, prickly pear cactus) or waterways (e.g. water hyacinth, water fern).

Independently of the nature of the pest and the damage it causes, the nature of the crop and how and where it is grown affect the prospects for biological control. Conditions favouring stability and therefore the long-term persistence of pest populations, favour long-term control measures. Conversely, where the crop environment is unstable, only short-term measures may be feasible. Greathead & Waage (1982) attempted to provide a guide to strategy based on the scoring of four indicators.

(1) *Climate*. Semi-desert areas (e.g. the Sahel) with a short growing and erratic climate at one extreme, and areas such as much of the west African coast, with no marked dry season, at the other.

(2) *Crops*. Short-term crops such as rice and many vegetables contrast with tree crops growing for many years.

(3) *Scale of planting*. At one extreme, there are fruit trees grown individually or in small groups in gardens, and at the other estate crops and irrigation schemes where the same crop is grown over a large area.

(4) *Husbandry*. Here, intense cultivation, as in vegetable gardens, with frequent hoeing to suppress weeds, contrasts with unweeded plantations with little disturbance except during harvest or occasional slashing of ground cover.

Table 2 summarizes results worldwide and suggests that high scoring crops do indeed have above average success rates in application of long-term biological control. They are also ideal for application of natural enemy conservation measures. At the other extreme, inundative releases and bio-pesticides are most likely to be useful. Under these conditions, there may also be opportunities to avoid pest attack by alteration of planting dates or cultural practices.

Table 2   Crop suitability assessment for biocontrol (modified from Greathead & Waage 1982).

Crop type	Score	"Classical" biological control		
		Attempts	Successes	%
Cereal	1–6	21	8	38
Root	3–10	3	2	67
Vegetable	0–7	88	27	31
Oil seed	3–8	2	0	–
Plantation	7–12	152	50	33
Fruit	4–12	185	107	58
Other cash	2–8	26	3	12
Forage	4–12	23	9	39
Total		506	208	41

Future needs

The effort which has been employed on major cash crops is required to solve problems on food crops, especially those being grown on small farms where the cost of chemicals is beyond the means of the farmer. Here, there is a basis on which to build as traditional practices such as mixed cropping, rotations and the juxtaposition of several crops within a small area can enhance the level of activity of natural enemies by providing necessary shelter, alternative hosts, sources of nectar and other food for adult parasites which enable them to maintain effective levels of activity. When this is insufficient, multiplication and periodic application of natural enemies may be employed, but to be effective, this requires a thorough knowledge of the pests. Development of simple monitoring systems, preferably usable by farmers, so that applications can be timed to have a maximum impact, a simple, cheap and effective system of breeding or culture is also required. Where only small numbers are needed for the inoculative release of insects, a simple insectary is needed and technicians to maintain the culture. Large scale breeding for inundation may require a heavy investment and is usually impracticable for larger species and those which develop slowly or have a low reproductive potential. In all instances, it is vital that the material is available at the precise time it is required and that the supplies are reliable. This tends to rule out insects, except the smallest and most easily handled, such as egg parasites, notably *Trichogramma* species, which are in use chiefly in China, USSR and India (INRA 1982) and are being evaluated on an experimental scale in South Africa against false codling moth and American bollworm. However, micro-organisms are more promising, especially those that can be cultured *in vitro* and stockpiled as spores until required. These can be manufactured in quantity relatively cheaply and, most important, can be stored until needed. Manufacture need not require special facilities since many micro-organisms can be rapidly cultured, for example by using fermentation plants similar to those in breweries.

Production of biological control agents is carried out in Chinese communes using simple techniques. A similar strategy might be developed in Africa with units set up in villages supervised by technicians from the extensional or research services who would also monitor production to ensure that the product is both effective and not contaminated by possibly dangerous organisms.

For biological control to achieve its potential in its service of African agriculture, the first requirement is plant protection personnel with an awareness of the alternatives to chemical control and the ecological background to enable them to develop soundly based management strategies. It is premature to expect most developing countries, short of trained scientists, to set up specialist units, but expertise is available, through the CIBC and other agencies, to advise and assist in the procurement and handling of biotic pest control agents by non-specialist plant protection staff.

References

Bardner, R. (1978) Pest control in coffee. *Pesticide Science* 9, 458-464.

Bedford, E.C.G. (1978) Citrus pests in the Republic of South Africa. *Bulletin of the Republic of South Africa, Department of Agricultural Technical Services* 391, 1-253.

Canada (1971) Biological control programmes against insects and weeds in Canada. 1959-1968. *Commonwealth Institute of Biological Control, Technical Communication* 4, 1-266.

Clausen, C.P. (ed.) (1978) Introduced parasites and predators of arthropod pests and weeds: a world review. *United States Department of Agriculture, Agriculture Handbook* 480, 1-545.

Cruickshank, S.; Ahmed, F. (1973) Biological control of potato tuber moth, *Phthorinaea operculella* (Zell.) (Lep. Gelechiidae) in Zambia. *Technical Bulletin of the Commonwealth Institute of Biological Control* 16, 147-162.

Evans, D.E. (1968) Recent research on the insect pests of coffee in Kenya. *Annual Report, Coffee Research Foundation, Kenya 1967-68*, 7-12.

Greathead, D.J. (1971) A review of biological control in the Ethiopian Region. *Technical Communication, Commonwealth Institute of Biological Control* 5, 1-162.

Greathead, D.J. (ed.) (1976) A review of biological control in western and southern Europe. *Technical Communication, Commonwealth Institute of Biological Control* 7, 1-182.

Greathead, D.J.; Waage, J.K. (1982) Opportunities for biological control of agricultural pests in developing countries. *World Bank Technical Paper* 11, 1-44.

INRA [Institut National de la Recherche Agronomique] (1982) Les trichogrammes. I[er] Symposium international. Antibes, 20-23 avril 1982. *Les Colloques de l'INRA* 9, 1-307.

Julien, M.H. (1982) *Biological Control of Weeds. A world catalogue of agents and their target weeds.* Farnham Royal; Commonwealth Agricultural Bureaux, 108 pp.

McLeod, J.H.; McGugan, B.M.; Coppel, H.C. (1962) A review of the biological control attempts against insects and weeds in Canada. *Technical Communication, Commonwealth Institute of Biological Control* 2, 1-216.

Melville, A.R. (1959) The place of biological control in the modern science of entomology. *Kenya Coffee* 24, 81-85.

Papavizas, G.D.(ed.) (1981) *Biological Control in Crop Production.* Totowa, New Jersey; Allanhed, Osmun, 461 pp.

Rao, U.P.; Ghani, M.A.; Sankaran, T.; Mathur, K.C. (1971) A review of the biological control of insects and other pests in South-East Asia and the Pacific Region. *Technical Communication, Commonwealth Institute of Biological Control* 6, 1-149.

Wilson, F. (1960) A review of the biological control of insects and weeds in Australia and Australian New Guinea. *Technical Communication, Commonwealth Institute of Biological Control* 1, 1-102.

Chapter 38

Post-Harvest Problems in the Tropics

T. JONES

Deputy Director, Tropical Development and Reasearch Institute, College House, Wrights Lane, London W8 5SJ, UK.

Introduction

Post-harvest problems from the farmer's field to consumption and end-use are multivarious and inter-reactive but all involve deterioration in quantity or quality of stored agricultural produce.

This may result from the activities of such "external" agencies as insect pests and rodents, fungi and bacteria or "internal" agencies such as enzymic processes and chemical reactions within the tissues, or a combination of both.

"External" Agencies

Insects and rodents cause direct losses by consuming or damaging produce and indirect losses by facilitating the activities of micro-organisms. Direct damage especially by insect pests is often spectacular and prevention and control of them relies on manipulation of the environment and/or the use of insecticides. Fungi and bacteria cause spoilage of food by effecting undesirable changes in appearance, colour, odour and flavour of produce to the point where it is unacceptable to the consumer. Other microflora constitute public health hazards by producing toxic metabolities or causing enteric diseases. Control of microflora commonly involves de-hydration but this can also cause deleterious effects in quality, and alternative methods of preservation such as chilling may be necessary.

"Internal" Agencies

These are essentially of two types – autolytic processes and chemical reactions. Autolysis, due to the activities of the tissues own enzymes is particularly important in meat and fish products but it also occurs in oil seeds and oil seed products and in fruit. The process can be controlled in part by chilling or by heat treatment to denature the enzymic proteins. Oxidation caused by reactions between chemical components of foods can cause browning of produce and off-flavours. Control lies in part in controlling the composition of produce in the pre-harvest stage and thereafter by manipulation of processing and storage conditions.

In reviewing the effects of these various agencies on specific commodities it is convenient to consider the problems of *storage* and those of *processing* under separate headings whilst recognising that they are closely inter-related.

Problems of Storage

Commodities are conveniently classified into durables such as grains pulses and oil seeds etc and perishables such as fruit and vegetables, meat and fish.

Durables

(a) Cereals and seed grains

The principles of storage of cereal grains and similar commodities concern moisture content and temperature and preventing or controlling the development of micro-organisms, insects and mites. Generally grains will store for long periods with negligible deterioration at low temperatures which inhibit pest and pathogen development. At higher temperatures with adequate moisture activity of micro-organisms and pests causes direct losses and produces heat and moisture which exacerbates further deterioration of the grain reducing the acceptability of the product and its fitness for human and animal feed.

Problems with cereals in store involve *first* the effect of temperature which determines the availability of water to micro-organisms associated with the grain and *second* – pest damage, injury and breakages (rupture of tissues and integument) which allow invasion of the grain by micro-organisms. These two factors alone are responsible for most of the losses incurred when grain is transported.

Water activity is critical because micro-organisms develop rapidly in environments of 75% relative humidity or higher. The "safe storage" mositure content of grain is therefore that having an equilibrium relative humidity of 70% or less. This varies with temperature; maize grains with a moisture content of 15.3% have an equilibrium relative humidity of 70% at 0°C – but this rises to 77% at 25°C.

In seed storage the major concern is to minimise losses in viability caused primarily by incorrect storage conditions and insect attack in open stores. Optimal storage conditions can however be prescribed as the loss of viability of a chosen seed correlates with the temperature and relative humidity and can be expressed by a simple graph. The relationship between storage life of maize seed, moisture content and effective seed temperature can be expressed as a simplified mathematical equation. This is important in store design and in determining the need for and feasibility of in-store drying. TDRI has constructed a mathematical model which can predict the storage conditions required for different intake and climatic environments. Effective seed temperature, safe moisture contents, hours of drying required, and costs can be computed for a wide range of seeds, building materials and climatic conditions.

Insect pests reduce viability directly by feeding, allowing the entry of micro-organisms and causing rises in temperature which may kill the seed. Control of infestations is usually by fumigation with methyl bromide or phosphine gas and despite some pest resistance to them these compounds can still be highly effective. Cotton sheets placed under the fumigation sheets, sealed and left in place after fumigation can prevent re-infestation. Storing stacks under sheets purged with carbon dioxide or nitrogen has shown promising results which may be applicable in developing countries. Insect aggregation pheromones, sex pheromones and development of grain varieties resistant to pests are other promising lines.

(b) Pulses (beans and peas)

Beans and peas are very susceptible to attack by insects, especially bruchid beetles, and the high cost of pre-harvest pest control measures requires the use of more economic measures in storage. Resistance to the bruchid *Callosobruchus maculatus* in cowpeas (*Vigna unguiculata*) has been shown by IITA in Nigeria, the University of Durham and TDRI in the UK to be conferred by high levels of trypsin anti-metabolite.

(c) Oil seeds

Particular problems relate to the production from these seeds of high quality oils, fats and waxes the flavour and constituents of which may be adversely affected by the feeding activities of pests, the metabolic processes and by-products of micro-organisms and by the seeds' own innate enzyme systems.

TDRI in collaboration with CEPEC (Centro de Pesquisas do Cacau) in Itabuna in Brazil has developed a sex pheromone technique to control heavy infestations of *Ephestia cautella* on dried cocoa beans normally done with expensive pesticides which can cause residue problems. This reduces frequency of mating of the moths and the combinations of pheromone concentrations and moth population densities which will give acceptable levels of control are now being determined.

The Problem of Mycotoxins in Cereals and Oil Seeds

Certain fungi produce highly toxic metabolites (mycotoxins) which can affect various organs in animals and man, having mutagenic, carcinogenic and teratogenic effects and lowering resistance to infectious diseases. Acceptable mycotoxin levels, particularly of aflatoxin, are increasingly being limited by legislation causing high losses to developing countries by the banning of imports into developed countries. Heavy losses also occur within developed countries where unacceptable aflatoxin levels develop in storage.

Control includes the use of resistant varieties, improved harvesting and drying; physical separation by hand selection and colour sorting; detoxification processes and solvent extraction. TDRI has developed a low pressure ammoniation process for destroying aflatoxin in groundnut cake which is now being assessed.

Perishables

(a) Fruit and vegetables

As sources of food these are second only to grain cereals and some such as cassava and cooking bananas are important staple crops. The post-harvest characteristics of these commodities vary enormously from highly perishable leafy vegetables and soft fruit to the more sturdy potatoes. Losses are generally very high but are difficult to estimate accurately (and difficult to define).

Perishable plant foods are living organs of the parent plant and their continuing metabolic activity is an important cause of loss. Perishable implies an inherent limited storage life, but this can be extended by maintaining the physical and physiological integrity of the produce, by provision of optimal environments or by manipulation of the physiological state of the material and by choosing healthy material at the ideal state of maturity. The two main factors in deterioration are physiological and phytopathological. The *physiological* factors are of two types – the normal endogenous metabolic processes essential for survival which cause unavoidable losses and other abnormal physiological processes caused by exposure to adverse environmental conditions and which are avoidable. Only the latter is of concern here. Abnormal physiological effects may result from mechanical damage or exposure to extremes of heat and cold or of relative humidity. Tropical produce tends to suffer chilling injury at temperatures well above zero and sometimes as high as 14-15°C. At high temperatures, including exposure to insolation, deterioration results from oxygen deficiency in the tissues causing a change to anaerobic respiration.

Phytopathological processes probably constitute the biggest single factor in post-harvest losses of perishable plant produce. Quantitative phytopathological losses result from initial infection by specific pathogens followed by secondary saprophytes which cause rapid, extensive breakdown of the host tissues. Qualitative phytopathological losses result from blemish or surface disease which render produce less attractive and hence less marketable.

In general losses of plant perishables can be reduced to an acceptable level by simple precautions, such as, good pre-harvest field hygiene, sound and appropriate harvesting techniques, the use of curing techniques and appropriate chemical treatments; the avoidance of mechanical damage and thermal stress. Techniques such as refrigeration, controlled atmospheres, hypobaric storage may be used for high quality produce fetching high market prices. The problems of plant perishables particularly demonstrate the need to produce primary products properly suited to the handling and storage conditions to which they will inevitably be subjected.

(b) Meat and fish

Livestock ownership is a traditional part of life in many developing countries. Slaughtering facilities available are usually minimal and as the meat is normally consumed within a day of slaughtering cold storage and refrigeration facilities are unnecessary. Standards of hygiene are usually low but have been improved by the education and training programmes of TDRI and others and the introduction of facilities and working techniques aimed at improving particular aspects where hygiene is important. Safe fly control is possible using pyrethrum/piperonyl butoxide or malathion sprays on walls and working surfaces but it is rarely used.

Commercial abattoirs with integrated boning and other processing facilities and with international certificates for export are amongst the major successes in some developing countries – including East Africa.

The introduction of new fishing vessels and new catching techniques to the Malawi Central Lake Fisheries Development Scheme in which TDRI has been closely involved, have made available fish stocks hitherto unexploited by traditional dug-out canoe fishing. This has been followed by the construction of a completely new infra-structure and establishment of new communities.

Much of the catch is sold fresh through new chilling facilities but traditional patterns of consumption and the need for some long distance transport require a proportion of the catch to be dried. Peak catches coincide with the wet season and difficulties have arisen with blowfly infestations and dermestid beetle attack on the dried fish. A novel insecticide treatment to control these pests by dipping or spraying the wet fish with pirimiphos-methyl before drying has been developed and awaits international approval.

Problems of Processing

Sun-drying

Is a universal and old-established method of preserving agricultural produce but it is not universally applicable to a common standard of efficiency and reliability.

Sun-drying can be improved if done on clean firm and smooth surfaces such as rendered plastic, cement, concrete, wood, plastic or metal sheeting. These improve hygiene and absorb solar radiation more efficiently than soil, become hotter and provide more energy for drying the produce. Black materials provide greater efficiency. Woven matting or mesh trays allow ventilation which improves drying and over 100 000 tons of grapes are dried annually in Australia by this simple process. TDRI has developed tray-dryers for coffee and cassava in Brazil using optimal angles for radiation and wind which have proved highly effective (allowing night drying also) and much cheaper than blackened concrete.

Solar dryers

Rely on enclosing the produce to facilitate moisture removal by ventilation with sunheated dry air. Fan driven forced convection dryers developed by TDRI for use by small farmer communities have more than halved the time for drying red peppers and coffee. In collaboration with the Kenya Industrial Research and Development Institute TDRI is studying the performance of dryers using fans powered by photovoltaic cells.

In-store drying

By aeration is the principle behind the success of maize cribs in the hot and humid tropics. Recent studies on developing model structures to maximise aeration and increase drying in peasant farmer structures have investigated drying under conditions of no change in temperature/humidity or moisture content. Contrary to general opinion – results have shown that long-term preservation of high moisture paddy rice in a flat bulk store by aeration with ambient air is possible in south-east Asia giving slow drying and only slight loss of quality.

Freeze-drying

Is an effective long-term preservation technique for perishables and high quality produce destined for high value markets, e.g. prawns and coffee. There are two basic techniques. Pre-freezing followed by ice evaporation with high vaccuum gives products with very good reconstitutional properties and is used for example for storing enzymes. More commonly the product is frozen and put under vacuum on traps which are heated enough to assist evaporation without melting the ice. This is used to produce high quality/low weight commodities, e.g. coffee, prawns and defence rations.

Chilling and freezing

Reduce micro-organism activity and autolysis to a very low level and are particularly suitable for preserving some tropcial perishables because the basic raw materials and their spoilage microflora are adapted to high temperatures (and their autolytic systems operate optimally at high temperatures); for instance fish in northern waters frequently live at 6–7°C whilst those in tropical waters operate at temperatures of some 25–30°C. Therefore in the tropics chilling is generally more effective and produces greater benefits than in temperate regions because greater temperature differences are involved. Cheap chilling transport systems rather than expensive cold storage facilities may be equally effective for African lake fishing systems.

Osmotic preservation

Involves impregnating the tissues of perishable products with high concentrates of salt and/or sugar which provide a hostile environment or substrate for micro-organisms which are themselves desiccated by osmosis.

Canning

Involves high temperature sterilisation followed by air-tight durable packaging and has been highly successful in the developed world. Pineapple and meat canning are examples of successes in developing countries but there have been some spectacular failures due to the cost of cans in many non-industrialised countries.

Extraction and refining

Oil seeds grown by farmers in developing countries are exported either out of the country or to some central processing facility remote from the growing area. There is considerable interest in the potential of cheap small scale expelling equipment which allows processing at or near the point of primary production. So that farmers could have a local source of cooking oil and proteinaceous feed for their livestock, TDRI recently developed such a machine in collaboration with a British manufacturer. Tests in Zambia show it to be suitable for sunflower, soybean, palm kernels and to a lesser extent groundnuts. A small expeller to extract macadamia oil from broken kernels and used for frying high grade kernels and making soap has also been developed. The seeds of the indigenous sal tree of India contain a good substitute for cocoa butter but traditional harvesting techniques result in a poor quality product. TDRI's studies showed that an acceptable butter substitute could be obtained if the seeds were sterilised immediately on harvesting to prevent natural biochemical activity. TDRI have also investigated the composition of oil from the winged-bean, a potential source of edible oil and animal cake or meal, and has developed a small-scale expeller for a solvent extraction system to obtain crude oil which can be readily refined by chemical and physical processes.

Conclusions

It can be concluded from this very brief review that whilst considerable improvements have been achieved by new technologies and new approaches to storage and processing at the small farmer level and at the commercial level there is still a sad lack of awareness of the need to ensure that primary produce is suitable for its proposed end-use.

Systems have been evolved through the ages which have allowed individuals and communities to cope generally and adequately with their post-harvest problems. In recent times, new high yielding crop varieties have been introduced which require new agronomic practices for optimum yields and better and bigger capacities for storage. The food which they provide is often very different from the traditional product and may require different preparation. They have not always been successfully incorporated into the traditional sociological structures and feeding patterns in rural communities, and have sometimes caused as many problems as they have solved. The "miracle" rice varieties in some south-east Asian countries and new bird and pest resistant varieties of sorghum in Africa have not been fully accepted because of human resistance to the quality of the grain.

Similar situations have arisen in developed countries where new high yielding wheat varieties have proved unsatisfactory for making bread. New varieties of high cropping potatoes have proved to be of very limited acceptability. New tetraploid varieties of dessert bananas bred and selected specifically for export to European markets were unsuitable due to their short shipboard and shelf-life and their proneness to finger-drop.

These examples serve to show how little attention is paid to matching primary produce to its end-use. Whilst successful plant breeding programmes against insect pests and disease abound and those for increasing yield and nutritional content of produce are equally numerous, there has been little if any effort made to breed and select for post-harvest properties in plant produce. The quality of field produce is determined initially by factors – such as varietal characteristics, fertiliser and water regimes, growing season and time of harvest and all associated operations concerned with pest and disease control. This quality more than any other single factor determines the pattern and degree of all subsequent losses. Low quality means low suitability and acceptability for consumption, storage or processing, and high susceptibility to the whole range of destructive agencies leading to spoilage and wastage.

Agricultural research and production, handling, storage, processing and marketing can be considered as a single system in which the individual elements are inter-reactive. There is a real need to relate agricultural research and production to end-use on the market.

I wish to thank the following Heads of Department of TDRI, Mr D.J.B. Calverley, Mr J.G. Disney, Mr D. Adair and Dr J.C. Caygill, and their staff, for their contributions to this paper, and especially Dr N. Jones, Deputy Director, who provided much information and read and approved the manuscript. The paper is published by permission of the Director, TDRI.

Chapter 39

Integrated Crop Protection

M.J. WAY and G.A. NORTON

Silwood Centre for Pest Management, Department of Pure and Applied Biology, Imperial College, Silwood Park, Ascot, Berkshire SL5 7PY, UK.

Introduction

In his introduction to the conference (Chapter 1), President Nyerere emphasized the crucial need for African countries to expand agricultural production by developing traditional techniques "without capital which we do not have" and without presupposing "the ready availability of chemicals'. This immediately implies that scientists need to learn much more about empirically developed cropping systems developed by farmers in order to provide a more enlightened basis for the conscious improvement of traditional practices.

It is salutary to remember that all the World's domestic animals and all but one of the World's crops were discovered by prehistoric farmers who owed nothing to science. Similarly, integrated crop protection practices have been developed and practised by farmers throughout the World for centuries before scientists invented the terminology! For example, traditional rotational cropping not only helped to control weeds but also successfully controlled severely damaging nematodes and soil pathogens which were unknown until comparatively recently. Even today, many mixed and successional cropping systems developed by small farmers in the tropics are now being recognized as containing important pest control components (e.g. Way 1973, Perrin 1977, 1980).

In our opinion, pest control specialists have a major responsibility for defining the control mechanisms that exist in these traditional practices and for developing ways of improving them before they are lost in the drive towards simplification. We cannot emphasise this need too strongly, particularly when it is recollected that small farm families account for 65-70% of the population in tropical countries, amounting to nearly 1400M people, most of whom are in the traditional farming category referred to by President Nyerere.

Terminology

Although farmers have for centuries used empirically, and often unconsciously, developed control practices against plant diseases, weeds and insect pests, the term Integrated Pest Control (IPC), which emphasises a conscious approach to crop protection was first defined by entomologists in the early 1950s. An original and vitally important feature of IPC is the need to use insecticides and other pesticides in ways which do minimal harm to indigenous natural enemies, the importance of which had previously been underestimated, particularly for insect pests of many major annual crops.

Although the term integrated pest control has remained an important theoretical concept, in practice, it has had limited application. This is partly because IPC has been confined primarily to insect pest control. A more important reason, however, is that IPC has reflected scientists' goals and has frequently not been compatible with farming technology and economic reality. We prefer the term Integrated Crop Protection (ICP), which is also the title of this paper. We envisage ICP as a sub-system constrained by and integrated as part of overall crop production requirements. ICP also avoids misleading interpretations of what is meant by a "pest", as well as encompassing the need for integrating the control of animal pests, plant diseases, and weeds — something which has not been adequately achieved by research workers in the past.

Control Methods in an Integrated Crop Protection System

There are four fundamentally important methods that can be employed in integrated crop protection, as follows:—

(1) *Biological control*, which involves the preservation, augmentation and introduction of natural enemies, including competitor species which may be said to act indirectly rather than by directly killing or harming the damaging organisms. Biological control is vitally important for insect pests, is limited but locally important in the control of exotic weeds, and in the form of competitor species is potentially important against some pathogens, notably soil pathogens.

(2) *Host plant resistance*, that has been established in the past for controlling many plant pathogens. Despite a few striking exceptions, the use of host plant resistance against insects has in general been rather disappointing. This is partly because resistance characteristics, notably tolerance, have frequently been incompatible with other desirable crop characteristics, notably improved yield and eating quality. Similarly, yield is normally decreased by preservation or enhancement of characters which enable crop plants to compete with weeds. Another problem with host plant resistance is the tendency for pests, especially pathogens and insects, to adapt and overcome the resistance mechanisms.

(3) *Cultural control*, which is achieved by the manipulation of cropping practices, particularly in ways which dislocate the life cycle of insect pests, diseases or weeds; it also provides a means of preserving or enhancing biological control. Cultural methods are widely important for controlling many different kinds of damaging organisms, particularly in arable crops, and they are much the most widely practiced form of control.

(4) *Pesticides,* that, apart from chemical pesticides, include biological pesticides, such as viruses and bacteria, which are applied and behave like synthetic chemicals.

Other methods for crop protection, such as the use of pheromones and sterilizing techniques against insects, are potentially important but at present have relatively limited application. Of the four key methods, the first three (unlike pesticides) mostly operate in ways which prevent pests reaching important pest densities. Thus, ideally, the pest becomes a non-pest and the farmer or extension worker is hopefully relieved of the burden of tedious pest assessment. In contrast, the use of pesticides is based either on the assumption, or knowledge, that economic losses will otherwise occur. Accordingly, the farmer either applies the pesticide as a routine or he makes his decision to apply chemicals based on pest forecasting and monitoring (Walker, Chapter 33). Since pesticides cost money they can form a major proportion of variable costs and of capital expenditure, particularly for a small farmer. On the other hand, cultural control normally involves the use of labour rather than cash, as does the preservation of natural enemy action. Similarly, the cost of host plant resistance and of the introduction of natural enemies in biological control does not fall directly on the farmer, if at all.

Contrasting approaches to crop protection in temperate and tropical countries

Predominant dependence on pesticides with limited consideration of biological control and decreasing reliance on cultural controls and host plant resistance is the philosophy behind most high yielding temperate farming systems. Such systems depend on high inputs resulting in increasing costs for pesticides and their application. In the UK for example, pesticide application accounts for about 40% of the annual costs of high value apple production (ADAS 1980): in oilseed rape the cost of pesticide treatment has risen from about 8% of the total cost of production in 1974 to 26% in 1982 (Murphy 1983). Reliance on pesticides, in part needed to compensate for lost host plant resistance which has been traded for high yielding potential (Chapter 34) and in part needed to replace cultural controls such as mechanical weeding and long term rotations, is therefore an increasingly important cornerstone of many high yielding annual cropping systems in temperate regions. A key feature is that temperate farmers have been able to depend on a climate that permits a reliably high yield every year, guaranteeing the ability to more than repay the large financial input needed each year to grow the crop to maturity.

In contrast, there are relatively few parts of tropical Africa where comparable climatic stability would permit dependable yields from capital intensive farming based on arable monocultures and pesticides: though this may not necessarily apply to tropical perennial crops. Nevertheless, amongst arable crops it might be said that irrigated rice should provide a reliable local climate for high technology pesticide-dependent crop production in the tropics. If so, why the early catastrophies and continuing serious problems in green revolution rice? This is surely because the tropical climate is ideal for pest multiplication in all-year-round successional irrigated rice. It provides a unique opportunity for pest and disease build-up without a winter which, in temperate environments, often dislocates the pests' life cycle. Hence the contrast between green revolution rice and the relatively very successful green revolution in temperate wheat or, in part, the success of temperate rice crops in California and Spain where a single crop per annum produces a higher yield than one to three successive crops per year in tropical irrigated rice.

No doubt, the problems of green revolution rice in south-east Asia are now being alleviated by greater emphasis on integrated crop protection and less on a single-minded drive for high yield potential. Nevertheless, if the objective of decision makers is to aim for the high potential yields that tropical irrigated systems permit, all-year-round rice production will always be faced with the spectre of catastrophic pest and disease problems, despite and perhaps because of pesticides. Put simply, tropical farmers are unlikely to emulate what is being done with equivalent temperate crops because of a higher order of weather-related or "pest"-related instability in much of the tropics. This fundamental difference is not just due to the contrast between developing and developed country technology and economics, as is often assumed. Indeed, the catastrophy experienced in cotton production in the Ord Valley of tropical Australia is a salutary example of complete failure, despite all that developed country technology had to offer!

We have laboured this contrast between crop and pest dynamics in tropical and temperate systems because it adds emphasis to the clear message of many participants in this Conference, that the development of arable agriculture in tropical Africa must, for a combination of reasons, be built up from traditional methods. We need to understand and then apply relevant knowledge of pest, disease, and weed dynamics in modified traditional systems in order to improve pest control without, as stressed by Odhiambo (Chapter 6), using pesticides as the centrepiece.

The knowledge basis for improved integrated crop protection in small tropical farm systems

The traditional farm system in the tropics has only recently been accepted by scientists as a respectable and sound basis from which agricultural improvements can be developed. Much of this change can be attributed to Norman (1974) and his co-workers who carried out the economic surveys of inter-cropping systems in northern Nigeria. Their interpretation of their findings challenged the "conventional wisdom" which had been responsible for prolonged and unsuccessful efforts to develop and encourage single species rotational cropping. For example, Norman concluded that, under the prevailing conditions, gross returns from certain mixed cropping systems were as much as 60 % higher than from single crops grown in rotation. He went on to state that until extension workers can suggest changes that have a convincing return and yet do not involve big changes in farming methods it is unlikely that they will ever be truly implemented.

While the lessons learnt from this earlier work have been incorporated into later research approaches, often under the title of the "farming systems approach" (Collinson 1984), the emphasis is still largely socio-economic. We believe that more emphasis is still required on the ecological corollary of this approach: namely, that a major research objective should be to assess carefully the effects of mixed crop complexes on pest, disease, and weed incidence. This was emphasised by Way (1973), and the subject comprehensively reviewed by Litsinger & Moody (1976) and by Perrin (1977, 1980) for insects in particular. Further evidence is given by Mumford & Ballidawa (1983) and Trenbath (1977). They, along with Norton (1975) and Way (1977), list the evidence for decreased insect pest incidence in particular crop mixtures, and outline possible reasons for this — notably enhancement of natural enemy action, deterrent or antibiotic effects of non-hosts of the pest, presence of diversionary hosts, camouflaging effects of other plants and loss of emigrants among the non-host plants as they disperse to other host species in the mixed crop.

Much of the information is qualitative however, and the mechanisms operating to reduce pest incidence are often a matter for speculation. In particular, the lack of knowledge of disease and nematode incidence in mixed cropping systems is abysmal. Existing knowledge on pathogens is mostly summarised by Sumner *et al.* (1981) who indicate how non-related crops can alter the life cycle, epidemiology and survival of both aerial and soil borne pathogens, thereby greatly reducing disease incidence on crops at risk even when grown continuously in a mixed cropping system on the same land. For nematodes, Olowe (1978) has shown much greater nematode infestation in a cowpea monoculture than in traditional mixed cropping systems in Nigeria. For weed control, there is no doubt that the practice of covering the soil with crop plants must greatly reduce weed incidence compared with iconventional monocropping. This is borne out by the notable demonstration of weed control using live mulches (Akobundu, Chapter 36).

However, where monocultures have led to increased pest problems, reintroducing diversity *per se* is not the answer (Way 1976, Mumford & Ballidawa 1983). In some situations mixed cropping systems may exacerbate pest and disease problems, as can the introduction of different crops in the rotation. For instance, Pearson (1958) has collated evidence of how, in different circumstances maize and other hosts of the bollworm, *Heliothis armigera,* can variously act as sources of bollworm infestation in cotton or act as diversionary hosts, keeping bollworms away from a cotton crop at risk. More recent evidence of this, and of the role of host weeds, is described by Nyambo (Chapter 24) in relation to integrated control of cotton pests in Tanzania, and by Way (1973) for cotton in the Sudan where inappropriate cropping sequences have, in retrospect, been shown to have created serious new pest problems that have disastrously affected the cotton crop in that country.

Knowledge of pest interactions for cropping systems and in particular for mixed cropping systems is essential if problems are to be avoided or at least minimised through appropriate cultural and biological control practices. Such understanding is not only important in relation to small farm systems but also to larger systems, as in the Sudan Gezira and also in the vast areas of the Sudan savannah where large scale mechanised systems are used for production of sorghum and other crops, the pests of which all originate from 'bush fallow'. If, as Odhiambo (Chapter 6) suggests, fallows will be replaced in the future by appropriate arable crops, using systems that are designed for local conditions, then unquestionably an appropriate choice of crops and design could greatly reduce many pest and disease problems in all parts of tropical Africa where traditional fallows are a source of pests.

Whilst the potential for developing cultural practices and host plant resistance is inevitably much less in perennial than in annual crops, the opportunities for biological control are generally much greater for perennial crops. This is well exemplified by a serious pest problem of coconuts in Tanzania and Kenya. The coconut bug, *Pseudotheraptus wayi* is known to decrease the yield of coconuts by about 50 % in the region as a whole. Therefore it constitutes a serious hazard to the enlightened work on crop yield improvement at present being undertaken by the National Coconut

Development programme in Tanzania (Way 1983). Fortunately there is already sufficient understanding of the system to allow control practices to be elaborated, particularly in new plantations of potentially high yielding hybrids. Briefly, *P. wayi* is very effectively controlled by the ant *Oecophylla longinoda* in palms which the latter occupies (Way 1953). However, this beneficial ant is destroyed by several competitor species which themselves are not predators of the coconut bug. Consequently, the objective of any effective control programme should be to create conditions unfavourable to the harmful species which would then allow the beneficial ant to colonise the palm. Secondly, there remains the question of improving the plantation conditions to favour vigorous colonies of the beneficial ant. Tactics include localised pesticide treatments to kill one harmful ant selectively, and manipulation of the ground vegetation against others; also interplanting certain perennial crops in new coconut plantations in order to encourage the beneficial ant.

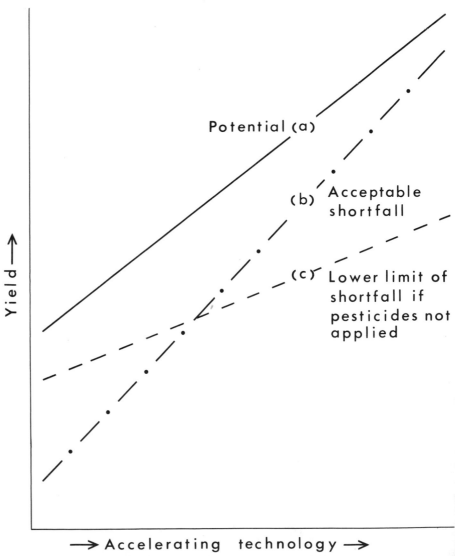

Fig. 1   Diagram illustrating how yield increase associated with technology is related to pesticide usage. The extent to which yield can fall below the level of (a) and remain economic to the farmer is indicated by (b). The widening gap between (a) and (c) indicates increasing potential for pest damage and consequently increasing dependence on pesticides needed to attain or exceed the economically acceptable yield represented by (b).

Conclusions

We hope that a case has been made for highlighting a different developmental path for tropical small farmers than that which has been adopted in most temperate systems. We suggest that the emphasis in crop protection should be placed on avoiding damage through integrated cultural and biological control together with appropriate host plant resistance, with a minimal dependence on costly pesticides used as curative or prophylactic treatments. This is not to deny the vital importance of chemical controls which must play a crucial role, particularly against insects and weeds of some crops. However, even where chemical control practices are desirable their use by small farmers in the tropics is often limited by financial and other constraints.

Against this background, what role do research and development programmes have to play in the future? In our view priorities should be placed on field studies and the development of selected systems in chosen villages where there is close collaboration between farmers, extension workers and scientists to build upon tradition systems. With the collaboration of pest control specialists, emphasis should be placed on the first three components of the IPC armoury, cultural controls in particular but also on the neglected role of indigenous natural enemies and partially resistant cultivars, all within the context of an integrated crop protection programme.

The adoption of such a strategic approach to the development of crop production /protection systems is rare or absent in high technology agriculture. Here a single radical technological development can often end up by driving the development of the system, to which pest control practices must adapt. Almost invariably this involves the routine use of pesticides as virtually the sole pest control measure. The implications of this stategy are shown diagrammatically in Fig. 1. Accelerating technology provides a combination of cropping practices which increase the potential yield of the crop. However, as capital costs and overall instability increases, less and less risk can be taken by the farmer in relation to crop loss caused by pests, diseases, and weeds. Hence, the increasing emphasis on pesticide use as a means of risk-aversion. If lower potential yields can be accepted, greater tolerance in the event of pest attack can be achieved with relatively low technological inputs. This not only means less capital cost but also a much greater in-built "resistance" to yield loss from insect pests, diseases, and weeds. We would suggest that the challenge for research and development work in the tropics is to identify and develop appropriate farming strategies which are intermediate between "safe", low yielding, subsistence systems and "unsafe", high yielding, capital intensive systems as practised in many temperate countries.

The farming strategy that is attained in any particular country and farming system depends on a sequence of conditions and actions beginning with decisions on national goals, e.g. food self sufficiency, followed by the policy makers' decisions on how to make political and social conditions sufficiently favourable for farmers to respond appropriately. As scientists and economists we can identify and design farming systems that are desirable in the context of the possible range of constraints and opportunities that are provided by national policies. There is therefore a crucial need to identify likely agricultural developments and to design integrated crop protection strategies appropriate to them. In our opinion such strategies must emphasise the modification and adjustment needed to maintain or incorporate desirable features of "natural" controls supplemented where possible with chemicals. This will need to be done within the framework of an overall farming systems approach.

References

ADAS (1980) *Enterprise Gross Margins*. Agricultural Development and Advisory Service, MAFF.

Collinson, M. (1984) *Farming Systems Research: diagnosing the problem*. [Annual Agricultural Symposium, 1984.] Washington DC; The World Bank, 22 pp.

Litsinger, J.A.; Moody, K. (1976) Integrated pest management in multiple cropping systems. *American Society of Agronomy Special Publication* 27, 293-316.

Mumford, J.D.; Ballidawa, C.W. (1983) Factors affecting insect pest occurrence in various cropping systems. *Insect Science and its Application* 4, 59-64.

Murphy, M.C. (1983) *Report on farming in the Eastern Counties of England 1981-82.* Cambridge; University of Cambridge.

Norman, D.W. (1974) Rationalising mixed cropping under indigenous conditions: the example of Northern Nigeria. *Journal of Development Studies* 11, 3-21.

Norton, G.A. (1975) Multiple cropping and pest control – an economic perspective. *Mededelingen-Faculteit Landbouwwetenschappen, Rijks Universiteit, Gent* 40, 219-226.

Olowe, T. (1978) Progress Report from the National Cereals Research Institute, Ibadan, Nigeria. In *Proceedings 2nd Research Planning Conference on Root Knot Nematodes*, Meloidogyne *spp.* International Meloidogyne *Project Abidjan, Ivory Coast*, 58-62.

Pearson, E.O. (1958) *The Insect Pests of Cotton in Tropical Africa.* London; Commonwealth Institute of Entomology, 355 pp.

Perrin, R.M. (1977) Pest management in multiple cropping systems. *Agro-Ecosystems* 3, 93-118.

Perrin, R.M. (1980) The role of environmental diversity in crop protection. *Protection Ecology* 2, 77-114.

Sumner, D.R.; Doupnik, B.; Boosalis, M.G. (1981) Effects of reduced tillage and multiple cropping on plant diseases. *Annual Review of Phytopathology* 19, 167-87.

Trenbath, B.R. (1977) Interactions among diverse hosts and diverse parasites. *Annals of the New York Academy of Sciences* 287, 124-150.

Way, M.J. (1953) The relationship between certain ant species with particular reference to the control of the coreid *Theraptus* sp. *Bulletin of Entomological Research* 44, 669-691.

Way, M.J. (1973) Applied ecological research needs in relation to population dynamics and control of insect pests. In *Proceedings of the FAO Conference on Ecology in relation to Plant Pest Control*, 283-297. Rome; FAO.

Way, M.J. (1976) Pest and disease status in mixed stands vs. monocultures: the relevance of ecosystem stability. In *Origins of Pest, Parasite, Disease and Weed Problems* (J.M. Cherrett; G.R. Sagar, eds), 127-138. Oxford; Blackwell Scientific Publications.

Way, M.J. (1977) Integrated control – practical realities. *Outlook on Agriculture* 9, 127-135.

Way, M.J. (1983) *Report on Consultant's visit to Tanzania to examine and advise on the* Pseudotheraptus wayi *problem on coconuts.* National Coconut Development Programme and GTZ, Tanzania, 82 pp.

Chapter 40

Crop Production, Protection and Utilization

J.M. HIRST FRS (Rapporteur)

Long Ashton Research Station, Long Ashton, Bristol BS18 9AF, UK.

All of us who participated in this meeting are more aware of the problems of Africa and its agriculture than we were at its start. Although some nations can claim to be close to self-sufficiency in staple foods, many are great importers and almost a third of the rice in world trade is bound for Africa. Efforts that have increased food production have been outstripped by population growth, so that food production per capita has not risen.

The first half of this symposium concentrated on surveys of the state of knowledge and the problems associated with various crop types, for example, cereals, root crops, and cash crops (both annual and perennial), concluding with more emphasis on strategies. Inevitably, division into concurrent sessions and the limitations of time meant that the programme did not pay sufficient attention to the fragility of tropical soils, cultivation practices, the conservation of soil moisture or the management of irrigation, the kinds, uses and abuses of agricultural chemicals, and in particular perhaps the nutritional values and dietary roles of the crops discussed. I am sure that those attending this symposium would not provide answers about agricultural practice until we had some factors defined as terms of reference.

Amongst *natural factors* we would need to know: (1) The needs of our market, what crops and to serve what purpose. (2) Information on soil types and their cultivability and fertility. (3) Climate and seasonal patterns, together with the frequency and severity of deviations from it. Crisis weather often determines the success and the future of cropping.

We would also need to define *human factors*: (1) The intensity of land settlement and its distribution; the size of holding and its purpose (self-sufficiency or cash crops). (2) Incentive rewards for the cultivator, if he is not to flee the land. (3) Factors such as family size, age distribution, the convenience of markets, and fuel reserves.

Finally, we would wish to know how these affected, and were affected by, *Government policies*: (1) The price, wage and tax structures. (2) The stability of incentives. (3) The improvement of transport and services to improve collection, marketing and distribution and to decrease the present heavy charges. (4) The policy decision between support of (a) a low-input, labour-intensive agriculture based on the aim of self-sufficiency, accepting very partial control of pests, diseases etc., and products not infrequently impaired in quality; (b) a high-input, mechanized agriculture possibly resulting in a very erratic distribution of wealth requiring costly inputs of fertilisers, plant protective chemicals, forecasting services, and supporting agribusiness; or (c) some more desirable compromise between these extremes.

Somewhat similar lists are to be expected from other symposia (Chapters 55 and 77). I am sure that we would expect the process of definition and decision to be an iterative one. As ground rules became established we would be able to define our cropping system, its costs and rewards, and the research problems that this would raise. We discussed such problems during our sessions, but it would be impossible and unhelpful to repeat these here. Although we had many excellent presentations with a proper defence of professional interests, we endeavoured to see all viewpoints.

In his contribution on weed control strategies, Akobundu (Chapter 36) introduced us to that all important member of the silent majority, "the small farmer". He was hoeing and apparently spent 40% of his life doing that irrespective of season or crop. When he finished one round it would be time to start again, and if ever he stopped, or even if he did not, it was certain that he would miss the timeliness of the first hoeing. Akobundu said the small farmer did this all the time thinking how he could save his sons from a similar fate. He was probably unfamiliar with single-spray residual herbicides, and most of my colleagues were sure he never could afford them. However, I am quite sure he would if we told him of them, spend the remaining 60% of his time dreaming how he might use them so that he could spend some of his time chasing the greater grain borer (Chapter 12).

If you think life may be difficult for the scientist and the small farmer, consider also the potato with 268 registered pests and pathogens existing in an unnumbered complexity of strains, biotypes, etc. (Chapter 20). To continue to present such important and intriguing facts, would contribute little to the primary purpose of the session, so I will concentrate on a few topics grouped under the headings of Problems of intensification, Biological virtues and sins, and Awareness, with a final section of Needs for development.

Problems of Intensification

Population density, distribution and wealth constantly interact with the agricultural system. I have already specified some of the controlling factors, but it is appropriate to mention some of the possible consequences of aggregation and intensification.

If we were African farmers we would wish to improve our position. However, greed, although a powerful motive, should be restrained if it were conclusively shown to damage the inheritance of future generations. Many ecosystems, both wild and agricultural, are in delicate balance, but also I believe that they have a marked ability to repair themselves. Intensification usually implies: (1) the cropping (and possibly irrigation) of a greater percentage of the cultivatable land; (2) the search for greater yields per hectare; (3) often in the past reducing the diversity of cropping and administering greater inputs of fertilisers and crop protective chemicals; (4) it usually requires an improvement of the national infra-structure service and communications which inevitably requires capital inputs; and (5) if production is for a particular market, there would often be a tight requirement for quality and for the future storage, transport and processing of the product(s).

Environmental concerns have led to the highlighting of presumed deleterious agricultural results of such changes, for example: the exclusion of wild life or its extinction; the precariousness of high yields based upon more demanding and perhaps untried varieties; the increased ease of spread of pests and pathogens in large areas of single variety crops, and the possibility of further increase if the cropping is repeated; and sensible use of fertilisers is of course most helpful but their careless use can be wasteful, and if very generous, even polluting. There are safety hazards with most crop protective chemicals, and there has been an increased incidence of insensitivity with a number of the recent organic materials active against specific enzyme systems.

However, the significance and the biological processes and consequences are similar and no more dangerous than the frequent "breakdown" of inherited mono-genic resistance, which does not attract such media cover. There is little doubt that the greater the crop and yield the more there is to lose if things go wrong. Conversely, the benefits are greater in good years, and intensification can better provide the costs of monitoring, survey and research on pests and their forecasting. Central processing can have enormous advantages for the products, indeed it can make their cultivation possible. Together with the much more convincing reasons of shortage of money, these factors constitute the basis of "back to nature movements", of which we have heard so much under a variety of names.

However, our sessions did produce two extremely interesting papers on compromise production systems: the Mumias Outgrower Scheme for sugar cane production (Chapter 26) and the Kenya Tea Development Authority Scheme (Chapter 28). Both provided capital benefits and central processing to small-area producers who were obliged to follow strict procedures to intensify production. These seem to be models of great interest, offering a feasible way of advance, yet minimizing social and other agricultural change.

Biological Sins and Virtues

Few would not subscribe to the conservation of our biological heritage, and it would be folly to pretend there had been no mistakes, no learning process. This must be a continuing part of the means by which we learnt which insects stung or transmitted diseases, or which plants were poisonous. It is well to heed the warnings and worries of "ecologists", but it would equally be inexcusable not to note and cater for the present population increase, a new London every month (Chapter 5). Of course I support the classical biological control model, and I am sure that CIBC (Chapter 80) will, through its liaison with the Institute of Invertebrate Virology in Oxford, carefully monitor introductions to see that introduced insects do not bring with them mammalian viruses. The risks are considerable, as could be the misbehaviour of the control agent in a new environment. I do not think we can assume constancy of behaviour in introduced agents. Greathead (Chapter 37) reminded us that man domesticated the cat to control the rat, but ornithologists are infuriated when it exhibits a preferential appetite for birds.

The figures presented showed the difficulty of managing classical biological control, but few results were produced. Of special interest was one set showing suppression of cassava pests for two years after the introduction of a pathogen. We had been similarly encouraged by early results showing that codling granulosis virus was as effective as the best insecticide in controlling codling moth on apple; in later years the trees suffered from several different tortricid moths previously unwittingly suppressed by the insecticide programme. This virus was too specific to be a useful biocontrol agent.

I believe that perhaps most of those present support the proposals for radical changes in control strategy proposed (Chapter 39) but unqualified support for mixed cropping is not justifiable. I would recommend the closest monitoring of root pathogenic fungi, nematode infestation, and virus infection before such practices are introduced. The continuity of an inoculum source is a cardinal sin that plant pathologists have long recognized; and farmers have known about rotation for millenia. Certainly I have no objection to the use of cultural, environmental, or chemical measures in the practice that began as Integrated Pest Management (IPM). However, I believe that in Integrated Crop Protection (ICP), weeds and diseases must also be considered. The concepts of IPM and ICP differ enormously in crop applicability, time and space.

In this symposium, there was a general recognition that plant breeding has contributed considerably to crop production, not only through increased resistance to pests and diseases, but also through making plants more efficient. Innes (Chapter 32) admitted that he would like to see better matching to African environmental factors, and of course to further increase the durability of resistance to pests and diseases. I think we agreed with him that at present the intriguing tricks of genetic engineering are best left to sophisticated laboratories in developed countries. Within the tropics, the primary objectives were rather the incorporation of multiple resistances into plants and the improvement of screening methods. There were suggestions that more attention should be paid to the storage capabilities of new varieties, but Innes stressed that this, or incorporating particular nutrients, would greatly add to the cost of breeding programmes.

Notable among the biological sins is the introduction of exotic pests or pathogens. We were given many examples, such as coffee rust, ground nut rust, the greater grain borer and the cassava mite and mealy bug. Despite the difficulty and inconvenience caused, there was general support for strict quarantine and, for example, amongst rice nematodes there were very useful and important aims to be achieved in limiting the spread to or within the African mainland. Health precautions are of course a trying delay for the plant breeder, but one that is ignored at the peril of their reputations. Fortunately the International Centres working mostly with vegetatively propagated crops now appreciate the importance of virus-tested meristem culture as a means of safe international exchange. Seedborne pathogens, especially viruses in legumes and nematodes, still remain problems for which safe procedures are difficult to define.

Awareness

The symposium contained much evidence of the importance, and often the speed, of the CAB Identification and Information Services (Chapters 79 and 81). There was an enquiry about the future development of the Services with reference to their use of newer techniques. As these Services are described separately, they will not be emphasised again here, except to note that they were fully supported. It is also important that, in partnership or directly, CAB should encourage the subsequent phases of assessment and prediction. The latter could concentrate on biological, climatic, or economic factors.

The need to ensure the sustainability of developed systems and practices was also stressed. Training is important in this context, and the need for specialized training and communication arose in discussion many times. We recognized the problems in some countries in arranging PhD programmes partly in developed and partly in developing countries. Equally, incentives were required to ensure the adequate staffing of remote stations; it was said that in Nepal in such circumstances seniority was gained at twice the calendar rate.

We also recognized a strong need to improve the information flow to farmers through teachers and officials. It was hoped that CAB might contribute to this by further improving the attractiveness of its training publications, and the quality of its description sheets and distribution maps.

Some Needs for Development

Stabilising Supplies

It is extremely difficult to obtain accurate measures of yield on the properties of small-area farmers.

However, there seems little doubt that this is only a fraction of that at which we should aim, and that this is much less than that attainable on research stations. However, I think we agreed with Odhiambo (Chapter 6) that it would be better for agriculture to attempt to stabilise on reasonable yields than to strive for erratic records. Apart from weather, the pests, pathogens and weeds create some of the most important variables.

We also applaud the efforts that CAB and FAO have made towards the improved assessment of damage and believe this to be a study worthy of expansion. Continuous monitoring of the costs of control are necessary, and are often heeded if expressed as "yield equivalents". Widespread adoption of biological control would of course make such information more difficult to collect.

Another very important factor that can contribute to the stability of yield is the seed industry and its associated responsibilities for seed health. The development of such facilities, either privately or with national support, is most important. There still seems room for doubt and investigation about the degree of uniformity of stocks. From an epidemiological standpoint, there is some advantage in diversity, and this may be the optimum situation until marketing to quality standards or processing become important, when homogeneous quality is almost certain to be required. This does not necessarily detract from the importance, which we recognised, of local selection for local adaptation.

The selection of African crops that we could accommodate during our sessions was quite large, but remains insignificant compared with the variety available to farmers. We considered the serious problems facing individual commodities such as coconut, cashew and oilseed plants, but found no common factor to present as a solution. Diversity of crop types is essential, not least because of dietary variety and nutritional complementation, but also because variety offers insurance against pests, disease or drought, and increases the flexibility to feed the human and animal populations or to grow some cash crops.

Scientific Disciplines Needing Reinforcement

It is difficult, perhaps invidious, to attempt to select some areas from among so many that could benefit from increased resources. I intend to mention post-harvest storage, nematology and weed science in particular, because each is important, neglected, and has an important component of location specificity.

(a) Stored products research

It is difficult enough to produce foodstuffs and cash crops in Africa; to lose much of it in subsequent storage is both an insult and a hazard to the producer. Some forms of loss are serious as well as unpalatable (35% of tobacco leaf is lost in curing and handling). Serious losses arise from insect and mould infestations; the special and particular hazards to man and animals of a wide variety of mycotoxins in grains and fodders merit increased attention.

The accumulated resistances of cultivars long native to a region was repeatedly commented on. Often these advantages were lost if they were replaced by unadapted but high yielding (and more demanding) cultivars bred elsewhere. This is one more evolutionary lesson that we have to learn, that plant breeders, stored product scientists, and national agriculturalists must become increasingly engaged in ensuring the incorporation of storage capability alongside yield capability.

(b) Nematology

Nematodes are relatively inconspicuous parasites, partly because of their small size and frequent attack underground, but also because their effects begin inconspicuously but become serious as their populations rise. Once established, they are great survivors and very difficult to kill, so that control usually involves prohibitively expensive nematicides or long rotation breaks. Elsewhere, they have been shown to be important not only through direct damage but also as virus vectors, a topic so far scarcely tackled in Africa. Their spread in some seeds, soil or vegetative planting material can be decreased significantly by reasonably simple routine treatments that merit adoption on a wide scale.

(c) Weed science

To return to the small farmer of Nigeria, that country is well endowed with weed specialists compared to most in Africa, but the ratio is only one such specialist to 900 000 small farmers. Considering the ubiquity of weeds and the time spent in their control, the investment in education about them is very poor. There is an urgent need to provide training courses for future teachers of weed science and weed control practices.

Chemical Crop Protection

I also wish to add a plea for a more mature consideration of the benefits of chemical crop protection and instruction in its careful and rational use. It is accepted as a component that may need to be adopted into integrated control systems, but there is an attitude propagated that chemical manufacturers are grasping men not interested in benefiting agriculture. As in any walk of life, there may be some that fall into this category, but my experience is that most, and particularly the larger firms are very responsible traders who rely upon satisfied customers who will come back for more. It seems foolish to deny a tropical market that currently absorbs about 12% of world pesticide output. Not to welcome pesticide firms into our thinking and discussions, could force them towards practices that we would wish to avoid. They should be accepted as partners in the safe and rational use and development of pesticides in African farming, perhaps first with respect to seed dressings and herbicides.

The Face of Farming

Many contributors mentioned the hard life of the African cultivator, and the flow of the population to the cities, with all its consequential problems. As a result, the farmer tends to have a low standing in public esteem. Changing this attitude is an absolutely fundamental requirement of improving African agriculture. The farmer must have incentives, and his rewards must come quickly after his sales. This, together with the gradual introduction of labour-saving practices, will improve the farmer's life and so reverse the flow of labour from the land.

The policies of Governments and the will of the people will decide future success. It is not for visitors to interfere in either of those processes, but I believe we would all agree that in the light of the most constructive discussions with scientists from Tanzania and other African countries we have now had, we are confident in their skill and determination to effect improvements.

Chapter 41

Trends in Sub-Saharan Africa's Livestock Industries

A. ANTENEH

Livestock Policy Unit (LPU), International Livestock Centre for Africa (ILCA), P.O. Box 5689, Addis Ababa, Ethiopia.

Introduction

The performance of livestock production in sub-Saharan Africa was poor in the 1970s. Many reasons have been pointed to for this: technological and technical constraints, a shortage of trained manpower, and a lack of well-conceived projects, have all been indicated as major ones. Constraints arising from disease and poor nutrition contribute to increased mortality and poor productivity. Constraints imposed or introduced by Government policies, such as pricing, internal marketing and external trade, are increasingly being recognized as important in supporting or discouraging increased production. Livestock production is an important economic activity in many African countries. For sub-Saharan Africa as a whole, livestock's share in agricultural GDP averaged over 15% in 1981. Out of 31 countries for which data are available, in 16 this share was over 15% ranging from 16% to 86%; in 10 the share of livestock production in total GDP was about 10%, with a range from 9.8 to 30% (Jahnke 1982).

The main purpose of this contribution is to describe the situation over the last two decades in order to serve as a background to subsequent discussions at this Conference.

Table 1   Growth and regional distribution of livestock units[1] in sub-Saharan Africa, 1963–1980.

Region[2]	Livestock units (1980) (M)	Growth rate (% y^{-1}) in livestock units		% of sub-Saharan African total no.	
		1963-70	1970-80	1963	1980
West	38	2.0	0.3	30	25
Central	9	4.7	2.7	4	6
East	80	3.6	1.4	51	55
Southern	20	1.2	1.5	15	14
Total	146	2.8	1.2	100	100

[1]LSU conversion factors: cattle 0.7, sheep and goats 0.11, camels 1.4.
[2]**West** (Benin, Chad, Gambia, Ghana, Guinea, Guinea-Bissau, Ivory Coast, Liberia, Mali, Mauritania, Niger, Nigeria, Senegal, Sierra Leone, Togo, Upper Volta); **Southern** (Botswana, Lesotho, Madagascar, Malawi, Mauritius, Mozambique, Namibia, Reunion, Swaziland, Zambia, Zimbabwe); **Central** (Angola, Burundi, Cameroon, Central African Republic, Congo, Gabon, Rwanda, Zaire); **Eastern** (Comoros, Djibouti, Ethiopia, Kenya, Seychelles, Somalia, Sudan, Tanzania, Uganda).

Changes in Size of the Livestock Population

A study of the growth and distribution of livestock units in sub-Saharan Africa (Table 1) indicates that the East Africa sub-region remains dominant in the share of the total stock of animals; its share (in total LSU) increased from 51% in 1963, to 55% in 1980. Central Africa's share, although it increased during the same period, still remains small at 6-7% in 1980. Both West Africa's and Southern Africa's shares (but not absolute numbers) have declined during the period.

Between 1963 and 1980, sub-Saharan Africa's total ruminant livestock population, in terms of LSU, increased at 1.9% y^{-1}. The rate of growth between 1963 and 1970 was higher than that between 1970 and 1980 reflecting the adverse effects of the major drought of the early 1970s on the major livestock producing areas in the Sahel and parts of Eastern Africa. West Africa, which includes the major part of the Sahel, showed the lowest growth rate among the regions during the whole period, while the highest was for Central Africa, perhaps because of the small initial base of livestock numbers compared to other sub-regions. East Africa recorded a growth rate of over 2% y^{-1} during the 17 years. Overall, these annual rates show that livestock populations increased at a slower rate than human populations (about 2.8%); only sheep populations have grown at about the same as or higher rate than that of the human population (except Southern Africa). The West and Southern African sub-regions show consistently lower growth rates than the average for sub-Saharan Africa, except for goats in the South.

It is difficult to be precise about the causes of these trends. Increasing war and conflicts as well as drought and more frequent outbreaks of disease, due to a breakdown of control measures must have contributed to the deceleration in the growth of the livestock population between the two decades. Poorer nutrition, resulting in decreased fertility and still higher mortality, has been another contributory factor (ILCA 1983). These factors are likely to have been common for many countries. Drought occurred in parts of Eastern Africa at the same time as in the Sahel, but seems to have had relatively less effect there than for the sub-region as a whole.

The tsetse and trypanosomiasis problem primarily affects the West and Southern African sub-regions, but it cannot be considered a phenomenon affecting livestock populations only during the last two decades. However, tsetse eradication and the control of trypanosomiasis, or programmes for breeding trypanotolerant animals in the infested areas would have to be significantly larger than they were to have an effect on livestock numbers.

Trends in Livestock Production and Productivity

Output of meat and milk

In 1980 sub-Saharan Africa produced about 4% of the world's beef and only slightly more than 1% of the world's cow milk. East Africa has the highest share of total livestock output, corresponding to its high share of the total livestock population (Table 2). Output growth rates over the decade have also contributed to this. Between 1963 and 1980 beef production for sub-Saharan Africa as a whole grew at slightly over 2% y^{-1} while mutton and goat meat production grew at 3.4% y^{-1}. Sheep and goat, however, contributed no more than 30% of total meat production. In the 1970s production of meat (all types) grew only at 1.8% while population growth in sub-Saharan Africa was close to 3% (ILCA 1983).

Although a trend of declining growth rates is evident for the three major livestock producing regions, Southern Africa recorded the lowest rates in meat production during the whole period, 1.3% for beef and 2.0% for mutton and goat meat, consequently reducing the sub-region's share in total production. West Africa seems to have maintained its share of the total production (for both types of meat) at almost the same level throughout the period. Eastern Africa's share has not only continued to be the largest but has grown slightly during the period. Central Africa's share, although small in absolute terms, increased between 1963 and 1980. The rate of growth of per capita beef production during 1963-70 was positive for sub-Saharan Africa as a whole, as well as for all the sub-regions except perhaps Southern Africa, while it was positive for all as regards mutton and goat meat output. This situation changed dramatically during the 1970s with per capita growth rates for both types of meat becoming negative for sub-Saharan Africa as a whole and all the sub-regions in it except Central Africa.

Table 2   Regional distribution of livestock output and human population in sub-Saharan Africa 1980. % of sub-Saharan African total (in Mt M^{-1} people).

Region[1]	Beef	Sheep and goat meat	Cow's milk	Human population
West	27	40	20	42
Central	8	4	5	16
East	43	50	64	29
Southern	22	6	11	13
Total (absolute values)	(2.05)	(0.87)	(5.65)	(350)

[1]See Table 1 footnote.

Data on sub-Saharan African production of milk of all kinds in the 1960s are thought to be particularly inaccurate (ILCA 1983), and this applies to the production of cow's milk; my discussions of trends therefore only relate to 1970-80. Overall, cows' milk production in sub-Saharan Africa increased from about 5M t to 5.7M t during the past decade, a growth rate of only 1.3% y^{-1}. The annual growth rate of sub-Saharan Africa's milk output in the 1970s has been much below the estimated rate of population growth, resulting in a negative per caput change in production much higher than that for meat. Cow milk data includes dry milk production and the major producers of dry milk, Kenya and Zimbabwe, have experienced substantial declines in dry milk production. Other sources which only considered fresh milk, have shown a positive rate of up to 1.3% y^{-1} (ILCA 1983). The annual growth rate of sub-Saharan Africa's milk output has therefore been below the estimated rate of population growth, resulting in a negative per capita change in production much higher than for meat. West and Central African milk output increased in absolute terms during the period; both sub-regions were able to increase their shares by about 2 and 1 percentage points, respectively. On the other hand, Eastern and Southern Africa's share declined from 66% and 19% in 1970, to 64% and 18% in 1980 respectively. Per capita rates of growth, however, were negative for all the sub-regions, Eastern Africa showing the highest negative value.

Factors Affecting Meat and Milk Output

Over the last two decades, total output of meat and milk has increased. In considering the factors affecting this increase, we can look at two aspects, if and how much of this can be attributed to *extension* rather than to *intensification* of existing systems, and whether one can identify the factors affecting the rate and direction of the changes.

(a) Yield per animal

Yield figures per animal can be expressed as yield per directly productive animal, or per every animal in the whole herd. If flocks/herds lead to one product (i.e. milk or meat) and if the technology and management were unchanging, so that calving, mortality and other technical coefficients remained constant, the two ways of expressing yield would be the same. However, since herds are multipurpose (e.g. producing milk, meat and draught power) and technical coefficients change over time, changes occur in herds' structures so that yields expressed in these different ways may change in different directions at the same time (Table 3).

For beef the general picture is of modest, if any, increases in yield per directly productive animal, i.e. almost no increase in carcass weight, but some evidence of higher beef off takes from cattle herds as a whole. In West Africa where there has been a fairly steady rise over both decades in the productivity, in respect of beef, of the herd. For sheep and goat meat in the region as a whole, there has been little change in yield by either way of expressing it. In Eastern and Southern Africa yields seem to have declined, in West Africa there was an increase in the 1960s but not much consistent change in the 1970s, and Central Africa the two ways of expressing yield give contradictory results. For cow's milk the picture for sub-Saharan Africa as a whole, and for the Eastern and Southern sub-regions, is of substantial increases in the 1960s, followed by a sharp decline in the 1970s which reduced by about 50% the gains of the 1960s. In West Africa yields increased in both decades.

Table 3   Changes in yield per animal in sub-Saharan Africa 1963–1980 (index 1963 = 100)[1].

Region[2]	Yield basis[1]	Beef		Sheep and goat meat		Cow's milk	
		1970	1980	1970	1980	1970	1980
West	productive animals	101	102	114	117	106	109
	total herd	114	131	123	126	108	128
Central	productive animals	109	110	107	107	145	146
	total herds	99	111	100	104	73	66
East	productive animals	104	103	94	97	138	117
	total herd	108	109	93	99	139	109
Southern	productive animals	101	99	94	95	271	221
	total herd	109	99	87	98	238	189
Total	productive animals	102	102	103	106	147	127
	total herd	109	112	104	110	143	121

[1]Calculations by E. Seyoum & S. Sandford based on FAO *Production Yearbooks* data.
[2]See Table 1 footnote.
[3]See text.

(b) Attribution of changes

The results of calculations attributing increases in total output to extension of numbers and intensification of herd productivity respectively are given in Table 4. In the case of beef in the 1960s, 60% of the increase in output in sub-Saharan Africa as a whole was due to increases in numbers and 40% to increases in yield per animal. The proportion of the total increase due to numbers was highest in Central Africa where it accounted for all of the increased output, and lowest in Southern Africa where it accounted for only 40%. During the 1970s over sub-Saharan Africa as a whole more than 80% of the total increase in output was due to increased numbers and less than 20% to increased yield. Only in West Africa, where cattle numbers actually fell in the decade, significantly less than two thirds of increased output was due to increased numbers.

In the case of sheep and goat meat in sub-Saharan Africa as a whole, nearly 90% of the increase in output in the 1960s was due to increased numbers and 70% in the 1970s. In the 1960s West Africa and in the 1970s Southern Africa are the only sub-regions to show a substantial proportion of the increased output as coming from increased yield.

Table 4   The attribution of changes in total outputs to increased herd numbers and to increased herd productivity in sub-Saharan Africa (% of total change attributed to the cause indicated)[1].

Region[2]	Caused by increase in	Beef		Sheep & Goat Meat		Cows' Milk	
		1963-70	1970-80	1963-70	1970-80	1963-70	1970-80
West	numbers	44.9	−9.2	44.5	81.7	59.48	−7.35
	productivity	55.1	109.2	55.5	18.3	40.52	107.35
Central	numbers	102.3	65.6	101.8	85.5	1196.62[3]	175.45
	productivity	−2.3	34.3	−1.8	14.5	−1096.62[3]	−75.45
East	numbers	64.6	94.1	122.3	64.4	27.91	211.44
	productivity	35.4	5.9	−22.3	35.6	72.09	−311.44
Southern	numbers	39.4	247.7	181.0	6.6	3.96	285.85
	productivity	60.6	−147.7	−81.0	93.43	96.04	−385.85
Total	numbers	60.6	80.2	87.1	69.3	23.52	322.01
	productivity	39.4	19.8	12.9	30.7	76.48	−422.01

[1]Calculations by E. Seyoum & S. Sanford based on FAO *Production Yearbooks* data.
[2]See Table 1 footnote.
[3]The high percentage is due to the very small value of the denominator, i.e. the change in total output.

With respect to milk, in the region as a whole over 75% of the increased output in the 1960s was due to increased yield, while in the 1970s absolute yields fell, as did, to a small extent, total milk output, as increased output due to growth in cattle numbers failed to compensate for the decline in yields. West Africa went against this trend with a high proportion of the total output increase coming from improved yields in the 1970s compared to the 1960s.

(c) Causal Factors

The rate, and occasionally direction, of change have altered from time to time and sub-region to sub-region (Table 4). Some countries in sub-Saharan Africa have performed appreciably better than others in increasing the supply of domestic livestock products. A study to see whether it was possible to identify causal factors by using national-level data and simple correlation techniques (McClintock 1983) involved comparisons of output changes between 1965 and 1980. Only two strong and significant correlations emerged from this exercise: that between increased output of livestock products (except cow milk) and growth in the livestock population, and that between increased livestock output and increased cereal output. The first corroborates Table 4, and the second has important implications for policy (see below).

External Trade in Livestock and Livestock Products

Overall indicators

Over the past two decades, sub-Saharan Africa has therefore shown declining growth rates in livestock output in the face of a rapidly growing total population, and even faster rates of urbanization, as well as, in some cases, a rising per capita income. One result will be lower nutritional levels for the human population; this has serious implications for food policy issues in the region. One quantifiable consequence is, however, the growing volume and value of imports of livestock products into sub-Saharan Africa. ILCA (1983) analysed 1970 and 1980 data on net external trade in livestock, meat, dairy and poultry products for about 32 countries. In 1970, 16 of these countries were net importers; that number increased to 21 in 1980, with the net trade position of five having been reversed; these additional net importers included important livestock countries such as Ethiopia and Kenya.

Examination of sub-Saharan Africa's net external trade in meat and milk as a ratio to its total value of exports of agricultural (including fishery and forestry) products, shows that overall the net imports in 1980 has a ratio of about 7% to the total value of these exports. While West and Central Africa's value of net imports came to 11% (Nigeria's share of West Africa is 54%) and 8% respectively of their total exports of agricultural products, East Africa's stands at 2.5% and Southern Africa's net milk imports and net meat exports balance out at 3% each. Although it is unsatisfactory to make such comparisons, it is interesting to note that the ratio of sub-Saharan Africa's net imports of meat and milk to total exports of agricultural products compares favourably only with Asia (11%), that of all developing countries being about 4%.

External trade in live animals

Two general patterns can be observed in the external trade of live animals. In West, and to some degree in Central, Africa trade in live animals, particularly cattle, takes place between individual countries within the sub-regions. In this case, exports from the countries of the Sudano-Sahelian zone, i.e. Mauritania, Mali, Upper Volta, Niger, Chad and Sudan go to their coastal neighbours in West and Central Africa (Shapiro 1979, ILCA 1979). Nigeria and the Ivory Coast are the major importers. Exports from Mauritania, Mali, Upper Volta and Niger approximately equal the imports of the former two; for example in 1980 580 000 head were imported by Nigeria and the Ivory Coast, as against exports of about 550 000 from exporting African countries.

The second identifiable pattern is that of the surpluses from East African countries being exported to the Middle East. Almost 70% of the sub-Saharan African export trade in bovine cattle, and more than 85% in sheep and goats, originate from Ethiopia, Kenya, Somalia and Sudan. About 60% of the live animals exported from these markets are sheep and goats. In both cases official and illicit trade in live animals are important features of exports going to the Middle East.

Intra sub-regional trade in live cattle in West and Central Africa face increasing competition from cheaper-priced beef imports from Europe and Latin America. Live sheep exports from East Africa to the Middle East face similar competition from Australian and Indian exports, despite the advantages of short distance (low freight costs) and consumer preference by the importing countries. Although now recovering, the Sahel drought has reduced exports from surplus producers in the West African sub-region. In East Africa, except for Somalia which has been successful in its livestock exports (Reusse 1982), countries have largely failed to exploit a relatively "captive" market. The short-term prospects for capturing the whole potential offered by these large and lucrative markets is not as bright as one might have expected.

External trade in meat and milk

Sub-Saharan Africa experienced a fundamental reversal in its net trade position in meat from having a virtual balance in 1963, a slight export surplus in 1970 (almost 15 000 t) into becoming a net importer of almost 60 000 t in 1980. This has meant that in 1980 sub-Saharan Africa spent almost $150M more for meat imports than it earned from meat exports. The development in the position of the sub-regions differs markedly from that of sub-Saharan Africa as a whole. During the two decades considered, Central Africa has remained a net importer throughout, Gabon and Zaire accounting for 50% of the sub-region's value of net imports of meat in 1980. Of the eight countries in the sub-region, only the Central African Republic and Cameroon have reversed their position from net importers in 1963 to net exporters in 1980. At the other extreme, Southern Africa has remained a net exporter, five of the nine countries for which data are available being net exporters in 1980. Botswana, Zimbabwe and Madagascar have always been the major net exporters, while Reunion and Mauritius have become the two major net importers of the sub-region.

Southern Africa owes its success partly to the preferential quota arrangements accorded by the EEC to the four meat exporters of the sub-region, Botswana, Zimbabwe, Madagascar and Swaziland; prices received by them for beef exports to the EEC are above world prices (von Massow 1983). Many of these countries could continue to meet the rigorous health requirements of the EEC. They could also meet their quotas, although variable from year to year because of foot and mouth disease or droughts, at a reasonably high rate; Botswana on average exported 66%, Swaziland 80%, and Madagascar 39% of their quotas during 1976–81. In contrast, because of veterinary problems, Kenya never utilized its small quota. West Africa's position in this trade has fluctuated between a net importer position in 1963, to a net exporter in 1970 and the largest net importer of the sub-region in 1980. This change is mainly accounted for by the inflow of oil revenues in Nigeria, which accelerated in the mid-1970s, and by the reduced supply of live animals from the Sahelian countries as a result of drought which forced importers, such as Nigeria and the Ivory Coast, to find other sources of supply. In 1980, the sub-region's value of net imports of meat almost equaled that of the sub-Saharan total. In East Africa, meat exports from Kenya and Ethiopia, which accounted for almost the total net export figures of the sub-region in 1963 and 1970, gradually declined to make the sub-region as a whole a net importer. Despite this East Africa remains in a relatively better position vis-a-vis the other two net importer sub-regions in 1980 (0.01 kg caput^{-1} as against about 0.45 kg caput^{-1} each for West and Central Africa). It may also be worth noting that East Africa's better performance in live animal exports may be indicative of its comparative advantage as against meat exports.

With regard to milk, the picture is even more gloomy: net import values for sub-Saharan Africa as a whole have risen from $39M and $81M in 1963 and 1970 respectively to $575M in 1980 (all at current prices), an increase of over 600% in the past decade. Even if one takes into account that about one third of this rise is attributable to price increases, the remainder still accounts for the more than a 400 % increase in the volume of net imports. Sub-Saharan Africa's livestock sector has failed to take advantage of the domestic market opportunities available to it. West Africa accounted for the largest increase in the value of net imports of milk, its share of the sub-Saharan African total rising from 54% in 1963 to 68% in 1980. As with meat, Nigeria and the Ivory Coast account for about 82 % of the sub-region's value of total net imports of milk in 1980. For the same year, the other three sub-regions more or less equally share the remaining 32% of the total sub-Saharan Africa figure. There is no success story in any of the sub-regions of a favourable net external trade position in milk and milk products. Despite the relative advantage that its high potential areas possess an increasing domestic milk output, even the East African sub-region has failed to surpass its Central and Southern African counterparts. In 1980 the value of East Africa's net imports of milk was only slightly lower than that of Southern Africa.

Factors affecting external trade

Some factors affecting external trade in livestock and livestock products have been mentioned above. Natural factors such as the drought in West Africa have affected intra-regional as well as extra-regional trade, influencing livestock supply and leading to subsequent rises in imports or falls in exports. The oil boom in Nigeria has substantially contributed to the enormous increase in West Africa's high imports in the 1970s. Some sub-regions have not been affected, or at least not to the same extent by these factors; other factors, perhaps with a longer lasting effect, have contributed to the present situation.

Internal prices for meat and milk and their ratios with respective world market prices determine the extent of external trade in those products. In the last decade, in many sub-Saharan African countries domestic prices for food commodities in general were set at relatively low levels. Therefore, producers could hardly be expected to cope with food demands of a fast growing population with increasing income at its disposal. At the same time, world market prices were at depressed levels, largely due to protectionist policies of developed countries (Tangermann & Krostitz 1982). Thus, the growing gap between production and demand of livestock products in sub-Saharan African countries could be filled by imports at relatively favourable prices. Additionally, for many countries there was food aid of milk powder and butteroil available as well as commercial imports. Both categories of imports had adverse effects on domestic production, generating further demand for imports.

Apart from setting domestic prices at relatively low levels, Governments have used various policy instruments and thereby directly or indirectly influenced external trade; these include direct effects by way of import and/or export taxes, quantitative restrictions and exchange rate policies, as well as indirect influences that may result from marketing activities, subsidies, and domestic policy measures. These factors can alter the basic effects of the price mechanism, and therefore require increased attention.

Conclusion

The livestock output situation in sub-Saharan Africa is very serious, not only in the last two decades has the rate of growth of output declined, but also the percentage of output growth attributable to productivity (i.e. yield growth).

In terms of production per capita, only Central Africa seems to have improved its performance vis-a-vis population growth in the 1970s. It is the only region that shows a positive rate of growth in meat output per capita and at the same time has the lowest negative rate of growth in milk output. However, this sub-region had in 1980 only 6% of the livestock population, contributed 8% of the beef output, and 4% each of the sheep/goat meat and milk output of sub-Saharan Africa as a whole.

The situation in the external trade position in livestock and livestock products has become alarming, particularly as regards milk. Overall gross imports of livestock and livestock products into sub-Saharan Africa has reached approximately US $2 billion per year, about twice the figure of all foreign aid to the livestock sector over the last 15–20 years. In the 1970s in all cases of livestock output, increased livestock numbers contributed most of the change in total output (Table 4). With the non-tsetse-infested land area available to livestock decreasing due to population pressure, this suggests that there is a need for a move towards finding the appropriate technology and policies to increase the yield per animal.

Perhaps the most important finding from correlation analysis is that livestock output is positively correlated with crop output. This appears to allay fears that livestock and crops compete with each other for scarce resources, and support the view of complementarity between crops and livestock. This has important policy implications in that it indicates that livestock should be considered simultaneously with crop outputs to improve food production in Africa.

This contribution is a joint effort of the LPU staff at ILCA. Besides commenting on the whole papers, S. Sandford and V.H. von Massow contributed major portions of some sections. I also gratefully acknowledge the assistance of E. Seyoum (ILCA Documentation Services) for the compilation of data, and S. Mbogoh (LPU) for editorial comments.

References

International Livestock Centre for Africa (1979) *Livestock and Meat Exports after the Drought in the Sahel.* [ILCA Bulletin No. 3.] Addis Ababa; ILCA.

International Livestock Centre for Africa (1983) *Annual Report 1982.* Addis Ababa; ILCA.

Jahnke, H.E. (1982) *Livestock Production Systems and Livestock Development in Tropical Africa.* Kiel; KWV.

McClintock, J. (1983) *What causes Supply Levels from African Livestock Sectors to Change?* [Livestock Policy Unit Working Paper no. 2.] Addis Ababa; ILCA.

von Massow, V.H. (1983) On the impacts of EEC beef preferences for Kenya and Botswana. *Quarterly Journal of International Agriculture* 22, 216-234.

Reusse, E. (1982) Somalia's nomadic livestock economy – its response to profitable export opportunity. *World Animal Review* 43, 2-11.

Shapiro, E. (ed.) (1979) *Livestock Production and Marketing in the Entente States of West Africa.* Ann Arbor.

Tangermann, S.; Krostitz, W. (1982) Protectionism in the livestock sector with particular reference to the international beef trade. *Gottinger Schriften zur Agrarokonomie* 53.

Discussion

Dr Ellis opened the discussion and, along with other participants, agreed that this paper presented an interesting new approach to the analysis of trends in livestock production, which can indicate the constraints to development. The database on which the work is based is far from perfect, and the analytical techniques need further development, but the value of the approach was amply demonstrated.

The evidence for the decline in productivity seems to be clear; any growth in output can be explained by increase in livestock population. Even more disturbing is the fact that the declines were apparently more rapid in the 1970s than in the previous decade, causative factors suggested being pricing policy and the Sahelian drought. Several participants suggested that some of the apparent decline could be explained by the increase in unrecorded offtake. The only way to test this contention would be to improve data recording. Improved data would also allow changes in the distribution of livestock ownership and the importance of livestock in subsistence to be examined. It was noted that the statistics for pigs and poultry were not examined, and these should be considered in future appraisal systems.

Although participants had some doubts about the significance of the apparent correlation between expanding cereal production and animal population density, all agreed that closer co-operation and integration of agriculture and livestock production systems was desirable. A prerequisite to improved animal productivity is the control of numbers in animal populations to reduce overgrazing and land erosion. Contributors also noted that in many countries the resources allocated to research and development in the livestock industry have not been commensurate with its contribution to the gross domestic product; a disproportionate emphasis had been placed on crops.

Chapter 42

The Problems of Marketing – Serving the Consumer

E. G. CROSS

General Manager, Cold Storage Commission of Zimbabwe, P.O. Box 953, Bulawayo, Zimbabwe.

Introduction

It has long been held by market-orientated companies and countries pursuing free market strategies, that the object of all production is to meet the consumer's needs. Even under systems which reject free market economics, there is a growing acceptance that if production is not guided by consumer demand, serious misallocation of resources and inefficiencies rapidly become a feature of the system. The question of how best to meet consumer needs in Africa is, therefore, of fundamental importance to production planning and policy formulation.

All production planning should begin by identifying consumer needs. Once these are established, we should then proceed to determine; (1) how much of a defined good or service is required to meet total demand; (2) what will be needed in the way of raw materials and manufacturing capacity in order to provide adequate supplies; (3) where the production of the goods or service should take place; (4) what system should be used to distribute the product or service to the final consumer; and (5) how to balance the relationship in the market system between the different economic institutions and between producers and consumers. In free market systems, these decisions are taken on the basis of supply and demand and Government intervention is restricted to ensuring competition between independent agencies. This may not be acceptable in more rigidly controlled systems.

In addition, those of us who work in agriculture and are concerned with marketing agricultural products, must deal with a number of additional factors which are peculiar to the industry. Marketing arrangements which do not take cognisance of these characteristics will fail, with dire consequences for both producers and consumers in the long-term. The most important features are: (1) Both production systems and consumer markets for agricultural products are characterized by the existence of large numbers of small independent units of production or consumption and this results in conditions of near perfect competition on either side of the market chain. (2) In contrast, market chains in agriculture are frequently characterized by major concentrations of economic power, created either by market forces, the character of the product itself, or by Government. (3) Production is affected by season, climatic and disease factors beyond the control of the producers and consumers and leading to frequent and wide fluctuations in supply. (4) The demand for agricultural products tends to be inelastic so that small shifts in supply can result in substantial price movements. (5) Because of the numbers of people involved in production and consumption and the political importance of urban workers, agricultural marketing systems are subject to political interference which results in the distortion of prices and supplies.

In the livestock sector, the above features are complicated by the character of the products which often require a high degree of processing, and both storage and distribution is difficult. The majority of livestock products are perishable and can only be converted into a less perishable form by sophisticated technology; processing and distribution systems can require special facilities.

In designing production and distribution systems, all these aspects must be taken into account. Each product and market has its own character and systems must bear this in mind. This contribution outlines steps which can be taken under most circumstances to facilitate the development of market systems which effectively meet consumer needs.

Market Requirements

The first step, and one which should be ongoing, is basic market research. This involves identifying consumer preferences and isolating elasticities of demand. The latter should include income elasticities, as well as those between competing products and price/demand relationships. Included in market research programmes should be the determination of population and income data, as well as the distribution of the population between regions and socio-economic groups. It must also consider religion and traditional habits and tastes, which might affect requirements.

Fundamental to all efficient marketing systems is the development of a database which accurately describes the market and which is constantly up-dated. One of the fascinating aspects of marketing is the dynamic character of such activity, which requires constant attention to changing demands. In the case of Africa, the majority of countries do not have a satisfactory database which can be used for decision making. Until this situation has improved, it will be necessary for market-orientated corporations to undertake a considerable amount of database investigation on their own.

Using market research findings and an up to date database, estimated market requirements can be calculated. In certain circumstances, the approach adopted for such calculations might be minimum nutrition levels. This approach can be adopted where national restrictions on supply inhibit the ability of producers or importers to meet total consumption demand, and controlled distribution would be necessary to ensure equity under such circumstances. Where supply constraints are not a consideration, needs should be based on prices, income and consumer preference. Even a small differential between supply and demand can lead to substantial price premiums.

A programme of market research is not complete until those responsible have actually defined products for development and distribution. Such product development requires an intimate knowledge of the capabilities of the production and market system, the availability of raw materials, delivery systems, and an understanding of storage needs as far as consumers are concerned. Consumers place a high degree of preference for both convenience and basic availability.

Meeting Market Requirements

Once the extent and character of demand has been defined, attention should be paid to how production can be organized so as to meet the consumers' needs. This is a complex process which requires attention to detail and an intimate knowledge of factors affecting the product. Raw material requirements must be analysed and quantified in depth, including attention to price and foreign exchange needs, as well as stock holdings to ensure that production is not inhibited by delays in delivery. A careful analysis of raw material requirements will give early warnings of potential difficulties in meeting market demands. Raw material requirements will depend, to some extent on the technologies used. Decision-makers need to understand the technical aspects of production to ensure that input requirements are satisfied. Even small errors in technical descriptions for required inputs can lead to a serious loss of foreign exchange and the disruption of production.

As investigations proceed, it will become apparent whether or not manufacturing capacity is adequate. As it takes time to establish new capacity, it is necessary to ensure that total production covers existing and projected market needs; forecasting therefore plays an important role in determining investment priorities.

A feature to which inadequate attention is often paid, is the selection of packaging. While the need for economy is important, the consumer requires a convenient pack priced at a level appropriate to his cash flow. At the same time, care should be taken to ensure that packaging standards found in free-market economies should not be automatically accepted for developing countries. Developed countries may sell a system of packaging which is too sophisticated for a developing country to maintain. A further feature which should receive serious attention is the re-cycling of material within the market. Plastic-based packaging can be re-cycled to the extent that over 60 % of the raw material can be recovered resulting in substantial savings.

The character of the distribution system will, to some extent, determine the character of the product and type of storage required. If a perishable product is to be distributed without adequate storage and sophisticated distribution facilities, it may be necessary to change the character of the product itself in order to give it a longer shelf-life. This can permit simplified storage and distribution with substantial resulting savings. Inadequacy of home storage can also force adaptations.

The Location of Production

The location of manufacturing capacity is an important decision which can affect production and distribution costs. All too often, the location is selected due to political pressures or to satisfy loosely defined objectives such as de-centralization. Such decisions can impose long-term burdens on both producers and consumers. Decisions on location should take into account political and policy constraints, but the primary issues which should be used, should be the supply of services and labour, the distribution of market demand, transport, and the need to secure economies of scale. All too often appropriate technology is interpreted as being "simple and small" when, in fact, developing countries can utilize sophisticated technology and should consider building larger centralized plants. For example, small abattoirs incur almost double the level of overheads of beef produced over large abattoirs with a minimum throughput capacity of 500 head or more per day. However, care must be taken to ensure that economies of scale are not secured by opting for far too large a plant instead of smaller plants located closer to production or consumption centres; the product and distribution system, will have a bearing on the decision.

Distribution Systems

The selection of a system for the distribution of an individual product or commodity will depend on the nature of consumption, the location of demand and the product. There is an urgent need to develop policies which will guide future development by encouraging the development of certain types of retail or wholesale institutions and can bring about substantial savings. By encouraging supermarkets instead of small specialist retail stores, retail margins can be reduced by up to 70%.

In addition, Governments should give attention to assisting in institution-building throughout the market chain, emphasizing the development of human skills and management; the greatest single obstacle to efficient and effective distribution. By encouraging certain types of institutions through state assisted schemes, substantial structural improvements in distribution systems can be secured. Also important is the need to evaluate alternative transport and distribution systems. Long-term savings might be secured by substantial infra-structural investment in railways or river transport. Such investment should always be influenced by projected requirements for the movement of the product within a country in order to meet market demand. The reliance of most African states on road transport may not be in the long-term interests of the consumer, although convenient and relatively inexpensive in terms of basic investment.

Finally, the design of distribution systems or their selection, will depend on the character of the product and its packaging. Long-life products packaged to facilitate easy transportation and distribution, are much less expensive to handle and require less infrastructural development than where the product is highly perishable and packaged so as to require careful handling. Markets where producers and consumers can exchange products directly is also a vital part of development; these are inexpensive to create and are largely self-administered. However, despite the absence of sophisticated processing and packaging, such markets tend to be expensive both in terms of retail margins demanded and prices paid to producers.

Competition and Control

A balance is needed between permitting a free competitive situation, which may be to the detriment of both producer and consumer, and intensive Government control which stultifies initiative and freedom of action. The ideal situation is one where Government acts as a referee, leaving producers and consumers to fix prices within certain guide-lines and intervening only when rules are broken; the referee role might include ensuring that there is competition within the market chain.

In agriculture this is seldom sufficient as it is almost always necessary for Government to assist in stabilising production prices and in ensuring the availability of raw materials. Ways of achieving such stability should be so managed as to be self-balancing over time, with no element of subsidy. In time, producers and consumers should be required to meet real costs as subsidies distort economic relationships and artificially raise consumer demand, so complicating decision making to the detriment of sound management. Governments may feel that this is insufficient and enter the process of production and exchange as a participant. Such decisions should be made only to remedy inadequacies in the market chain and to ensure that longer-term interests are protected. Wherever possible, Governments should not attempt to become sole operators or producers. There are too many examples of state-financed failures in Africa. Where parastatals are created to undertake such functions, Governments should insist that they behave, as far as possible, as business corporations required to balance their accounts and satisfy the demands of producers and consumers.

Policy

Policies should be geared to deal with the question of producer prices on a basis which will give producers confidence and which will encourage investment. This requires sophisticated management to be successful. Production policies can deal with fluctuations in world prices and domestic production. For example, in the long-term it may be less expensive to pay farmers drought relief than leave them to their own devices in adverse weather conditions while paying higher long-term producer price levels to enable them to build up resources to cover climatic vagaries. It will also be necessary to protect the consumer and producers from extensive fluctuations in prices due to world market conditions but such measures should not protect consumers and producers from long-term shifts in markets.

Policies need to be established to guide project development, to ensure the development of appropriate processing technologies, and guide the development of distribution systems. Where adequate competition exists, less attention should be paid to controlling consumer prices as this leads to artificial distortions and empty shelves. Consumers prefer dealing with producers directly providing that products are in free supply.

Given the application of these principles, African countries can give the consumer democracy in the market place, a great need of the continent.

Milk Products

Improved collection systems for small scale producers have resulted in a substantial upsurge in milk production for delivery to processing plants. Small farmers, given a suitable production and economic environment, can meet demands for milk and dairy products in consuming centres. Equally important is the selection of technology for processing. The alternatives include the conversion of milk into cheese, powder or acidulated products, as well as the development of long-life products such as sterilised and UHT milk in long-life packaging. In addition, distribution systems for milk and dairy products offer many alternatives. Zimbabwe's urban workers have a daily door-to-door distribution system operated by vendors which has proved less expensive than distribution through retail outlets. In India, the distribution of milk in low income communities through public vending machines has proved equally successful. Lastly, the use of small retailers and markets for the distribution of dairy products can provide a perfectly adequate infrastructure, providing processing and transport arrangements are efficiently managed.

Beef

Abattoirs must be convenient to production as abattoir-based processing is becoming increasingly important; developing countries should also be more active in developing products such as sausages, canned meat products, pre-packed products, and cooked or processed beef which can be long-life, refrigerated or effectively distributed using simple technologies. Of fundamental importance is the choice between traditional butchers and other retail establishments. Retail outlets such as supermarkets and hypermarkets are taking an increasing share of the traditional meat trade and in some countries the traditional butcher has almost disappeared. There are advantages, both in terms of costs and hygiene, for centralized meat processing.

Goats and Sheep

Goats and sheep are widely used to satisfy meat requirements in urban and rural areas, and greater attention needs to be paid to establishing live animal markets in urban areas. Although these often exist on an informal basis, without appropriate organization they can constitute health hazards. Abattoir arrangements can be less complex than for beef, but where goat and sheep meat is distributed chilled or frozen, the same requirements exist.

Poultry

The fastest growing single item of consumption in Africa is probably poultry, because of its convenience and ease of production, which is now widely marketed as a live product and in a frozen or chilled form. The main need is to develop adequate refrigerated distribution systems for poultry and organized markets for eggs. There is no real requirement for Government intervention here, apart from facilitating the development of infrastructure. The provision of live poultry markets is also important for consumers who do not have refrigeration.

Other Considerations

Africa currently imports 10–15 % of its total food requirements, and could absorb at least the same amount again before consumer needs are satisfied. The dangers in food aid schemes are well-known but it needs to be emphasised that where food aid is provided or sought, it should not be distributed in a way that undermines markets for local producers; it should be sold at full internal prices based on the production priorities of the country.

Advertising is often criticized as unnecessary and expensive but one problem of marketing in Africa is the paucity of product-and price-awareness by the consumer which can lead to monopolies and exploitation. In the long-term, consumer welfare is best protected within marketing systems which are free of unnecessary control and where there is competition. Governments should facilitate the development of such systems without becoming involved in the market place.

Money acts as a means of exchange between producers and consumers, and also provides, in well-ordered societies, a store of wealth. It is vital, for the sake of consumers in African states, that money exchange values (either for goods and services or for hard currencies) are at a realistic level. In many African countries, rates of exchange are out of line with estimated "real" values. Street values can vary enormously from official rates. Such discrepancies destroy efficient market systems to the detriment of producers and consumers and parallel market systems take over. Corruption and smuggling become the options for survival, and confidence in Government and in money is destroyed. Unless currency values are set at realistic levels, it is difficult, if not impossible, to operate soundly-based and effective market systems meeting consumer needs at the lowest cost.

Discussion

Dr C.E. Williams (Senior Lecturer, Department of Agricultural Extension, University of Ibadan, Nigeria) focussed on the problems of serving consumers in Africa and outlined the attributes expected of a market by society: (1) goods and services to be produced and offered in the right quantities and qualities at the right time, in the right places and the desired forms; (2) fair prices charged for goods and services; (3) full information about uses, usefulness, strengths, special attributes, weaknesses of goods and services; (4) dynamic institutions performing marketing functions – modifying the character of offerings in response to changes in consumer needs; (5) special arrangements such as discounts/rebates, credit facilities, after-sales service, spare parts, exclusive facilities; and (6) markets regulated by law to prevent abuses and protect consumers. Generally, consumers want information and convenience, equality and satisfaction, and value for money. In a competitive situation the seller tries to attract consumers to his goods and away from his competitors' goods by providing things critically demanded by consumers, especially things his competitors do not or cannot meet.

Unfortunately, sellers' markets predominate in most African countries and consumer desires are ignored. Marketing companies, middlemen, and retailers do not know their products well and cannot explain their qualities and uses and so cannot service the consumer. In fact, a consumer often feels lucky to obtain anything at all and cannot talk of quality, or raise issues of price or additional convenience.

Longer-term aims should therefore be to create competitive conditions where anyone can buy or sell, where information is not restricted, and where price depends on the ability of the purchaser to negotiate. As more competitive conditions emerge, the purchaser will want more information on conditions at source, through packing and processing technique, to storage, preservation, preparation and use. Increasing attention will be paid to consumer needs, the size of packages for family use, and different forms of incentive will be offered to encourage purchases. As the chain becomes more complex, credit facilities will be required to help wholesalers, middle-men, and retailers.

Dr Williams also drew attention to some of the special problems experienced in beef marketing in Nigeria. She attributed the high cost to the fact that while most cattle are owned by Fulani, they are marketed through middle-men from the Hausa tribe who double or triple the price in selling on to butchers. Butchers have to pay local government taxes and may also have to pay stall fees in the markets. While the butchers pay cash to the middle-men they often have to give credit to the various sellers who handle different parts of the carcass. Only when direct sales are made to consumers does the butcher receive immediate cash. The cost of credit and losses from debtors add to the cost to the consumer. Seasonal patterns can affect prices as butchers may have to divert their capital to small ruminants for festival periods and increase the price of beef to help cover additional outlays. Butchers also become involved in undesirable practices. They may buy, slaughter and sell old, sick, or even dead, animals to an unsuspecting public. Scales may be inaccurate, storage facilities unhygienic, and contamination may occur in slaughtering or cutting up carcasses.

Short-term steps

Short-term steps considered in relation to meat marketing included:

(1) Consumers should be organized into groups for the equitable distribution of animal products.

(2) Quantity purchase by consumer groups will reduce prices, and improve the quality of meat to members.

(3) The existence of selective consumers will gradually build up consumerism.

(4) Education is needed on the acceptability of frozen meat as African consumers often want freshly slaughtered meat or even live animals to slaughter themselves.

(5) Sellers should be compelled to observe hygiene and quality standards for meat and offal. A system of "Best Butcher" prizes could be an incentive for improved quality.

(6) Meat from small ruminants could be promoted to overcome localized scarcities of beef.

Long Term Steps

(7) Consumer education on the nutritional values of livestock, and the quality and uses of livestock products is necessary. An educated consumer is weight and quality conscious.

(8) Provision must be made for marketing information by Government agencies through mass media.

(9) The introduction of pre-packaged meat for consumers must be considered.

(10) Provision should be made for better slaughtering facilities, more thorough inspection, for effective disposal of waste such as bones and dung, and movement and storage of meat should be improved by providing refrigerated vans and cold rooms.

(11) Those who handle meat should be educated in all aspects of meat hygiene.

(12) Building standards and fly-screening should be enforced in sales markets.

(13) Butchers and retailers must learn how to provide good consumer services.

Cattle owners, middle-men, butchers, and even consumers, are likely to resist change until they can see its benefits. Many complex abattoirs have been built but remain unused in Nigeria. Butchers boys who used to trek live cattle to slaughter slabs for one naira per head once started a riot when central slaughtering removed the need for trekking. However, many now support the new system, having adjusted to the change.

The highly critical views of livestock policy and marketing in Africa, presented by Mr Cross and Dr Williams met with a large measure of agreement. In particular, the point that consumer protection through price control or subsidy does not equate to long-term consumer welfare was repeatedly underlined; underpricing leads to shortages. Frequently, inefficient marketing systems add a disproportionate amount to the consumer price, damaging the interests of consumers and producers. Unrealistic foreign exchange rates discourage the production of exports and make imported commodities artificially cheap.

It is appreciated that practical and political constraints make it unlikely that consumer subsidies, inefficient marketing boards, and unrealistic exchange rates can be abolished overnight. However, improved efficiency in the marketing chain to reduce the differences between producer and consumer prices and secure greater quantity and quality of produce was essential. The problems of sellers' markets, which predominate in many African countries, were emphasized by several speakers who endorsed the ideas of Dr Williams concerning the usefulness of consumer groups and education on the use of new products. They also felt that advertising should receive much more attention from producer and marketing organizations.

Chapter 43

Pricing Policy

R. DONALD

Dairy Economist, FAO, Box 2, Dar es Salaam, Tanzania.

Pricing policy is often an important determinant in the pattern of agricultural development. Actual prices can influence the production mix purely through the profit they imply, but where administered prices are used, price movements and changes in price relativities can also serve as an indicator of the Government's priority areas for development, and thereby provide a secondary stimulus. The importance of price and the method by which it is fixed is, however, generally less in the livestock sector than in the agricultural sector as a whole. Empirical studies have clearly shown that climate, sociological factors, and general economic conditions are of greater importance than price in determining off-take in the traditional sector. The sociological and biological factors which underlie this feature are deep-seated, and therefore, while some scope exists to increase productivity and off-take of traditional cattle keepers, the main scope for influencing the production system through pricing policy is in the "modern" sector.

The profit margin is affected to a greater extent in the modern sector by price changes, due to its greater reliance on purchased or higher value inputs and greater response to improved nutritional status and health care, etc., but the sociological constraints to change are fewer. The adoption of new technology brings with it the opportunity for redefining values and objectives. Price policy is therefore very important in influencing the rate of uptake and degree of sustained interest in new production systems.

The effect of pricing decisions in the livestock sector, particularly where ruminants are concerned and where lower quality feeds are being relied upon for both maintenance and production, as is generally the case in Africa, is often marked by the long time horizon being faced by producers. Unlike annual crops, where timely price changes can alter the product mix dramatically, the ruminant sector is generally slow to change in response to price. The influence of price is therefore as much if not more an interpretation by the producers of the Government's resolve to support the sector in the future, rather than being short-term profit related. This may have serious implications for ruminant owners. Pricing authorities, in endeavouring to curb the effect of price increases on the consumer, are more likely to hold prices of commodities where short run responses to price changes are known to be low; consequently the prices of livestock products may, particularly during periods of general economic hardship, lag behind those of other agricultural products. As a measure to help overcome a short-term problem this may make macro-economic sense, but if used, compensating price increases, larger than would normally be justified, need to be accorded to these products when the emergency subsides.

A decision not to intervene in the market is as much a pricing policy as adoption of an administered price regime. The main objectives of pricing policy must be to:−

(1) Foster the orderly development of the sector to which the prices apply; order not being regarded as synonymous with slow change or no change. This will involve providing both some degree of price stability and an assurance of payment.

(2) Provide protection for the consumer against exploitation arising from short-to medium-term supply shortages. In the absence of a sole supplier or monopoly producer, this implies a general under supply to the market which may then necessitate some form of rationing.

(3) Enable the pricing authorities to guide the direction and rate of development of the sector.

The general rules for effective pricing policy apply equally to livestock products and other agricultural commodities:—

(1) The policy must be able to achieve a more efficient production system than can be achieved without it. Pricing policy must, with its associated structures for implementation, add value and not just costs.

(2) To be effective, administered pricing systems have to be relatively straight-forward to understand and implement. (Policies that require numerous administrators to enforce are generally a waste of effort or unenforceable.)

(3) Negative effects, particularly in relation to other markets and products, must be kept to a minimum.

The possible combinations of administered pricing policies that can be considered are legion, ranging from panterritorial prices for both the producer and the consumer through to a system of regional, or even smaller zonal area prices, with quality, transport and seasonal premiums and discounts operating. While the latter offers the ability, in theory at least, to tailor the pricing system to meet the specific objectives of profit control, resource allocation and welfare considerations, it suffers in terms of its complexity which will frequently make its administrative burden untenable. At the other end of the scale, panterritorial prices should require minimal administrative overheads, but where production conditions or population densities vary, they will almost certainly produce some anomalies in terms of resource allocation, particularly in terms of production and marketing locations. In situations where transport is high cost and/or unreliable, the problems caused by the use of panterritorial prices are likely to be exacerbated many times. Products are only worth what the final consumer will pay and where the cost of getting the product to the consumer varies greatly from place to place then a common price at the farmgate is inconsistent with the attainment of efficient production patterns.

There is extreme danger in linking livestock commodity pricing directly into a general pricing strategy for agricultural commodities, particularly grains. In general the outputs are so dissimilar in terms of their marketing requirements that a common pricing structure is unlikely to be suitable. Pricing systems which ignore perishability must be treated with caution; unless meat, milk and eggs are produced close to both the market and time of sale (consumption) then marketing costs, in terms of storage, processing, or wastage will form a significant part of the final market value. In contrast, for most cereal and industrial crops, although the time of harvest may be critical, transport, storage and deterioration costs need not be particularly high. Livestock products, more particularly from monogastrics, meet the high storage and possibility aspects of their marketing problem by having a year-round rather than seasonal harvest system; ruminant meat production also overcomes this in part by on-the-hoof storage, achieved at the expense of lower total production.

Whereas the simplicity of panterritorial pricing may be a sufficiently appealing feature to warrant its adoption in the cash cropping sector, it is unsuitable for most livestock products, particularly if the livestock sector is diverse in geographical location or production systems. The classic result of panterritorial pricing in the wrong situation is to strangle the supply to those markets which are difficult to reach, and which therefore may realistically be expected to pay a premium, and to remove the market from low-cost but isolated producers who could generally be expected to sell at a lower price. The net result is that part of the market is destroyed and total trade with its concommitant benefits declines.

Free market forces are allowed to determine the prices for livestock products in many places and often provide the best method of allocating resources and promoting development. However, particularly in situations where input supplies are temporarily unavailable or are not free to move easily, then managed markets and prices have a definite role in the effective use of resources. Free market pricing under these conditions normally results in substantial gains or losses to individuals which, rather than promoting long-term development, can give rise to stagnation and retrenchment as windfall gains are squandered and the possibility of losses is protected against by the adoption of low input/low output production systems.

Pricing policy which can effectively reduce the uncertainty of fluctuating prices by providing a more stable price regime requires not only a price setting mechanism, but also a marketing organization that can, either from its own resources or by Government assistance, meet price shortfalls through market intervention or price support. Sophisticated marketing organizations may be able to support prices and still be self-funding by creaming off price highs, but in developing countries this frequently proves difficult to supplement. The stability of prices which an administered price system can offer should not be over-rated for livestock products produced for the domestic market under low-cost feeding systems; the inherent stability of the production system often requires little under-pinning from short-term price declines.

The problems of effectively operating administered price schemes over a long period where there is very little competition should not be under-rated. Price adjustment systems easily become a means for perpetuating inefficient practices and allowing cost plus pricing systems to be adopted by default.

Almost without exception, official foreign exchange rates adopted by developing countries are above those which could be expected to prevail if rates were not controlled. One result is that prices received by exporters are lower than they otherwise would be; the prices for imports are also correspondingly lower. There is therefore a danger that by holding these rates artificially high, the overall foreign exchange position of the economy will be further weakened by discouraging the production of commodities for export. The traditional solution is to directly subsidise some part of the production or marketing chain for export products, thereby rendering their export profitable to producers and marketing bodies. The possibilities for this are, however, frequently constrained by the low tax base and high demand for Government revenues to fund social, health and educational programmes in developing countries. Care needs to be taken when setting prices for domestic products to ensure that the minimum of disruption to the production of exportable products occurs.

The problem of how to get farm prices high enough to protect those who are nutritionally at risk relates to food crops and not to livestock products. Meat, milk and eggs are expensive to produce, and, while they are highly nutritious, their role in providing higher nutrition levels for needy groups must be part of a welfare programme and not pricing policy.

Examples of the above issues can be found in the livestock sector in Tanzania. The benefits from a properly administered price system for milk produced on the state farms are obvious. The local milk supply, even when supplemented by quite large commodity aid programmes (4000 t SMP^{-1}) is totally inadequate to match the local demand and free market prices are well above the costs of production in all but the most difficult areas. The large profit to be made selling milk at the free market price attracts resources into the milk production sector and is causing it to grow quite rapidly, but the constraint to growth is not the profit level but the supply of inputs such as improved cattle, feed and skills. The same volume of resources can be attracted to the industry at a much lower price and therefore the administered price system provides some protection to the consumer during resource scarcity while supplies are building up.

The difficulty of enforcing official prices when free market prices are significantly higher has been experienced in the milk market in Tanzania. The unequal struggle has now been lost in the small-holder sector and only the large-scale farms are bound by the official pricing structure. Problems of inflexible pricing policies have also been experienced, but are now being overcome, through regular price reviews. Unaltered milk prices from the mid- to late- 1970s sent many small and medium sized producers out of business in the high-cost production area around Dar es Salaam; many of the large-scale farms also suffered a marked set-back in their development programs due to this non-revision. Temporary seasonal meat shortages appear to be experienced in areas where controlled prices make it more profitable for beef to be retained on the hoof to obtain higher carcass weights when the feed supply improves with the onset of the rains.

The negative impact of panterritorial pricing can be seen most clearly in the milk sector. In general conditions for milk production in Tanzania are not favourable, but in some areas surpluses exist, especially in the north-west of the country near Lake Victoria and in the southern highlands near Mbega. The major markets for milk are in urban centres where milk production conditions are difficult; the logistical problems of ferrying milk, other than as processed dairy products, to Dar es Salaam, Moshi and Arusha would prevent this trade occurring, even if the price system did not mitigate against it. In Mwange, the panterritorial pricing system, geared to transporting milk short distances from farm to factory and then to the consumer, prevents the profitable development of a collection system based on centres in the Mara area and the relatively long, but technically quite possible, cartage of pasteurised milk to the Mwanga market. The impact of the introduction of panterritorial milk pricing, with its resultant massive increases in the farmgate price for milk, can be held in large part responsible for the situation. The losers are the Mwanga consumers who have an unsatisfied demand for milk, and the cattle keepers in Mara who have been priced out of the market. The winners are the few producers sufficiently close to the milk plants for their milk to warrant being picked up.

The impact of international prices and the exchange rate dilemma in Tanzania is largely a matter of debate rather than substance, although milk production is in some areas beginning to compete with coffee for land. The debate occurs in the area of project preparation and evaluation, where, as a result of the production and export subsidies accorded to the dairy industry in Europe, the international prices for these commodities are below even the costs of production of the most efficient international producers. This means that when dairy products are assessed using border prices they are invariably shown to be poor investments from the national point of view. The argument in defence of further dairy development therefore has to be that even though for the last 10 – 15 years Europe has had surplus milk to pass to the developing world, the milk surplus entering the world market represents only a small part of the total production; fairly minor changes in demand or supply may consequently take the international dairy product market from a chronic surplus to a deficit situation. As dairy development by its nature has a long time horizon, to protect Tanzania's milk supply against possible reversals in the international market development has to be proceeded with.

In conclusion, it should be re-emphasised that the potential benefits of introducing an administered price scheme have to be large to warrant the administrative overheads and undesirable disruptions it may inadvertently cause. Panterritorial pricing for livestock products should be introduced with great caution, and only when production and marketing conditions are similar over most of the area to be covered. Special consideration must be given when setting the prices for domestic products to their impact on export crops, and lastly the scope for considering nutritional welfare aspects in the pricing of livestock products is limited.

Discussion

While the mood of the participants was in general against price control, it was agreed that there are circumstances where it may be desirable. Where price control is to be used, it must be effectively enforced, it must take account of regional differences in supply and demand, and it must not cause consumer shortages, unless rationing is feasible.

The controlled price is very difficult to determine as it involves balancing the interests of producers and consumers, a difficult political problem. It may also cause a long-term imbalance of supply and demand. The usual tendency is to over-protect consumers, leading to shortages or the necessity for subsidies, which can become a serious drain on Government resources.

It would be desirable to give long-term guarantees for animal product prices to secure confidence among producers for the long time scales involved. However, this raises the danger that market conditions could change and make the policy an unacceptable drain on finances. It was concluded that, in the long run, Governments should act as referees to ensure equitable pricing for both consumer and producer, rather than as a controller of prices and purchaser of commodities.

Chapter 44

Movement and Marketing Systems for Live Animals in Sub-Saharan Africa

S. BEKURE[1] and I. McDONALD[2]

[1]Agricultural Economist, International Livestock Centre for Africa (ILCA), P.O. Box 46847, Nairobi, Kenya; [2]Agricultural Economist, Ministry of Agriculture and Livestock Development, P.O. Box 68228, Nairobi, Kenya.

Introduction

Ruminant livestock production in sub-Saharan Africa is unevenly distributed in relation to the effective demand. The bulk of the slaughter stock is produced in the arid and semi-arid zones which contain about 60% of the ruminant biomass (Jahnke 1982); these zones are far from major population centres and so slaughter stock has to be moved over great distances to centres of consumption or ports of export. The efficiency of the livestock marketing system determines both the income of livestock producers and hence the level of offtake, and the consumer price of meat and hence the level of consumption. The more efficient the system is in minimising the costs of moving animals, the better it can stimulate consumption and production.

Livestock Markets

Auction Market

Auction markets were introduced into eastern and southern Africa by Europeans and still operate in Malawi and Zimbabwe, but only some variants of these remain in Botswana, Kenya, Swaziland and Zambia. Animals classified by age and sex, and graded for quality are often segregated into similar lots, price per unit of liveweight for each lot being determined by bidding. Buyers deposit money to guarantee cash payments for anticipated purchases and sellers can withdraw animals if they are not satisfied with prices offered. Auction sales normally exist in conjunction with a state-owned or parastatal organisation; in Malawi and Zimbabwe a parastatal organization acts as a buyer of last resort. The sales are attended by traders, butchers farmers and cold storage commission representatives; the latter guarantee floor prices, which are the starting point of the bidding. In Botswana and Somalia, auctions without Government or parastatal attendance have an established place, and in Upper and Lower Juba districts of Somalia, auctions are often conducted by private auctioneers in an open area, buyers and sellers forming a ring around animals being sold. In the eastern districts of Botswana, monthly auctions are organized by licensed auctioneers in major villages; in western districts, auctions are held less often and conducted at the request of Farmer's Associations. The auction system exists alongside and competes with other marketing channels. Approximately 10 000 cattle per year are sold through auctions, about 5% of the export offtake (McDonald 1978).

Traditional Markets

In the traditional livestock marketing system, which is operative in most African countries, animals are brought into the open and sold on an individual basis. Livestock producers may choose to sell in the bush (or at the farm gate) to small traders, who frequent watering points or walk from hut to hut, or depending on location, sell at a local primary (or collection) market, where market days are held on specific days well-known to both sellers and buyers. The principal purchasers are other producers looking for either breeding stock, oxen for ploughing, or immature steers for fattening; and local butchers and traders who purchase mainly slaughter stock for shipment to secondary (regional) or terminal (national or export) markets.

Secondary (or regional) markets, to which cattle are supplied by a number of primary markets, are usually located in large towns. In addition to supplying the local need for meat, they are staging points for assembling and transporting slaughter stock to terminal markets. The principal sellers at the secondary markets are traders, but producers from the immediate hinterland may also supply animals. The buyers are individuals purchasing animals for personal consumption, local butchers or slaughter-houses, and parastatal agencies and major traders operating in national and international markets.

Transportation

An efficient livestock transport system minimizes costs by (1) reducing mortality and weight losses, (2) shortening the time between primary and terminal markets, (3) increasing offtake by facilitating the movement of animals, and (4) enabling movement of animals throughout the year.

Trekking

In Africa the most important and cheapest means of live animal transport over long distance is trekking. In many instances there are no alternatives but mortality rates *en route* are very low (Reusse 1982). In many countries trek routes have been established by custom, not law, and are not well-marked so that conflicts over rights of way arise, especially when trek cattle damage crops. The ensuing controversy and litigation can cause delays which increase costs. Well-defined trek routes can both solve this problem and facilitate the monitoring of livestock movement. In Togo traditional trek routes were officially confirmed by a decree in 1937 which is still in force and in Botswana there is a policy of leaving a corridor of at least 1 km along trek routes within which no permanent settlement is allowed.

Inadequate grazing and water along trek routes can be limiting during dry seasons and result in a severe loss of condition by the time animals reach their destination. The problem is being gradually alleviated by development projects, but there is considerable room for improvement, especially in West Africa, in providing holding grounds at the end of trek routes. The proper management of watering points and holding grounds established is essential. In Botswana, individuals given permission to use a trek route borehole, used it to graze and water their own animals, limiting or denying access to trek cattle.

Rail Transport

Where available, rail transport is invariably used for moving live animals in Africa. Its potential advantage lies in that it is quick which enables traders to make rapid turnovers of capital, and to deliver animals in relatively good condition. With the exception of Tanzania, rail transport has been effectively used for moving live animals in eastern and southern Africa, but in West Africa their rail transport is fraught with problems arising from poor management and inappropriate handling procedures. Cattle suffer from heat and humidity when boxed for more than three days, resulting in weight losses and mortality (Staatz 1979). Railway directives could ensure that livestock are given priority in rail movements and unloaded for feeding and watering on journeys lasting more than 48 h to minimise losses arising from stress, bruising, and death.

Botswana Rail provides a major service to the cattle industry in its movement of cattle to the terminal market. In a normal year, 65–70% (c. 150 000 head) of all deliveries to the Botswana Meat Commission arrive by rail. The line is 700 km long and at each of 16 loading points adequate holding pens, sorting crushes, and watering troughs have been built and are maintained, and cattle are inspected by veterinary staff before loading. Railway operating procedures ensure that livestock movements have priority and most journeys are less than 16 h and normally take place at night; cattle are rarely on a train for 36 h or more. In a normal year less than 40 cattle die during the train journey to the Commission.

Truck Transport

The use of trucks for transporting small stock is gaining importance as road networks improve and penetrate more remote areas. Small stock cannot easily be trekked over long distances, and as a large number can be carried on trucks road transport is competitive. Under normal circumstances, cattle trucking is expensive and cannot compete with trekking or railing but cattle are mainly trucked as back-loads on multi-purpose lorries. There are few routes where the combination of good roads and a high volume of livestock movement justifies the purchase of specialised livestock trucks.

The shortage and price of oil in many African countries will continue to militate against the wide use of trucking for cattle. Unofficial levies paid at roadblocks also add to the cost. The costs have meant that only under special circumstances is trucking used on a large scale, for example: (1) When disease control regulations are such that other forms of transport, especially trekking, are not possible. (2) At seasonal peaks in demand when terminal market prices are high enough to compensate for the increased costs. (3) When other forms of transport, especially trekking, are made impossible or expensive by weight losses and mortality, as where trypanosomiasis is prevalent.

Boat Transport

Boats are used to transport livestock in the navigable portions of the Congo River and its tributaries. Live animals are exported by boat from the Sudan, Ethiopia, Somalia and Kenya to the Arabian peninsula, and boats are also used to transport livestock along the African coasts.

Market Organization

If this part of the marketing system is to be efficient then:– (1) the resources used by livestock traders and merchants (i.e. capital, management and labour) should realize a return comparable to that earned in the rest of the economy; and (2) information about changing supply and demand conditions and hence prices received and paid should be quickly transmitted throughout the marketing chain from primary to terminal markets and vice versa; impediment results in either higher costs in the form of higher risks assumed by traders or, where information is known to only a restricted number of groups or individuals, in higher costs in the form of monopoly or monopsony profits. Judged against these two criteria, the available evidence indicates that the traditional livestock marketing systems in Africa are fairly efficient within the framework in which they operate, and that the scope for increasing their efficiency does not lie in attempts to regulate and control the market participants, but rather in facilitating their operations and instituting measures which reduce their costs.

Concentration of Market Power

As pointed out above, livestock producers normally have a number of options on where and to whom to sell their animals. Livestock traders, who canvass at watering points in the bush and at homesteads, are numerous and operate in competition; these are often young producers trying to build their own herds and flocks with income from trading. These part-time traders increase the level of competition in the bush and the primary markets.

If producers feel that prices offered by itinerant traders or at their closest primary market are too low, they commonly take their animals to other markets in the region. It is also common for one or more producers to be entrusted with the task of selling on behalf of colleagues. Producers also frequently visit markets to obtain market information for themselves and their colleagues and use this in decisions on where to sell. In Botswana livestock producers can sell animals directly to the Commission terminal market on their own, or through co-operatives, or agents who charge commissions for arranging transport and delivery. They can also sell to private traders, or to a parastatal, the Botswana Livestock Development Corporation. Prices at the terminal market are well publicised.

At primary and secondary markets there are normally so many competing traders that prices are competitive, especially in West Africa. This may be because it is only recently that traditional producers in East Africa have been drawn into the market economy. Its absence prompted the estalishment of parastatal agencies to undertake livestock marketing functions. Few serious attempts have been made to measure the efficiency of livestock marketing systems in eastern and southern Africa. In the Sudan secondary and terminal markets are dominated by a few big merchants who restrict price competition, and Bekure *et al.* (1982) concluded that traders' margins at a secondary and a terminal market in Kenya were higher than what could be considered reasonable. On the other hand in Somalia Reusse (1982) found little evidence of unreasonably high margins.

In West Africa the tradition of inter-regional trade in livestock across several countries is long-established and well-developed. Since the logistics of livestock trading in volume across great distances is difficult and ties up large amounts of working capital, individual merchants cannot handle more than a small segment of the trade. This ensures competition, and a sophisticated and complex system of market relationships has evolved with specialized groups of intermediaries ("landlords" and *dillalis*) with well-defined functions (Staatz 1979). Ironically, the presence of a large number of participants, which makes the market more efficient, has been the target of attacks by those who allege that the proliferation of intermediaries must invariably increase costs. However, the various intermediaries perform important economic functions, (1) providing market information, (2) guaranteeing credit, and (3) concluding sales (Staatz 1979). They perform the functions of brokers in western markets, and have information on the demand in terminal markets and supply conditions in the hinterlands so enabling them to gauge a fair price. They also have a knowledge of the capacity and integrity of traders and merchants, are acquainted with and have the confidence of suppliers, are reasonably sure that animals offered for sale are not sick, and through discussions form a good idea of the minimum price sellers will accept. They are consequently able to conclude quick sales and offer guarantees of payment when credit sales are made.

Studies conducted in West Africa (Ariza-Nino *et al.* 1980) lay to rest allegations that traditional livestock markets in Africa are inefficient. After intensive studies in Upper Volta, Ivory Coast, Mali, Benin, Ghana, Liberia, Togo and Niger the overriding conclusion was that:

> "The current, traditionally organized cattle and beef marketing system is rather efficient – given the institutional and infrastructural framework within which it operates. Marketing of cattle between the Sahelian and coastal states is costly, but the costs are attributable mainly to high transportation costs, and export taxes, not monopoly profits of traders and butcher. ...Cases of market power being concentrated in a single entity are found in areas subject to government intervention".

Market Information

In Africa information throughout the livestock market chain is primarily transmitted by "word of mouth"; it usually travels as quickly as the fastest means of communication between the various links in the chain, i.e. by road, rail, telephone, telegram, or telex. Within each market, price information is readily available. In the auction markets of eastern and southern Africa, prices are openly announced. In traditional markets, although sales of individual animals are negotiated privately, the information is made available when sellers, intermediaries and buyers compare notes. It is transmitted to producers in the hinterland by bush traders and producers visiting livestock markets. Published and broadcast price information have only limited places in traditional livestock marketing systems at present. Newspapers and farming magazines tend to be distributed slowly, and many owners and traders are functionally illiterate. Broadcasts have a greater potential than its present usage would suggest. The major problem associated with expanding its use is the speed at which the information can be gathered, collated and presented; information is often obsolete before being broadcast. Both published and broadcast prices are, however, useful when the terminal market price is stable for long periods as in Botswana, where the Botswana Meat Commission changes its buying prices every 3 months. On the basis of such transmission of market information, the traditional marketing systems in West Africa have shown a remarkable capacity to respond to changing conditions in production, consumption and prices (Ariza-Nino *et al.* 1980).

Other Marketing Costs

(a) Taxes, Licenses and Cesses

Since the administration of personal income taxes is poorly developed, and poll taxes are difficult to administer and enforce, it is understandable that central Governments wish to extract revenues from this important subsector of their economy through indirect taxes, fees, cesses and levies.

However, caution should be exercised in their application as their net effect is to increase the cost of marketing. In West Africa taxes accounted for 24 % of the costs of marketing and processing live cattle from Ouagadougou in Upper Volta to retail meat markets in Abidjan (Delgado & Staatz 1980), all other individual costs represented a lower proportion. There is a need for Governments to periodically review the effect of taxes on marketing costs and competitiveness in export markets. This is particularly important in view of the growing importation of large quantities of frozen beef from Argentina, Australia and New Zealand into the coastal countries of West Africa since 1974.

Governments also issue licences for livestock traders. While licencing is required for taxation and other purposes, there is some tendency to use this instrument to control the number of traders. This can promote monopsonistic practices in the system.

(b) Veterinary and Export Permits

Most Governments in Africa impose veterinary regulations on the movement of livestock. They are introduced and are administered to protect livestock industries and public health and normally take the form of quarantine periods, compulsory vaccinations, etc. The direct costs paid by the livestock owner are reflected in lower prices to farmers and higher prices to consumers. Cost-benefit analyses are rarely carried out before these regulations are introduced, but they are normally justified on the grounds that benefits in the form of higher herd and flock productivity and better public health are greater than the costs of administration.

The enforcement of veterinary regulations in Africa is not easy. Most livestock is owned by pastoralists who are often nomadic and suspicious of central Government; this means that veterinary control measures are normally only partially applied. The effectiveness of veterinary services in enforcing disease control regulations depends almost entirely on the conditions required by the terminal markets. For countries whose national economy is to a large extent reliant on the export of meat or live animals, the incentive to build and maintain effective disease contol systems is great. In countries where most livestock is consumed internally there is usually neither the political will nor sufficient resources available to ensure more than partial controls.

Over 90 % of the throughput at the Botswana Meat Commission is in the form of chilled and frozen meat, and the Veterinary Department has successfully introduced a range of disease control measures designed to satisfy the standards imposed by her export markets. The most notable of these have been those instituted to identify and control foot-and-mouth disease (FMD). Cordon fences have been erected which divide the country into FMD control zones and prevent the uncontrolled movement of cattle. All cattle moving from one zone to another must pass through a quarantine camp; those travelling across several zones must pass through several quarantine camps or remain in quarantine during transit. That Botswana has managed to maintain access to the EEC market for chilled and frozen meat testifies to the success and efficiency of her veterinary services. The costs of building and maintaining fences, quarantine, and vaccination campaigns are met from central Government. The indirect costs which this system imposes on the marketing system, i.e. the extra costs of labour and trading capital, are, however, borne by the livestock owner.

In most African countries the veterinary services are weak and poorly co-ordinated, and the vaccination and quarantine which should be carried out on all stock exported is not adhered to. The continual issuing of veterinary movement permits without proper enforcement only serves to increase the costs of livestock marketing.

Permits are usually required for the exportation of livestock. These facilitate the application of foreign exchange control, taxes, and the compilation of statistical data. However, the procedure for obtaining the permits is in many cases cumbersome and costly (Delgado & Staatz 1980). As well as the payment of official and unofficial levies, it is necessary in some cases to visit up to eight different Government officials, so that the process cannot be completed in less than one week.

Government Roles in Marketing Systems

Price Controls

Many African Governments attempt to control live animal and meat prices by (1) fixing minimum prices per unit of liveweight which slaughterhouses and butchers can pay, and/or (2) fixing wholesale and retail meat prices which they can receive. The presumed intention of these price controls is to limit the margins of traders and butchers and so protect both producers and consumers from exploitation. Although there are legal provisions for the frequent review of these prices, this is seldom done. Few Governments in Africa have the analytical and administrative resources or perhaps the will to alter gazetted prices as market conditions change. As a result, prices remain fixed despite radical shifts in market conditions. For example, in Zaire maximum producer, wholesale and retail prices for livestock and meat were fixed in February 1973 and remained unchanged until May 1976, meanwhile, actual producer prices had risen by 40% and retail prices by 100%.

In most countries controlled prices are totally or partially ignored. Where price controls are enforced, the result is often a shortage of meat which leads to black market operations, and in the end consumers pay higher prices than would otherwise be the case (e.g. Farris & Stokes 1976). When fixed retail prices are maintained below market prices, an income transfer from the farmer to the urban consumer takes place. It also discourages the farmer from improving productivity or expanding production, and encourages illegal exports; livestock movements across borders in response to price differentials often take place.

Parastatal Organisations

In many African countries, parastatal organisations are actively engaged in the livestock marketing system. These parastatals are normally abattoirs with a monopoly over the export of meat or in the wholesale sector of the meat trade. Other parastatals have also been established to stimulate livestock trade and promote the stratification of the industry, or regulate the livestock marketing systems by offering competition to private traders.

The experience of parastatals in Africa has been mixed. Firstly, there are well-managed examples which have fulfilled their objectives. Almost invariably such parastatals have held monopoly powers in some part of the marketing chain and, while they may not incur financial losses, the extent to which a lack of competition allows them to operate at higher costs than they otherwise could, represents an additional cost to the system. Secondly, there are parastatals which have accomplished their objectives but with colossal inefficiency and cost. Finally, there are parastatals which have failed to achieve their objectives and have in addition incurred substantial losses, even destroying the livestock marketing system in their country (Sullivan & Josserand 1979). The potential damage caused by failures outweighs the successes. The major reasons for failure in parastatal agencies are (1) bad management (including pilferage and corruption), (2) political interference, and (3) price controls.

Conclusions and Policy Implications

Movement of Live Animals

Livestock marketing systems have developed to move live animals long distances from areas of surplus production to those of deficit. For most countries the cheapest method of transport is trekking and consequently this remains the method most widely used. Movement by rail is only important regionally, and trucking is only used on a large scale under special circumstances. Water transport is used along navigable internal waterways and along the coasts, but in the context of internal African trade it is relatively unimportant. However, much can be done to reduce the costs of transporting livestock. Most countries have poorly developed trek routes, not properly demarcated, and with inadequate facilities for watering, holding and feeding. The provision of holding grounds at terminal markets with adequate water and feed should be given particular attention. User-fees can be charged to provide these services at cost. The transport of cattle by rail, particularly in West Africa, can be substantially improved. Cattle cannot be handled in the same manner as inert commodities; their movement should be expedited and adequate provision made for holding, feeding and watering at loading and unloading points. Journeys of longer than 48 h should be broken to allow the cattle to be rested, fed and watered, and the wagons used properly ventilated.

**Market Efficiency and
Government Roles**

The livestock marketing systems which have developed in sub-Saharan Africa are by and large efficient within the framework in which they operate. In the past Governments have intervened in various ways, aimed at regulating and increasing the efficiency of the marketing system, and ranging from the control of prices to the purchase and sale of animals and meat. Experience, however, shows that the scope for increasing efficiency lies neither in attempts to regulate and control the market participants, nor in efforts to control prices, nor in the creation of parastatals; but rather in facilitating the operations of the private sector and instituting measures which reduce their costs.

The effect of Government interventions in the forms of taxes, licences and cesses, procedures required for the movement and export of livestock, controlled prices, and the policies of parastatals in livestock and meat marketing, need to be reviewed periodically and streamlined with a view to stimulating the industry. Prices must also remain competitive with imported frozen beef. Using licences to control the number of participants in the market should be avoided as it tends to decrease the level of competition and so increase traders' margins. Care should be taken that taxes and cesses imposed do not unduly increase prices at terminal markets. In some countries procedures for obtaining permits for the movement and export of livestock are cumbersome and costly. Streamlining to reduce the time that traders have to spend obtaining permits will reduce marketing costs, and efforts to erradicate unofficial levies will also help improve efficiency.

Controlled prices for live animals and meat do not seem to be effective in protecting the interests of producers and consumers. They have often introduced distortions into the market, spawned black markets, and directed livestock away from established markets.

When parastatals or other Government agencies engaged in livestock and meat marketing are run efficiently, they can increase competition and stimulate the marketing system. Unfortunately, success stories are the exception rather than the rule. As stated above, livestock marketing systems in Africa are fairly efficient, except under certain circumstances such as drought, or when there are large seasonal fluctuations in supply. The evidence also indicates that the inefficiencies are most severe in situations where Governments have directly intervened. This implies that Governments would be well advised to refrain from direct interventions in livestock marketing systems and that they should concentrate efforts and resources into effecting measures which will relax constraints that participants in the system cannot remove. These include (1) improving the infrastructure for livestock marketing, (2) streamlining procedures for the movement and export of livestock, (3) the provision of market information through mass media, (4) facilitating the provision of credit for traders in primary markets, (5) regulating the standards of products and services, (6) negotiating favourable trade agreements in export markets, and (7) aligning property taxes and foreign exchange rates to promote exports.

Views expressed in this paper are those of the authors and should not necessarily be interpreted as reflecting those of the International Livestock Centre for Africa (ILCA) or the Kenya Ministry of Agriculture and Livestock Development.

References

Ariza-Nino, E.J.; Herman, L.; Makinen, M.; Steedman, C. (1980) *Livestock and Meat Marketing in West Africa*. Vol. 1. *Synthesis Upper Volta*. Ann Arbor; Center for Research on Economic Development, University of Michigan.

Bekure, S.; Evangelou, P.; Chabari, F. (1982) *Livestock Marketing in Eastern Kajiado, Kenya*. [ILCA/Kenya Working Document no. 26.] Nairobi; ILCA.

Delgado, C.; Staatz, J. (1980) *Livestock and Meat Marketing in West Africa*. Vol. 3. *Ivory Coast and Mali*. Ann Arbor; Center for Research on Economic Development, University of Michigan.

Farris, D.E.; Stokes, K.W. (1976) *Tanzania Livestock – meat sector. Consultants Report*. Vols 1-4. TAMU/USAID.

Jahnke, H.E. (1982) *Livestock Production Systems and Livestock Development in Tropical Africa*. Kieler Wissenschaftsverlag Vauk.

McDonald, I. (1978) *A Report on Cattle Marketing in Botswana.* Gaborone; Ministry of Agriculture.

Reusse, E. (1982) Somalia's nomadic livestock economy. *World Agriculture Review* 43.

Staatz, J. (1979) *The Economics of Cattle and Meat Marketing in Ivory Coast.* Ann Arbor; Center for Research on Economic Development, University of Michigan.

Sullivan, G.; Josserand, H. (1979) *Livestock and Meat Marketing in West Africa.* Vol. 3. *Benin, Ghana, Liberia, Togo.* Ann Arbor; Center for Research on Economic Development, University of Michigan.

Discussion

Mr McDonald, in referring to data provided by Mr Vahaye and drawing on his own experience, outlined the role of the Botswana Meat Corporation which successfully exports fresh chilled and canned meat to EEC standards. The Corporation provides the main outlet for beef production and has fostered the development of an efficient delivery system for stock through agents and co-operatives. Grading and seasonal pricing are designed to encourage quality meat production, and confidence of the producer is maintained by payments being guaranteed within five days of sale. Stock movements by the company are integrated with Government animal health procedures aimed in particular at controlling foot and mouth disease, a prerequisite for access to the high priced EEC market.

Much emphasis was placed on the problems of seasonality in marketing and the possible advantages of incentives, such as reduced transport charges and taxes during seasonally low periods of production to stimulate offtake. Arrangements for accommodating the marketing of surplus stock in drought conditions may require specific planning to ensure the timely provision of facilities such as emergency feed reserves, cattle banks, subsidized movement to unaffected areas, and the encouragement of the market to absorb surplus stock. It was accepted that the export of stock from some countries, including Lesotho, had not been treated in this contribution due to their representing a small proportion of the total volume of the trade of the region.

The efficiency of traditional marketing systems was questioned by several speakers because it failed to provide any mechanism to stimulate improved quality and productivity. It also allowed wide differentials in producer and consumer prices to persist in some areas. There appeared to be a need for stratification although little evidence of progress in this direction in Africa was forthcoming. The purpose of stratification was seen to be to encourage the offtake of young and growing stock as well as mature animals from the primary producers, thus reducing the overall population pressure at critical times while improving the general supply of meat. It was noted that the zonal stratification of production systems had evolved on a satisfactory basis in South America. It was recognised that marketing strategies could provide the mechanism for transition into a stratified system which could contribute to discouraging the practice of building up livestock numbers as a safeguard against periods of drought. Price incentive was also considered to be a vital factor.

Chapter 45

Problems of Marketing: Appropriate Policies and Organizations

L. G. K. NGUTTER

Principal Economist, Ministry of Finance and Planning, P.O. Box 30028, Nairobi, Kenya.

Introduction

The problems of advancing agricultural production in the developing world in general, and in Africa in particular, may be viewed as the problem of providing: (1) the necessary public sector services such as extension, research and training; (2) essential inputs such as credit, agricultural chemicals and improved crop varieties and livestock types; (3) incentives to the producers so that they may improve farming methods – incentives include adequate prices and market services; and (4) the necessary policies and organizations to deliver (1) – (3). This contribution is concerned with the essentials of marketing and its development with reference to the policies and organizations which would encourage farmers to produce more and better products. The value of a reliable, well-developed and low cost marketing system, particularly for foodstuffs, has often been underestimated in Africa. Yet, such a system is crucial to the specialization of production in the rural areas and to the distribution of food and non-food agricultural products in the towns. Reliable markets would encourage producers to move away from subsistence production to commercial production. Additionally, incidences of political instability being either caused, or at least aggravated, by food shortages and/or rising food prices are too common in Africa. Africa is a continent of wide diversity; ecologically, politically and indeed as far as the level of development is concerned. I will therefore limit my discussion to the Kenyan situation, but hope that most of what is true for Kenya can be adapted for other countries in Africa.

The Marketing System in Kenya

The general marketing system in Kenya has been widely discussed and perhaps the most useful summary is that of Heyer (1976). Institutionally the domestic market works through three principal channels, viz. the Marketing Board/Authority (or parastatal) channel; the co-operative marketing channel; and the private channel.

The Marketing Board Channel

The Board of Authority receives produce from the producer, either directly or through appointed agents who may be individual produce buyers, private traders, large farms/firms, or a producer co-operative. The Board then supplies processors who may be either other co-operatives or private entrepreneurs and/or wholesalers. The last organizations in the chain are retailers who deal with the consumers. In the case of export commodities, the Board may receive the produce after it has been processed to whatever level is necessary before the Board sells it to an agent representing overseas firms. Kenya has a long history of statutory Boards. Some are deeply involved in either the production or marketing or both of the commodities they deal with, while others are less involved. The National Cereals and Produce Boards (NCPB) deal in maize, wheat, rice, and to a lesser extent with pulses and similar agricultural products. On the other hand, the Coffee Board of Kenya simply oversees the industry, leaving production and marketing substantially to another Board, the Kenya Planters Co-operative Union.

The major problems associated with the parastatal marketing system are: (1) management problems resulting in high marketing costs; this is aggravated by (2) the tendency (where the price of the commodity is fixed by Government) to squeeze the Board's margins in an effort to force the Board to be more efficient and cost-conscious, and hold down consumer prices; and (3) the net effect of the above, delayed payment to producers, who in turn devise ways to evade this channel in search of ones where payment is more prompt. This can have two consequences: (a) where the Board relies on constant throughput of produce to process, diversion of the produce reduces the throughput below break-even points, increases unit processing costs, and places the Board in a worse financial position; and (b) where the Board is the sole official dealer in the commodity, diversion of produce means that the consumer markets are satisfied other than by the Board (except in times of shortages) and that the Board must continue to hold produce in their stores at substantial insurance, pest control, security, and interest-on-capital costs, again placing the Board in financial problems.

There are about 50 statutory Boards involved in agriculture in Kenya which have different types of origins. Old Boards established prior to independence, basically to protect colonial settlers' produce from competition with African production, have survived, some with minor modifications. After independence, it became the practice that each time a new crop was established in Kenya, a new Board was introduced to handle the crop. Finally, some Boards have been established specifically to deal with production from areas where the marketing system was judged under-developed, as with the Kenya Meat Commission (KMC). The Livestock Marketing Branch of the Ministry of Agriculture and Livestock Development, though not a parasatal organization in the strict sense, also fits into this mould.

The Co-operative Marketing Channel

Under the Co-operative Marketing Channel, produce moves from the farm to a primary society and thence to a processor, who may be a larger co-operative or a private firm, or to a wholesaler and then to the retailer, consumer or export agent. Two major types of co-operatives may be recognized. Old pre-independence producer co-operatives which have grown to the status of national co-operatives and behave as parastatal organizations sponsored by the Government, and smallholder-orientated organizations, usually formed at the request of Government; there are now nearly 1000 such co-operatives in Kenya. Major problems of the co-operative sector include: (1) Weak management, smallholder co-operatives in particular lack the financial control, accounting and capital to run efficiently, reflected in the high but variable deductions made from the producers' proceeds, reducing the producers' take-home returns. (2) Primary societies are required to affiliate to district level unions, and while the costs and benefits of such affiliations have never been fully analysed, it is likely that the affiliation requirement simply increases marketing costs and so reduces farmers' incomes without corresponding benefits. (3) Delayed payments are common, and may be as much as three or more months; the incentive for farmers to by-pass the co-operative and sell directly to consumers is therefore high.

The Private Marketing Channel

Under this system, produce moves directly from farm to the consumer, trader, or processor. Payment is usually on the spot, even though the absolute price level may be lower than that nominally paid by the Board or Co-operative or the gazetted price. Even for those commodities which legally fall under the purview of a state-established monopoly, the private trader is deeply involved, often through what is known locally as "*magendo*" (illegal trade). The volume of agricultural produce traded through the private channel is increased further by sales in local rural markets. These, while allowed even for maize, may be in small quantities only per seller, but the total volume of trade can be substantial, depending on the commodity. The problems in this channel include: (1) The local traders are often harassed through such mechanisms as the need to be licensed, produce movement control, and market regulation for commodities falling under statutory Boards; attempts to circumvent these controls through, for example bribery, night journeys and evasive routes increase marketing costs and lead to the existence of higher consumer prices side-by-side with very low producer prices. (2) There is disdain against a freer participation of the private trader in such commodities as maize in the country as a whole, and milk in urban areas; this limits the market channels available to the producer as, for example, only 1–10 (90 kg) bags of maize may cross district boundaries without a permit. (3) Small-scale traders lack the managerial skills, business acumen and capital base to increase their volumes of trade; an educational and credit programme would facilitate expansion of this market.

The general marketing system in Kenya therefore appears to be unnecessarily restrictive for some major commodities such as maize; is high-cost in attempts to circumvent such restrictions; fails to offer sufficient market services in such key areas as milk; and results in the subsidization of certain classes (e.g. urban dwellers) by others (e.g. rural producers) in case of beef, or the consumer by Government in the case of major food grains.

Problems of Livestock Marketing in Kenya

Some of the problems of livestock marketing, especially as regards livestock movements and their costs are considered in Chapter 44. I therefore propose to limit my discussion to beef livestock and dairy produce only.

(a) Beef

Beef in Kenya is produced by large-scale ranchers, pastoral societies under range conditions, and by smallholders. Little is known of marketing problems in the higher potential smallholder areas, compared with large farms and pastoral areas (Aldington & Wilson 1968, Chemonics International 1977). Beef consumption is highest among the higher and medium income peoples in central Kenya and in urban areas. The problems of beef livestock marketing include: (1) Facilities to move livestock from major areas of production to major fattening, slaughtering and consumption areas are inadequately developed. (2) The large number of legal and veterinary restrictions to livestock movement has given rise to debate in the country as to their usefulness. (3) Historically low beef, and all red meat prices, which are controlled, have stifled the development of beef substitutes such as pork, poultry and mutton; this policy has also acted to prevent the full stratification of the industry as described in Chapter 44. (4) Additionally, the price differential as to the quality of meat is very narrow, so that nearly all large-scale beef feedlots have had to cease operations; traditionally, export sales have been used to subsidize local sales and except for low quality carcasses sold in the bone-in bone-out markets, gazetted prices are often flouted by butchers whose retail prices for high quality cuts may be 40% above the gazetted prices. (5) Finally, the main official organizations in beef and livestock marketing are high-cost institutions, squeezing suppliers' margins and so reducing supplies.

(b) Dairy Produce

In Kenya about 60% of marketed milk production now comes from smallholdings, and the remainder from medium and large farmers. Most of the pastoral output is used for domestic consumption, but there is a dual milk market with Kenya Co-operative Creameries (KCC) taking nearly 100% of all official commercial sales, while substantial quantities exchange hands directly in rural markets. This is understandable considering that milk is produced all over the country while KCC depots, collection points, and a majority of the milk co-operatives are concentrated in a strip, running north-west to south-east through the country. Organizationally, the inability of the system to collect all milk available (particularly evening milk) and preserve it until it reaches processing plants is the main marketing problem. Other problems in dairy marketing are: (1) Until the last two or three years the returns to producers who market through the co-operative system have been so low and payments so delayed that this channel was unattractive to the producer. (2) Until two years ago there was no inter-seasonal price differential, resulting in over-supply during the wet seasons and under supply in the dry; a dry season bonus has recently been instituted but has yet to be evaluated; a similar situation applies to inter-zonal differentials. (3) Weak and inefficient milk marketing co-operatives, and the monopoly status of the KCC in milk processing and urban sales.

Key Issues in Marketing

Objectives in Pricing and Marketing

Government pricing and marketing objectives may be intended to achieve either revenue or welfare objectives. Under the revenue objective, the Government would seek to obtain the maximum revenue from taxes and cesses, whereas under the welfare objective, it would seek to improve the situation of producers and consumers through measures such as price and income stabilization, the provision of secure and reliable outlets for sales and supply, stabilization of agricultural supplies in both deficit and surplus areas, maintenance of strategic reserves of staple foodstuffs, and protection of producers from exploitation.

The Kenyan marketing and pricing policies appear to be geared more towards welfare than revenue objectives. This appears to be appropriate for the many African countries with underdeveloped agriculture. Excessive taxation in early stages of development tends to discourage producers. One shortcoming in Kenya has, however, been a failure to re-orientate attention away from large farm production (for which it was originally designed), towards smallholders who now dominate agricultural production in the country.

Level and Flexibility of Prices

Price fixing, an integral part of the pricing system for major commodities in Kenya, appears to be a common practice in African countries. The take-home price needs to be high enough to encourage the producer to keep producing the commodities consumers desire and are willing and able to pay for; this is particularly so for basic foodstuffs. Delayed payments by marketing Boards and/or Co-operatives amount to interest-free loans to the Board or Co-operative by the producers. The result is often for the producer to devise ways to evade the formal market system. Further issues are the extent to which prices are differentiated so as to reward quality produce, and the costs of enforcing a fixed price structure out of tune with market-discovered prices. Finally, there is a need to differentiate prices according to season and zone of supply in relation to the location of the market.

Role of Marketing Boards

The performance of nearly all Boards in Kenya has been rather poor and the problems of them have already been indicated above and in Chapter 44. We are forced to conclude from the Kenyan example that the granting of monopolies to Boards, especially in smallholder production situations, often aggravates marketing problems when Boards fail to accomplish their objectives. This frustrates both producers (who cannot find official market outlets) and consumers (who have to pay increased marketing costs), resulting in illegal trade, regional supply imbalances, and related costs.

Market Development

An alternative to market regulation through enforcement by Marketing Boards, imposing movement controls and granting monopolies, is a full development of the marketing system, including both physical infrastructure such as market sites, transportation (especially feeder roads), and communications. It also includes market services such as product grading and inspection for hygiene standards; and institutional development of market channels, organizations and traders. National policy in Kenya encourages private initiative in all fields of economic endeavour. This initiative is fairly well-developed, as evidenced by the existence of dual marketing systems even for Board-controlled commodities. What remains is to give private traders a freer hand in agricultural marketing, and to help them in improving their capital base and formal business skills.

Conclusions and Recommendations

Advancing agricultural production is closely linked to the structure of incentives, of which pricing and marketing policies, institutions, and services form major parts. High degrees of Government intervention represent a major and important commitment of administrative and managerial skills, scarce resources in developing economies. It is therefore important for each country to re-examine its present objectives, articulate them and assess the costs and benefits of pursuing present policies. In doing so, it is important to bear in mind that it takes time to re-orientate institutions. Organizations and institutions which have outlived their usefulness often continue to be cherished for no logical or economic reasons.

To determine appropriate policies and organizations, each country is recommended to:

(1) Clearly identify the nature and scope of its own problems. Background economy-wide or sector-specific studies may be necessary for this problem-identification phase. In Kenya, a re-examination of the usefulness of the Marketing Boards is currently underway.

(2) Design alternative policies. A sector by sector approach appears the best, but care should be taken to minimize conflicts in objectives, for example income stabilization versus high agricultural taxation versus the need for Government revenue. When conflicts remain, it will be necessary to rank the objectives. In Kenya, the welfare objective has been chosen over the revenue one.

(3) Policy objectives needing emphasis in the African context include: (a) flexibility in marketing and pricing; (b) improved market services, especially to smallholders; (c) least-cost provision of services; (d) nutritional security; and (e) adequate price incentives.

(4) Organizationally, the most economically efficient market channels should be allowed to develop. In Kenya, the private sector now seems to have advantages over the parastatal one, at least in the marketing of foodcrops.

(5) Full market development. Improving marketing systems in developing countries with large subsistence-orientated rural populations poses several problems. The system must accommodate a large number of producers, scattered over a large area, each with only a little surplus for sale which may vary substantially according to weather and prices. Cost may also be high. Market development policy must therefore be long-term.

The views expressed above are those of the author and should not be interpreted as necessarily reflecting those of the Government of Kenya.

References

Aldington, T.J.; Wilson, F.A. (1968) *The Marketing of Beef in Kenya*. [Occasional Paper no. 3.] Nairobi; University of Nairobi Institute for Development Studies.

Chemonics International (1977) *Livestock and Meat Industry Development Study*. Nairobi; Ministry of Agriculture.

Heyer, J. (1976) The marketing system. In *Agricultural Development in Kenya: An economic assessment* (J. Heyer; J.K. Maitha; W.M. Senga, eds). Nairobi; Oxford University Press.

Republic of Kenya (Ministry of Agriculture) (1978) *Summary of Major Technical Studies and Recommendations for Agricultural Marketing Development*. Nairobi; FAO/UNDP Marketing Development Project.

Discussion

In opening the discussion, Dr Anthonio (Economist, University of Ibadan, Nigeria) drew attention to the need for clear policy objectives. First among these was consumer satisfaction at minimum but equitable prices. Effectiveness and efficiency of marketing systems, with minimised costs but reasonable incomes for handlers and distributors was essential. Equity and optimum incomes for producers of the commodities concerned were equally important. While these basic requirements had to be secured for satisfactory near-term operations, it was also important to accelerate growth and development of appropriate outputs, and to emphasize quality as well as quantity as income levels and total demand expand. Furthermore, the overall objective of economic development of specific areas, sub-regions, countries and even sub-continents had to be borne in mind. Dr Anthonio identified a number of policy areas which required attention if these objectives were to be achieved:– (1) *Research*: Governments have assumed the right to intervene in marketing and therefore need basic information to avoid doing damage; this involves continuing research into how the market is performing to identify bottlenecks, prescribe alternative remedies, and determine how best to minimize costs. (2) *Education and extension*: For the producer this should cover cost-reducing technology, efficient use of inputs, and consumer preferences for different products; for distributors it should focus on handling, price, margins, and consumer preferences. Meanwhile consumers must be informed of input and output considerations faced by the industry in the formulation of prices, as well as the tasks of handling commodities through educational and advertising activities. (3) *The improvement of marketing services*: These must include the collection and delivery of live animals and products to assembly and processing centres, processing techniques, hygiene and inspection in handling, and consideration of types and size of product units for distribution as well as packaging and delivery to the consumer must also receive attention; price information at each stage must be collected, analysed and disseminated and the marketing system must be assured of adequate financing, insurance, brokerage and credit. (4) *Marketing livestock, feeds and medicine*: Adequate supplies and orderly deliveries must be assured at appropriate costs, together with the necessary technical advisory backup. (5) *Market facilities*: Policy must include location of physical sites and provision of transport. (6) *Basic needs*: Control of epidemic diseases, over-grazing and environmental hazards must be included.

Dr Anthonio pointed out that while it was impossible and clearly undesirable to try to design an ideal organization for all cases in all African countries, there were certain issues that had to be recognized:– (1) *Characteristics of the commodity basis*: All are highly perishable and expensive to produce, but the main items are highly nutritious, including meats, milk and milk-based products, eggs, and by-products including skins, bones, and fats. (2) *Variety of outlets*: Animal products may go for export, domestic consumption, or industrial processing. (3) *Ownership and authority over marketing*: A variety of systems are invariably involved and must be harmonised. (4) *Nature and degree of integration*: This may be vertical as in the case of meat processing and canning, or horizontal as in the case of egg-producer co-operatives. (5) *Market control needs*: Essential requirements include legislative facilities for arbitration, relative price and income parity and equity within and between industries, health guarantees, environmental protection and national security.

There was a strong consensus that the role of Marketing Boards should be carefully considered and periodically re-appraised. It was accepted that while they had provided a valuable service in many countries, there was a danger of them drifting into efficiency with a concurrent loss of farmer-confidence. The principal role of such Boards was considered to be the motivation of farmers through price incentives and by securing adequate equity income. While it was important to accommodate consumer preferences and recognize consumer price limits, the overall objective for the foreseeable future in most countries was seen as stimulating supply to a level at which consumer demand can begin to control prices. It was generally agreed that Boards should operate on a commercial basis, detached from political influence, and with a strong element of producer control.

Highly skilled personnel required for marketing organizations had to be rewarded with appropriate salaries if efficiency was to be encouraged. This would also reduce the drift of skilled personnel to other industries and discourage corruption and malpractice.

The capability of co-operative organizations to develop the marketing system was also discussed; there are several successful examples in the dairy industry. Their ability to accommodate fluctuations in supply and to develop a variety of products enables smallholders to gain benefits previously only enjoyed by large-scale producers. The widespread adoption of the co-operative system in India was described as an advanced example of their potential. Not only has the co-operative system helped to improve the production and distribution of milk in India but now also secures for producers a comprehensive range of veterinary, advisory and support services, in particular the provision of reliable credits facilities. Success in dairying is now linked to the similarly successful co-operative marketing of vegetable oils, thereby providing important residues for animal feed.

It was concluded that one vital deficiency was the lack of adequate information and research into all economic aspects of production, marketing and consumption. This is essential for all organizations concerned, and especially policy-making units whose overall concern should be the stimulation of producers to provide adequately for market needs at appropriate times. Only in this way can the requirements of the consumer and producer be balanced and contribute to overcoming the common problems of overstocking.

Chapter 46

The Adjustment and Integration of Livestock and Cropping Systems

S. NURU

Director, National Animal Production Research Institute, PMB 1096, Zaria, Nigeria.

Introduction

Concern to increase livestock production through the optimum utilization of resources has directed the attention of agricultural scientists in developing countries towards developing strategies that give not only maximum return to the farmer but also optimum land and labour utilization.

Dual and tri-commodity integration has been practised in Asia for centuries. In Africa, the concept is also well-established and often referred to as "mixed-farming". However, it is pertinent to note that the concept of mixed farming, even in Nigeria, varies in meaning. In Kaduna State, for example, it may refer to the keeping of two bulls for traction on the farm, whereas in Kano State, it may apply to the keeping of animals for manure, particularly small ruminants. In all cases in Africa the integration involves crops and livestock in varying proportions, the determining factors being socio-economic, cultural, and climatological/ecological situations. Whatever the relative proportions may be, integrated farming systems are based largely on traditional ideas and customs. Recently, however, attention has focussed on research to make such systems more economically viable for small farmers with limited land and usually surplus, family labour.

In Africa, and particularly in Nigeria, the need to implement dual commodity farming systems is now more urgent than before. A few decades ago, land was abundant, sparsely occupied, and often freely-used by livestock rearers. Cropping systems are now well established and understood, but today, the role of livestock, although significant, is not well-defined. That reform of economic livestock production in Nigeria was both inevitable and desirable was realised by van Raay (1975). This became inevitable because of the requirements of the livestock industry, i.e. access to pasture, water supplies, and markets, could no longer be harmonized with population growth and the extension of arable cultivation. Shortages of animal protein can only be met by exploiting the full economic potential of the cattle population and the possibilities of integrating livestock into the cropping system to solve ruminant feed problems. With explosive population growth between 1956 and 1982, and geopolitical activities from 1967 which led to the creation of more states and local governments, there has been an unprecedented growth of urban communities thereby reducing the land formerly available for grazing. Meanwhile, developments in arable crop production technology, particularly the use of chemicals and tractors to open up large tracts of land, has hastened the great reduction in fallow lands and permanent pastures hitherto used for "global grazing".

Advantages of the Integration of Livestock into Nigerian farming systems

(1) A multi-commodity farming system presents several advantages over single cropping, among which is the economic benefits of two sources of income. Moreover, if one commodity fails the farmer may be able to benefit from the other. In the 1983 drought in Nigeria maize and sorghum failed to produce grain, but the standing crop residue was able to be harvested to feed cattle and so provide more meat and milk.

(2) Nomadic livestock farmers may be perusaded to settle near large Government farm projects and rural irrigation schemes and adapt new husbandry methods, alongside new technology in crop production, which could not be utilised by hitherto landless wandering people.

(3) Fodder crops, such as grain legumes (cow peas, groundnuts, soybeans), and leguminous fodder crops such as *Stylosanthes*, lab lab and siratro, grown in rotation as sources of cattle feed, can lead to an improvement in soil fertility and possible reductions in the need for nitrogen fertilizers.

(4) Soil conservation is facilitated.

The Traditional Pastoralist System

The traditional livestock producers in Nigeria, the Fulanis and the Shuwas, own or herd over 95 % of the ruminant livestock population of the country, and are often landless. Different social groups amongst the Fulanis practice different combinations of livestock and crop farming. Semi-nomadic pastoralists have large herds which move in search of feed and water for their animals, but also have a home base where millet is grown in the wet season; their settlement is transient. Semi-settled pastoralists, to whom both cropping and livestock rearing constitute essential aspects of life, usually have smaller herds than the semi-nomadic groups and stay longer in one place. Settled pastoralists grow crops on a much larger scale than the other groups, but also derive income from the small herds they keep. There is increasing evidence that many Fulanis prefer to settle in the midst of farming communities (van Raay 1975). Among the most important reasons for the growing link between livestock rearing and cropping are the availability of crop residues as dry season fodder, fallow land grazing, the use of animal manure for crops, and market channels for milk. Whether pure nomadic, transient, or settled, Fulanis need bush grazing to provide 80 % or more of the total feed resources to their animals. Large dams are additional inducements to settle.

Settled Fulani crop farming is able to provide 30 − 60 % of the family need for grain, but the majority still find it necessary to move to new sites every 3 − 5 years. With increased pressure on land, these traditional herdsmen are forced either to restricted areas of grazing in the wet season, or move further into unoccupied lands away from arable farmers. If they have a home base they still send 60 − 90 % of their herd away from the farming community in the wet season to avoid crop damage and resultant litigations, but a few lactating cows are kept at the home-base to supply daily milk needs and food for aged family members. It is partly to cater for the former needs that the 1965 Grazing Reservice Law of Northern Nigeria was promulgated. The inclusion of arable cropping in livestock systems within grazing reserves had never received adequate attention in spite of the fact that the majority of traditional herdowners cultivate as a major or subsidiary activity. The International Livestock Centre for Africa's (ILCA) research work in a large reserve in the sub-humid zone of Nigeria is aimed a boosting livestock productivity in these circumstances and the success of recent introductions of fodder legumes among the arable crops of settled Fulanis in the reserves has been amply documented. In this ecological zone a 10 000 ha grazing reserve can carry 4600 LVU's or 138 kg live wt ha^{-1} assuming that 15 % (1500 ha) of such land is cultivated.

Traditional Arable Farmers in Urban and Suburban Areas

In large cities such as Kano, Katsina and Sokoto in Nigeria, agriculture is the main traditional occupation, most households keep either cattle, sheep, goats, or a mixture of these. However, in such areas rapid urbanization and related developments have placed pressure on the land available for crop and livestock production.

An essential component of the farming system is manure from livestock, indeed the current intensity of land use could not be maintained without it. While manure is a major reason for owning livestock, another is the demand for animal products; they also provide a means of accumulating wealth through herd expansion. Over one third of the family farm units in Kano own sheep, goats, a donkey, and some poultry, while some larger farm units also have cattle (Hendy

1977). In an urban farming community like Kano, where over 80 % of the available land is cropped, the system concentrates on cereal and grain legumes. Crops are commonly interplanted and livestock relies heavily on crop residues for fodder. Growing groundnuts and cowpeas provide the farmer with security and some flexibility; should one fail, the other provides fodder for the livestock.

In the sub-humid zone, some farms are solely arable with no livestock. In these circumstances the traditional cattle owners operate as a separate group, although both live together in a mutalistic way; the Fulani cattle are allowed to crop residues on farmers' plots in return for manure dropping on the farm. In so doing they improve soil fertility and reduce the costs of organic fertilizer that farmers would otherwise require. In this zone, the possibility of intercropping or inter-row planting of leguminous fodder crops in the traditional farmer's plots exists, in addition to crop residues from planted sorghum, maize and soybeans. These add to natural grazing feed resources; and alleviate the major constraint of nutrition in animal production. Problems remain, however, as the cultural and social attitudes between Fulanis and the Kaje differ as do their economic goals.

Potential Large-scale Integration Schemes

Large-scale development schemes are gaining momentum and revolutionizing agricultural production. There are at present 11 River Basin Development Authorities (RBDA), and over twice this number of World Bank assisted agricultural projects (ADP) at state levels. In the Chad RBDA area 4000 ha was under wheat in 1981, while 10 000 ha was planned for 1983. The US $500M Bakolori Dam, part of the Sokoto-Rima RBDA, commissioned in 1982, irrigates 30 000 ha, converting or transforming a semi-desert area into a potentially fertile modern agricultural region; in 1983 3500 ha of rice, 250 000 tons of fodder, 130 000 tons of cereals, 50 000 tons of vegetables, and 25 000 tons of industrial crops e.g. groundnuts, cotton, tobacco, tomatoes, were produced. The Bakolori Scheme is also expected to raise 30 000 fattened cattle in the period 1985-90.

All rain-fed and irrigated cereal crops (i.e. sorghum, maize, millet, rice and wheat) produce straw and stubble that could be grazed and/or stall-fed after the grain harvest. Because of sequential harvesting, these fodders are available for feeding nearly the year round, with millet residues in late July, maize in August, rice and sorghum straws early in the dry season while wheat straws are available from February; it has been estimated that 3-5 tons ha^{-1} dry matter could be available. If grain legumes such as cowpeas and groundnuts are planted, the additional quantity of crop residue will be less, but the quality and protein content will be higher. Further increases in fodder crops and pastures may be possible by using recommended adapted species of grasses and legumes. This example illustrates the unexploited potential for livestock within cropping systems in Nigeria.

Factors to be Considered for Adjustment and Integration

Adjustment by practising farmers to the concept of integration must be on a reasonable economic scale as well as an adaption of both systems. In the case of each of the three types of farming systems outlined above, human (socio-cultural and socio-economic) as well as scientific and technological factors need to be considered. The choice of specific components to be integrated depends on the ecological areas, but even more importantly on the priority needs of the farmer concerned. Predominantly subsistence arable farmers place greater emphasis on crops to supply their customary dietary requirements while the Fulani's priority is for milk and the social prestige of large herds. In a close-settled zone, or urban farming community as at Kano, the primary objective of integration is manure (mostly from small ruminants). To adjust these divergent needs into a dual-commodity farming system, the commodity mix must fit not only a particular farmer's capability, but also his resources, needs and socio-economic and environmental forces.

Socio-cultural factors are particularly important. The Fulani tribe enjoy social bondage between clans, and because of their social background and affinity for livestock are more likely to adjust to integration than purely arable farmers. Many already grow food crops for survival. On socio-economic and ecological grounds, sheep and goats are the preferred livestock in urban and sub-urban farming areas because purely arable farmers, who predominate, can easily manage these animals and find them cheap to purchase; sufficient fodder is conserved during harvesting to supplement the dry season food needs when the animals are free ranging.

Social interaction through marriage is very rare between Fulanis and members of other tribes. This can be a disadvantage when one considers the need to integrate with purely arable Kaje farmers in the sub-humid zone on whose land the Fulanis depend for crop residues. The co-existence of these tribes can be a delicate issue, especially as the Fulanis own no land in the area.

Conclusions

Factors of paramount importance for success in adjustment and integration, apart from the social issues raised above include:–
(1) Education in the concept of integration. Cultural change is essential for the Fulanis who must settle to practice animal husbandry in the modern way. All concerned must be made aware of the economic and food security advantages of the systems, and need to learn more of the management problems likely to arise and how to cope with them, for example, farm chemical applications, the management and conservation of swards, fodder and feeds resources, and animal health care. Whether the typical arable farmer will ever incorporate livestock keeping on a commercial scale, especially in large scale farming in RBDA's, is yet to be seen.
(2) There is a great need for on-farm studies on the integration of livestock-cropping systems and the development of appropriate technologies for particular ecological zones. In Nigeria, work is in progress on the use of leguminous plants such as *Glyricidia* and *Leucaena* in alley-cropping systems in the humid zone where small ruminants are socially and economically important; whereas in the Northern States, the use of fodder legumes such as *Stylosanthes* and lab lab and grain legumes is being investigated.
(3) Effective extension services are of paramount importance.
(4) Governmental assistance and favourable policies are also essential. These must include marketing, procurement of inputs and development loans. Settlement of the landless Fulanis with secure tenure is a pre-requisite for meaningful and successful integration in parts of Nigeria.

References

Hendy, C. R. C. (1977) *Animal production in Kano State and the requirements for further study in the Kano Close Settled Zone.* [Land Resources Report no. 21.] London; Overseas Development Administration.
van Raay, J. T. G. (1975) *Rural Planning in the Savanna Region.* Rotterdam; Rotterdam University Press.

Discussion

Mr P. N. de Leeuw (Research Scientist, International Livestock Center for Africa, Kenya Country Programme, Nairobi) opened the discussion and pointed out that there was a wide range of multi-commodity production systems combining cropping and livestock; mixed farming represents a continuum from transhumant pastoralists with limited millet/sorghum farming in their home base in the Sudan zone, to settled farmers who keep a few livestock. To understand the Nigerian situation fully and place this in a West African or pan-African context, it should be realized that

Nigeria is unique and that comparisons with other countries in Africa are difficult to make. In particular: (1) Nigeria contains half the total human population of West Africa has a density of 40 rural persons km^{-2} compared with 20 km^{-2} in the rest of West Africa. (2) Livestock ownership is mainly confined to the Fulani, who own 95 % of the cattle, although smallstock are more equally distributed among the tribal groups. (3) Throughout Nigeria the predominant cattle breed is zebu (95 %) in contrast to the rest of West Africa, where in the tsetse-risk zone about 75 % of the cattle are either trypanotolerant (Ndama or West African Shorthorns) or crossbreeds between these breeds and the zebu. (4) Nigeria is a rapidly developing country because of the oil revenues ranging from $15 000 – $20 000M y^{-1}. Hence, farming and livestock production systems are in a state of flux, with traditional systems co-existing with modern capital-intensive rainfed and irrigated farming systems. Due to the high population density, mechanised farming is much less common in Nigeria than in the remainder of West Africa. In the rainfall belt 800 – 1500 mm, cash crop cultivation of cotton with animal traction is a major farming system, providing a large proportion of the foreign exchange to francophone West Africa. Animal-powered millet farming is also very important in the 500 – 700 mm zone in the northern Sudan.

Although herds are small, that "mixed" farmers have invested cash crop revenue back into cattle and herd ownership accounts for a very large share of the total. In Nigeria, only in Gombe has a comparable farming system based on animal power and cotton cropping developed due to suitable soil (vertic luvisols) and a relatively low population density. Smallholder fattening is common in Nigeria and is actively sponsored by the Livestock Production Unit; similar schemes are widespread in francophone West Africa.

Large-scale irrigation is a new development in Nigeria and is mainly in floodplain areas. Traditional livestock exploitation systems have been disrupted and adjustment to the new situation by these traditional livestock owners has given rise to conflicts. As Professor Nuru pointed out, while byproducts from irrigated cropping are plentiful, the integration of livestock and irrigated cropping has still to be accomplished. Floodplain grazing combined with rice cropping (either in controlled irrigation schemes like those of the "office du Niger", established in 1928) and used in the Delta floodplain for centuries are common in Mali. Exploitation strategies associated with them could guide Nigerian approaches to this problem.

Stratification of the livestock industry is slowly developing. There is a high demand for work bulls for training (2-3 y old), which are retained as traction animals for 4-6 years and then sold as mature slaughter animals. Smallholder fattening during the dry season is becoming commoner both in Nigeria and elsewhere in West Africa. This trend may produce an upward price adjustment for young stock generating a greater willingness of pastoralist cattle breeders to sell surplus immatures.

High-protein legume fodder banks, as developed by agropastoralists in the subhumid zone in Nigeria, are located in a dual-tribe economy. The principal techniques used in introducing such banks are similar wherever a high-quality fodder base is needed within the climatic conditions of the subhumid zone (1000–1500 mm rainfall). It should be realized that Nigeria has about 2–2.6M cattle in its subhumid zone, while there are another 9M of mostly trypanotolerant cattle in this zone in the rest of West Africa.

Finally, in north-east Nigeria there is no doubt that livestock ownership has enabled Fulani farmers to become wealthy cotton cash croppers. It can be concluded that livestock functions as a financial buffer, providing cash in times of need for financing cropping, and as an investment for surplus cash crop revenues.

There was a concensus that the settlement of nomadic peoples such as the Fulani creates special problems in mixed farming because of the conflicts between indigenous farmers and livestock owners, although there is much evidence for the benefits of integration. In dry lands fodder crops can often be planted between food crops without affecting their yield; some legumes can supply 70-90 units of nitrogen in such circumstances. In other areas, fallows provide a considerable livestock carrying capacity. The conservation and storage of crop residues and forage crops can contribute to the successful integration of crop and livestock production.

The settlement of nomads to participate in integrated production improves the prospects for preventing land degradation. Traditionally, livestock have been unpopular in irrigation projects because they were thought to damage the banks of canals; however, there are now several irrigation projects where livestock have been successfully introduced. The provision of credit to landless livestock owners has been dificult because of the lack of collateral. However, schemes providing inputs rather than cash have given rates of recovery of 80-90%.

Africa seems to be at an intermediate stage in the development of land use. Virgin land is usually exploited first by livestock, moving to crop production, with eventual integration of the two. It was felt that it was pointless to try to arrest this trend and that the sensible policy was to facilitate it by promoting research to overcome the problems of developing integrated systems.

Chapter 47

Delivery of Animal Health and Production Services – General Aspects

D. M. CHAVUNDUKA
2 Kenilworth Road, Highlands, Harare, Zimbabwe.

Introduction

I have interpreted the title of my contribution as "How to provide extension advice and technical assistance to smallholders which will enable them to improve the health and production of their stock." There are many similarities in peasant communities throughout Africa. As a rule, they depend for their livelihood on a combination of crop production and the rearing of animals; some depend on animals more than others. Nevertheless, there are parts of Africa where the rearing of animals has not been possible due to the presence of tsetse-fly; in such areas, cattle have never been institutionalised in the lives of people. During the last decade or so, considerable research, extension and financial resources have been committed to livestock improvement programmes throughout the continent. The thrust of this effort has been towards animal health and disease control, breeding, and pasture improvement.

Animal Health and Disease Control

During the 19th and early 20th centuries disease was a major factor limiting animal production in Africa. Outbreaks of rinderpest, contagious bovine pleuropneumonia, East Coast Fever, and foot and mouth disease occurred from time to time, decimating animal populations. Considerable advances have been made in research on these diseases, to the extent that most states have either eliminated them or are in a position to control them. As a result animal populations have increased. I believe that the control of specified epizootic diseases and zoonoses is the responsibility of the state. In addition, there are numerous non-scheduled diseases that the stockowner himself must control. The latter consist mainly of parasitic problems associated with tick and worm infestations, and bacterial infections causing sudden deaths or debilitating ailments.

The smallholders are fully aware of the occurence of disease in animals. Indeed, a wide range of traditional remedies exists for their treatment; many of which are still used. What was not understood was the cause of disease. Extension services need to dispel myths and impart the modern concepts on aetiology, therapy, and prophylaxis. Media through which this knowledge can be disseminated, depending on the degree of literacy, include radio, leaflets, booklets written in the vernacular, training courses, lecture tours, and demonstrations.

Accepting that fully qualified veterinarians will be in short supply for sometime, and recognising the need for a greater extension effort among communal stockowners, the Government of Zimbabwe now intends to establish "Animal Health Centres" in all districts of the country. A training programme of cadres to man these centres has already started. These veterinary assistants will reside in the communal areas and be available at all times to assist stockowners in problems they encounter.

Overstocking and Overgrazing

The increase in human population and the concomitant increase in stock numbers on a finite resource base, the land, has brought about an imbalance in the man − stock − land relationship. Because of the excessive stocking rates, the animals tend to exist in a state of chronic malnutrition and with low conception rates; calving rates of 30-35 % are common. For a variety of reasons it has hitherto not been possible to control stock numbers to any great extent. Most of the land in Africa is highly fragile sandveld, with a low and erratic rainfall and occasional mid-season droughts. The high stock concentration has resulted in widespread erosion, the silting of dams and rivers, and gully formation. In Zimbabwe, desert-like conditions have started to develop in over 40 % of the land with a rapid replacement of grasslands by xerophytic bushes, so reducing the carrying capacity of the land.

Overstocking was formerly attributed to a lack of knowledge of good animal husbandry and an inherent belief in "large herds" rather than the quality of the animals. In Zimbabwe several attempts have been made to "correct" the situation. In 1929 Alvord (Chief Agriculturist) produced a "Scientific Land Use Plan" which set aside large areas within the peasant sector for grazing. He sent "agricultural demonstrators" to persuade peasants to limit their stock numbers; they did not succeed. In 1952 the "Land Husbandry Act" apportioned land belonging to the peasants into villages, cropping, and grazing areas. Draconian "destocking regulations" were introduced but there was so much resistance that by 1960 the Act had been repealed. The "Tribal Trust Land Authority Act" of 1969 gave powers previously held by central Government over the use of land to chiefs and headmen; it was "passing the buck" without offering any real solution.

This lack of success was because the authorities had not started to tackle the root of the problem, but had only tried to treat the symptoms. While enlightened farmers measured progress in terms of calving rates, quality of the animals, and 'offtake', the smallholders view was very different. The peasant herd is not simply a beef herd, it is multipurpose within the agricultural complex of the peasant sector. Productivity is considered by owners more in terms of the provision of draught power, milk, and manure than in beef output. A pair of oxen will work $2-3$ ha y^{-1}, the indigenous cow will yield up to 800 kg of milk per lactation (in addition to producing a calf), and enough manure in a year to fertilize 0.4 ha of cultivated land. These outputs are in real economic terms much greater than if the animals were sold for cash.

The problems of overstocking will remain until the subsistence economy has been transformed to such an extent that there is less reliance on animals on the hoof to satisfy basic needs, or until there has been a complete reorganization in the general pattern of life. Further study is required on the different economic uses of animals in the rural sector, and what the alternatives could be to present usages.

Land Tenure

In most African countries the system of land tenure contributes to the overstocking situation. The individual stockowner does not own the land, but has rights to communal grazing. His pride of ownership therefore resides in his stock holding; the animals are his only form of security. Furthermore, he must express his status in society through cattle numbers, regardless of the condition of the land on which the animals graze. A system needs to be evolved which places collective responsibility on the people for the conservation of land.

Animal Breeding

One of the most serious errors of judgment during the colonial period was in respect of cattle breeding. Administrators and agriculturalists of the day considered the slow growth rate and small size of the indigenous animal as a disadvantage, and to overcome this, they crossed indigenous cows with exotic bulls. They argued that the offspring would have the properties of fast growth and be a large framed animal which would yield much meat on slaughter. However, (1) because of the limited food resources the first generation offspring did not reach the full-growth potential that could have been realised; (2) the inter-breeding of subsequent generations produced animals of inferior quality to either of the parent stocks; (3) these animals were less adapted to the environment and they were the first to show signs of debility and to die in a drought; and (4) the policy was a threat to the continued existence of true indigenous breeds.

In Zimbabwe we now consider that the introduction of exotic blood into local herds was retrogressive and plan to reverse this situation by using proven indigenous bulls on phenotypically indigenous-type cows, and by selection and culling to "breed-out" traits of exotic animals from smallholders' herds.

Livestock Co-operatives

On gaining Independence, most African countries opted for socialism; such policies are by definition more responsive to the the needs of the poor. In Zimbabwe the last two years of drought have seen the Government undertake various relief measures. For instance, tractor tillage units were introduced to alleviate shortages of draught power, and millions of dollars have been spent in providing food to destitute families. This type of response by the state to natural disasters means that the vital role of stock as a "cushion" in times of need has somewhat diminished, and that a more rational approach to livestock husbandry can now be pursued. Socialism also offers a framework for closer co-operation in the means of production, and there is no reason why this should not be extended to livestock. Moreover, peasants have traditionally herded their stock together, and the concept should not be strange to them.

In a livestock *co-operative*, grazing land belonging to a group of families is clearly demarcated and divided into paddocks. Each participating family is allocated grazing rights for a given number of livestock units, taking into account the carrying capacity of the grazing. The animals are managed as a single herd, a code of conduct being drawn up and administered by a Livestock Management Committee.

An alternative to a livestock co-operative is a *collective*. In this system stockowners lose possession of their individual animals but have shares in the communally owned herd. There should, as far as possible, be an equitable distribution of the shares amongst the families involved. While a co-operative appears to be more acceptable to peasants, offering as it does direct ownership of animals, a collective has merit in that it is easier to rationalize the herd composition, and make more economic use of the animals, for example with respect to the draught power needs of the community as a whole, the organization of marketing, and maintenance of the correct stocking rates. Stockowners should be allowed to select the form of co-operation they prefer.

Short Duration Grazing

In the last 15 years we have learned a great deal about short duration grazing on sandveld soils using multi-paddock grazing systems. Studies undertaken on growth patterns in grasses and the effects of defoliation show ways to more efficient grazing management. The trend was towards rapid rotational grazing involving 4-6 paddocks; the aim in short duration grazing is to obtain maximum carrying capacity while maintaining a high level of individual animal performance and maintaining or improving the grass cover. Results to date show that there could be a 20-30% increase in productivity using this system; claims of a carrying capacity 2-3 times that achievable under continuous grazing appear false. Livestock numbers still have to be controlled, regardless of the system of grazing used.

In low rainfall areas, stock rearing must be based on veld, and assessed carrying capacities have to be adhered to. In moderate to high rainfall areas, various other strategies exist for improving carrying capacity and animal productivity. These include: (1) Schemes to integrate crop and livestock production, crop residues being used as supplementary winter feed or cattle fattening. (2) Preparation of portions of the grazing to include fertilized planted pastures; the nutritional value of the pasture can be enhanced further by the inclusion of legumes such as siratro and silver-leaf *Desmodium*. (3) The natural veld can be seeded with pasture legumes; in Zimbabwe a 30% increase in beef production has been achieved by overseeding veld on granite sands with fine stem stylo.

Conclusions

I have endeavoured here to give an overview of a very complex issue. There are "grey" areas where more precise information is required before we can be more confident of our approach to the peasants. It is necessary, after reviewing the literature, to conduct trials with constraint removal tests and modifying techniques until a model of livestock production which can easily be understood has been established under a given set of circumstances. Whichever system is adopted, it will require constant monitoring and evaluation.

Even where solutions appear to have been worked out, in the final analysis their effective implementation will depend on the capability of Governments to formulate, legislate and enforce policies embodying corresponding land-use principles. Successful implementation will mean a gradual improvement in the quality of life through agricultural production. In this context, extension becomes an important tool in motivating people to adopt the necessary measures and offering technical advice. An extension effort which does not enjoy substantial backing from the political and law enforcement agencies is unlikely to succeed.

Discussion

Dr M. Moteane (Director of Livestock Services, Lesotho) opened the discussion with comments on ways of strengthening health and production services, focusing attention on needs at the producer level. The roles of livestock assistants must be expanded and their numbers generally need to be increased. The main tasks are to ensure that technical advice reaches and is applied by farmers, that necessary supplies are made available, and that avenues for marketing are developed. In Lesotho encouragement is given to the formation of farmer associations at village level to facilitate these developments. Given the wide variation in climatic conditions, research and demonstration centres are needed to provide appropriate information, but it is equally important to strengthen communication both between research and extension workers and between extension workers and farmers. Veterinary services also need to extend their scope to emphasize production methods and preventative medicine as well as clinical care. The training of vets needs to be modified, and their numbers need to be increased along with those of supporting staff.

In general discussion participants re-emphasised many points. As the control of the major epizootic diseases of livestock progresses, more attention must be concentrated on other constraints to production. The problem of land degradation caused by overgrazing becomes increasingly serious. The problem of overstocking will remain in many areas until there are fundamental changes in the attitudes and lifestyle of livestock owners. Common land ownership gives individual livestock owners no disincentive to overstocking. Possible solutions to the problem are individual land holding, or collective responsibility for conservation.

The supply of appropriate breeding stock is important. For high milk production, some exotic blood seems to be necessary in most circumstances, while African breeds can usually supply the necessary genetic material for beef production. Where cross-breeding is practised, it is important to control the breeding of the progeny, or very inferior stock can result. Pasture improvement can also dramatically improve productivity, but requires grazing for long-term success.

Veterinarians must concern themselves more with the less dramatic production diseases which severely depress offtake. To control these diseases requires a knowledge of the whole production system: drugs and vaccines cannot cure problems by themselves.

All these problems require the efforts of research and extension workers, and there is a shortage of both in quantity and quality. Only additional training facilities can overcome these deficiencies. Communications between research workers, extension workers and farmers must also be improved.

Chapter 48

Delivery of Services – The Case of Tick Control

I. E. MURIITHI

Director of Veterinary Services, Ministry of Agriculture, Kilimo House, P.O. Box 68228, Nairobi, Kenya.

In the tropics amongst the most important animal health problems which constitute a major constraint to the small livestock holder is the control of external parasites, especially ticks. Tick infestation causes local irritation and discomfort, leading to some loss of production of milk, meat, wool and eggs. Contamination by tick mouth parts may result in abscess formation and consequent damage to the hides or transmission of infectious diseases. When ticks are present in large numbers, they remove substantial quantities of blood. Attempts by the animal to rid itself of the ticks may produce bleeding and sores which in turn are subject to secondary infection.

In addition, ticks are vectors of diseases including anaplasmosis, babesiosis, hearwater, Nairobi sheep disease and East Coast Fever. They also cause tick paralysis and one species cause sweating sickness.

Tick control is therefore not only worth-while but essential if an individual or community, in this case small-scale livestock holders, wish to keep improved cattle which will give more milk or meat by creating conditions where such cattle prosper. However, when a tick control programme is begun one factor must be borne in mind. The adult cattle, especially the indigenous zebu, have already contracted whatever tick-borne diseases are prevalent in the area and are immune to such diseases. There is therefore little to be gained by tick control measures unless the objective is the introduction and maintenance of more productive but susceptible improved cattle.

The standard methods of tick control include Machakos dip, hand spraying, motorised spray pumps, spray races and the use of plunge dips. Each method has its own merits. The Machakos dip has been used in the past in Kenya but its use has been discontinued because the operator is excessively exposed to the acaricide. Hand spraying in the hands of an experienced person is most efficient. It has the disadvantage of being wasteful as it is not possible to collect and re-use the acaricide once sprayed on to the cattle. For a smallholder with less than ten cattle, a well supervised operator should do an efficient job. Motorised spray races have similar disadvantages to hand spraying but in addition, have the disadvantage of possible mechanical failure. A spray race, if not beset with mechanical failures, unavailability of spare-parts and maintenance problems, gives excellent tick control. The spray race allows rapid treatment of cattle and in addition, the acaricide is freshly made so there is a problem of fouling. The most generally used tick control method is the plunge dip, where the animals are totally immersed in an acaricide. To enable correct acaricide concentration, the dipping tank must be calibrated so that the total volume is known.

In many developing countries, there is a general awareness that the demand for livestock products is outstripping the domestic supply. Therefore, policies are being designed and implemented to improve livestock production. This is being done especially in countries where there is a productive livestock industry, where land potential is high and where prices of livestock and their products (meat, milk, eggs, etc.) are such that there is an economic advantage in raising livestock.

With the above in mind, a Government may consider offering certain special services to improve livestock production. Services rendered by the state to small-scale livestock producers might include artifical insemination services, clinical and diagnostic services, credit facilities and tick control. Throughout East and Central Africa, tickborne diseases, and specifically East Coast Fever, are considered to be amongst the most important diseases inhibiting the development of the cattle industry. At present the only method of controlling these diseases is by close-interval application of acaricides to cattle to kill the tick vector. Susceptible cattle not protected, especially in the case of East Coast Fever, experience morbidity and mortality approaching 100%. In the small-scale livestock production system (farmers with up to 5 ha and keeping few cattle), it is very difficult for an individual to raise the capital necessary to construct a dip or spray race for his few animals. The alternative is a hand or simple mechanical spray on the farm. There is also a tendency for him to use communal dips or spray races.

In order for a tick control programme to be effective, there is a need for over 60% of the small-holder farmers in a given area to dip all their cattle regularly. For this to happen, a large majority of the farmers must be committed to dipping and the country must have the political will to enforce and support the tick control programme. This would necessitate drawing up suitable legislation and establishing mechanisms for monitoring and enforcement. Mechanisms for monitoring include the provision of the laboratory services.

Laboratory Services

It is necessary to test regularly the strength of acaricides in all dipping tanks to ensure that each dipping is done in a solution lethal to ticks. The same laboratory service is required to monitor the development of resistance by ticks, so that changing from one acaricide to another can be carried out economically and where necessary. Laboratory services are also needed to study and evaluate the various acaricides on the market so as to grade their useage in such a manner that they can be used for long enough periods without causing tick resistance. Laboratory services extend too, to other needs of a thriving livestock industry, including feed analysis and disease diagnosis; the costs of all these activities are best met by the state.

Funding

With dips and spray races, there is the problem of initial capital available for purchase or construction, together with the initial filling with acaricide. It is possible to obtain this capital through contributions from individuals, as a proportion of what is required, and to borrow the rest from commercial and other banks. It is also possible that donor countries can be approached to provide construction money and maybe the initial filling of the dip or spray races. However, recurrent costs of dipping, such as the topping up of acaricide have to be met by participating small-scale farmers from the sales of their produce. The same is true of the funds for the repair and general running of the dip.

In the preparation of this paper I have been indebted to Dr D. P. Kariuki who offered valuable criticisms. I also drew freely from the knowledge and experience of other officers available in Ministry files.

Discussion

Dr R. J. Tatchell (FAO/UNDP Adviser, Ministry of Agriculture and Livestock Development, Kenya) opened the discussion by congratulating Dr Muriithi on having shown how the need for tick control enabled the veterinary department to create an infrastructure providing valuable support for the livestock industry. Not only were ticks controlled, but a base was provided from which other problems could be attacked, but now the base is established, is there still a case for the intensive tick control? At a recent FAO workshop on tick control he noted that participants from countries lacking large-scale intensive tick control were very keen to establish it, whereas those from countries where it was already practised were trying to find ways of stopping it. Interest in stopping control arises mainly from the fact that tick control is now very expensive in terms of capital and recurrent costs, acaricides are now many times more expensive than earlier arsenicals and require foreign exchange, high wages for staff involved in supervising dips and collecting and analysing samples, not to mention the cost of test chemicals. This activity brings additional logistic problems, keeping sufficient vehicles and equipment operational, distributing acaricide, etc. In Kenya, where many highly productive farms dip twice weekly, the total costs could easily amount to US $15 per animal per year. With a herd of 1000 that is a substantial cost, in a national herd of 10 000 000 it is obviously impossible to provide "perfect" tick control.

Interest in starting, and carrying on, is due in part to tradition and in part to a misunderstanding of the nature and extent of the problem. Settlers importing exotic breeds into East Africa were forced to control ticks in order to ensure the survival of their cattle; they were concerned with controlling tickborne diseases (TBD) through the vector, while in Southern Africa intensive dipping was necessary to eradicate the recently introduced East Coast Fever (ECF). Elsewhere, TBD's are controlled by encouraging enzootic stability by allowing small numbers of ticks to be present, by treatment or by immunization but, in the case of ECF, these were not possible. We now appreciate that what was essential for the survival of a few susceptible introduced animals has been unnecessarily extended to include indigenous animals in a state of enzootic stability to TBD. Where this programme has been applied too vigorously, enzootic stability, a priceless natural resource, has been lost and all animals are at risk. Zimbabwe lost at least 1M head during World War II and Tanzania some 300 000 cattle in the Southern Highlands when intensive tick control was relaxed just prior to independence.

One reason for the extension of intensive tick control was the tendency to ascribe an unrealistically large number of deaths to ECF. Thus, indigenous calf mortality in enzootic situations was believed to be as high as 20 %. We now know that in the Trans Mara, in a highly enzootic situation complicated by ticks from the many buffalo present, calf mortality to ECF is negligible. Another reason was belief that ticks must be bad as they look unsightly and so their control was viewed as a form of progress. Also, it was not recognized that while certain ecoclimatic zones are very suitable for the vector of ECF, others are completely unsuitable; these are the extensive rangelands which could be the heart of a cattle industry, and yet control programmes are advocated in these areas as they are in those where ECF is a problem. Because of these misunderstandings, intensive tick control is being applied in areas free of ECF where all TBD are curable and enzootic stability is well established. Of course, tick worry does cause production loss, but really heavy infestations are needed to make it economically worthwhile to dip; the current economic threshold in Australia is 158 *engorged* ticks day $^{-1}$ (the total adult tick count would be of the order of 3500). The economic threshold for Africa is raised to an even higher figure due to the 3-host life cycle of the ticks, but is also lowered by reduced labour costs; it is likely to be at least as high in Australia. In Kenya it has recently been shown that there is no benefit from controlling ticks on rangeland cattle in an area where total tick numbers would fluctuate up to around 100-250. Similarly, in the Sudan the normal burden of 50-60 adult ticks caused no significant reduction in weight gain compared with treated cattle. There is a case for minimal tick control, however, to reduce udder damage.

In the future enzootic stability must be preserved where it remains. Where it has been lost it should be re-established by immunization and treatment; there is no TBD for which immunization and/or treatment does not exist. Susceptible imported cattle can be dealt with similarly. We can then dip only when it is really necessary to prevent production losses, and we can save the cost of expensive dips where minimal hand spraying is all that is required. We can also be sure that a reduction in dipping intensity will reduce the risk of acaricide resistance. In this connection we must educate farmers to tolerate a few ticks. For example, it is normal in Africa for a dip to be changed when *Boophilus* species become resistant even though the real vector is still being controlled. In African dipping regimes there are unlikely to be enough *Boophilus* species to reach an economic threshold. It is much better to preserve the life of the old acaricide as long as it is effective against the true problem tick species.

In tick control, as any other agricultural activity, we must be sure that there is an economic benefit. Governments also have to be sure they can afford new acaricides. One country recently changed from the affordable arsenic to a more effective but much more expensive, chemical with the result that the acaricide budget for the year was inadequate; acaricide purchasing ceased and hundreds of cattle died from babesiosis, an unnecessary disease if enzootic stability had been preserved. Perhaps our first step must be not only to think critically about why and how we control ticks, but also to design and carry out large-scale experiments to measure the damage done by ticks and determine economic thresholds, such as those currently underway in Zimbabwe, Zambia and Kenya.

Many participants shared Dr Tatchell's optimism that alternative control strategies for TBD could be developed to replace dipping in many livestock production systems. However, some doubts remained and the difficulties of making a transition should not be under-estimated. By and large, improved animals, especially dairy types, are susceptible to ticks and tick-borne diseases. It is debatable whether some or all of these can be protected by immunization or preservation of pre-immunity, but the economic advantage of being able to reduce the intensity of dipping would be enormous.

Chapter 49

The Large Livestock Unit: its role in development

N. G. BUCK

1 Market Place, Bedale, North Yorkshire DL8 1ED, UK.

Introduction

Recent reviews of the livestock industries of post-independence Africa emphasize increases in cattle numbers and static or declining individual animal productivity (Jasiorowski 1976). Reasons cited for these population increases are improvements in watering facilities, and prophylactic health measures; of probable significance in respect of reducing production has been investment in livestock of capital generated outside the livestock economy (Baker 1981). Such investment has been economically sound, particularly when costs were minimal as cattle were grazed on communal pasture, but then resultant over-stocking has proved detrimental to poorer producers whose livelihoods became threatened. Cattle populations have now reached the point where each subnormal rainfall year becomes a drought, and range degradation is continually more progressive. In Botswana the cattle population, which grew from 1.3M in 1965 to 2.9M in 1980 with a comparatively level offtake during this period, is reported to have collapsed to 1M during the current drought.

To fulfil national expectations of productivity and reverse the trend of urban drift, agriculture must be attractive in social and economic terms; continuing urban migration suggests these requirements have not been met. However, for the vast majority of Africans, agriculture remains the only opportunity for gainful employment. The development of urbanization places the rural populations in a position of disadvantage as they are often illiterate, without access to public communications, or lack a political voice. This situation is likely to continue if agriculture, and particularly animal agriculture, are considered subsistence and "peasant" occupations. The development of a sector of large-scale animal agriculture with the investment of capital, the development of service and processing infrastructure, and of political influence, is essential to strengthen the broad base of production.

The Development of Animal Agriculture Services

Strong veterinary departments were among the major institutions which African countries inherited at independence. Their principal role had been the control of epidemic disease, and noteworthy achievements were recorded. Eradication of rinderpest and pleuro-pneumonia were attainable, but have received recent set-backs. Tsetse-fly control was the other main area of activity, but trypanosomiasis is still limiting animal production in the continent. In spite of criticisms levelled at veterinary services, the control of major epidemic and endemic diseases in Africa is still a prerequisite for livestock improvement and production increases.

The veterinarian is not trained or equipped to undertake all the roles needed in animal production, extension and development. In western Uganda, as a District Veterinary Officer, I was responsible for the importation of livestock, marketing of poultry, eggs and milk, advising on pasture improvement and farm planning; the response and enthusiasm of smallholder farmers was a most rewarding experience. Their failure has not been due to technical incapability or unwillingness to adapt. The main constraints at the time were: (1) The difficulty of adequately servicing and advising farmers scattered over a wide area. (2) The rapid build-up of a surplus perishable commodity beyond local demand, but below that which would justify transport to a distant market. (3) The frustration of the efforts of successful farmers by the jealousy or prejudice of neighbours. The consolidation of such ventures into an area which could be adequately supervised, with a sufficiently large nucleus production unit to ensure year-round production and processing, would help meet these constraints.

The paucity of technical manpower in the disciplines serving development remains a severe limitation. In the time that educational programmes have received the priority they deserve, only a limited cadre of such manpower has been produced. Many were educated in foreign institutions, or in domestic universities where expatriate teaching staff often had limited experience of the country in which they were teaching. The contribution of these first-educated scientists to the development of policy within their own countries, based on their own experience, is yet to be fully realised (Eicher 1982). Early employment opportunities for such graduates have been almost entirely in administration; expansion of bureaucracies has created overburdened budgets which absorb much of the available domestic funds. Such staff frequently work with little experienced supervision, in isolation, and under limiting conditions. Due to that inexperience, practical capabilities and contributions have been limited, and positions are often overmanned and underutilized. Inflation has exacerbated this situation, leaving trained manpower unable to travel, with limited mileage allowance, and inadequate supplies of drugs, vaccines and pesticides.

Land Tenure

Increasing human and animal population pressures require a re-appraisal of traditional land tenure in Africa. Generalizations are unlikely to be adequate, but individual security for at least a temporary period is usually available for cultivation, while communal open grazing is the norm for cattle. The absence of security is an impediment to any conservation or improvement of land, and individuals or communities who have traditionally used such areas feel threatened by the operations of outside investment or neighbours who have capital to invest from another economic sector.

The degree of dispossession incurred by colonial settlement in those countries where it was permitted is subject to debate; as will be those areas which have more recently been allocated for development. If such areas are approaching their potential in productive agriculture, fragmentation may well not serve the long-term national interest. In contradistinction, there are recognized land holdings, which for diverse reasons are currently used extensively, which may have greater potential if farmed as smaller units. As far as livestock production is concerned, an essential requirement is the recognition of land capability and the limitation of livestock numbers accordingly. This requires administration of land allocation at the local level, by and with the compliance of the local community.

A stratification of land-use, and with it the evolution of arable and animal production systems, will be dictated by both environmental and strategic considerations. Three broad categories and systems are likely to emerge:− (1) High potential/peri-urban areas; where land is of high productivity or by its proximity to urbanization with a ready market justifying irrigation; vegetable growing and dairy farming production systems. (2) Middle distance areas; cereal growing under rainfed systems, dual purpose cattle production with seasonal milk sales. (3) Peripheral/semi arid areas; beef cattle production under ranching systems phasing into the arid semi-nomadic and nomadic production systems. It is against this background that a large-scale animal production sector is proposed.

The Peri-urban Areas: Dairy Production

Successful dairy farming is the animal production system with the greatest potential for high return from land area and for employment creation. Exotic dairy cattle have been productive in a sufficiently wide range of climatic conditions to indicate that they can contribute much more widely in Africa. To keep such cattle adequate nutrition and the control of disease are required. Transition from the informal hawking of raw milk to the provision of a reliable town supply of processed and packed milk, requires a sufficient year-round supply to economically justify a dairy plant. The size of such facilities will be determined by potential market, but to ensure economic throughput commercial dairying on a sufficiently large scale must be justified at the onset. Kenya provides a prime example of the emergence of successful dairying by smallholders built on the established Kenya Co-operative Creameries. In the absence of any dairy production, initiative needs to be taken by Government. Major dairy projects are, however, likely to attract domestic finance if encouraged and given access to the land required. The potential benefits from research, teaching and extension suggest that at least some units should be in the care of research institutions or universities. Fodder production and energy feed source research are of high priority, but the relative performance of the different strains of black and white cattle in Africa, sire progeny performance testing, embryo transfer for breed expansion, and appropriate crossbreeding evaluations are also needed.

In the foreseeable future, the over-production of milk by large-scale units seems unlikely. Such units can also be seen as a nucleus for the satellite development of smallholder dairying, with the sale of surplus cattle, exposure to technology, and in areas where fodder production is difficult, its central production and sale. Research in Botswana indicates that a 50 cow 50 ha^{-1} intensive unit could be economically viable, relying on non irrigated arable silage production, with economies of scale operating up to 150 cow 150 ha^{-1} units (FAO 1982). Such schemes seem preferable to the importation of European surplus milk powders.

The Middle Distance Areas: Mixed Farming

The production of adequate supplies and reserves of staple grains remains the greatest need of many African countries. Countries which have achieved this have relied on capital-intensive mechanized agriculture, with minimal labour requirements, but these solutions are not readily applicable to most areas of Africa. The desirability of developing from the hazard of subsistence farming to that which, still providing employment, provides an attractive income from the sale of surplus produce is well recognized.

At present arable and animal production are frequently in conflict with grazing animals encroaching on growing crops and rested or fallowed areas providing poor grazing whilst recovering from a series of grain crops. The evolution of mixed farming with the use of a leguminous pasture to restore fertility requires further land-use reorganization and enclosure; legumes are now available for even the most unfavourable areas. Large-scale animal production enterprises in these areas are difficult to visualize. As the land has potential for cereal growing, the possibility of large-scale forage production units for feedlot finishing of cattle justifies consideration. In higher rainfall areas the mechanized production of elephant grass, sugar cane, and maize for ensilage may serve to finish cattle from both within the mixed farming area and from more peripheral areas. Although considerable isolated research on fodder crop production has been conducted, too few applications have been demonstrated.

Information on the development and economics of feedlot finishing in Africa is limited, although the experiences in Kenya were initially reported on favourably (Squire 1976) and the use of high forage percentage diets with limited cereal or cereal by-product ratios is appropriate for African economies. In Botswana in the early 1970s a number of commercial feedlots fed both young cattle and cull cows to a more acceptable level of finish than on grain-based rations; this practice declined with increases in cereal prices.

The development of feedlot finishing systems where forage crops can be grown for ensilage or where arable crop residues, waste grain or by-products are available has been frequently advocated (Preston 1981); more favourably watered countries particularly Uganda, Zambia and Zimbabwe, have considerable potential for this. Mechanization for the production, harvesting and ensiling of forage crops in a timely method seems essential, and so feedlotting becomes a commercial exercise with indirect benefits to small-scale producers. This is again likely to be more successful as a private sector venture; sugar-growing organizations also have potential for collaboration here.

The Peripheral/Semi-arid Areas: Beef Cattle Production

Beef cattle production will, as land pressure increases, become confined to semi-arid areas. Such areas have a limited potential for improvement and their conservation will remain a matter of judicious stocking with the structural opportunity for the adjustment of cattle when climatic circumstances dictate. Potential profitability of beef production in such areas is dependent on the achievement of an adequate cow reproductive performance. Reconception and calf survival are largely functions of adequate nutrition. The opening of new areas for communal grazing expansion should be approached cautiously if this will provide only temporary relief to population pressure. The potential impact of any proposed development should, however, be related to the beneficial effect on neighbouring communal areas and not considered in isolation. Relief of grazing pressure by the removal of progressively younger males and barren females will only occur if there is commercial incentive.

Ranching remains a most logical land-use method for semi-arid areas; its failure in many areas of Africa cannot be blamed on the system so much as its application. Selection of potential ranch holders, inadequate managerial capability and overburdening lease-holders with debt repayments have been major factors. This should not be a barrier to the more capable educated younger generation of potential farmers who will have a greater incentive to succeed.

If the natural evolution of the stratification of beef cattle production from calf producer to finished product is to occur, artificial restrictions such as price control should be removed. The large-scale producers capacity to accumulate, hold and market numbers of stock will benefit the smaller producer. The competitive buying activities of independent graziers is more likely to elevate prices than monopolistic organizations; an organization adhering to civil service salaries and benefits may prohibit the transfer of full value to the producer.

The Influence of Large-scale Production on Livestock Improvement

Those African countries with a history of colonial settlement agriculture benefitted by the influence of introduced production systems and genetic resources. The improvement of the boran and sahiwal in Kenya, and of the tuli in Zimbabwe, has been conducted on commercial farms and is having increasing influence in livestock breeding. In southern Africa, the importation of the brahman (*Bos indicus*) from the USA and numerous examples of *Bos taurus* breeds have contributed to a greater potential for beef and dairy production. Few of these breeds in the pure strain would be advocated as improvements on the indigenous cattle, particularly under the harsher environment of communal grazing conditions, but their use in crossbreeding systems has advantages (Trail *et al.* 1977). Commercial or large scale farms have traditionally supplied breeding bulls for the communal sector in Botswana, Kenya and Zimbabwe. As information accrues on the most suitable breeds for crossing systems, producers and buyers will become more selective in their choice. The controlled breeding environment available on such ranches makes them invaluable as a future stud sector of national herds. Such herds are also potential sources of research information. Trail & Gregory (1981) were able to produce an evaluation of the sahiwal breed in Kenya based on commercial farm data, and in Botswana a Commonwealth Development Corporation (CDC) ranch collaborated in progeny testing of Africander bulls. Where large-scale research facilities exist they should be carefully guarded; when managed as "ongoing production units" they are invaluable as demonstration centres, provide improved genetic resources, and if allowed to operate without restraints should be self-financing.

References

Baker, P.R. (1981) Sociological implications of intensifying animal production in developing countries; the case of the African pastoralists. In *Intensive Animal Production in Developing Countries* (A.J. Smith; R.G. Gunn, eds.), 11-21. British Society of Animal Production.

Eicher, C.K. (1982) Facing up to Africa's food crisis. *Foreign Affairs* 61, 151-174.

FAO (1982) *International Scheme for the Coordination of Dairy Development: Botswana, Final Report.* Rome; FAO.

Jasiorowski, H.A. (1976) The developing world as a source of beef production for world markets. In *Beef Cattle Production in Developing Countries* (A.J. Smith ed.), 2-18. Edinburgh; Edinburgh University Press.

Preston, T.R. (1981) The use of by-products for intensive animal production. In *Intensive Animal Production in Developing Countries* (A.J. Smith; R.G. Gunn eds.), 145-150. British Society of Animal Production.

Squire, H.A. (1976) Experiences with the development of an intensive beef feedlot scheme in Kenya. In *Beef Cattle Production in Developing Countries* (A.J. Smith, ed.), 150-153. Edinburgh; Edinburgh University Press.

Trail, J.C.M.; Gregory, K.E. (1981) *Sahiwal Cattle: an evaluation of their potential contribution to milk and beef production in Africa.* Addis Ababa; International Livestock Center for Africa.

Trail, J.C.M.; Buck, N.G.; Light, D.; Rennie, T.W.; Rutherford, A.; Miller, M.; Pratchett, D.; Capper, B.S. (1977) Productivity of Africander, Tswana, Tuli and crossbred cattle in Botswana. *Animal Production* 24, 57-62.

Discussion

Dr J.E.U. Mchechu (Ministry of Livestock Development, Tanzania), brought out most of the main issues:–

(1) *What constitutes a large scale livestock unit?* In most cases this would be one that is capital intensive, economically manageable, and ideally it should integrate multi-disciplinary systems of two or more of the following: dairy cattle, beef raising (and where necessary a small beef-lot), pig raising, poultry, sheep and goats, rabbits and guinea pigs. It should also practice mixed farming systems. This approach, however, has limitations, particularly in arid/semi-arid areas of Africa where the bulk of the livestock are found, most in the hands of nomadic pastoralists.

(2) *Land tenure systems and land-use in general.* These are necessary, and must be an essential component of national policy guidelines. Lawful land ownership is necessary to guarantee security to both the large- and small-scale livestock units including family-holdings, communally owned land, and reserves gazetted as grazing areas. To ensure adoption of proper land-use practices by the major land users in Tanzania a "Land Use Commission" was set up by an Act of Parliament in February 1984. The Commission, with units at National, Regional and District level, has the primary objectives of planning and co-ordinating the effective and economic use of land by the major sectors of the national economy (Ministries of Agriculture, Livestock Development, Natural Resources, Lands and Urban Settlement, Water, and Energy and Minerals).

(3) *National livestock policy.* Some countries in African did not experience land takeover and settlement during colonial times and still have potential areas for livestock development. In these planning for large-scale livestock units can be undertaken as an essential component of overall national policies. Tanzania has planned the development of several large-scale livestock units over a ten-year development period (1981–1990) and involving private and public investors. Other African countries are taking similar steps.

(4) *Zoning and stratification of land-use.* Tanzania has, as part of its livestock policy, grouped potential areas into ecological zones with land stratified for various economic uses, such as "growing and fattening ranches." Use of feed-lots has not been very successful to date due to the high cost of grain.

(5) *The role of the large-scale livestock unit in development.* The points raised by Dr Buck are commendable. However, these units in Tanzania's case also serve as training schools for farmers, extension staff, trainers and research workers in a peripheral way. In short, large-scale livestock units offer back-stopping support facilities and services to small-scale livestock units.

(6) *Constraints to large-scale livestock units.* The major constraints unique to this development in Tanzania are:– (a) Shortage of improved breeding stock in the case of dairy cattle, this can be over-come by the establishment of crossbred dairy multiplication farms and continued importation of stocks; (b) shortage of skilled and trained management particularly in the short and medium

term; and (c) shortage of inputs, particularly those which depend on imports (e.g. farm machinery, equipment and accessories, seeds, fertilizers, spare parts, drugs, vaccines and chemicals).

Dr Mchechu concluded that large-scale livestock units are necessary in offering supporting facilities and services to small-scale livestock units which in return must be managed more as economic units and less as "subsistence" units in order to achieve some form of surplus production for marketing.

In subsequent discussion of the roles of large holdings in the stratification of production, it was recognized that separate facilities for growing youngstock for a year or more may be required in addition to breeding and fattening facilities. This raised again the desirability of having stratified production systems to improve the overall efficiency of livestock industries.

Experiences with the feedlot system, originally established in Kenya to utilize surplus maize in fattening beef for the export market, were discussed. They had proved difficult to operate following outbreaks of foot and mouth disease which prevented further export. The consequent fall of the home beef produce rendered them uneconomic. In Zimbabwe, although the establishment costs were relatively low, the running costs of feedlots were high. This has resulted in them being used only in periods when cheap surplus stock had to be accommodated. Feedlots also provided opportunities to use crop by-products, such as citrus pulp.

Chapter 50

Delivery of Health and Production Services: The Producer's View

W. A. WILSON

Kilifi Plantations, Kilifi, Kenya.

Introduction

My task is to present a producer's ideas on the adjustment and servicing of large scale Production Systems. In inviting me to contribute Mr Ellis expressed the hope that the conference would prove a source of creative ideas which would foster new initiatives in livestock development. I submit that creative ideas and new initiatives are not what is needed, but a return to basics, which can be summed up in the word "management".

The Roles of Government

Governments must manage their affairs so that there are adequate and available supplies of fuel, spares, and servicing, drugs, fertilizers, acaricides, weedicides, vaccines and controlled credit. Governments must arrange for diagnostic and analytical services not only to be available but to function quickly and effectively in the fields of animal diseases, mineral deficiencies, and plant diseases. If a farmer has dying or non-breeding animals, it is very little help to him to have to wait six months or more for an answer.

Governments must also manage their communication systems and maintain them so that what is produced by the farmer is able to get to market, and those inputs required by a farmer are available close to the point of production. They must also manage disease control, vaccination campaigns, and stock movement controls so that disease is not spread indiscriminately. Once Government has managed to arrange that the essential ingredients for production are in the country, then distributors (parastatal or private) must manage to distribute them so that producers can use them at the proper time.

The Roles of the Producer

Once the producer has the essentials it is up to him to manage them. To see that his stock is properly fed, that the correct dipping or spraying routines are carried out, that inoculations are done on time and that fertilizer is correctly applied with the correct seed rate. There is nothing inherently difficult in this, but the farm manager must be motivated to pay attention to detail, 365 days of the year and 24 h each day when necessary. There is a proverb, "the best manure is the farmers boot"; it is very true.

To turn to the problems of a producer in securing the technical support he needs to maintain and promote productivity in his livestock, if the above resources and services are made available promptly and efficiently, I believe that nearly all large-scale producers would willingly contribute to the costs. I would suggest that prepayment of a fee annually is a method which could be considered. A farmer would then be able to budget without receiving sudden shocks. It would also be far easier for providers of the services to budget realistically and Government to know its liabilities.

Service Structures

In the developing countries, where I believe that the large-scale stock unit should be the exception rather than the rule, it should be the Government who should operate the clinics and laboratories. Clinics must be as close to the producer as is possible, and offer a seven a day week service. Private practice should be allowed if there is a demand.

It would be highly advantageous to have the farmer more closely involved in policy and problems at grass roots level in the form of District Agricultural Committees. A farmer as chairman, elected members who are also farmers, the Agricultural Officer and District Officer as ex officio members and, where warranted, an Executive Officer paid by the Government. This would be a forum where farmers could discuss mutual problems, exchange ideas, and perhaps also control credit. Above all, the greatest stimulant would be a fair net price to the producer and prompt payment. The Kenya Breweries Barley Scheme is a good example of what can be achieved; a fair price is offered for barley, inputs and credit are available, and payment is prompt.

New Technology?

Finally I should like to comment on the apparent belief that new technology is the only answer for increased production. This is something that worries me. First of all, what is the definition of technology; the dictionary defines it as "The Science of Industrial Arts". Looking at the Kilifi Plantations which produced over 2 000 000 kg of milk, 320 steers, 300 female stock for sale, and 520 t of sisal from 2600 ha in 1983, I ask what technology is available to me or used by me that is not available to other farmers. Apart from the fact that I use a computer to analyse my cow records, I can see none. To me the key to our production successes is the management team which has been built up over the last 20 years. This should be linked to the fact that I, as part-owner and manager, have been resident on the farm, where my day starts at 0545 h and often finishes at dusk, or later. I believe that what is needed to increase production is not new technology, bright ideas or large conferences, but hard work, fair prices, discipline and the availability of inputs.

For Discussion of this contribution see below Chapter 51.

Chapter 51

Delivery of Health and Production Services: The Veterinarian's View

S. KAMVASINA

Department of Veterinary Services, Ministry of Agriculture, P.O. Box 30372, Lilongwe, Malawi.

Introduction

In the various reports from national institutions and reputable international organizations involved in livestock production in Africa, it has been shown and accepted that livestock production in Africa as a whole has deteriorated, although there are a few small islands of progress.

Livestock production in most African countries depends on the traditional/indigenous sector and to a limited extent, the small commercial sector. Whereas the small commercial sector is independent from state subsidies in terms of routine animal health and production inputs, the major traditional livestock sector is dependent on them, and it has been assumed that this should be the trend. Unfortunately, economic development during the past two decades has not been compatible with a two-sector approach, and neither has flourished. This situation calls for a reformation in our approach to animal health and production delivery services, bearing in mind that economic viability is the ultimate aim for all sectors, and will dictate the trend of events.

Problems with the Present Services

The present animal health and production delivery services have not provided the level of service to meet the full requirements of the traditional livestock producer. Limited financial resource allocation, due to a lack of recognition of the importance and potential of the livestock sector, has greatly diminished the effectiveness of services. Organizational structures for the departments responsible for the delivery of animal health and production service are generally imperfect. The faulty design of livestock operations and inadequate, sometimes faulty, policies within the delivery system has countermanded its effectiveness. Further, wars, conflicts, and a lack of co-operation within and between countries have dramatically reduced and sometimes eliminated certain types of assistance to farmers.

Model Services

The aims of the livestock producer must be recognised and accepted as important. Farming systems should be clearly defined, and the association between livestock production, and crop-farming, traditional and cultural factors, social advancement, and the aspirations of the livestock producer must be assessed. Extension methods to be adopted should be targeted towards achieving the farmer's aspirations, and must be adapted to the farmers' circumstances and environment.

Necessary inputs for all livestock sectors should be assessed and areas for subsidies separated from those provided at cost to the farmers. Areas of operation must be defined, and thereafter Government or other responsible institutions, and the farmer, should be committed to honour corresponding obligations. Organizational structures for ministries and departments responsible for the delivery of animal health and production services require a definite line of command backed by sound policies and adequate and efficient extension personnel. Although the ratios of veterinarians and livestock officers to producers may be low, auxiliary technical staff can have great impact provided the concept of functional veterinary centres is adopted. Veterinary centres located within easy reach of livestock producers and provided with the necessary inputs for disease control and livestock production extension services, can utilize professional capabilities to full advantage. The assessment of manpower requirements/projections, and training required to balance the operational needs is important. Sound policies on marketing, i.e. price structure for livestock and livestock products, good forward planning and cost effectiveness of the operations, must all be considered critical in the system.

Although wars and national conflicts are beyond the control of participants in the animal health and production delivery system, strife should not be used as an excuse for abandonning help to farmers.

Discussion

There was general agreement in discussing the above contribution and that by Mr Wilson (Chapter 50), that efficient management was the over-riding factor in determining the success of livestock production systems, and that services should focus on this. Extensive discussion followed on the Government's role in the supply of services. There was a concensus that certain responsibilities, such as the control of major diseases and the formulation of overall policies must remain with Government. However, it is clear that, in the present economic climate it is difficult for Governments to meet all the farmer's needs. Large farms could meet most of the costs involved directly, provided that the necessary human and material resources are accessible. Meanwhile, groups of medium and small-scale producers could progressively assume a share of the costs through the formation of co-operatives and associations which set up their own veterinary, artificial insemination, and advisory services. Such schemes already operate in Zimbabwe and farmers groups there even contribute directly to problem-orientated research.

It was felt that when Government does maintain services and implement programmes, realistic time scales should be required to ensure greater efficiency. The need for economic evaluation of these and other kinds of activity was emphasized. It was pointed out that recent Kenyan studies had shown that benefits of foot and mouth disease control alone exceeded the total cost of the entire veterinary service, in spite of the fact that the average cost of employing a veterinarian was equivalent to US $40 000. It was generally agreed that the Government cannot continue to be the main employer of support staff. If farmers were prepared to purchase services, then private enterprise should be encouraged to take over.

The proven success in India of a comprehensive service to dairy farmers funded entirely by farmers co-operatives was cited as a model. In this connection, it was also pointed out that extension work needs to be developed with staff drawn from the communities concerned, and that everyone receiving benefits should contribute to the cost of the service in some way. If farmers are required to contribute, however, they must be involved in the management of the service and ensure that those employed by it are suitably motivated by job satisfaction.

All the developments mentioned were considered to be dependent on a realistic commercial orientation of the industry.

Chapter 52

Generating Information from the Small-scale Farmer

S. R. MAGEMBE

Deputy Chief Veterinary Officer, Ministry of Livestock Development, P.O. Box 9152, Dar es Salaam, Tanzania.

Introduction

Most developing countries of sub-Saharan Africa produce insufficient animal protein to meet the demands of their expanding human populations. In Tanzania, for example, the Tanzania Food and Nutrition Centre (TNFC) record that the daily *per capita* animal protein consumption was about 14.6 g in 1974-75; it had declined to 12 g by 1980. At the same time, the contribution livestock can make to draught power, soil fertility, and domestic fuel supplies, is under-exploited.

There are many technical reasons for inadequate progress in livestock development in sub-Saharan Africa. Pressure on land in traditional grazing areas has forced livestock to less favourable areas, poor communications have limited access to markets, and the marketing system does not appear to have stimulated the producers to improve productivity. Moreover, despite the importance of livestock to most national economies the development expenditure on livestock is minimal. In Tanzania investment in livestock services and support activities accounted, on average, for only 2-3% of the national development budget from 1972-73 to 1979-80, contrasting with the contribution of 10 − 15% of livestock to the gross domestic product (GDP).

The traditional sector needs to devise appropriate methods for the circumstances which have resulted from the re-structuring of populations. This sector has enormous potential, and could do much to help overcome the current and increasing deficits of foods of animal origin. The over-riding constraint, however, appears to be poor management of the resources involved. To remedy these deficiencies and imbalances, authorities at all levels, and the producers themselves, need information on how production systems operate in the country's differing ecological environments. They should then be able to remove constraints and design realistic steps to improve animal productivity. An efficient livestock information system should facilitate the flow and analysis of data, and provide appropriate information to different parts of the livestock industry and the institutions and services associated with it.

Sources of Information

General questions concerning information are:−

(1) What information/data on livestock do we need?; (2) where and how can we get this information?; (3) what form should it take and what should be the mechanism of dissemination?; and (4) how reliable is the information/data we recieve? We all know the answers in general terms, and detail varies with circumstances. Pertinent sources in the present context include small and large-scale farms, veterinary diagnostic laboratories and practices, Government Ministries and institutions, and universities, etc. All these keep some form of livestock information to suit their particular needs, but the quality varies widely. In order to effectively plan and control livestock development, the quality of the information utilised must be of a high standard and reliable; if not planning will be faulty or unrealistic.

In most sub-Saharan African countries, increasingly reliable information can be obtained from sources such as veterinary laboratories and practices, universities, large-scale private and public farms, and from survey reports in relevant ministries. These organization all need information for the successful management of their activities. It is, however, disheartening to note that little progress is being made with respect to small-scale farmers, who own over 90% of all livestock in most of our countries. Some small-scale farmers do keep records, but the data kept are often of little value even to the farmers themselves. This poses the question of whether such farmers need information, or whether they are unaware of the potential value of better information.

Most small livestock owners in the Arusha, Kilimanjaro, Tanga, Iringa, Mbeya and Rukwa Regions of Tanzania, especially those who keep beef animals, do not maintain any records (Tyler *et al.* 1981). In most cases they memorized dramatic events occuring within their small herds. When asked why they were not keeping records, some said they were never told to do so, others questioned their usefulness, and others argued that even if they did keep them nobody would collect the data. These responses indicate that it would be possible to generate information from these communities if their enthusiasm was aroused.

Generating Information from Small-scale Farms

As in other farming enterprises small-scale livestock owners are faced with a control problem (Pugh 1977). First they have to decide on objectives, on the best strategy to achieve these objectives, on the way to implement this chosen strategy, and if necessary how to modify either the objectives or the strategy. An essential feature of this control process is that firstly it involves decision-making for the long, medium or short term. Secondly, reliable information is essential for this decision-making activity, and consequently appropriate data analysed in an approriate fashion are essential for the production of this information.

Therefore, in attempts to generate information from small-scale farms, it is important to consider the farmer's own information needs; both from within his unit and from outside his farm. This consideration is essential since all small livestock holdings are family units and they have varying goals depending on their degree of market orientation, and the information need not be detailed. In a subsistence setting, for instance, a family strives to produce enough to satisfy its own needs. At this stage, the farmer does not seek detailed information from his farm or outside it. If, however, he recognizes a potential for greater production, the farmer will usually attempt to maximise first his subsistence production, secondly his capital reserve, and thirdly output and income. He will then begin to require information on management practices, disease control, production, and marketing, and it will be our responsibility to provide such information.

Once the information needs of the farmer have been identified, then he must be encouraged to keep data. Farmers are reluctant to change habits, so they must be convinced that the exercise is for their own well-being. The only way to do this is to show that there are economic (i.e. monetary) benefits from embarking on such a programme. As some of the smallholders may be illiterate, ways and means should be devised to facilitate simple recording. However, the recording should be in a format that permits simple analyses which can be interpreted to advise the farmer.

This brings us to the question of the dissemination of information. The flow of information is a two-way system. Those who keep records must have feed-back, and this must be prompt, maintaining the enthusiasm of the smallholders. It is also important for the disseminators of this information to realize the importance of the duty they are performing. Furthermore, facilities to accomplish this whole undertaking need to be provided.

In conclusion, from our experience to date, it is possible to generate information from small-scale farmers, provided that they are not being asked to do much (particularly paperwork). There must be a feedback of information to those who in the first place collected or recorded data, and there must be follow-up on the part of the disseminators of information. Most importantly, there must be a system and a service that facilitate the process.

References

Pugh, C. L. (1977) Farm management information systems. A practical approach. *Farm Management* 3, 247-255.
Tyler, L.; Kapinga, C. B.; Magembe, S. R. (1981) *Disease Control Zone Project. Final Report Section II.* Dar es Salaam; Ministry of Livestock Development.

Discussion

It was clearly recognized that in order to help the small-scale livestock owner it was essential to have good data to understand his problems, the constraints affecting his operation, and his potential for improvement. However, the difficulties in obtaining this information were also obvious. Surveys must work to a system as initially there may be little understanding of a situation. The small farmer may react with suspicion to an approach for information, and it is essential that clear explanations are given to obtain true answers to questions. It was also emphasized that the relevance of the questions must be clear and, if possible, the data provided should generate help and advice quickly; in this way continued interest and motivation to assist can be assured. To this end, microcomputers are a useful tool enabling the rapid processing of data and feedback to the farmer. Although it is helpful to be able to provide facilities to aid data collection, it was also possible to devise simple recording systems capable of being used by illiterates. There were also techniques to ensure continued interest, such as comparing the best and worst producers in a situation, and by studying the best farmers to detect effective local management techniques. It was felt that many surveys collected too much data and that use of the Systems Approach could enable the key points to be recognized and studied.

The value of research was recognized, but basic problems such as how to use farm waste were not being studied. A service to digest the results of research and pass them on in an understandable form to extension workers was seen to be valuable. In many instances extension services, particularly to livestock owners, were not given adequate support and finance. Despite the lack of finance for expensive surveys, there was often duplication of data collection by different bodies which could result in irritation to the farmer.

It was pointed out that although mortality statistics alone were of little help in disease studies, other statistics (such as those on hides) could generate useful additional information such as the national offtake for sheep and goats. Much assumed information may be incorrect, and studies may reveal surprising results (for example, the high potential fertility of indigenous Zimbabwean cattle).

It was also emphasized that small-scale farmers were not all similar. Household economic studies could reveal differences and enable help to be given to the various categories in a specific way. Family cash-flow distributions could also assist in developing pricing policies. There was agreement that data were urgently needed from the small-scale farmer. To obtain them, final emphasis was given to the need to make studies simple, immediately beneficial, and to have good handling and storage facilities for the continued usage by extension and research services of the data.

Chapter 53

An Approach to the Information Needs of a Ministry of Livestock

S. CHEMA and A.D. JAMES

Livestock Development Research Division, Veterinary Research Laboratory, P.O. Kabete, Kenya.

The main users of information in a Ministry of Livestock are: (1) Development policy decision-makers who need information to make rational decisions on such questions as "Should our priority be to control diseases, or to provide credit for smallholders to buy improved animals?" (2) Extension workers who need to know whether they should be encouraging farmers to adopt new production systems, and if so what problems may arise. (3) Research workers who must be made aware of new problems emerging in the field; they must also be able to refer to the work of other workers nationally and internationally to avoid duplication of effort. (4) Administrators who require information to ensure that the resources of the ministry are optimally allocated; if additional staff and equipment are required in one area, they should be able to determine that there are no under-utilised resources in other areas before purchasing new equipment or employing new staff.

There has been a tendency for information systems to be designed from the starting-point of the data that can be collected, rather than considering the information that is actually required. The legacy of such systems is to be found in archives of files that either have no use, or which cannot be used because of data processing constraints. This leads to a decline in the quality of the data available, since field staff are less likely to be conscientious in collecting information if they know that it will simply be filed and forgotten. Unless the collectors of data receive some feedback, the quality of the data will deteriorate. Well-designed computer systems can help to improve data quality by providing timely and useful feedback to field staff. Furthermore, they make it easier to detect inconsistencies which may be due to apathy on the part of field staff or to defects in the design of the information system.

A common problem of computerised information systems is the inaccessibility of the computer to non-specialist staff. To obtain a piece of information, it is necessary to explain what is required to a computer operator who, after other higher-priority tasks, may produce nearly, but not quite the infomation requested. The lack of accessibility of computers used to be a result of their high price. However, US$ 6000 will now buy a computer equal to the demands of a Ministry of Livestock Development; the economics of information handling have clearly been revolutionized. Indeed it is now cheaper to run a computer than a car. The programs for the new generation of computers are designed for non-specialist staff. The computer asks questions about the task in simple English and waits for replies in English, explaining when the replies do not make sense to it. During such conversations, the computer is only using a fraction of 1 % of its capability, but the time used is justified by the benefits of direct access to the data.

With the assistance of donors, mainly the West German Agency for Technical Cooperation (GTZ), the Kenyan Ministry of Livestock Development has been investigating the use of computerised data management techniques to improve information systems. Four microcomputers are now being used for a number of applications. The section responsible for the computers has a staff of three, who assist and advise users. Each user is taught to operate the programs unaided, most people taking about a day to learn how to enter data and produce simple reports. More complex techniques are acquired as needed. About 20 staff have already acquired some familiarity with the machines, and about six more have learned most of the facilities offered by our present programs. The use of non-specialist staff makes the project much more reliable, as it is not entirely dependent on one or two key people for its continued operation.

The applications already developed, or under development, include:–

Administration	Staff Records
	Orders and payments control
	Current budgets
	Forward budgets
Monitoring	Vaccine and drug issues and use
	Cattle dip performance
	Diagnostic laboratory investigations
Research	Analysis of experimental and survey data
	from various sources

As routine monitoring activities develop, the database available for research and planning increases in size and quality. Technical staff are aware of the data accumulating because they manage it. This gives them more interest in its potential uses for planning, and it is to be hoped that this will result in more planning activities originating from technical staff directly concerned.

One example of an information system using new data management techniques is the dip-recording system. The Government of Kenya is responsible for about 4000 cattle dips. Dip attendants are trained and paid, acaricides are supplied, and dipwash samples are tested for acaricide concentration. Some of these functions may be taken over by the supplier of the acaricide, but the Government retains a supervisory role. There is no simple method of determining the concentration of most acaricides; sophisticated laboratory procedures are necessary. There is a laboratory at Kabete equipped for this purpose analysing about 2000 samples each month. There is a considerable information problem in the dip testing section, because after analysis of the sample, a report has to be prepared for the dip attendant to specify action to be taken to correct the strength of the dip; this requires access to background information about the dip, such as its capacity. Although the calculations are relatively simple, the consequences of errors could be serious so the work must be carefully supervised. Much more management information could be obtained from the data held by the dip-testing section, but the clerks are fully occupied in preparing dip-attendants' reports. Employing more clerks would not solve the problem, because of the physical difficulties of paper records; time is already wasted when two clerks need to use one card or ledger simultaneously.

A previous attempt to use a main-frame computer to handle dip recording failed, mainly because data entry had to take place at the computer centre using cards. Errors would not be detected until the job was processed, normally several hours and sometimes days later. The erroneous data would then have to be corrected, and the job re-submitted to the queue. This involved so much time in travelling that the system proved impractical; moreover, the costs of data entry staff, computer time, and travel were quite high. It was therefore decided to design an independent computerised dip-recording system. The advantages of using microcomputers are that non-specialist staff can enter the data themselves, using an interactive program. Errors are detected as they are entered, and pointed out to the user so that they can be corrected. Reports are printed immediately, so that they can be checked and despatched on the same day. The system maintains three data files the administrative unit's reference file, the acaricide reference file, and the dip reference file.

The administrative unit's reference file keeps a record of all the provinces, districts, divisions and locations; this is essential as reports are prepared on a geographical basis with dips in one location grouped together. When a dip is introduced to the system, the district, division and location must be specified. The system checks that their names exist; there are many variations in the spelling of place names in Kenya and this file ensures that one location is not given two names with different spellings.

The acaricide reference file maintains a list of acaricides known to the system, which again helps to avoid errors resulting from spelling mistakes, and important parameters are recorded for each. With this information, the system can calculate such values as the quantity of product needed to bring a dip to the correct strength. Upper and lower reportable concentration limits define the range of concentrations within which it is practical to produce a recommendation for the dip-attendant to correct the strength of the dip. If the strength is outside this range, then the attendant will be told to consult a supervisor. These limits are not to be confused with the acceptable range of concentrations for the acaricide. In fact, the system considers any value within 10% of the ideal acceptable as this is the accuracy range of the laboratory tests.

The dip reference file is the main data file of the system. It stores background information about each dip including the name, postal address for reports, and capacity. It also keeps a record of the last 36 dip tests for each dip. Before dip test results can be entered for a dip, the dip must have been introduced in to the system so that it has a record in the dip reference file.

When a dip test result is to be entered, the following information must be provided:–

> The date that the sample was taken.
> The acaricide in use.
> Whether the dip has been refilled since the last sample.
> If so, the date of refill.
> Quantity of acaricide added since the last sample.
> Number of cattle dipped since the last sample.
> The date of receipt of the sample at Kabete.
> The date of the test.
> The concentration recorded.

All these values are checked for inconsistency during the data entry process. One of the two GLC machines in the dip testing section has an integrator which allows it to print concentrations directly; the other is necessary to calculate the concentration from four peak heights on the plotter, and the standard concentration. The system allows the peak heights and standard concentration to be entered instead of the sample concentration, which is automatically calculated. When the data have been entered they are stored in the dip reference file, and a report for the dip attendant is printed. The basic arrangement of the report is specified by the user in a format file prepared by a word-processing system. Report formats can be tailored to particular requirements without altering the program; report variables can be arranged to suit pre-printed stationery. If it was not possible to perform the analysis, the reason is stored, and a report printed so that the attendant will know why he has not received a result.

Reports are designed so that they can be folded for posting. The first third shows the address, the second the report, and the last is a form for return with the next sample to ensure that the information required is always provided. Each month a report for supervisory staff is prepared for each district, printed in triplicate with one copy for headquarters, one for the district veterinary officer, and one to be split and the relevant portion sent to each divisional office. The details shown on the report are as follows:–

> Code and name of dip
> Current acaricide code
> Date of last refill
>
> Date ⎫
> Concentration ⎬ for the latest three recordings
> Indicator flags ⎪
> Percentage under/overstrength ⎭
>
> Other warning indicators

The indicator flags for each recording are a " + " or "–" sign to indicate whether the dip is overstrength or understrength, preceded by a "*" if the concentration is outside the reportable limits as defined in the acaricide reference file. Other warning indicators are "MO SA", which indicates the number of months since a sample was received; "MO LO", the number of months for which the dip has been consistently understrength; and "MO HI", the number of months for which the dip has been consistently overstrength. If the dip was toxic at the last recording, "TOX" is shown on the right-hand side of the listing. An "ERR" on the right-hand side means that the concentration recordings have been erratic. This may mean that the dip attendant does not understand the replenishment policy, or that he or she is selling acaricide and trying to disguise the fact that the dip is understrength by adding concentrate to the dipwash samples. These warning indicators allow divisional staff to easily identify problem dips, and district veterinary officers to see at a glance the situation area by area.

At the end of the district report, a graphical representation of the situation in the district is presented as a histogram which showing the number of dips 0–10% understrength, 11–20% understrength, etc. Only dip samples taken in the relevant month are included; numbers in the bars indicate the code numbers of the dips, and letters the acaricide used.

The system has been in use in the Kiambu district on a pilot basis for 18 months. This has provided time to overcome problems which only became apparent in practical use. The staff of the dip testing section, and those in Kiambu district, have been enthusiastic and co-operative. It has been essential to hold meetings with all staff, including dip attendants, not only to explain the purpose of the system and the use of the reports, but also to receive comments and criticisms which have resulted in improvements.

The most important problem, which has not yet been overcome, is that the return slips in the report are rarely returned with the next sample. This means that data on the refilling of dips, the number of cattle dipped, and acaricide usage are not received. This information would allow analysis of the efficiency of the dipping programme, both in terms of the percentage of the population dipped, and the quantity of acaricide used per animal dipped. The main reasons for this are that the importance of the information had not been sufficiently stressed, and that the attendants often do not receive the report before they take the next sample. The companies supplying acaricide have not been collecting the samples from attendants as they should, and difficulties with transport have led to delays in the receipt of samples at Kabete. It is hoped that both of these difficulties are now resolved. Despite these problems, the system appears to have improved dip management in Kiambu district, and it is now intended to apply it to all Government-controlled dips.

Discussion

It was pointed out that appropriate technology did not always mean simple technology, and that computer usage was clearly essential to cope with the bulk of some statistics, particularly where rapid processing and feedback was essential. Computers could also be programmed to detect errors in entry and fraudulent samples. The computer also permits the analysis of data at any time once it has been entered; something particularly valuable in following trends. In Botswana the Beef Cattle Information Service gave essential information enabling rational culling, replacement and bull evaluation to support farm management. In addition, breed comparisons gave highly valuable information for future programmes. The Nigerian farmers credit card system was another computer-based effective aid to development.

Another example of relatively sophisticated technology is the use of aerial livestock and crop surveys, which are particularly valuable as they are rapid and enable exact comparisons to be made, seasonally and often at lengthy time intervals, and with other remote sensing and ground-truck data.

Despite the advantages of the computer, it was also recognized that their usage was not great in Africa at the moment because of a lack of appreciation of their user-simplicity and trained manpower. This could be remedied by the incorporation of appropriate material in undergraduate courses, together with training in the methodology of data handling. Much data remains in Government files and these should not be neglected. There was also a need to integrate field data with research data and ensure that they reach the extension worker and the farmer. Again, it was recognized that there was overlap and duplication of effort by different bodies and agreed that the eventual aim should be a central integration of data.

Chapter 54

The Role of International Bodies in the Provision of Information for Livestock Development

M. HAILU

Library and Documentation Services, International Livestock Center for Africa (ILCA), P.O. Box 5689, Addis Ababa, Ethiopia.

Introduction

Information has a very important role in the whole process of agricultural research and development. It is only through the smooth flow of information that the results of research can be transferred to its end-users, farmers and decision makers. Moreover, if research itself is to be cost-effective and successful, it must have a continuous feedback from its end-users and must be supported by a strong information and documentation service which will provide researchers with up-to-date information in their areas of specialization. This is not the case in most tropical African countries. Low priority is often given to documentation in the planning and management of agricultural research and development, and very little money is allocated to libraries and documentation centres. Increasing prices of books and diminishing library budgets make the problem more acute.

Considerable efforts have been made by the international community to help developing countries strengthen their national information systems. Most of the international institutes supported by the Consultative Group on International Agricultural Research (CGIAR) have strong information/documentation units specializing in the commodities they are dealing with. FAO has established two international co-operative information systems, AGRIS and CARIS, where developing countries can both contribute to the inputs and benefit from the outputs.

This contribution describes the activities of international organizations, particularly the International Livestock Center for Africa (ILCA), in providing information services in the field of livestock research and development in Africa.

Information Needs for Livestock Development

Everyone involved in livestock development, policy makers, planners, researchers, extension workers, and farmers/pastoralists, need access to information that will enable them to perform their duties most effectively. The objective of any information service is to get the right information to the right person at the right time. This is not easy, mainly because of the diverse requirements of different classes of users and the unique nature of agricultural information in general (of which livestock information is only a part). Such information is unique because of its interdisciplinarity, which leads to a wide scatter of literature of potential interest, its wide range of relevant forms of document, its international character, and the need to disseminate information to a diverse community having widely varying educational backgrounds. It is therefore important to identify classes of users and develop information services accordingly.

If livestock development efforts are to be successful, they must receive appropriate support from decision makers. Inappropriate policies have had adverse effects on the performance of several livestock development projects in Africa (ILCA 1984). This may be due to either the total lack of information on policy options, or the absence of information services that put together such information in a form palatable to policy makers. Decision makers do not usually use primary sources of information. They need digests and repackaged information ready for action or implementation which requires the specialized input of consultants in various disciplines.

The livestock development planner, on the other hand, should have a wide array of information on the biological, socio-economical and technical aspects of the livestock sector in the country where he or she is working. The range of "relevant information" is wide: from aggregated "background" information concerning the place of livestock in the national economy, to detailed measurements concerning specific livestock systems (Hallam *et al.* 1983). However, there is a problem of inaccurate data on African agriculture in general, and that on livestock numbers, slaughter, production, consumption, trade and prices tends to be fragmentary, varied in source, and of limited usefulness. This has been one of the major problems facing livestock development planners in Africa.

In another context, livestock development planners in Africa have been accused of paying little attention to the technical base for their interventions, except in the case of veterinary innovations. This may be due, among other things, to the planner's lack of information on modern techniques in such areas as range management, nutrition, forage agronomy, etc. The livestock development planner should also be provided with the factual data that are invaluable in the planning process. Information is required on such aspects as the current and future demand for livestock products, the productivity of different breeds under various agroclimatic zones, the availability of feed resources, and the output of crops, to mention but a few (Hallam *et al.* 1983).

Agricultural scientists engaged in research, education, advisory services and administration require periodic information on new developments in their disciplines and, from time to time, retrospective information about specific problems. The major sources of information for scientists are scholarly journals, abstract journals, conference proceedings, and personal communication with colleagues. In Africa, where most research results do not have the opportunity of conventional publication as journal articles, "grey literature" is a valuable source of information for the scientist. This includes reports, theses, conference proceedings, official documents, consultants' reports, etc. The livestock researcher in Africa has a serious problem of access to both conventional and non-conventional literature. Because of limited funds for libraries and severe foreign exchange restrictions, most research institutes in the continent are not able to subscribe to most of the professional journals they desperately need. Neither do they easily have access to locally produced literature, due to a lack of bibliographic control in most countries. Livestock research is a diverse field involving the whole complex of animal, plant, environmental, social, and economic factors affecting production. The information requirement of the scientist in the livestock field is no less broad, and should be supported by a multidisciplinary information service.

Research work will be meaningless unless its results are communicated to the ultimate end-users, farmers and pastoralists. It is not, however, often practicable for international bodies to meet the information needs of farmers and pastoralists directly (Thorpe 1980). National extension agents should therefore play an important role in communicating the results of research to the farmer or pastoralist. Their local knowledge of ecological and social constraints is invaluable in selecting the appropriate information and presenting it in the appropriate way. Extension agents should be provided with the necessary information support, including audio-visuals which if prepared in the right language can also be used directly by farmers.

ILCA's Role in Providing Information Services

ILCA is the only international institute which has developed its information services specifically to meet the user needs in the field of livestock research and development in Africa. It provides information services through a Documentation Centre, publishing activities, and research networks it has helped establish. Services offered by the Documentation Centre include the collection and dissemination of non-conventional livestock literature on microfiche, computer searches on specific topics, and the selective dissemination of information (SDI).

Collection and dissemination of non-conventional literature

Much research has been conducted on the problems of livestock production in Africa, by Africans themselves, foreign experts, consultants, post-graduate students, and station annual reports; masses of data are also generated by survey and research work throughout the continent. However, most of this useful work is never published in the form of journal articles or books for wide circulation, and hence appears in a non-conventional form. This leads to a situation where work is duplicated, simply because officials and scientists are not aware of the results of previous work in comparable areas. ILCA, with funding from IDRC, consequently launched an exercise to microfilm non-conventional literature on animal production and health. The purpose of this exercise is to identify and retrieve "grey literature" of value from experimental stations, Government departments, educational institutions and libraries for use by livestock researchers, planners and educators. Each participating institution is provided with a copy of the complete set of microfiches collected from that country, a microfiche reader, and copies of the national bibliographic catalogues printed at ILCA. Under this project, Botswana, Ethiopia, Ghana, Kenya, Malawi, Nigeria, Sudan, Tanzania, Zambia and Zimbabwe (in anglophone Africa), and Burundi, Cameroon, Ivory Coast, Mali, Senegal, Rwanda, Upper Volta and Zaïre (in fracophone Africa) have been covered. Further countries will be covered during 1984.

Searches on specific topics

References pertinent to livestock production in Africa are collected from various sources, catalogued, indexed, abstracted, and then entered into ILCA's computerized database, on which specific literature searches are conducted on request. Requests can be sent to ILCA's Documentation Centre by mail, telex, or filling in request forms. Print-outs of the results are returned to the requestor, who can scrutinize the list and request photocopies or microfiches of titles of interest; these are provided free to African users. ILCA also has contacts with the world's three biggest agricultural databases, AGRIS (FAO), CAB (Commonwealth Agricultural Bureaux) and AGRICOLA (US National Agricultural Library); searches can also be made on these databases as required. A new title, *African Livestock Abstracts*, is planned which will have a regular and wide distribution.

Selective dissemination of information (SDI) services

The aim of the SDI service is to keep key African professionals in the livestock field abreast of new developments in their particular areas of interest by regularly scanning the world-wide agricultural literature. Every month, user "profiles" are matched against updates of CAB ABSTRACTS and AGRIS records, using ILCA's computer. The resulting list of abstracts ranges from 10–50 per month depending on the topic. These abstracts are then printed out, along with the name of the user for ease of distribution. The service started in early 1983 and has already become popular.

Publications

ILCA produces a series of publications reporting the results of research by its own staff or other African researchers, and the proceedings of major conferences and workshops, organized by the centre. The series include *Research Reports*, *Bulletins*, and the *ILCA Newsletter*. ILCA distributes its publications free of charge within Africa.

Services through networks

ILCA has taken a leading role in co-ordinating various networks in sub-Saharan Africa. Even though the primary objective of these networks is to facilitate joint research between ILCA and national institutions, they also play an important role in enhancing information exchange.

Through the *trypanotolerance* network and livestock productivity studies, ILCA has been involved in data analyses, in co-operation with national institutions or private producers, with the aim of building up production information on important livestock groups in Africa so that decisions can more easily be made when breed has been shown to be a bottleneck in a particular production system. The ARNAB network, established to deal with the utilization of crop residues and agro-industrial byproducts, and the *forage germplasm* network, initiated to increase contacts among African researchers working in the field, both run quarterly newsletters which facilitate the exchange of ideas and communication of research results among a wide family of researchers. ILCA also plans to establish an *African Livestock Policy Analysis Network* (ALPAN), to serve policy makers and individuals in official, commercial and academic circles (ILCA 1984). Through a bilingual newsletter and a series of ALPAN Discussion Papers, members of the network will be provided with information on the issues facing livestock industries in Africa. The network also plans to provide a periodic review of African livestock statistics by assembling data available elsewhere so as to make it easily digestible by policy makers and planners.

Training and conferences   ILCA's training and conference activities also play a role in facilitating communication among scientists, planners and administrators from different African countries. For example, the "Biennial Conference for Leaders of Livestock Research and Development in Tropical Africa" brings together officials responsible for livestock research and development programmes from various countries, and gives them the opportunity to establish personal contacts and exchange ideas and experiences.

The Role of other International Organizations

Other international bodies, particularly FAO, have set up active agricultural information systems providing both primary and secondary information services. AGRIS (International Information System for the Agricultural Sciences and Technology), launched in 1974, is a co-operative information system dealing with published literature, with 110 countries and 15 international institutes participating as input centres. The AGRIS database carries a substantial literature on animal production, and makes its output freely available to all participating centres in published form (AGRINDEX) and on magnetic tape. CARIS (Current Agricultural Information System), also co-ordinated by FAO, deals with information on ongoing agricultural research projects in developing countries. Even though both these systems try to cover agriculture in general on a world-wide basis, they also provide information relevant to livestock production in Africa.

FAO's periodical publications, for example the *Animal Health Yearbook, Production Yearbook, Trade Yearbook*, and *The State of Food and Agriculture Commodity Review and Outlook*, are invaluable sources of agricultural statistics for planners and researchers in Africa despite the fact that problems of data collection at the national level raise doubts as to the accuracy of the data provided.

The Commonwealth Agricultural Bureaux (CAB) provides a world-wide coverage of literature on various aspects of agriculture through its specialist abstracting units (Chapter 81). A number of CAB journals cover the livestock literature, and the extensive abstracts provided have made the service popular amongst researchers.

Conclusion

User needs in the field of livestock research and development are diverse. Due to problems such as limited budgets for information and documentation activities, national institutions in Africa are not often able to provide the necessary information services to their professionals. International bodies such as ILCA have developed services to help them do so. Full use of and active participation in these services should therefore be encouraged.

References

Hallam, D.; Gartner, J.A.; Hrabovszky, J.P. (1983) A quantitative framework for livestock development planning: Part 1 — the planning context and an overview. *Agricultural Systems* 12, 231-249.

ILCA (1984) *ILCA Livestock Policy Unit Programme Review and Protocols, 1984.* Addis Ababa; ILCA.

Thorpe, P. (1980) Agricultural information service for the Third World: Problems, developments and prospects. *IAALD Quarterly Bulletin* 25, 2-3.

Discussion

The role of international bodies such as FAO (through the AGRIS and CARIS services), CAB (through the abstracting journals), and ILCA was recognized as vital. Co-operation between these bodies seemed good and enabled ILCA to concentrate on the important task of extracting data from Government and institutional files through its microfiche service. However, it appeared that co-operation with Governments and institutions varied. It was clearly desirable for ILCA to be able to obtain the maximum information possible. For Africa to benefit to the full, it was necessary for all workers to be aware of the ILCA services and to provide feedback to ILCA in terms of publications and important unpublished material and comments. In addition, it was felt that the current *ad hoc* training given by ILCA to librarians could usefully be formalised on a more regular basis in co-operation with other bodies such as CAB.

There were some suggestions to improve the services of data dissemination by citation listing instead of keyword analysis and by including new data sources. In addition, the FAO FARMAP form data gathering service had worked well in Zimbabwe. However, it was felt that no single body could hope to cover more than 85% of the available material. It was also pointed out that although the current usage of databases by telephone was expensive, there were good prospects for cheaper and more convenient satellite systems in the future.

Chapter 55

Constraints on Livestock Development

P.R. ELLIS (Rapporteur)

Director, Veterinary Epidemiology and Economics Research Unit, University of Reading, Early Gate, Reading RG2 2AR, UK.

Introduction

Extensive consultations had exposed the need for a comprehensive discussion of processing, marketing and consumer preferences, as well as the problems faced by producers. The programme of this Symposium, which this contribution reviews, had been designed, therefore, to bring together leading authorities on different phases of the production and supply chain so that a consensus might be formed on the full range of factors that are impeding livestock development in Africa. It would have been tempting to deal, in depth, with some specific themes such as the control of major diseases or the problems of feeding animals in unpredictable climatic patterns. However, many specialised topics continue to be the subject of major conferences at other venues. This Symposium consequently aimed at evolving a balanced view of the many constraints involved and attempting to identify those which require priority attention.

The main purpose was therefore to discuss the reasons for the failure of the livestock industries of Africa to meet increasing needs for animal production. As Anteneh (Chapter 41) showed, growth in output over the past decade has been far less than in the period 1963–70. Although new technology was available to increase herd productivity most of the limited gain has come from increased numbers of animals. This implies a further threat to the environment through overgrazing. Some doubt was cast on the quality of data used, but the analytical techniques themselves represented an important step forward. Evidence from other sources, particularly aerial surveys, supported the basic findings. These sources indicated an increasingly close association between livestock and crop production. It was also noted that resources allocated to livestock development and research had been much lower than those allocated to crops and much less than the contribution of livestock to Gross Domestic Product justified.

Prices and Incentives

There was a strong consensus that a key constraint on livestock development was the underpricing of animal produce. Low prices were decided with the aim of consumer protection. The common result was a sellers market which operated to the disadvantage of the consumer. Shortages have led to clandestine marketing at exorbitant prices and to little interest in the quality and type of products. A starting point for help for the consumer could be buyer co-operatives to promote their interests. Consumer education and advertising should follow. It was the opinion of several leading authorities that price control was effective when there were no significant shortages, or when fair rationing was feasible. For exporting countries, unrealistically low exchange rates prejudice food exports and allow cheap imports to depress internal production. From these discussions it was concluded that a reversal of present problems depended on sustaining producer confidence over the long time scales involved in efforts to increase productivity. Only then will supply begin to approach real demand and real free market or tolerable controlled prices emerge. Better information was needed at all stages.

Marketing

Consideration of the effect of marketing systems (Chapters 42–45) led to the conclusion that while the trader and the middleman operating in the traditional marketing system provided a satisfactory service for pastoralists they added to the difference between producer and consumer prices. Better alternatives were described, which integrate co-operative sales groups, trekking routes with feed and water resources, mechanised transport and reliable payment systems. Similarly marketing co-operatives were proving to be successful means of organizing dairy product supplies for the benefit of both consumer and producer. In both fields control can be exercised over hygiene and quality and the producer can be stimulated to relate type and quality of the product to consumer preferences. Marketing Boards can facilitate orderly marketing and fair pricing in the same way as co-operatives but they can develop bureaucratic inefficiency and need to be reappraised at regular intervals. As single marketing channels, Boards are also subject to breakdowns and dishonesty when staff are not paid adequately, transport and storage are seldom sufficient, and payments may be so delayed that producer confidence is lost.

Here too the need for information at all stages of the chain was stressed to ascertain real costs for policy and price determination as well as for management of the chain.

Changes in Production Systems

Participants all stressed that co-operation and integration with crop production was inevitable and, in fact, highly desirable. Nomads were being required to settle by pressure on grazing land. Given secure title to land they seemed willing to do so and it was noteworthy that several countries were taking steps to provide secure land tenure. Traditional stockeepers seemed more inclined to become cultivators than cultivators were to move into animal keeping. Advantages include the better use of draught power, the provision of manure, and utilization of valuable byproducts.

Particular emphasis was placed on the need to improve the productivity of small holder's animals since a large proportion of stock was owned by smallholders in most African countries. Much can, and is, being done, particularly through co-operatives. Nevertheless, it was agreed that there were important roles for large settled herds, for balancing supplies of products, providing breeding stocks for improvement purposes, for the training of managers and for research and investigation. In fact, they are needed as key parts of stratified production systems which allow specialized units to improve overall productivity. Although breeding, rearing and fattening can be carried out in the same holdings and areas, it can be advantageous to have specialized rearing units and feedlots to take surplus animals from areas that come under pressure of land shortage or drought. Organised marketing chains can help to evolve appropriate stratification of this kind. Several contributors pointed out that the general pre-occupation with cattle was leading to a neglect of the potentials of other species that could also make valuable additions to animal production. Sheep and goats have important niches and roles, while pigs, poultry, and rabbits deserve more attention, particularly as users of crop byproducts and in smallholder farming systems.

Adjustment of Services

If development is to be accelerated, support services must be strengthened. The consensus was that management techniques should receive very high priority. This means evolving advisory services to complement clinical and disease control services already in operation. However, it was recognized that no Government can afford comprehensive services of the desired quality out of tax revenue. Large farms can meet almost all costs if the necessary specialists and material resources are accessible. Groups of smaller farmers and co-operatives representing smallholders can share some of the costs with Government, but it was felt that in these circumstances farmers must also have a share in the design and management of their services. Governments must, however, retain responsibility and authority for national problems, such as major diseases, and for overall development policy.

In studying major disease control needs, with ticks as an example (Chapter 48), the group felt that economics should be brought into planning activities to determine the utility and priority of different schemes. Economic investigation accompanying surveys of production systems also reveal the importance of less dramatic production diseases which seriously depress offtake but are currently overlooked. Applied research on systems will enable us to give new impetus to promoting herd health and productivity.

Information Systems

Each of the earlier sessions had identified important deficiencies in information. It was clear that new initiatives in data gathering and information management are needed to help the small-holder, the manager of larger and specialized units, for the disease control planner, herd adviser, research worker and, of course, Government regulators and policy-makers.

Several successful initiatives were reported. Chema & James (Chapter 53) described a microcomputer system now being used to process a wide variety of livestock survey data and to monitor veterinary schemes. It also contributes enormously to the management of research data and to administrative tasks such as budget and stock control. Representatives of ILCA (Chapter 54) reported on the extensive investigations they have been fostering, conducting and analysing by computers. In fact it was clear that most participants were involved in or aware of the need for such information generating activities.

In the wider field of reference materials, important progress is being made. ILCA is developing a documentation network within Africa while CAB and FAO are bringing in such information from world-wide sources. Communications are still a major constraint on the accessibility to these services but advances in electronic technology should soon improve the availability and use of these valuable materials.

Conclusions

We conclude that the greatest need is for the creation of producer confidence to overcome shortages. Realistic pricing and exchange rates, organization of marketing chains, improvement and re-orientation of services with producer support and participation, and strengthening of information systems, are the essential ingredients.

Chapter 56

Climate, Weather and Plant Production in Sub-Saharan Africa: Principles and Contrasts with Temperate Regions

F.J. WANG'ATI

National Council for Science and Technology, P.O. Box 30623, Nairobi, Kenya.

Introduction

As the developing countries of Africa approach the third decade of their political independence, the challenge of producing enough food to sustain their growing populations is increasing in magnitude and, apart from energy, is posing the most severe developmental crisis in virtually all countries of the region. A multitude of factors are responsible for this situation but only those related to climate are discussed in this contribution.

Climate determines what crops the farmer can grow. Weather influences the annual or seasonal yield and, as we have seen repeatedly in tropical Africa, how much food there is to eat. For most crops the relation between weather and yield is extremely complex but three main aspects can be distinguished. First, individual elements such as rainfall, light and temperatures have a direct effect on physiological processes such as photosynthesis, leaf expansion and plant growth and development in general. Second, weather controls the spread of fungal diseases, insects, pests and weeds which can affect crop growth. Third, weather governs the farmers' day to day programme by setting variable limits to field operations.

Since climate is the summation of weather over time, the yield fluctuations commonly experienced in a given climate are a consequence of all three sources. It is therefore unlikely that crop yield variations can be linked exclusively to one climatic parameter although one or two such parameters may have a dominant influence. It is significant that in almost all situations of food shortages in Africa agricultural drought is considered the principal cause for deficiency in agricultural production. The following sections of this paper therefore explore some of the climatic principles and factors which may be contributing to inadequate food production and include some suggestions on remedies which could be applied to help the African farmer cope better with the effects of climatic uncertainties.

The Climate in Sub-Saharan Africa

Temperatures

In contrast to the temperate countries, low temperatures are not a serious problem in agriculture except in a few situations where crop production has been extended to high altitudes. The base temperature for tropical crops is, however around 10–20°C and low soil temperatures at the beginning of the season may adversely affect young seedlings and final yield. Maturity of cereal crops such as maize and sorghum is also delayed and incidence of night frosts may result in crop losses. The majority of agricultural enterprises are, however, carried out in the humid, sub-humid and dry tropics and sub-tropics where ambient temperatures are generally favourable to plant growth. Temperatures above the optimum levels of 32–37°C do however occur but the duration and effects of such high temperatures on crop growth in the tropics is not well documented. Incidentally, high temperatures and humidity also favour the maintenance of high insect pest populations and multiplication of plant pathogens.

Rainfall

The annual movement of the Inter-tropical Convergence Zone (ITCZ) creates two rain seasons separated by dry spells near the equator and one rain season followed by prolonged drought in the sub-tropical zones (Sirakumar *et al.* 1983).

In contrast to the temperate environment where winter rain and snow accumulates in the soil and is supplemented in most cases by frequent showers throughout the growing season, the tropical rainfall is characterised by heavy rain storms of short duration followed by prolonged dry warm periods. High agricultural potential is therefore dependent on the availability of deep friable soils, which can store large quantities of water, as well as crops with a good rooting system capable of extracting moisture from the entire soil profile. Water availability is therefore the most important climatic factor in agricultural production in most tropical and sub-tropical countries of Africa.

Evaporation

Potential evaporation in the tropics is high, reaching about 2200 mm a year; this compares with 400–500 mm in temperate countries. Potential evaporation in most areas also exceeds precipitation except during the height of the rainfall season. Evaporation therefore reduces drastically the effectiveness of rainfall and has a profound influence on crop production potential as will be seen later.

Interaction between Climate, Soil and Crops

Evapotranspiration

Plant growth and production is dependent on two important processes, both of which involve interactions between weather factors and the plant. The first essential process for growth is transpiration, that is the movement of water from the soil through the plant and into the atmosphere. This process is essential for cooling the plant in the warm tropical environment, and supplies the plant with water needed for biochemical metabolism within the plant. Evapotranspiration also provides a mechanism for movement of plant nutrients from the soil into all parts of the plant and enables the plant to maintain turgidity and hence open stomata for photosynthesis. The scientific principles involved in the evapotranspiration process are well understood and the rate of evapotranspiration can be reliably estimated through field measurements of vapour pressure gradients above the crop, soil water balance, lysimeters or derived from energy balance, temperatures, windspeed and humidity in the crop environment. All these methods are described in literature and it is not necessary to repeat them here.

It is however, important to stress that although every farmer in the tropics is aware of rainfall as an overiding climatic factor in agricultural production, the real significance of evapotranspiration and its influence on effectiveness of rainfall has not been widely recognized. Even in the highly developed temperate countries the significance of evapotranspiration was apparently not recognized until the begining of this century and serious efforts to quantify evapotranspiration have only been made in the past four decades.

Evapotranspiration from a field crop is closely related to potential evaporation and can similarly reach very high levels in the tropics depending on soil water availability. The top layers of the soil profile which are normally the most fertile therefore dry up quickly and crop growth and yields suffer even before the entire soil profile is exhausted. Rainfall alone is therefore not an adequate guide to agricultural potential and it is necessary to work out soil water balance and select crops which will grow and mature in the short periods of favourable soil water regime. Under these conditions, timeliness of agricultural operations becomes critical for successful crop production. The definition of timeliness, however, requires detailed information on the reactions of various crops to various levels of water characteristics and ability to predict early in the season whether the rainfall season will be above average, average or below average.

Photosynthesis

From a climatological point of view, the relevant parameters in photosynthesis are light intensity and duration, ambient temperatures and carbon dioxide concentrations. Dry matter production is, however, the balance between gross photosynthesis and the fraction metabolised in the process of respiration. The latter fraction increases with ambient temperatures and values close to 50 % can be expected in the tropics. Solar radiation and the associated light energy which is a principal factor in photosynthesis deserve mention. Although generally the tropics have higher intensities of solar radiation than temperate countries, the longer daylengths in the summer cropping season in temperate countries result in higher yields per unit area per season than can be obtained in the tropics. Tropical countries, however, have the advantage that provided water is available, the land can support crop production all year round and could thus offset the yield differential.

One important principle, often not sufficiently emphasized in tropical crop production, is the advantage of having a high plant population in utilising the water and solar radiation particularly in the early stages of growth of the crop where water is not limiting. Leaf Area Index is a principal determinant of light interception in crop canopies (Monteith 1965), and hence the rate of dry matter production. This principle was well utilized in traditional agriculture in the tropics where cereal crops were usually interplanted with short term legume and root crops so ensuring that both available light and water were utilized to the optimum. With the replacement of this system by widely spaced monocrops, this advantage is often lost and production even further reduced when the sole crop fails to germinate evenly. It is consequently not surprising that inter-cropping has recently gained acceptance by agronomists in the tropics; although competition between intercropped species can depress yields, in favourable situations careful choice of the crops to be intercropped can result in higher levels of production per unit area of land (Prasad 1983, Marimi 1983). Several models of light interception and photosynthesis in field crops have been developed (de Wit 1966, Monteith 1970) and have been shown to produce realistic estimates of crop production in the tropics provided the various agro-climatological and crop structural parameters are quantified (Wang'ati 1970). More effort is needed to measure light interception characteristics of canopies of various intercrop systems. The principal components of solar radiation should also be recorded regularly throughout the region, where cloud cover and dust storms are known to influence the intensity and spectral composition of radiation reaching the ground.

Contrast with Temperate Climates

Although scientists from temperate regions have made a contribution in the development of basic scientific principles regarding crop – weather interractions, there is limited scope for transfer of this experience to the tropics. The basic problem is that the agroclimatic systems in the temperate countries are different in many respects from those prevailing in tropical and sub-tropical Africa; experience gained in temperate environments must therefore be evaluated through local experiments before it can be applied. An example of this situation is the climate–soil interaction. In the temperate environment, the cropping season starts in spring with the soil cold and thoroughly wetted by winter rain. The potential evapotranspiration is also low as a result of low temperatures and shorter daylength. The cropping season is further preceded by a long period during which tillage operations can be carried out comfortably on relatively soft ground. The cold winters also suppress weed germination and the newly planted crops have little initial competition with weeds. The season then gets progressively warmer and drier and there is therefore little loss of nutrients from top layers of the soil through leaching. This situation may be contrasted with the cropping season in the tropical and sub-tropical regions.

In the tropical environment the cropping season starts with the soil hot, dry and the surface is often hard to till. The rain then arrives in heavy storms usually far in excess of the infiltration capacity of the soil. The effect is often severe soil erosion, capping and rapid evaporation. Build up of soil moisture is slow due to run off and evaporation unless special measures such as tie-ridging are taken (Dagg & MaCartney 1968). The initial warm soil environment also primes the weeds for rapid germination. The farmer in the tropics is therefore faced with serious problems of a multitude of farming operations – land preparations, ridging, planting, weeding, side-dressing of crops with fertilizers; all of which require labour and other resources before the crop is well established. Considerable leaching of nutrients also takes place and the farmer is often unable to utilise the flush of nitrogen which occurs through mineralization of organic nitrogen when the warm dry soil is quickly wetted by rain. The situation is even more acute in the drier tropical regions where the rainfall season lasts for only two to three months. In these circumstances, soil and water conservation are not only essential but vital for any crop production and opportunity for cropping large acreages is limited without appropriate technology and energy for rapid land preparation, planting and weeding.

Role of Agro-climatic Data in Agricultural Planning

From the above, it will be evident that although the experience gained in temperate agriculture may not be directly applicable in the tropical environment, the principles of agroclimatic modelling have much to offer if the parameters are quantified *in situ*. Three basic strategies could be adopted in the application of such models to agricultural planning. First, is the overall planning of what crops should be grown in which areas. This can be greatly facilitated by a systematic delineation of the agroclimatic zones in each country and the use of known crop requirements to match the crops to the zones offering optimum production potential. Several efforts have already been made to establish the ecological or agroclimatic zones in the tropical African region (Brown & Cochéme 1973, Woodhead 1970, Sombroek *et al.* 1982). Unfortunately, it has often been necessary in such maps to interpolate climatological and soil parameters due to scarcity of actual measurements. The resulting maps consequently present difficulties when it comes to detailed planning. These difficulties can, however, be overcome through intensification of observation networks, especially in the mountainous regions where local climatic variations can be substantial and in the semi-arid areas where the coverage of observation stations is poor due to difficulty of access. Technology in data telemetry is now highly developed and could be introduced in some situations with economic advantage. Higgins *et al.* (1980) describe the efforts being made to utilize agroclimatic information in land use planning and the potential value of such maps in increasing agricultural production*. It is, however, rather unfortunate to note that even with good agroclimatic maps, it is unlikely that crop varieties are readily available to match climatological requirements for optimum production in a given area.

The second strategy is therefore to encourage close interaction between plant breeders, agroclimatologists and agronomists so that the required characteristics are incorporated in breeding programmes, usually very long term processes.

The third strategy is the recognition of the basic concept of response farming which predominates in subsistence agriculture. This is simply a systematic evaluation of the agroclimatic parameters which goes on intuitively in every farmer's decision-making process, but this time introducing into this process the fundamental physical and biological principles and models which have been shown through research to stabilize or increase crop yields through better assessment and prediction of seasonal opportunities. Attention is drawn to recent findings in Kenya (Stewart 1980) and the possible impact of application of these results in the process of response farming.

*See also Chapter 8.

The highly erratic and short duration rainfall in the semi-arid areas of Kenya has led farmers to accept a high frequency of total crop failure. For similar reasons, response to inorganic fertilizers in semi-arid areas is considered normally uneconomic and such inputs have therefore been little used in the past.

Table 1   Farm yields of maize in a good rainfall season, comparing results of following normal practices in the area to practices adjusted in response to actual rainfall. Machakos District, short rains, 1982.

Farmer	Farmers' Normal Practice[1]		Practices Responding[2] to Rainfall Events		Increased Nitrogen	Yield due to Fertilization (kg grain	Nitrogen[3] Efficiency
	Population $(000ha^{-1})$	Yield $(kg\ ha^{-1})$	$(000\ ha^{-1})$	$(kg\ ha^{-1})$	$(kg\ ha^{-1})$	$kg\ N^{-1})$	(%)
A	30.0	443	30.0	2843	2400	40.0	91
B	31.7	676	46.0	1216	540	9.0	21
C	39.3	887	46.0	2208	1321	22.0	50
D	31.0	1155	41.0	3512	2357	39.3	90
E	38.3	1165	70.7	2139	974	16.2	37
F	37.7	1429	24.0	2703	1274	21.2	48
G	34.7	2434	28.7	3122	688	11.5	26
Mean	34.7	1170	40.9	2535	1365	22.75	52

[1]Normal practice is neither to use commercial fertilizer, nor to thin the stand if rains are weak. [2]When early rains are strong, the recommended response is to leave plant populations high and boost N to 60 kg ha^{-1}. [3]100% efficiency is assumed to result in an increase of 43.75 kg ha^{-1} grain at 15.5% H_2O content per kg N. Reproduced from Stewart & Kashasha (1983).

Recent analysis of effective rainfall covering a large part of the semi-arid areas in the Eastern Province of Kenya has, however, revealed that the rainfall seasons are not necessarily haphazard but could be put into three specific classes, good, medium and poor. It has also been shown that the date of onset of the rainfall season, as defined for example for maize by a total of 40 mm in a period not exceeding 10 days with no dry period exceeding one day at a time, is correlated with total seasonal rainfall expectation and with intensities of early season rains (Stewart 1980, Stewart & Hash 1982). Using such analyses, it is possible to draw up guidelines to help farmers decide early in the season the type of season likely to result and hence the type of crop, plant population and fertilizer inputs required to optimise yield. Table 1 presents findings of a comparative study on response farming carried out with maize in a low rainfall area in Kenya (Stewart & Kashasha 1983). These yields were obtained on farmers' plots using traditional technology and demonstrate what can be achieved even in dryland areas through the application of response farming based on the water production function of a given crop and the effective rainfall.

Unfortunately, few countries have managed to establish reliable and regular programmes of agroclimatic data collection, and the study of agroclimatic patterns has therefore been difficult if not impossible in many agricultural areas. Where such data exist, experience shows that considerable progress can be made not only in matching crops to the agroclimatic situation (Dagg 1965) but also in optimising yields of existing crop varieties and agricultural inputs. It is also necessary to ensure that the extension staff are not only exposed to the use of agro-climatological data but that they are convinced that the pains taken in recording such data are worth the possible gains in agricultural production.

Conclusion

Climate is a major factor in agricultural production in tropical Africa where, except in the very humid regions, water availability is the principal determinant of both growing season and yields. Success in increasing agricultural production will therefore be strongly dependent on the extent to which agroclimatic factors are quantified and the information used to optimise selection and management of crops. This requires a determined effort to improve recording and systematic analysis of agro-climatic parameters, validation of agro-climatic models in the local environment and the use of these models to develop detailed information on the agroclimatic potential of each region. It must, however, be recognized that collection and analysis of data can be an expensive undertaking. There is therfore a continuing need for agro-climatologists to work closely with agronomists, plant breeders and other agricultural scientists in order to determine the critical parameters to be recorded and to develop simplified agro-climatic models which avoid collection of non-essential data.

Finally the application of agro-climatic knowledge to agriculture requires a joint effort between agroclimatologists and extension staff. The latter are not usually familiar with these techniques and effort is required to orientate the extension services to record and use such information.

References

Brown, L.H.; Cochéme, J. (1983) *A study of Agroclimatology of the Highlands of Eastern Africa.* [Technical Note no. 125 (WMO 239).] Geneva; World Meteorological Organization.

Dagg, M. (1965) A rational approach to the selection of crops for areas of marginal rainfall in East Africa. *East African Agriculture and Forestry Journal* 30, 295-300.

Dagg, M.; MaCartney, J.C. (1968) The agronomic efficiency of the NIAE mechanical tie ridge system of cultivation. *Experimental Agriculture* 4, 279-294.

Higgins, G.M.; Kassam, A.H.; Shah, M.M. (1980) *Report on the second FAO/UNFPA Expert Consultation on Land Resources for Populations of the Future.* Rome; FAO.

Marimi, A.M. (1983) Increased crop production through inter-cropping – Kenyan experience. In *More Food from Better Technology* (Holmes, J.C.; Tahir, W.M., eds), 323-330. Rome; FAO.

Monteith (1965) Light and crop production. *Field Crops Abstracts* 18, 213-219.

Prasad, R. (1983) Increased crop production through intensive cropping systems – Indian experience. In *More Food from Better Technology* (Holmes, J.C.; Tahir, W.M., eds), 323-330. Rome; FAO.

Sirakumar, M.V.K.; Virmani, S.M.; Reddy, S.J. (1983) *Rainfall Climatology of West Africa: Niger.* [Information Bulletin no. 5.] Hyderabad; ICRISAT.

Sombroek, W.G.; Brown, H.M.H.; Pouw, B.J.A. van der (1982) *Exploratory Soil Map and Agro-climatic Zone Map of Kenya, 1980. Scale 1:1 000 000.*

Stewart, J.I. (1980) Effective rainfall analysis to guide farm practices and predict yields. *Proceedings of the 4th Annual General Meeting of the Soil Science Society of East Africa.*

Stewart, J.I.; Hash, T. (1982) Impact of weather analysis on agricultural production and planning decisions for the semi-arid areas of Kenya. *Journal of Applied Meteorology* 21, 477-494.

Stewart, J.I.; Kashasha, D.A.R. (1983) *Rainfall Criteria to enable Response Farming through crop-based climate analysis.* [Cyclostyled Report USDA/USAID/GOK. Dryland Cropping Systems Research project (615-0180).] Muguga; Kenya Agricultural Research Institute.

Wang'ati, F.J. (1970) *A study of water use efficiency in field crops of maize and beans in East Africa.* PhD thesis, University of East Africa.

Wit, C.T. de (1965) *Photosynthesis of Leaf Canopies.* [Agricultural Research Reports no. 663.] Wageningen; Centre for Agricultural Publications and Documentation.

Woodhead, T. (1970) A classification of East African rangelands, II. The water balance as a guide to site potential. *Journal of Applied Ecology* 7, 647-652.

Chapter 57

The Zonal Climates of Africa and Biological Resources (and the collection and analysis of primary data in the AGRHYMET Programme)

D. RIJKS

World Meteorological Organization, 16 rue Gautier, 1201 Geneva, Switzerland.

Introduction

Agrometeorology can contribute to the advancement of agricultural production in Africa by enabling man to match, as realistically as possible, the requirements for solar radiation, optimal temperatures, water, humidity and wind of different crops, varieties and animal species to the characteristics of the climates of different regions of the continent. It can help man to increase his efficiency in the planning of agriculture and in day to day activities, and in particular in the use of energy in agriculture, be this in the form of his labour, mechanical energy, fertilizer, pesticides or heat for drying.

Agrometeorology can forewarn and so minimize the destruction of the environment, especially when it is fragile, and reduce pollution. It provides methods to monitor crop and animal production to facilitate nationwide marketing, transport and food-strategy planning.

The application of agrometeorological techniques will inform farmers to plan the agricultural season better and implement day to day operations. It will help the country to plan land use and agricultural systems, to monitor agricultural production and to take remedial action when production falls below expectation.

Agrometeorology and Levels of Food Production

There exists a considerable gap between actual average agricultural production in farmer fields and results obtained under generally favourable conditions such as at agricultural experiment stations. According to ICRISAT, average sorghum, millet and chickpea yields are 810, 590 and 510 kg ha^{-1} respectively, while yields at experimental stations are about 4900, 3050 and 7500 kg ha^{-1}. Theoretical calculations of potential yield of these crops are even higher. Average yield increases in Africa in the past 20–40 years have only been 3–10 kg ha^{-1}y^{-1}, depending upon the region. Figures exist for other crops, showing an essentially parallel pattern. These figures indicate that biologically and technically significant increases can be obtained if the necessary agricultural practices can be adopted. Data on yields in Europe and the USA, show a remarkably similar trend in time, but with a shift in phase of about thirty years. Initially wheat yield increases in these regions were 5–10 kg ha^{-1} y^{-1}, followed by a fairly sudden change to 50 kg ha^{-1} y^{-1}. This change coincides with the beginning of a period of increased use of energy in agriculture. Agrometeorological knowledge may contribute to a more economic use of inputs necessary to increase production.

De Wit *et al.* (1983) introduced the concept of potential production to indicate and quantify the role of optimal and non-optimal weather parameters and input factors such as fertilizer, additional water and crop protection measures in each stage of development of a crop. Principal climatic factors, that can not be influenced are solar radiation, temperature, rainfall and potential evaporation. Input factors that can be influenced are the water, fertilizer and pest control regimes.

Potential production can be calculated with fair precision (! 25% or better) from climatic data and fairly simple plant and crop data. An analysis of changes in the level of the input data can indicate measures to reduce the gap between potential and actual production, including the introduction of other varieties of a crop. A similar approach can be made for each of the different operations that a traditional farmer usually makes on a rain-grown crop (Fig. 1). The aim of this exercise is not only to calculate production under ideal conditions, but to indicate the loss of production in different farming operations, unless these are made to the extent required at the right moment and within specific time limits. Agrometeorological information can help define the need for, and the efficiency of, energy inputs necessary to complete these operations in time.

Fig. 1   Potential and actual production and the effect of agricultural operations. A, If water, fertilizer and pest regimes are optimal and operations are implemented in time. B, If water, fertilizer and pest regimes are non-optimal and operations not implemented in time. .

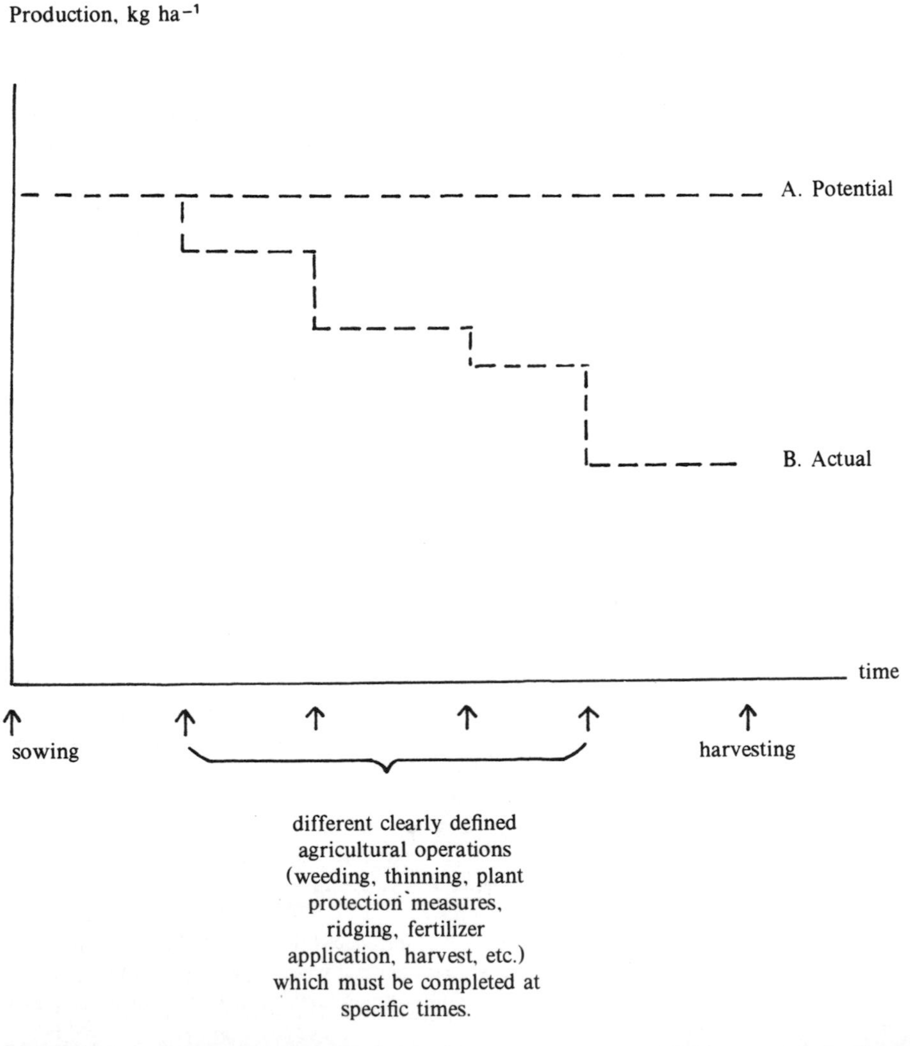

Production, kg ha^{-1}

A. Potential

B. Actual

time

sowing   harvesting

different clearly defined
agricultural operations
(weeding, thinning, plant
protection measures,
ridging, fertilizer
application, harvest, etc.)
which must be completed at
specific times.

Climate and weather

Although agrometeorologists may distinguish between climate and weather information, a farmer makes many decisions by combined use of the two:

Climate	Weather
choice of farming system, land layout	timing, extent of land preparation
choice of crops	date of planting
choicc of variety	choice of alternative variety
choice of farm equipment	actual use of equipment
choice of row width	within row distance of planting points
choice of irrigation method	timing and amount of water given
choice of pest control system	timing and extent of control

Temperature and Solar Radiation

In most of Africa temperature does not limit development and growth of crops in the rainy season. It may be a limiting factor for growth by being too low at higher altitudes or by being either too low or too high in the dry season at some distance from the equator ($>14°$ latitude). The wealth of observations of temperature in Africa is only partly exploited. The probability of high or low temperatures exceeding a certain value in areas where growth or development of crops might be limited because of the temperature can easily be given provided the agronomist specifies the threshold values and/or periods to be taken into account.

Potential photosynthesis, determined by solar radiation, is often higher in most of the dry season, contributing to high yields of many crops grown under irrigation and some flood retreat crops. However, the introduction of irrigated agriculture, and in particular the acquisition by farmers of effective water management techniques, is not considered as a large scale solution to advance agricultural production in Africa within the time span of less than one generation.

Potential production of crops grown during a rainy season is in several regions limited by insufficient solar radiation. In regions with a long rainy season it may be possible to shift planting dates and practice multiple cropping to reduce this effect.

Rainfall

In many areas the length of the effective rainy season limits the growth of crops. However, a better indication of water availability is given by the water balance regime, which includes water supply and requirements of crops and the possible soil water storage. In many areas near the equator ($3°S - 5°N$) and between $10-15°N$ or S (Fig. 2) the rainy season is very short (approx. 90 days) and even if suitable soil water storage is possible, the growing season rarely exceeds 105 days. The growing seasons start on an empty soil water profile. A common way to quantitatively describe the rainy season is the use of diagrams of the probability of receipt of specified ten-day rainfall totals at any time during the season (FAO 1976). Probability diagrams can be prepared for daily, ten-day, monthly and annual totals, but ten-day totals seem most adapted to use in agriculture, because of the "inertia" in soil and crop reactions to changes in water supply. From these diagrams can be derived the probabilities of the onset of the agricultural season. Using these diagrams only a very approximate estimate of the probable length of season can be made, because the statistical populations in successive ten-day periods are treated as independent. A water balance technique or a series of actual data on length of season permits more reliable estimates of season length to be obtained. There are also techniques that calculate the probability of within season dry periods of different duration and at different times in the season.

Atmospheric Humidity

Apart from its importance in the calculation of potential evaporation, high atmospheric humidity is a major factor in the development of many crop diseases and some crop pests. Low atmospheric humidity may inhibit fertilization during flowering.

Fig. 2   Schematic presentation of potential evaporation, rainfall regimes and lengths of season near the equator (a), at 6-10°N or S (b), and at 10-15°N or S (c).

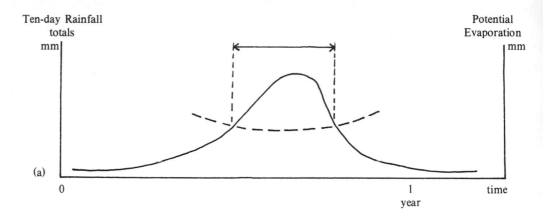

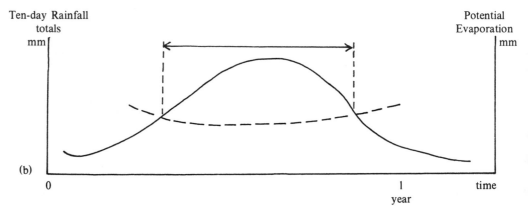

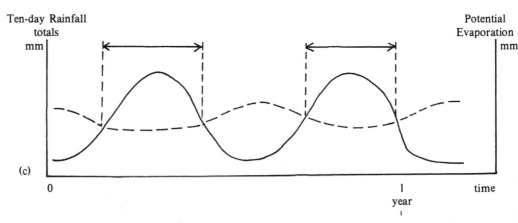

Wind Speed and Direction   Information on past, recent and forecasted wind regimes is important in farming operations such as ridging to prevent lodging, the timing of irrigation, and plant protection measures. Wind affects the flowering and fertilization of some crops and the development and movement of pests and diseases. Wind data can be used to orientate wind breaks efficiently and to warn of the danger of bush and forest fires. Wind data are available for many stations in Africa on a daily or three-hourly basis.

Potential Evaporation

Potential evaporation can be calculated with a precision that matches its spatial variation from data on solar radiation or sunshine, temperature, wind and humidity. The empirical constants used in the formula require verification, to make values reliable. In most cases the accuracy of the estimates is within the limits required for its application to agriculture.

Biological Resources

Most crops grown in Africa possess inherent genetic variability, which allows them to adapt to different climatic regions. In their phenology this may be shown by variations in crop architecture, rates of leaf area development, depths of rooting and responses to soil physical conditions, and requirements for and responses to solar radiation and water balance conditions. These characteristics permit the introduction of a multitude of inter-cropping and multiple cropping systems, that eventually should lead to optimal exploitation of land and energy resources. Different varieties have different lengths of season and adapt more or less effectively to dry periods within the season and to different temperature regimes. They may also have different responses to photoperiods, allowing them to flower at different times.

There is less quantitative information on responses of varieties to weather than many other factors and therefore on the magnitude of the loss of potential production in each stage of development. The formulation of such information to a large extent falls on national and international agricultural research institutes and needs to be actively persued.

Information for individual crops exists, but there is very little available for cropping systems as a whole, this also applies to forests and forest systems. Production-weather relations of natural pasture have lately received considerable attention, and a number of quantitative relationships have been established.

Relationships between animal production and the ambient temperature regime have been studied and the results have been applied in ways as varied as the provision of shelter and in animal breeding programmes.

Matching Climate, Weather and Biological Resources

Climate knowledge is used to formulate the agroclimatology of regions and to define farming systems, cropping systems, crop calendars, intercropping and crop associations, average number of work days and many other planning inputs. Weather data are used to inform farmers and others about the agricultural consequences of recent weather, such as the actual water balance of crops, the development and threat of outbreaks of pests and diseases, the need for, and probable efficiency of, various farm operations.

The practical value of this information to the farmer can often be increased if it is coupled to short-term weather forecasts (12-24-36 hours). Such information can help a farmer plan the correct time and extent of application of input of different forms of energy (manual labour, fertilizer, pesticide, fossil energy for mechanical operations) and to use this energy more efficiently.

Pre-sowing land preparation

In a region with a bimodal rainfall pattern, and so relatively short rainy seasons, sowing is sometimes delayed because land preparation can not take place until the soil has been sufficiently wetted. Therefore, longer season higher yielding varieties can not be introduced with a reasonable chance to complete their cycle. Climate information may inform planners about the probability of completing all or part of the land preparation at the end of the preceding wet season, and about the necessary outlays in the farming system, weather information will help the farmer to decide on the actual date of the land preparation.

Weeding

The timing of the first weeding is important, especially for crops grown on an initially empty soil water profile, when young seedlings are competing with weeds for water and nutrients in the top layers of the soil. Climate information can help a farmer plan the size of a holding that can be weeded with a certain probability with existing equipment aids, and the type of equipment needed, while weather information will help him to decide on the timing of the operation and the amount of outside help needed.

Thinning

Most farmers sow their crops with a very much denser stand than is retained later in the crop cycle. In the case of row crops, row width is usually chosen (wittingly or not) as a function of climate, recent weather and forecasts (whether by traditional methods or from a meteorological service) are taken into account when a farmer thins his crop to the desired distance or density within rows. Weather information may also help him to decide whether he will thin to the desired density in one or two operations.

Fertilizer application

Information about the water balance of the crop and the statistical probability or actual forecasts for the water regime expected over the next few days can help a farmer decide whether to apply fertilizer or wait, or apply in more than one operation, depending on the risk of leaching in wet soils or lack of uptake in dry conditions.

Supplementary irrigation

Supplementary irrigation systems often obtain water at a relatively higher cost than continuous irrigation systems. Climatic information will help planners decide on the need and the probability of profits from irrigation and on the system of providing water. Weather information can help a farmer decide when and how much water to give. Especially when water is pumped from water courses that lie below land level, the information may avoid unnecessary spending or assure spending to make the operation worthwhile.

Plant protection

Weather has a significant effect on the development of pests and diseases throughout the crop cycle. Recent weather and weather outlook may indicate the need for seed treatment, for protection during the growth of the crop and dictate measures of drying and storage. Good warnings require a certain knowledge of the physiology of the parasites. Current weather also affects the movements of migrant parasites.

Information on recent weather (sometimes coupled with forecasts) can help a farmer decide whether to remove sources of further infection and if so, before which date. Climate information can be used to choose a system and equipment for plant protection with mechanical or chemical intervention and recent weather and forecasts can be an aid in deciding whether and when and how much to treat. In several countries these techniques have been so well developed that farmers have increased production and reduced the costs of treatment significantly, to the extent that they are willing to pay to participate in the warning operations. Moreover, the techniques result in a reduction in pollution of soil, water and air. Wind, humidity, sunshine and to a lesser degree temperature, can also influence the efficiency of crop protection measures.

Ridging of row crops to prevent lodging

Some grain crops, such as maize, become top-heavy towards the end of the season when ears fill. If stems lodge during this stage, significant losses may occur. On the basis of recent weather (notably relating to the water balance) and wind forecasts for 24–36 hours, farmers may be informed about the danger of lodging and the need to rebuild ridges.

Harvest

Some crops are left on the stem to dry after harvesting, others are collected and dried away from the field. Climatic information may help a planner to select a sowing date or variety (other factors being equal), to select a suitable time for harvesting, or assess the need for outside sources of thermal energy. Recent weather information and forecasts may help a farmer decide on the actual practice to follow, in any one year, to avoid incomplete drying and subsequent losses to pests and diseases in the quality of his product.

Application of Weather Information

Erosive capacity of rainfall

Techniques exists to assess the erosive capacity of rainfall, using records of rainfall total, intensity and duration. This information, together with data on slope, land and soil management, are a tool in devising proper erosion control measures. The information can also be used in the choice of crop associations, farm layout and agricultural machinery.

Crop water balance

Crop water balance is calculated from water supply (rainfall and/or irrigation), potential evaporation, and a factor representing the stage of crop development. It takes into account the depth of rooting and the water holding capacity in the root zone. The formula employs empirical crop coefficients that can be obtained from experiments in the locality, or from the literature, with sufficient accuracy to be valid for practical use in day to day operations. The concept is employed in studies concerning raingrown, irrigated and flood-retreat crops and natural pasture, in the planning of the land layout, the farming systems, cropping systems, crop rotations, the agricultural calendar, the calculation of irrigation water requirements (optimizing profits from available water) and in the choice (as a function of the length of the available growing season) of crops and varieties. It can contribute to avoiding salinization of irrigated soils. It is used to monitor the water regime of crops and to monitor regional and national crop yields. Other uses include the assessment of dry matter production of natural pasture and of its capacity to carry grazing animals; early assessment of potential grazing land can permit a Government to influence the marketing of cattle and reduce pressure on grazing lands at the end of the dry season.

Biomass assessment

Weather information, often combined with remote sensing information, is used to assess biomass development in more or less remote and fairly homogeneous areas. This knowledge is applied to discern areas of potential development of cricket swarms, reducing the need for ground surveys and the assessment of potential grazing land in semi-arid zones with nomadic populations.

Agroclimatological mapping

The mapping of zones with distinct agroclimatological potentials (FAO 1978) uses well developed techniques and can aid long term planning.

Animal health

Research on the relation between environment and animal performance has led to practical operational techniques on weather and animal diseases and housing (World Meteorological Organization 1970, 1972, 1978).

Crop weather models

Recently, some crop weather modelling techniques have become operational (World Meteorological Organization 1983) but most are specific to the temperate zones. Techniques have been employed with apparent satisfaction in semi-arid (Penning de Vries & Djiteye 1982) and humid (de Wit *et al.* 1983) zones using standard weather data and provide an estimate of potential crop production. The relationships employed in the models are somewhat site specific and empirical constants will need verification when the method is applied to areas other than those for which it was developed.

The AGRHYMET Programme

This programme aims to contribute to food self-sufficiency of the Sahel countries through the provision of agrometeorological and hydrological information to farmers and Government officials. The Programme is composed of national projects in the eight members states of CILSS and a Regional Centre in Niamey. It aims to issue agrometeorological studies for planning and information and warnings for day-to-day operations, including daily and ten-day bulletins on the state of the weather and the crops issued by national meteorological services using data and information gathered in their countries and the region. The latter is relayed via the Regional Centre.

A daily bulletin may include: an analysis of the regional meteorological situation (based on the analyses received from the AGRHYMET Centre at Niamey), a meteorological analysis within the country, relying on the national forecast centre, a list of the amount of precipitation during the past 24 hours at selected stations, a list of maximum and minimum shade temperatures on the preceding day, also for representative stations, agrobiological information concerning the state of crops and a weather forecast for the next 24 hours (based on the forecasts issued by the national forecast centre).

It is also proposed that, during the rainy season, the various statistical probabilities of there being 10, 20 or 30 mm of rain in the ensuing 24/48 hours/five days should be included.

The ten-day bulletin, disseminated to responsible staff at the various levels of agricultural and rural development services, may include: the amount of rainfall during the past 10-day period at the main synoptic, agrometeorological and climatological stations and at a selection of rainfall stations, the average maximum and minimum shade temperatures during the past 10-day period, at the same stations, the average grass minimum temperatures during the past 10-day period at the same stations, the average daily duration of sunshine, the average daily global radiation, the saturation deficit, the soil water storage, the probability of the water deficit being made good during the next 10-day period and data from hydrological watch stations.

This bulletin includes general information on major crops. It is proposed to supplement this document with comments and a preview of expected weather and the development of crops for the next decade and hence the agricultural operation to be recommended. A review of the growing season is prepared as soon as possible after the end of the season.

An initial period of six years was devoted to training staff and the establishment of the infrastructure for the Programme. The Regional Training Programme for technicians and engineers had a syllabus covering the disciplines of meteorology, hydrology, crop phenology, crop and animal diseases, and agricultural practises. National training programmes were set up to train observers. About 180 students were accepted for training in the first eight years. The training activities continue, to strengthen national and regional services.

The tangible infrastructure consists of: a network of agrometeorological observing stations; a telecommunication network; a station and instrument maintenance service and national and regional data reception, verification and anaylsis centres. There is also an intangible infrastructure consisting of: the definition of a complete and uniform methodology; the establishment of inter-Ministerial contacts between data producers and data users; and the establishment of an information dissemination system, to planners, decision makers, and primary producers. To be accepted by farmers, information provided must be of a high quality, allowing them to make sound decisions.

The application of the information, as required by the users and in user-tailored language, formulated by an interdisciplinary team, consisting of a meteorologist, agrometeorologist and an extension officer, is tested in pilot projects. For example in Mali, farmers agreed to treat half of their land accordingly to traditional methods, and half according to the information received. After two years of operations, farmers indicated that they wished to continue to receive the information, were willing to pay a modest sum for it, and wished this information to reach them through the existing agricultural extension service. The pilot project provides a possibility for evaluation and the adaptation of the information provided and techniques used, which is fed back to the agrometeorological service.

A programme like AGRHYMET, with an ultimate spread of about 170 stations reporting on the full range of observations sought, can provide only a sample assessment of the actual situation in the Sahel. However, when the data provided by the system are combined with data derived from satellite observations, a more complete understanding of the physical environment in the Sahel with fully-defined limits of distribution of distinctly described conditions can be obtained. The value of each single system of observation is greatly enhanced by such a combination.

The AGRHYMET Programme can not become operational in only a few years, professional training of adequate numbers of staff is a multi-year activity. Started in 1975, the first six years have been mainly devoted to training. The next 6 years include a continued training programme and the establishment of the data analysis programme.

The Economic Benefits of Agrometeorological Information

The economic benefits of agrometeorological information take the form of reduced costs in farming and avoidance of losses in agricultural production, less often they clearly manifest themselves as increases in agricultural production. Therefore, the economic benefits are often unnoticed, as long as they occur, and noticed only when they are no longer available.

There are no quantitative figures to describe the benefits achieved in weeding, ridging and fertilizer application, and only a few relating to crop protection. However, savings in pumping costs of irrigation water through correct calculation of water requirements have been shown to have benefit/cost ratios of ten.

Action required to Promote Application of Agrometeorological Information

(1) The establishment of regular contacts between meteorologists and agriculturalists, perhaps informally at first, and institutionalized later.
(2) Joint definitions of requirements for information.
(3) Joint collection and transmission of observations and data.
(4) Analysis of data by teams of specialists.
(5) Formulation of the information for users in inter-disciplinary teams and in user-tailored language.
(6) Dissemination of the information through established and accepted channels.
(7) Better descriptions of the response of crop varieties, animals and forests to climate and weather.
(8) Formulation by the farming community as a whole of the climate and weather constraints for the timely and efficient implementation of agricultural day-to-day operations.
(9) Training in the formulation and use of agrometeorological information.

Conclusion

There is not one single solution to the questions raised, because climate and agriculture in the different regions of Africa are enormously diverse. However, techniques are available that have proved their value although they need to be adapted to local conditions when used elsewhere. Advancement of agricultural production in Africa will need the balanced contribution of many input factors, applied agrometeorology is one item for inclusion in this package.

References

de Wit *et al.* (1983) *Modelling of Agricultural Production: Weather, soils and crops.* Wageningen Agricultural University.
FAO (1976) *Development de la recherche agronomique dans le basin du Flure Senegal.* Agrometeorologie AGP: SF/REG 114. Rapport technique 2.
FAO (1978) *A Study of Agroecological Zones. Methods and results for Africa.* Rome; FAO.
Penning de Vries, F.W.T.; Djibeye M.A. (1982) *La Productivité des Paturages Saheliens.* Wageningen; PUDOC.
World Meteorological Organization (1970) *Weather and Animal Diseases.* [Technical Note no. 113.] Geneva; WMO.
World Meteorological Organization (1972) *Some Environmental Problems of Livestock Housing.* [Technical Note no. 122.] Geneva, WMO.
World Meteorological Organization (1978) *Weather and Parasitic Animal Disease.* [Technical Note no. 159]. Geneva; WMO.
World Meteorological Organization (1983) *Guidelines on Crop Weather Models.* [WCP-50.] Geneva; WMO.

Chapter 58

Variation and Change in Climate in Sub-Saharan Africa

L. OGALLO

Department of Meteorology, University of Nairobi, P.O. Box 30197, Nairobi, Kenya.

Introduction

The major factors which will determine agricultural production in any given region are the climatic variations, and in the tropics precipitation is the only climatic element with very high spatial and temporal variations. The climatic parameter which will limit the year-round agricultural production in the tropics is consequently water availability, the major source of which in many tropical regions is rainfall. This water resource also limits that available for domestic and industrial use, and for hydro-electric generation.

Historical, botanical, geological and meteorological records show that the global climate has experienced fluctuations ranging from regional anomalies to long-term changes of great magnitude. In certain regions of the sub-Saharan Africa, some have suggested that the climates have been deteriorating in recent years. Here the rainfall characteristics in sub-Saharan Africa are examined and compared with other regions of the tropics.

Here, the term "climatic change" refers to permanent changes in climate, while "climatic variations" or "climatic fluctuations" is used to refer to the climatic changes which are not permanent and have short time scales.

Climatic Fluctuations and Changes in Sub-Saharan Africa

The climatic conditions in the tropical sub-Saharan region range from dry to humid climates. The major source of food is rainfed crops, but there are very high interannual variations in rainfall, especially in the dry areas where the spatial and temporal variability patterns are both high and erratic. The dry regions are drought prone, as other dry regions of the world. Rainfall records from the sub-Saharan region indicate that severe droughts occured in some parts of the Sudano-Sahel region throughout 1968–73. During 1974–75 there was a substantial increase in precipitation in the drought-stricken parts, but droughts returned in 1976 and have continued in some parts to date. Studies on these fluctuations include those of Bunting *et al.* (1976), Ogallo (1979), Faure *et al.* (1981), and Nicholson (1983).

The available statistical and historical records show that catastrophic climate episodes like floods and drought have been part of the normal climate in Africa. Severe rainfall deficits were observed in many parts of sub-Saharan Africa between 1904–15 and 1939–49. Conditions similar to the currently severe ones in the Sahel region were also recorded about 150 years ago. Rainfall records also indicate that the Sudano-Sahel region received excessive precipitation in the 1950s, while the East African region had abnormally wet conditions in the early 1960s. Wet and dry episodes have been observed many times in the history of the region, but the persistence of droughts is higher in dry regions (Nicholson 1983). In general, no significant change has been recorded in the climate of the sub-Saharan region during the present century, and this is also true of other parts of Africa (World Meteorological Office 1983, Hare 1983).

Climatic Fluctuations and Change in Latin America

The records of rainfall in tropical America (see Stoeckenius 1981, Hastenrath 1983), indicate that severe floods and droughts have affected many parts of the region in the past. In the recent times, severe droughts have been reported in north-east Brazil, the Altiplano region of Bolivia, south-east Peru, north-east Mexico, Jamaica and other parts of Latin America. The recent Brazilian drought has been reported as the worst in 300 years, and caused severe food shortages to millions of people. The recent severe drought in the Altiplano region of Bolivia, has also been considered as the worst in 100 years. Catastrophic damage by excessive precipitation (floods) is also common in the region. North-east Brazil is one area with high interannual rainfall variability in tropical central and southern America; during 1900–1981 drought occured in 1915, 1919, 1932, 1951, 1952, 1967 and 1971, while extremely abundant precipitation was received in 1912, 1917, 1924, 1964, the 1970s and 1981 (Hastenrath 1983). Interannual rainfall graphs for Quito (Equador) and San José (Costa Rica) indicate that droughts and floods have also occured in some years during the present century, but no significant climatic trends are discernible. Rainfall records from other tropical areas of America confirm that no significant change in the regional climate has occured during the present century.

Climatic Fluctuations and Changes in India

India has an excellent distribution of rainfall stations, most of which have been operating for long periods. The study of rainfall fluctuations in India has attracted particular attention (see Ramasastri 1979, Mahodaya 1979, Das 1983), as most of the agricultural production is rainfed and is dependent on the monsoon rains.

Negative and positive departures from rainfall means have occured in some years, and in some cases anomalies have persisted for more than one year. The history of severe precipitation deficits in India is considered by Das (1983) who reviews drought conditions in terms of regions and the proportion of India affected from the last quarter of the previous century. The results from numerous Indian studies indicate that although catastrophic climatic episodes like floods and droughts have affected many areas of India, most interannual rainfall series have not revealed any significant non-random patterns. No significant change in the climate of India appears to have occured, to judge from available meteorological records.

Climatic Teleconnections

In some years spatial coherence has been observed in anomalous climatic patterns, which have ranged from regional to inter-continental or inter-hemispherical teleconnections (Fleer 1981, Stoeckenius 1981, Pittock 1983). Maura & Shukla (1981), Hastenrath & Heller (1977) and Hastenrath (1983) have all observed some teleconnection between anomalous climatic conditions in north-east Brazil and West Africa. Teleconnections between precipitation in the tropical Americas and changes in the general circulation parameters over other tropical areas, mid-and high latitudes have also been discussed, (e.g. Kousky & Gau 1981, Ramos 1975).

In sub-Saharan Africa teleconnections have been reported between anomalous precipitation in this region and several general circulation parameters over Asia, and the rest of the globe (e.g. Lamb 1978, Winstanley 1973, Kidson 1977, Kraus 1977). Nicholson (1983), for example, noted continental spatial coherence in droughts which affected many African countries in the early 1970s.

These anomalous climatic teleconnections, sometimes between widely separated landmasses, indicate that the causes of anomalous climatic patterns are global rather than regional. It is well-known that climatic fluctuations including extreme cases are controlled by fluctuations in the general global circulation patterns which are in turn influenced by complex interactions between the oceans, land, and the atmosphere. The dependence of climate on the global general circulation parameters clearly indicates why anomalous climate has been observed in all parts of the globe. Observed climatic teleconnections, together with the patterns discussed above clearly show that no region of the tropics has had a particularly favourable climate this present century.

Changes in Agricultural Production

While agricultural production in many African countries has deteriorated over the last two decades, significant advancements have been made in some tropical Latin American and Asian countries. Some nations in tropical Latin America have managed to minimise famine and deaths caused by climatic calamities through increased food production and good planning. However, in some areas, especially those dominated by civil strife and other crucial social, economical and political problems, abnormal declines in agricultural production have been reported. The situation is very severe in some of these nations as it is in some countries of sub-Saharan Africa.

In India significant advances have been achieved in agricultural output through the improvement of seeds, agricultural chemicals, fertilizers, irrigation systems, tools, and all other agricultural systems including traditional methods. All this has been achieved with local manpower. Although extreme climatic variations such as droughts and floods have suppressed agricultural output, as in the years 1965–66, 1966–67 and 1972–73, no clear link has been observed between droughts and famines in India since the last quarter of the 19th century (Das 1983). In some cases, severe weather anomalies affect too small a section of the country to have a significant influence on national production, but impacts may be very severe locally due to bad planning, such as a lack of transportation and inadequate food distribution systems.

From the preceeding sections, it is evident that change in climate cannot be taken as the only reason for the severe drop in agricultural output in sub-Saharan Africa. It is evident that in many developing nations man-made factors, including unstable economic, social and political systems, have in many cases played dominant roles in the decline of agricultural production. The contribution of agricultural constraints should also be highlighted. These have included a lack of skilled local manpower, to improve production through improved seeds, tools, pest and disease control, and agricultural advice.

The economic problems which are epidemic in many African nations are induced by both local and international factors. These range from low prices paid to farmer producers to the international marketing systems. Low producer prices result in low wages for farm workers, resulting in an influx of job-seekers into urban areas and neglect of agricultural areas. Low producer price can also lead farmers to reduce agricultural production, or change to other crops.

Exports from many developing countries have dropped significantly in recent years. While the international prices of many of these exports has not increased substantially in recent years, the prices of the imported items, including oil, fertilizers, agricultural chemicals, and tools has, forced many nations to constantly devalue their currencies, subsidise exports, and place quotas on imports. These economic problems can result in serious financial constraints to both policy makers and farmers, limiting the financial resources that can be invested in agricultural production.

Social problems which have aggravated agricultural production in Africa include increasing population and the influx of political and economical refugees. A population increase of $2-3\%$ y^{-1} is common in many African countries (FAO 1982) and this requires a similar trend in agricultural production from available resources. It also induces other financial constraints on policy makers as extra facilities will be required (education, health, etc.) which again restrict the resources that could be invested in agriculture. Many African nations attained independence within the last 2–3 decades, and in some of these anomalous climatic conditions have pre-dominated since independence. This has brought serious problems to some young nations, sometimes inducing unstable political systems ranging from civil strife, to unrest and wars. No increased agricultural production can be expected from any nation with crucial political instability even if all other factors are favourable; farms may be turned into battle fields and farmers will seek refuge in more peaceful locations.

Improving Agricultural Output

In order to improve the deteriorating agricultural output, the causes of the decline must be clearly identified. While environmental factors cannot be controlled, man-induced constraints on the agricultural systems can. This requires national agricultural systems which can absorb the shocks of unstable environmental, economic and social conditions; such systems cannot be achieved under unstable political conditions as they require long-term plans. Any plan of action for the improvement of agricultural production in the sub-Saharan Africa should therefore consider the:-

(1) Impacts of climatic variability on agricultural production which should be assessed using the available agro-climatological records. Agricultural systems which can endure anomalous climatic conditions should be identified since such systems can reduce the risk of damage caused by unfavourable climatic conditions. This will also require the improvement of traditional farming and livestock systems.

(2) Solutions to the agricultural constraints which might have also contributed to the severe decline in agricultural output that should be found. These solutions, reviewed elsewhere in this volume, include ways of obtaining high yielding varieties, effective fertilizers and chemicals, irrigation systems, and the control of losses due to pests and diseases.

(3) Availability of adequate skilled local manpower, training, and research facilities.

(4) Availability of an effective national or regional agro-climatological division with adequate early warning systems.

(5) Identification of local and international financial sources necessary in order for any plan to succeed; funding may be beyond the scope of economically depleted nations.

Finally, it should be noted that many of the political and social problems, together with other constraints contributing to the severe drop in agricultural production in many African countries, cannot be solved without the fullest co-operation and initiative of policy makers.

References

Bunting, A.H.; Dennet, D.M.; Elston, J.; Mildford, J.R. (1976) Rainfall trends in West Africa Sahel. *Quarterly Journal of the Royal Meteorological Society* 102, 59–64.

Das, P.K. (1983) Drought and famines in India. A historical perspective. *Mausam* 102, 59-64.

FAO (1982) *Famine in Africa*. Rome; FAO, 36 pp.

FAO (1983) *Special Report – Food and shortages. FAO Global Information and early warning system on food and agriculture.* Rome; FAO.

Faure, H.; Gac, J. (1981) Will the Sahel drought end in 1985? *Nature, London* 291, 475-478.

Fleer, H. (1981) Large-scale tropical rainfall anomalies. *Bonner Meteorologische Abhandlungen* 426, 1-114.

Hare, F.K. (1983) *Climate and Desertification*. [WMO WCP–44.] Geneva; World Meteorological Office, 149 pp.

Hastenrath, S.; Heller, L. (1977) Dynamics of climatic hazards in northeast Brazil. *Quarterly Journal of the Royal Meteorological Society* 103, 77-79.

Hastenrath, S. (1983) Towards the monitoring and prediction of northeast Brazil droughts. In *First International Conference on Southern Hemisphere Meteorology 31 July – 6 August*, 116-119.

Kidson, J.W. (1977) African rainfall, its relation to the upper air circulation. *Quarterly Journal of the Royal Meteorological Society* 106, 441-456.

Kousky, V.E.; Gau, M.A. (1981) Upper tropospheric cyclonic vortices in the tropical south Atlantic. *Tellus* 33, 538-551.

Kraus, E.B. (1977) Sub-tropical droughts and cross-equatorial energy transports. *Monthly Weather Review* 105, 1052-1055.

Lamb, H. (1978) Large scale tropical Atlantic surface circulation patterns associated with the sub-Saharan weather anomalies. *Tellus* 30, 240-251.

Mahodaya, M.M. (1979) A critical study of the drought conditions in Dhar district. In *International Symposium on Hydrological Aspects of droughts, 3–7 December*, 520-531. New Dehli.

Maura, A.D.; Shukla, J. (1981) On the dynamics of droughts in northeastern Brazil. *Journal of Atmospheric Science* 38, 2653-2675.

Nicholson, S.E. (1983) Rainfall and atmospheric circulation during drought periods and wetter years in West Africa. *Monthly Weather Review* 109, 2191-2208.

Ogallo, L.J. (1979) Rainfall variability in Africa. *Monthly Weather Review* 107, 1133-1139.

Pittock, A.B. (1983) Climatic teleconnections in the Southern Hemisphere. In *First International Conference on Southern Hemisphere Meteorology*, 283-286.

Ramasastri, K.S. (1979) On the desertification and trends of rainfall in West Rajasthan. In *International Symposium on hydrological aspects of droughts*, 47-66. New Delhi.

Ramos, R.P.L. (1975) Precipitation characteristics in the northeast Brazil dry region. *Journal of Geophysical Research* 80, 1665-1678.

Stoeckenius, T. (1981) Interannual variations of tropical precipitation patterns. *Monthly Weather Review* 101, 1233-1247.

Winstanley, M.B.C. (1973) Rainfall patterns and general atmospheric circulation. *Nature, London* 245, 190-194.

World Meteorological Office (1983) *Report of the expert group meeting on climatic situation and drought in Africa.* Geneva; World Meteorological Office, 27 pp.

Chapter 59

Weather and Insect Plagues in Africa

D.E. PEDGLEY

Tropical Development and Research Institute, London W8 5SJ, UK.

One contribution to the low agricultural productivity in Africa compared with other parts of the tropics is crop loss due to insect pests, both in the field before harvest, and in store. Such losses can be severe when insects appear as plagues. The causes of insect plagues are complex, but they need to be understood if rational control strategies are to be developed against not only crop pests but also disease vectors. Because the weather strongly affects the build-up, spread and decline of insect plagues, it is worthwhile to examine some of the consequences in relation to the development of control strategies.

We may define an insect plague as an exceptional increase in population density, leading to crop damage or spread of disease over some given area. When the area is large, the plague is more severe. The increase is exceptional in the sense that the density, although normally expected to fluctuate on time scales of a season or a generation, increases suddenly, over days or even hours, perhaps by an order of magnitude or more. Population density is defined as the number of insects per unit area, so an increase can be brought about in two ways: by increasing the number of insects in a given area, or by decreasing the area occupied by a given number of insects (i.e. by concentrating them). The rate of increase in number is the residual of two opposed rates: of gain (by birth and immigration) and of loss (by death and emigration). We take immigration and emigration to mean simply movement into and out of the area in which an initial population is living (i.e. feeding, sheltering and all the other activities that need only trivial movements). Thus, a plague can come about either by an exceptionally large rate of gain or of concentration, or by an exceptionally small rate of loss. Man affects these rates in various ways, for example by the ill-considered application of pesticides that preferentially kills predators and therefore increases pest survival rates, by increasing the pest status of a species through introducing it to an area with few predators, or by introducing a susceptible crop to an area where an insect species then has a chance to become a pest (e.g. cocoa, in West Africa, where indigenous mirids have become pests). Additionally, weather affects these rates in various ways.

Weather can be defined as the state of the atmosphere at some particular time; climate is the integration of the ever-changing weather over some given time interval, usually many years. We are really concerned with the weather as experienced by the insect population over some weeks or months, but because this is seldom measured it has to be more or less approximated by standard measurements made for other purposes, such as crop production or aviation. Much work has been carried out and continues on the effects of weather on the various stages in the life cycles of many insect pest species, mostly outside Africa, for example on feeding, growth, sheltering, flight, mating, egg laying, fecundity and survival. The generally higher temperatures of air, soil and vegetation in lower latitudes provide an optimum environment for physiological systems to function and therefore favour faster and longer development than in higher latitudes. Hence the main problems of crop damage and disease spread by insects are in the tropics. Where rainfall is seasonal, population density also tends to be seasonal if it depends on the presence of lying water for breeding, or on plant cover for food and shelter. During the dry season, insects may enter a resting stage or simply leave in search of a more suitable habitat. Even when the weather favours maximum birth rate and minimum death rate, any resulting increase in density can hardly be said to constitute a plague because it is not sudden. An exception is perhaps where two markedly different equilibrium densities are possible for each value within a certain range of some external constraint, such as food supply. If there is an increase in density above the lower equilibrium, as might be brought about by immigration, a rapid growth to the upper equilibrium may occur instead of a return to the lower. With a consequent decline in food supply, or an increase in predation, there is a collapse of the population density back to the lower equilibrium. Such upsurges and declines are typical of some forest pests of temperate latitudes, but there do not seem to be any comparable events among African species. Moreover, the population changes, although unusually large, require several generations to reach their maximum and can therefore be called sudden. Even an unexceptional increase in density may not be revealed until, for example, many individuals appear simultaneously after a seasonal rest in the ground or in vegetation.

Concentration of a population, by flying or crawling onto a limited and perhaps decreasing number of hosts, can be rapid and lead, for instance, to dense infestation on one crop from another nearby close to harvest. Individual weather events, such as rain or strong wind, may favour or inhibit such movement. But the mechanism that is most likely to lead to a sudden and exceptional increase in density, or plague, is immigration. The arriving insects add to any population already present; they have come from elsewhere, whether near or far. Whereas weather effects on local build-up are understood in principle, and have been applied to may species, even to the extent of incorporation in operational forecasting systems, the effects of weather on immigration, by contrast, and even the very existence of immigration, have tended to be neglected. However, in some species it has been shown that immigration dominates the population dynamics of a given area. In Africa this has been most clearly demonstrated with various grasshopper and locust species (notably the desert locust, *Schistocerca gregaria*) and African armyworm moth (*Spodoptera exempta*), but others behave similarly, such as the cotton stainer (*Dysdercus voelkeri*) and whitefly (*Bemisia tabaci*). It is likely that more species in Africa will be found to be migrants once their field ecology has been studied adequately.

It is a characteristic of migrants that individuals move hundreds of km in a lifetime, often in only a few days or nights. They are also well-known for their great fecundity and rapid growth in numbers, even with the appearance of wingless or small-winged generations when energy is concentrated into reproduction and not flight. Because food resources are often seasonal or patchy, and may be outstripped, there follows either widespread death or emigration, or both. These migrant species contrast with others that have lesser fecundity, slower rates of gain in numbers, more stable numbers and more static life styles.

Migrant crop pest species tend to be adapted to annual or seasonal hosts. Some such hosts, for example cereals, pulses and roots, provide the main source of food for man and his livestock. Consequently many field crop pests are likely to be migrants, or at least migrant individuals are likely to appear at times within the population. Where only a small proportion is migrant it is unlikely to lead to a plague because it will seldom add significantly to a population that has bred locally, but such migrants can reach new habitats where there may be few enemies, and so extend the normal distribution area, if only temporarily. The numbers involved will be affected by source strength, attractiveness of hosts in the new habitat, and weather met on the way.

Migration is affected by the weather in many ways, from take-off, through displacement and dispersion, to landing. Wind, temperature and rain, in particular, help determine the timing of these events, their magnitude, and the numbers of individuals taking part. Although these influences have been well-studied in the field for only a few species, they are of considerable significance in determining the strategies of monitoring, forecasting and control. With mobile populations it is advisable to direct the available control resources to the largest and preferably the most vulnerable populations, wherever they may be. This highlights the need for regular and continuous field monitoring over the whole area likely to be reached by migration, and for a system of communicating reports to a central analysis and forecasting centre in ample time for the issuing of advice to control services. Such systems already exist for the desert locust and the African armyworm. Alternatively, cultural techniques may be modified so as to reduce population build-up. For example, planting date might be altered to minimise overlapping of the times when hosts are most susceptible and when migration takes place; for all such measures to be effective, migration must be understood.

In summary, most insect plagues, as defined, whether of crop pests or of disease vectors, seem to be caused by immigration from more or less distant places where there have been increases in numbers by successful breeding over several generations. Population density increases suddenly and much damage may occur within days or even hours; or subsequent breeding by an initially smaller invasion may lead either to a progressive build-up or to a catastrophic jump from one equilibrium density to a much higher one. In the last two cases, however, build-up can hardly be said to be sudden. Both the role of migration, and the effects of weather on it, have been made clear for a few species in Africa, to the extent that they are now taken into account in regional control strategies, but they have probably been underestimated so far for other species.

Chapter 60

Weather and Plant Diseases in Africa

T.L. LAWSON and E.R. TERRY

International Institute of Tropical Agriculture (IITA), PMB 5320, Ibadan, Nigeria.

Introduction

Any consideration of plant diseases, particularly in relation to agricultural production, can be meaningful only in the context of the prevailing weather and climate. A significant body of information has been accumulated, but this derives primarily from studies in the temperate zone. Nevertheless, the current knowledge of the weather and climate of Africa and evidence of the influence of weather (or weather factors) on pathogens, enables working hypotheses to be developed. On the basis of such hypotheses, investigations may be conducted and/or deductions and tentative recommendations made. Formulated hypotheses contribute to a sounder approach to the study of the epidemiology of diseases in Africa, and lead to the development of disease management systems.

The subject of plant disease and weather interactions is complex and a significant portion of what is known is based only on circumstantial evidence.

General Characteristics of the Weather and Climate in Africa

Except for the extreme north and south, the prevailing weather and climate over much of Africa is tropical. Within the tropical climate, we may distinguish humid tropical, sub-humid, and semi-arid areas; a further differentiation may be made with respect to altitude-induced modified temperatures.

The Humid Tropics

The humid tropics include areas where the rainfall exceeds potential mosisture demand for more than half the year; it extends mainly from about 8°N to 9°S latitude, excluding the areas east of longitude 30°E, as well as the coastal areas of Togo, Ghana, and north-western Angola. The wettest core of this zone covers most of the Congo (Zaïre) basin, the Niger delta area extending into Gabon, and the coastal regions of Liberia, Sierra Leone and south-western Ivory Coast. In West Africa, the weather and climate in the zone is dominated by the interaction of two air masses, one continental and dry, and the other oceanic and moist. South of the equator a third air mass, of oceanic origin to the east of the continent, is involved. As temperature differences between the air masses are minimal, variation in day-to-day surface temperature is limited. Differences in moisture content do, however, induce significant day-to-day variations in rainfall and cloudiness. Zonal and local modification in the surface layers of the continental air mass ensures high humidity for most of the year, night-time and early morning relative humidity values generally rise above 90%. Surface wetness from dew and/or rain, as determined from dew balance, may average over 10 h day^{-1}; this has important implications for many pathogenic diseases, especially those caused by fungi. Insolation is relatively low due to cloudiness and typically averages 345–454 g-cal cm^{-2} day^{-1} at Yangambi, Zaire (Griffiths 1972), and 325–477 g-cal cm^{-2} day^{-1} in Ibadan, Nigeria, near the outer limits of the zone. The effect of low radiation regimes on temperature, relative humidity and evapotranspiration may be important to plant diseases. Surface winds in the zone are generally light, but gusts during thunderstorms may have a role in the dispersal of pathogens.

The Subhumid and Semi-Arid Zone

The sub-humid and semi-arid zones include areas with rainfall exceeding potential evapotranspiration for periods equal to or greater than 2 months, but not exceeding 6 months, in the year. It occupies the area between the humid tropics and the outer edges of the deserts (Sahara and Kalahari), and extends eastward from longitude 30°E to cover most of Eastern Africa. The weather here is tied to the displacement of the convergence zone between the respective air masses (Ojo 1977), which also accounts for the seasonality in rainfall. During the rainy season, the moist air conditions are similar to those of the wet tropics, except for the higher amount of insolation (Griffiths 1972). Temperatures are also more accentuated; the mean maxima for Bauchi, Nigeria (10°17'N, 9°49'E) for example are 28.6–36.7°C, and the minima 12.8–22.3°C; respective extremes are about 40.6°C and 6.1°C. Early morning relative humidity remains comparatively high during the rainy season, but can drop drastically during the day, particularly in the dry season. The pronounced and relatively long period of drought separating successive moisture cycles (cropping seasons) in this climatic regime appears functionally similar to the temperature-imposed break in the cropping cycles of temperate latitudes, and may well have a parallel role in the control or selection of disease organisms; ability to survive between cropping seasons assumes increased importance.

The Arid Zone

The arid zone represents the extreme of the weather regimes. Maximum temperatures routinely rise above 40°C, while the minimum may drop below 0°C. The mostly cloudless skies ensure very high levels of insolation; mean daily global radiation may exceed 650 g-cal cm^{-2} day^{-1}. Rainfall is very low and extremely variable, and relative humidity is also very low (10–20%); intense surface cooling at night results in an increase in early morning values which may occasionally reach 100% (Griffiths 1972). The harshness of this environment limits both hosts and pathogens.

The Mediterranean and Humid Subtropical Zones

The mediterranean weather experienced in the extreme north and southern part of the continent is rather limited in areal extent. It is characterized by mild and moist winters and hot and dry summers. There is a significant day-to-day variation in the weather associated with intrusions of remnants of middle latitude cyclones, especially in winter.

The south-eastern coastal areas of Africa share some of the above climatic features, except that the rains occur in summer rather than winter. Wind speeds are considerably higher than in the other zones, reflecting the mesoscale air flow associated with intruding cyclones.

Areal/Seasonal Complexes of Plant Diseases in Africa

Most of the crops grown by the small farmers in Africa, particularly tropical Africa, have reached some equilibrium with their endemic pathogens, both demonstrating a high level of genetic flexibility, that of the pathogens determining their ability to overcome spatial and sequential discontinuities of the hosts. Disease patterns in tropical Africa are considered here in the context of the agroclimatic zones recognized above.

The Humid Tropics Disease Complex

The duration of the rains in the humid tropics makes cropping possible virtually all year round. Crop production potential is therefore high, as is crop diversity. However, farming in much of the region has remained at the subsistence level. The cropping system primarily involves shifting cultivation with an emphasis on the reliable year-to-year production of modest yields in preference to fluctuating annual outputs.

The uninterrupted availability of crop hosts throughout the year significantly influences the level of available inoculum. The warm to hot and continuously humid conditions of the ambient environment are conducive to fungal growth and result in a predominance of fungal diseases.

The diverse subsistence crops grown in this zone, with their long history of evolution within the existing pathosystems and their resultant genetic flexibility, are generally able to endure diseases induced by indigenous pathogens, which are normally limited in incidence and severity. While diseases of annual crops grown at the subsistence level within the wet tropics may be amenable to control by cultural practices, plantation crops with varieties bred primarily for yields and grown over large continuous areas are prone to epidemics. Some measure of disease resistance and chemical control is required to protect them effectively against prevailing diseases.

The Sub-humid and Semi-Arid Disease Complex

Annual crops in these environments can usually be grown within precipitation periods which range from a minimum of 2 to a maximum of 6 months. The species and varieties of crops grown here are fewer in comparison with the humid tropics and the crops (with a few exceptions such as some palms and mango) are hard to grow. The range of crops includes sorghum, millet, groundnut, cowpea, and cotton. Maize has now assumed prominence in these agroclimatic zones; as have vegetables, sugarcane, rice, sweet potatoes, and tobacco.

The weather/climatic characteristics of this environment are evidently inconducive to the majority of plant pathogens. They lack epidemiological competence, i.e. the ability to survive and cause epidemics, and consequently may never be locally important in this environment. However, certain smut, rust and soil fungi (e.g. *Macrophomina phaseolina, Sclerotium rolfsii*) are fairly common. An important feature of the regions is the seasonally drastic reduction in inoculum in dry periods; pathogen survival thus assumes paramount importance. However, with increased use of irrigation in the dry season these pathogen curbing effects may be nullified. This consideration will be increasingly important in the future.

Weather and Disease Development

The link between disease and weather is not always simple as the optimum conditions for the life cycle phases of several diseases can vary with other factors. As indicated above, some of the dominant features of the weather pattern in the wet and semi-arid tropics can adversely or favourably influence the dominant crops and their pathogens. It is important to distinguish between the effects of weather on the pathogen and on the host, and further to determine whether the effects of weather tend to accelerate the development of the pathogen or adversely affect the host so that it is predisposed to infection.

The two most common weather elements that influence disease patterns are temperature and wetness. Winds, in addition, play a major role in their dissemination. As examples, the influences of one or a combination of these factors on the development of a disease of (a) a subsistence crop in the wet tropics (b) a plantation crop in the wet tropics, and (c) a subsistence crop in the semi-arid tropics, are presented.

(a) Effect of Temperature and Wetness/Aridity on Cassava Bacterial Blight Disease

In the disease cycle of bacterial blight of cassava (see Chapter 18) caused by *Xanthomonas campestris* pv. *manihotis* in Ibadan, Nigeria, an area with a five month dry season, new angular leaf spots appear on plants which had been in the field over the five-month dry season (November – March) within 1–2 weeks after first rains in the succeeding wet season (Persley 1978). This suggests that the new infections may have resulted from an epiphytic population of *X. campestris* pv. *manihotis* which had survived the dry season within that plant population. This was investigated further by monitoring leaf surface populations of the pathogen and correlating these with weather factors. The bacterium was detected on leaf surfaces of plants with no symptoms of bacterial blight on both resistant and susceptible varieties on symptomless leaves of infected plants, and on leaves with angular leaf spots. The pathogen was consistently detected on leaf surfaces during the rainy season, but less frequently and in lesser numbers at the beginning and end of the rainy season and during the dry season. The pathogen appears to survive the dry season as an epiphyte, increasing in number when moisture becomes available. The widespread and rapid development of new angular leaf spots as rains become more frequent at Ibadan during April/May results from such epiphytic populations which survived the dry season.

Temperature was the most important factor affecting the severity of the disease in the southern region of Brazil (Takatsu *et al.* 1978); susceptible and resistant cultivars when infected and kept under high temperatures (night/day temperatures above 20/30°C, respectively) normally recovered from the infection, but at low temperatures (night/day) temperatures below 20/30°C, respectively) even resistant cultivars develop the most severe phase of the disease.

These examples of the effect of seasonal changes in wetness and the effect of temperature on the development of bacterial blight in cassava demonstrate the complex nature of disease-weather inter-actions. The pathogen is more ecologically competent to survive after the cessation of an epidemic in the savannah, where there is a 5–7 month dry season, than in the forest zones with a short dry season of 1–2 months. We also have evidence that under high temperature conditions when the growth of the cassava plant is vigorous and the tissues mature rapidly that rate of infection is limited; an indirect effect of temperature on disease through its effect on the host.

(b) Effect of Seasonal Variations on the Coffee Leaf Rust Disease

Coffee leaf rust (*Hemileia vastatrix*), was one of the earliest major tropical plant diseases to be reported and studied. *Coffea arabica* and *C. canephora* probably originated in East Africa and this is the most likely home of the coffee rust pathogen. Seasonal periodicity of the pathogen in Kenya was found to be different in districts east and west of the Rift valley, the major factors influencing this difference being climatic (Bock 1962). In the east districts the outbreaks correspond to the monsoon rain pattern of the two wet seasons; there are thus two peaks in the annual disease cycle, increases in the disease level starting soon after the onset of each rainy season, and the maximum level extending into the following dry periods. During the dry season the disease incidence falls (Nutman & Roberts 1970) largely as a result of rust-induced premature leaf fall when most infected leaves abscise. In this instance there is a definite interaction between weather and disease incidence on the one hand, and an effect of the severity of a disease on the foliage density on the host which influences the subsequent rate of increase of the disease. West of the Rift the one extended epiphytotic is a result of the more or less continuous rainy season. Here there is one extended disease peak in any annual outbreak which is only broken by periods of heavy leaf fall.

Although temperature also affects the development of this disease, Nutman & Roberts (1970) found its effects to be complex; on leaf surfaces the germination curve is bimodal, with peaks at about 21°C and just above 25°C separated by a depression of germination between 23° and 24°C.

This example constitutes another clear cut relationship between weather and a plant disease. The course and severity of seasonal outbreaks are determined by the interaction of three factors, any one of which may be limiting; (1) the distribution and intensity of rainfall; (2) the degree of leafiness of the tree (i.e. the extent of the potential surfaces on which sporces can be retained and inoculum accumulated; and (3) the amount of residual inoculum at the end of the dry season. Temperature can also be a limiting factor in that exposure to suboptimal temperatures stimulate the process of urediniospore germination during subsequent periods of higher and more favourable temperatures. The degree of stimulation increased with the length of exposure to cool conditions (Nutman & Roberts 1963). The implications here are that under field conditions urediniospores dispersed by rain during the night may be subjected to low temperatures, which in Kenya can fall as low as 5°C, and then germinate at an accelerated rate when the morning temperature rises so that infection occurs rapidly.

(c) Effect of Drought on the Severity of Charcoal Rot in Sorghum

Sorghum production in West Africa is greatest where rainfall varies from 600–1000 mm y $^{-1}$ or more. The crop is mainly rain-fed, but some is produced on the flood plains of Lake Chad. Charcoal rot of sorghum is caused by *Macrophomina phaseoli*, a fungus which is widely distributed in the warmer areas of the world on a wide range of host plants. This pathogen can cause a complex of symptoms in sorghum including damping off, seedling blight, root rot, and dry rot of stalks (Tarr 1962). The fungus enters through the roots and thereafter advances into the crown and stem where rot occurs.

Colonization and subsequent rot are completely dependent on stress conditions; the most important being low soil moisture accompanied by high temperatures during seed development (King 1972). In grain sorghum plants with a range of maturities subjected to combinations of 35°C and 40°C soil temperatures, 40°C daytime air temperature, and 25 and 80% available soil moisture (ASM), for 4–7 days before inoculation there was no infection in plants kept at 80% or more ASM, while at 25% ASM, plants that bloomed 14–28 days before inoculation were killed within 5–7 days or 3–5 days at soil temperatures of 35°C and 40°C respectively (Edmunds 1964).

This suggests that total susceptibility occured in plants subjected to maximum stress, while total escape may be achieved in plants subjected to moderate stress. Here is an indirect effect of weather as a predisposing factor in the manifestation of a disease in a crop which, even though well-adapted to the drought stress conditions in the semi-arid regions, can succumb to an interaction between weather factors, one of which is moisture stress. Interventions to reduce the two most important weather-related stresses during seed development may help minimize the damage caused by this rot.

Strategies to Minimize the Weather Impact on Crop Productivity through its effect on Diseases

The basic prerequisite for disease development is that viable inoculum should interact with a susceptible crop under conditions favouring infection. Favourable conditions include the macro- or meso-climate which may define the areal or regional distribution of pathogens, as well as the ambient microclimate in the crop which controls reactions at the infection site. The strategies developed here apply mainly to subsistence farming because of their relevance to the African situation. An appropriate crop protection strategy for subsistence economies should emphasize the manipulation of existing interactions, while rigorously limiting inputs incompatible with the ecosystem in question. Intervention points in the behaviour pattern of diseases need to be identified and utilized to reduce disease severity without adversely changing the ecological realities of subsistence farming.

The alternation between endemic and epidemic phases of diseases has important implications for disease management. The general strategy for disease control consists of either reducing the amount of inoculum capable of initiating the disease, or alternatively reducing the rate at which the disease increases after it has started, or both.

Sanitation

In the case of cassava bacterial blight the strict enforcement of quarantine regulations restricting the movement of cassava vegetative material within Africa and prohibiting movements from outside the continent should at least have delayed the introduction and rapid spread of *X. campestris* pv. *manihotis*. The presence of the CBB pathogen in the xylem of stem cuttings is difficult to detect, and the propagation of cuttings carrying the pathogen from the previous year's crops is largely responsible for the continuity and dissemination of the pathogen. The pathogen was detected in debris on the soil surface approximately six months after planting in Nigeria (Persley 1978). The pathogen population fell sharply after the first rains as it cannot survive more than a few weeks in moist debris and soil at field capacity, but can survive several months if the soil remains dry. It should therefore be possible to break the disease cycle by harvesting with the first rains, removing all infested materials from the field, and allowing the field to fallow for three months before replanting with disease-free material. While damage due to CBB can be severe in the forest zone (wet tropics), if infected planting material is introduced into an established canopy it is difficult to establish an epidemic under these wet conditions from a few infected cuttings as such plants die quickly and the wet conditions and short dry seasons do not provide favourable conditions for pathogen survival.

Resistance to Climatic Stress

The evidence indicating that the low soil moisture which prevails during the dry periods in the semi-arid environment accompanied by high soil temperatures during seed development predispose sorghum to severe attack of charcoal rot (Edmunds 1964) suggests that measures taken to reduce these stresses during seed development may help to control disease incidence and severity. Resistance in sorghum to predisposing stresses had shown some promise for control. At ICRISAT, more than 2500 sorghum lines comprising germplasm and breeding materials were grown in the post rainy season under irrigation and screened for resistance to charcoal rot. Irrigation was withdrawn at 50% flowering to induce the moisture stress essential for disease development; evaluation after inoculation with *M. phaseoli* revealed 715 lines of low susceptibility. Consistent genotype reaction of stressed plants to the disease over locations or in replicates have been difficult to obtain, and the present ICRISAT strategy is therefore to emphasize the relationship between moisture stress and crop management.

Manipulating the Environment

Little can be done on the macro- and meso-scale, and so intervention is primarily limited to the micro-scale. In the case of charcoal rot, modification of the soil temperature and conservation of soil moisture through the use of mulches constitues a practical way of reducing stress, and consequently the impact of the disease. Moisture stress as a predisposing factor can also be relieved by supplementary irrigation.

Weather-based Control

As all commercially grown varieties of coffee are susceptible to coffee leaf rust, the possibilities of utilizing genetic resistance to reduce disease development are minimal. The recognition of the role of water in the dispersal of urediniospores of the pathogen has therefore become the basis for a control strategy utilizing weather forecasting to time copper fungicide application. Dispersal of urediniospores by rain is extremely short-range and generally takes place within an individual tree (Nutman & Roberts 1970). Movement of inoculum is virtually impossible until after the first "dispersal shower" of a rainy season (i.e. over 7.5 mm); the heavier this dispersal shower, the more effective the spore distribution. In parts of East Africa where it is possible to forecast the probable date of this shower with reasonable accuracy, there is a period of approximately 3 weeks during which a spray with a copper fungicide provides effective control. In the East Rift districts, where there are two cycles of coffee rust outbreaks each year, both require similar treatment, although successful control of the first invariably results in a low level of residual inoculum at the start of the second cycle.

General Conclusions

With respect to agricultural technicians mainly working with subsistence farmers, it is important to stress that sanitation in all its ramification is the key to minimizing losses in crop productivity due to diseases in general. However, the evidence presented here indicates that an integrated approach recognizing (a) the subsistence farmer's intimate involvement with his crops, (b) that over the centuries the African farmer has shown a remarkable ability to adapt to new conditions and a readiness to accept innovations of *proven* benefit, and (c) the role of environmental factors in disease manifestation and the scope of environmental manipulation to influence the course of diseases, is not only possible but could in the long run be most effective. A level of intimacy with the crop where some farmers harvest daily or at very short intervals, and often tend their plants individually, should aid the observations necessary to facilitate effective sanitation; ultimate success being dependent on its implementation at community level. The status of cassava as a crop in Africa provides evidence of farmers' ability to adapt to new conditions and accept innovations of *proven* value; farmers not only adopted it rapidly as a staple crop as they became aware of its advantages, but they were also able to integrate it into their cropping systems. The African farmer will utilize improved varieties with disease resistance when this is compatible with his circumstances.

Cultural practices such as mulching, shading, staking and manipulation of planting patterns which can all bring about changes in crop microclimate are well-known to farmers. No major obstacle can therefore be envisaged, *a priori*, in the further adoption of such practices, if they are proved to be an appropriate means of inducing conditions favourable to disease control. In an integrated approach to disease management, using a combination of sanitation and host resistance under subsistence farming conditions, the major obstacles are therefore socio-economic and not technological. Similarly, there is need for an increasing awareness that the logistic and economic requirements for implementing community orientated sanitation and the multiplication and distribution of new improved varieties are only components of the challenge for the development of effective disease management systems, to which the skills of meteorologists/climatologists and plant pathologists contribute.

References

Bock, K.R. (1962) Seasonal periodicity of coffee leaf rust and factors affecting the severity of outbreaks in Kenya Colony. *Transactions of the British Mycological Society* 45, 289-300.

Edmunds, L.K. (1964) Combined relation of plant maturity, temperature and soil moisture to charcoal stalk rots development in grain sorghum. *Phytopathology* 54, 514-517.

Griffiths, J.F. (ed.) (1972) *Climates of Africa.* [World Survey of Climatology vol. 10.] London; Elsevier Publishing, 604 pp.

King, S.B. (1972) Sorghum diseases and their control. In *Sorghum in the Seventies* (Roa, N.G.P.; House, L.R., eds). New Dehli; Oxford and IBH Publishing.

Nutman, F.J.; Roberts, F.M. (1963) Studies on the biology of *Hemileia vastatrix* Berk. & Br. *Transactions of the British Mycological Society* 46, 27-48.

Nutman, F.J.; Roberts, F.M. (1970) Coffee leaf rust. *PANS* 16, 606-624.

Ojo, O. (1977) *The Climates of West Africa.* London, Ibadan; Heineman, 219 pp.

Persley, G.J. (1978) Studies on the epidemiology and ecology of cassava bacterial blight, In *Cassava Bacterial Blight in Africa* (Terry, E.R.; Persley, G.J.; Cook, S.A., eds). Ibadan; International Institute for Tropical Agriculture.

Takatsu, A.; Fukudu, S.; Peria, S. (1978) Epidemiological aspects of bacterial blight of cassava in Brazil. In *Diseases of Tropical Food Crops* (Maraite, H.; Meyer, J.A., eds). Louvain La-Veuve.

Tarr, S.A.J. (1962) *Diseases of Sorghum, Sudangrass and Browncorn.* Kew; Commonwealth Mycological Institute, 380 pp.

Chapter 61

Weather and Animal Disease in Africa

R.J. THOMAS

Department of Agriculture, The University, Newcastle upon Tyne, UK.

Introduction

Agricultural production is a complex of soil, climate, crops, animals and people, and in looking at weather and animal diseases we are considering the interaction of only two of these variables, but it must not be forgotten that they all are interdependant. This can be illustrated by reference to the relationship between degrees of tsetse infection and cattle population as expressed in cattle biomass (Bourn 1978). In tsetse-free countries cattle biomass increases linearly with increasing rainfall, and there is clear evidence of land degradation due to excessive exploitation by increased cattle numbers. Countries with partial tsetse infestation show no such correlation. Tsetse control schemes are likely simply to extend this degradation unless carefully coordinated with other management controls.

Table 1   Cattle Production (Africa).

Output	Meat (kg)		Milk (kg)	
	1950	1970	1950	1970
Per animal	13.9	13.6	50.9	57.6
Per head population.	12.0	11.1	32.6	31.5

Overall, tropical Africa has one of the lowest per capita and per animal levels of meat and milk production in the world, and even more seriously per capita production is declining (Table 1). Mahadeven (1982), in a review of the constraints to effective use of livestock resources, pinpointed the major constraints in relation to African climatic zones (Table 2). The lowest production levels are found not in the arid and semi-arid zones but in the humid and sub-humid zones where animal diseases are the dominant constraint. Conversely, production is maximised by control of nutrition, environmental stress and infectious diseases. Since all three aspects are climate-related, climate is of major importance in determining production levels, and since all three are also involved in animal health, the importance of the interaction of climate and disease is clear.

Weather describes day to day fluctuations in atmospheric conditions, while *climate* is the sum of the daily fluctuations over a long period of time to characterize the annual and seasonal patterns of change. For example, the English climate is considered good since the seasonal variations are small around a temperate mean, whereas the English weather is notoriously bad since it fluctuates in a what often appears to be a totally unpredictable way.

Table 2   Factors limiting cattle productivity (Mahadeven 1982).

Area	Limiting Factor	Solution
Arid	Overstocking	Education in grazing management
Semi-arid	Excessive cropping	Improved pastures
Sub-humid Humid	Disease	Trypanosomiasis control (tolerant breeds, drugs & fly control)
Highland	None	Successful intensification

While the climate/disease interaction is clear, what is less clear is the extent to which an understanding of this interaction can be used to improve productivity. Agricultural development is the art of the possible, and where the requirements for effective disease control are not available on the ground, theoretically satisfactory plans cannot be implemented, even where all the necessary information is available; the ratio of veterinarians to cattle indicates the severity of this constraint (Table 3). In addition, a high proportion of the livestock population, of which cattle are the most important, is found in the arid and semi-arid zones where grazing is extensive and management is difficult to control.

Table 3   Distribution of veterinarians (1972).

Country	Veterinarians	Cattle (x10^6)	Ratio c:v
UK	5772	13.4	2310
India	10 800	176.6	16 360
Nigeria	225	11.4	50 600
Ethiopia	36	26.4	733 300

Weather is a complex of components including altitude, light intensity, radiation, sunshine, cloud cover, temperature, humidity, wind velocity, etc. However, the dominant factors in relation to both animal and crop production are temperature and moisture, both of these requiring simple measuring apparatus, and for which extensive records are available. The World Climatic Data Handbook compiles average monthly and annual data for temperature and rainfall from 19 000 stations, and many more records are available of rainfall than temperature.

Bunting (1961), reviewing meteorological factors related to crop production, concluded that in the tropics temperature tends to be less important since it is usually adequate or above, while rainfall is often highly seasonal and critical. Because of the importance of crops as staple human foodstuffs, and the direct correlation between climate and cropping, a good deal of analysis of meteorological data has been carried out, particularly in relation to evapotranspiration and soil moisture retention which are more crucial than simple rainfall in plant growth, and can be used to model soil moisture stress (Nieuwolt 1982), which might be a useful predictive aid. Much of this information (e.g. Beets 1978) is also directly applicable to the epidemiology of animal disease which is largely governed by the same factors, apart from the direct correlation between plant growth and animal nutrition, which while perhaps not strictly a disease problem, has a major bearing on losses both directly from starvation and indirectly in its influence on susceptibility to disease. This emaphasizes the value of a multidisciplinary approach involving meteorologists, biologists, crop and animal specialists, and veterinarians.

Nutrition and Disease

The effects of the Sahel drought are the most striking example of the importance of rainfall in animal production. This is one of the main cattle producing areas of Africa (Temple & Thomas 1973), but a series of seasons of low rainfall is estimated to have caused losses of over 3.5M cattle and a 50% decrease in numbers. This change in climatic pattern may well not be short-term, and if this is the case it has serious implications for the redevelopment of the area. Local populations, particularly agriculturalists, generally respond to experience of weather variations, but the changes are often not frequent enough or rapid enough for such experience to be useful, and better long-term forecasting is urgently required to tackle catastrophies on this scale.

From experience in tropical conditions in Australia, Vercoe & Frisch (1982) suggest that under difficult conditions attention should be concentrated on idigenous breeds, and exotic cattle should be avoided, since the exotic animals although potentially more productive have less resistance to climatic, nutritional and disease stress. This approach is based on an acceptance of climatic limitations and maximum utilization of land resources within these limitations, increasing output by reducing losses rather than by improved productivity. This may not be attractive in the long-term, but offers some scope for improvement by selection of local stock, and is probably a realistic holding operation until better methods are available.

Under more normal conditions, the seasonality of rainfall results in traditional migrations of stock over large areas of northern and central Africa which have implications for disease incidence and control. The movement of animals is an important factor in the spread of microbial diseases, as in the recent recrudescence of rinderpest in Nigeria (Nawathe *et al.* 1983), but does permit the utilization of grazing in tsetse-infested areas in the southern dry season (Hall *et al.* 1983).

Climate and Pathogens

The main influence of weather conditions is on the transmission of infection. It is therefore appropriate to consider the disease agents in this respect, particularly in terms of the effect of the external environment on the development/survival/infectivity of disease.

A general approach to this aspect of disease epidemiology is simple recording and data collection. For example, Smith & Olubunmi (1983) recorded all cases of disease on a 1250 ha university farm in an area with distinct wet and dry seasons and identified those conditions such as listeriosis which show no climatic associations, in contrast to helminthiasis which is strongly correlated with rainfall and worthy of more detailed investigation. Records of this kind are also valuable in monitoring changing disease patterns, such as longer term climatic relationships, and in pinpointing faults in managment. It seems probable that a good deal of such data could be unearthed and manipulated by computer.

Microbial infection   Bacteria and viruses require no period of development or maturity outside the body of the host. The effect of environmental factors is thus largely on the survival of the organism during the transmission period.

(1) *Short range pathogens.* Two of the three most crippling cattle diseases are Rinderpest and CRPP, transmitted mainly by large droplet infection. Such droplets persist only for a very short time and both these organisms are very susceptible to dessication (Hislop 1979), but the close association which normally exists between cattle ensures highly efficient spread irrespective of weather conditions. In practice the influence of weather and climate is limited to its effect on animal

movement, annual migrations being largely responsible for spread (Rweyemamu 1981). However, highly effective control of both conditions is obtained by mass vaccination, thus while the indirect association of climate and disease is well recognized there would seem to be little to be gained from further clarification of the relationship.

(2) *Longer range pathogens.* Other disease agents spread over much wider areas, and therefore tend to show more marked associations with climatic and weather conditions. In the UK, the progress of foot and mouth outbreaks in relation to prevailing weather has been extensively documented and a numerical model to predict spread based on source strength/windspeed/turbulance has been successfully tested. In Africa an increased incidence is similarly associated with cool conditions which would favour aerosol persistence. Although giving some indication of rate and direction of spread, more detailed weather correlation is probably not justified since again containment is largely by vaccination. However, Daborn (1982) reports some rather unusual disease/weather correlations which are a reminder that simple pratical effects should never be overlooked. In the Songwe Valley in Malawi, spread of foot and mouth disease has orginated from Tanzania during the dry season, not because of favourable conditions for the virus but due to extensive movements and large group grazing by cattle which is prevented by flooding during the wet season. The vaccination programme is therefore initiated at the beginning of the dry season to give maximum protection during the period of maximum challenge. The state of the roads due to the weather can also limit the ability to carry out control programmes!

Parasitic infection

Parasitic life cycles usually involve either periods of obligatory development in the external environment or in a second, usually invertebrate, host; in these cases climatic involvement is greater. As a result, more data is available on seasonal variations in parasite populations and the parasite/weather interaction.

In temperate regions epidemiological studies have resulted in the highly successful liver fluke forecast (Ollerenshaw & Rowlands 1959) and similar bio-meteorological studies (Ollerenshaw & Smith 1969). Mathematical models to simulate the population dynamics of a number of parasite life cycles may be used to both assess the effects of variation in the dominant climatic parameters, and test the effects of control strategies (Thomas 1982, Meek & Morris 1981). Even in Europe and Australia this work is still at a relatively untried stage, and its successful application to African conditions in the immediate future is unlikely. However, a simpler approach based on the direct correlation of parasite or insect vector populations with temperature and rainfall does offer considerable scope for improvement in the effectiveness or cost of control measures.

(1) *Helminth infection.* On the lines of European studies, an abattoir survey in Malawi of fluke infection in cattle and field survey of infection in the snail intermediate host (Mzembe & Chaudry 1979, 1980) offers a basis for efficient control measures. This data might also form the basis for an epidemiological model of the type used by Meek & Morris (1981).

Gastro-intestinal nematode parasitism is a widespread cause of lowered production in ruminants, and the identification of weather-related seasonal patterns of the parasite in the host and in the external environment (Michel 1976, Thomas & Starr 1978) have greatly improved the effectiveness of control measures, particularly in the integration of drug use and pasture management. Such data is probably most useful in the more intensive highland areas of Africa but the general value of this approach is well demonstrated by the work of Chiejena & Emehelu (1984) in Nigeria which has closely correlated gastro-intestinal parasitism in cattle with rainfall, to show a clear seasonal pattern, forming a basis for control programmes.

(2) *Protozoan infection.* Among the most serious and intractable problems are the arthropod-transmitted protozoan diseases, tsetse-fly and trypanosomiasis are by far the most important, but tick-borne East Coast Fever, babesiasis, etc., are also highly significant causes of loss. The major climatic influences on these diseases are on the arthropod hosts.

(a) *Trypanosomiasis.* The incidence of trypanosomiasis is directly related to the distribution of the insect vector, which is largely climate-controlled either directly or via the climate-associated habitat. Unfortunately, climatic conditions are favourable over a large area of Africa, some 7M km² being affected, and it is ominous that in an extensive review of tsetse control (Dame & Jordan 1981) no reference was made to climate or weather data. The most promising approach to reducing the impact of this condition is probably the promotion of trypanotolerant breeds of cattle (Murray *et al.* 1982). However, in a further 3M km² where a seasonal pattern of tsetse population occurs, an understanding of the climate/vector relationship must be of value (see Griffin & Allenby 1979). Such data are valuable in prophylactic chemotherapy and in timing insecticidal treatment in clearance schemes, and in the longer-term a more positive approach to the ecology of the insect must deserve support, such as that on malarial transmission which has led to the development of a promising model (Molineux *et al.* 1978).

Where the environmental requirements of the insect vector are more specific, climatic data are of more direct utility, as in the arbovirus disease Rift Valley Fever. Epizootics occur at relatively infrequent intervals associated with periods of heavy and prolonged rainfall, and it seems possible that analysis of disease records could pinpoint with some accuracy the potential risk conditions, which would make vector control measures cost-effective.

(b) *Tick-borne diseases.* East Coast Fever is limited to the area of the tick *Rhipicephalus appendiculatus*, and other tick-borne diseases show similar associations with tick distribution and seasonal activity. Like tsetses and mosquitos, ticks are widespread in Africa, but control is much more effective via the destruction of the arthropod on the host by repeated dipping, if necessary at twice-weekly intervals. However, this is expensive and time consuming and there is an urgent need to make dipping more strategic and more efficient (Keating 1983). Advances can best be made on the basis of a better understanding of tick ecology and this offers an obvious role for biometeorology. In contrast to trypanosomiasis, possibly because of the effectiveness of acaricidal treatments, there has been more interest and activity in this field in relation to ticks in Africa, and in addition extensive studies elsewhere, particularly in Australia, can be applied to African conditions. One model (Short & Norval 1981) predicts the seasonal occurrence of adult ticks and correlates well with field observations; this opens up opportunities to modify and test control programmes without the necessity of extensive field studies.

Conclusions

The most significant advances in the control of major disease problems in Africa have been through immunology, as in the development of vaccines for rinderpest, CBPP and FMD, and the work of ILRAD on trypanotolerance in cattle. However, the application of biometeorology can make a significant contribution in a number of areas, and has the great advantage of flexibility in that useful information can be derived from simple disease recording and analysis as well as from complex modelling. The climatic data required is widely available.

References

Beets, W.C. (1978) The agricultural environment of Eastern and Southern Africa and its use. *Agriculture and Environment* 4, 5-24.

Bourn, D. (1978) Cattle, rainfall and tsetse in Africa. *Journal of Arid Environments* 1, 49-61.

Bunting, A.H. (1961) Some problems of agricultural meteorology in tropical Africa. *Geography* 46, 283-294.

Chiejena, S.N.; Emehelu, C.O. (1984) Parasitic gastro-enteritis of beef cattle in Nsukka, Eastern Nigeria. I. Seasonal changes in populations of infective larvae on herbage and in soil. *Research in Veterinary Science*, in press.

Daborn, C.J. (1982) Foot-and-mouth disease control in the Songwe Valley, Malawi – a review. *Tropical Animal Health and Production* 14, 185-188.

Dame, D.A.; Jordan, A.M. (1981) Control of Tsetse Flies. *Advances in Veterinary Science and Comparative Medicine* 25, 101-119.

Griffin, L.; Allonby, E.W. (1979) Studies on the epidemiology of trypanosomiasis in sheep and goats in Kenya. *Tropical Animal Health and Production* 11, 133-142.

Hall, M.J.R.; Kheir, S.M.; Rahman, A.H.A.; Naga, S. (1983) Tsetse and trypanosomiasis survey of southern Darfur Province, Sudan. *Tropical Animal Health and Production* 15, 191-205.

Hislop, N. St G. (1979) Observations of the survival and infectivity of airborne rinderpest virus. *International Journal of Biometeorology* 23, 1-7.

Keating, M.I. (1983) Tick control by chemical ixodicides in Kenya: A review 1912 to 1981. *Tropical Animal Health and Production* 15, 1-6.

Meek, A.H.; Morris, R.S. (1981) A computer simulation model of ovine fascioliasis. *Agricultural Systems* 7, 49-77.

Michel, J.F. (1976) The epidemiology and control of some nematode infections in grazing animals. *Advances in Parasitology* 14, 279-366.

Mahadeven, P. (1982) Pastures and animal production. In *Nutritional Limits to Animal Production from Pastures* (J.B. Hacker, ed.), 1-17. Farnham Royal; Commonwealth Agricultural Bureaux.

Molineux, L.; Dietz, K.; Thomas, A. (1978) Further epidemiological evaluation of a malaria model. *Bulletin of the World Health Organisation* 56, 565-571.

Murray, M.; Morrison, W.I.; Whitelaw, D.D. (1982) Host susceptibility to African trypanosomiasis: trypanotolerance. *Advances in Parasitology* 21, 2-68.

Mzembe, S.A.T.; Chaudry, M.A. (1979) The epidemiology of fascioliasis in Malawi. 1. The epidemiology in the intermediate host. *Tropical Animal Health and Production* 11, 246-250.

Mzembe, S.A.T.; Chaudry, M.A. (1980) The epidemiology of fascioliasis in Malawi. Part 2. Epidemiology in the definitive host. *Tropical Animal Health and Production* 13, 27-33.

Nawathe, D.R.; Lamorde, A.G.; Kumar, S. (1983) Recrudescence of rinderpest in Nigeria. *Veterinary Record* 113, 156-157.

Nieuwolt, S. (1982) Tropical rainfall variability – the agroclimate impact. *Agriculture and Environment* 7, 135-148.

Ollerenshaw, C.B.; Rowlands, W.T. (1959) A method of forecasting the incidence of fascioliasis in Anglesey. *Veterinary Record* 71, 591-598.

Ollerenshaw, C.B.; Smith, L.P. (1969) Meteorological factors and forecasts of helminthic disease. *Advances in Parasitology* 7, 283-323.

Rweyemamu, M.M. (1981) Surveillance and control of virus diseases: Africa. In *Virus Diseases of Food Animals*. Vol. 1. *International Perspectives* (E.P.J. Gibbs, ed.), 79-94. London and New York; Academic Press.

Short, N.J.; Norval, R.A.I. (1981) Regulation of seasonal occurrence in the tick *Rhipicephalus appendiculatus* Newmann, 1901. *Tropical Animal Health and Production* 13, 19-26.

Smith, O.B.; Olubunmi, P.A. (1983) The incidence, prevalence and seasonality of livestock diseases in a hot, humid forest zone of Nigeria. *Tropical Animal Production* 8, 7-14.

Temple, R.S.; Thomas, M.E.R. (1973) The Sahelian drought – a disaster for livestock populations. *World Animal Review* 8, 1-7.

Thomas, R.J. (1982) The ecological basis of parasite control: Nematodes. *Veterinary Parasitology* 11, 9-24.

Thomas, R.J.; Starr, J.R. (1978) Forecasting the peak of gastro-intestinal nematode infection in lambs. *Veterinary Record* 103, 465-468.

Vercoe, J.F.; Frisch, J.E. (1982) Animal breeding for improved productivity In *Nutritional Limits to Animal Production from Pastures* (J.B. Hacker, ed.), 327-342. Farnham Royal; Commonwealth Agricultural Bureaux.

Chapter 62

Forecasting and Weather Services for Agriculture in Africa

C. J. STIGTER

Section Agricultural Physics, Physics Department, University of Dar es Salaam, Box 35063, Dar es Salaam, Tanzania.

Introduction

Smith's 1967 comment on weather forecasts for aviation applies increasingly and also to general weather forecasts in the First World, " much has been done to overcome the inherent deficiencies of a weather forecast by a highly organized system of *functional application"*. But he immediately added, "relatively little has been done of a similar nature for the more complicated requirements of agriculture". In the years since this statement was made, a slow but clear improvement has occurred in the First World but for Third World agriculture it remains true.

In the useful new *WMO-Guide to Agricultural Meteorological Practices*, a definition by Wang is used which indicates the difference between weather forecasting for agriculture and agrometeorological forecasting. The latter is defined as crop prediction without weather forecasting and it is necessary to distinguish between the two and later to unite them again in defining requirements for a comprehensive operational weather service, including functional advice, for farmers in Africa.

Predicting crop conditions influence yield outlook positively. Weather advisors should assist farmers here, but the present situation has been summarized by Arkin & Dugas as "Management oriented weather advisories are generally not available to farmers. Advisories currently available have little practical utility and therefore are not often incorporated into the management decision making process". When this is true for the developed world, it *must* also be true for Africa! Although in Africa most advice should be on a non-real time basis, the chances appear poorest there that this century will, even with this restriction, see any improvement in this situation. Essential improvement remains unlikely unless specific policy-decisions are made directed at making agriculture the focus of operational weather forecasting and towards developing operational research and services in agrometeorological forecasting and management. It is difficult to be optimistic on these aspects.

Pessimism

Some reasons for a pessimistic view are:

(1) Whoever has in the past few years entered weather forecasting units in African countries, will have shared some of my experiences and found services generally deteriorating through a lack of consumables, spare parts, transport, and other communication facilities. It has been pointed out to me that international aviation safety in parts of Africa would have been endangered if the services of these units was as important now as it was 20 years ago. Fortunately this is not the case due to changes in aviation practices.

(2) Unfortunately, this situation has led only slowly to changes in the training of African meteorologists. As Davy said of African climatologists "Work on the bench of an aerodrome office has been the foremost experience of the great majority of meteorologists who take up posts in climatology. The value of a background of synoptic meteorology is recognized but it is not essential". Some of these personnel will also be asked to provide inputs into agrometeorological services.

(3) Agronomists of all levels with an adequate knowledge of environmental meteorology are also very scarce in Africa. This relates to the general negligence of agricultural and environmental sciences and education in the developing world.

Remedies for these three causes of pessimism are known, but are necessarily of a long-term nature. Two further causes for pessimism will now be considered more closely.

(4) The difficulties decision-makers experience in obtaining, absorbing and applying new knowledge leading to a more comprehensive appraisal of yield fluctuations and limitations, and to improved management decisions. This is true, with a few exceptions for identical reasons, for Government ministries, extension agencies, and the farmer (or farming communities).

(5) Until recently, meteorological problems in agriculture have been rarely treated with the same degree of operational urgency as meteorological problems in aviation. Validated operational information of a kind that can be used on a relatively large scale is hardly available at Government level and is virtually non-existent at the farmer level.

Environmental Information Needs of the Decision-Maker

Holdgate recently looked at the application of knowledge to specific resource problems under this head. We interpret the weather and climate of agricultural lands here as a natural resource, and can therefore consider with him two issues: (1) Can we identify the types of decision in the agricultural enterprise that should draw on weather information, and (2) can we discover some properties of the system by which weather information is translated into management, or fails to be used? The answer to these questions must be given at the three levels of Government ministries, extension services, and farmers.

Agricultural strategies

Long-term production factors ("strategic factors") with agrometeorological/climatological components exist, land use, design and extent of farming systems, and production planning. These are also a function of political conditions. Whether one believes with Eicher in a new incentives relationship of African Governments towards farmers, or with Hyden that a growing conflict in relationships is a precondition for development in Africa, decisions for action in both cases are partly dependent on the interpretation and use of weather and climate information; whether the information is used and how. The following examples illustrate this point.

(a) Crop Monitoring (or Forecasting) and Early Warning System projects, as launched in some African countries with assistance from FAO, are examples of improved attempts of application, on an almost real-time basis, of knowledge as to how atmospheric demand and soil-water conditions determine yield. Local weather services are involved in such exercises, but also Ministries of Agriculture. If such information is of a high quality, the Government would be informed on the actual status, and provided with an estimate of possible yield, of the country's main crop(s) in different regions. However, the use of such knowledge in decisions and their implementation is often largely political in nature and because of their implications ministries can be reluctant to deal openly with such information. This may pose problems for those involved in collecting such knowledge.

(b) Country-wide food strategies as now introduced in some African countries are examples of modern production planning. For example, suitability for the growth of certain crops, or choice of methodology and places of bulk grain storage, are in such strategies at least partly determined by climatic factors. In this case agroecological zonification, for various purposes and on various scales, should be carried out with the assistance of local weather services. However, decisions based on that knowledge taken at governmental level may have adverse consequences for certain regions or groups of farmers.

In conclusion, identification of the types of decision at the Government level which require weather and climate information is not difficult. There is tremendous scope for weather services to agriculture in Africa. However, political factors external to such services, related to information being classified and to the absence of two-way communication on derived implications, should not remain unmentioned as a blockage to actual management.

There are problems in establishing operational weather services easier to recognise and acknowledge. These may of course complicate management decisions on agricultural strategies or their implementation. They include the organization of primary information collection; the availability and accessibility of the appropriate basic and derived information; decisions on priorities within the weather services; and the training and number of employees at weather services and in extension services for communication with the farmer.

Agricultural tactics and techniques

At the farmer level, seasonal production factors ("tactics") and daily avoidance, protection and improvement factors ("techniques") require particular attention. The types of decision bearing on weather information can again be identified reasonably easily, but the weather services needed by the majority of African farmers are quite different from those of Governments; they are developed to an even lesser degree at this time.

Assessment of the situation in deteriorating agricultural production and failing economies in general indicates that a majority of African farmers will for the time being have to be satisfied with no, or only very limited, external inputs in food cropping. As stated by Harwood "This [the farmer's] perspective may lead us to the conclusion that improved subsistence for the small farm family may be the most that can reasonably be expected, or even hoped for, in the near future... Subsistence may be the best the small farmer can do until off-farm development offers an alternative to his half-hectare for the support of his family."

Small farmers have come to cope with their environment by developing practices that result in a low but guaranteed minimum yield under severe climatic and poor soil conditions. In such a low-input agriculture, weather factors limiting production more frequently than in more modern farming systems. Interestingly, microclimate management and manipulation techniques are amongst those developed practices. Microclimate in agricultural lands is one of the natural resources on which there is little scientific information with regard to its traditional management and manipulation. As lamented by Bunting, "mine must be the only branch of applied science, incidentally, in which we let people loose to improve the working of a system when they do not know how it works now."

The types of management decisions in low-input agriculture that should draw on (micro-)climate and (micro-)weather information have been recently catalogued by Stigter. These range from shading and coverage against night frost, to the use of solar energy for field drying; from the degree of tillage and mulching to the use of shelter and (protection against) natural dew; and from wind break use to protection against the mechanical impact of rain and hail. The quantification of such techniques, the only way to obtain specific knowledge on cause and effect relationships, should develop as a line of operational research.

The role of weather services

In the tasks of weather services to provide environmental information to the African farmer, as a decision-maker we may distinguish two kinds: (1) To give, on a relatively short-notice basis, advice on the (accumulated) weather suitability, or a necessity caused by weather, to carry out certain management operations (that mean functional forecasting). (2) To provide input for and stimulate external research on quantitative advice on the optimum benefits of such management operations (i.e. functional advisories).

Farmers need advice not only when to mulch (for low-input agriculture), or when to irrigate (for high-input cash-crop agriculture), but also approximately the extent to which weather components determine that.

Forecasting for the African Farmer

Initiatives from strong agrometeorological sections of weather services, to produce simple operational and functional weather forecasting *and* agrometeorological forecasting methods, are required. This can only be achieved in relation to a world context and the World Meteorological Office (WMO) is an appropriate co-ordinating organization. The CARS-system to be established in the near future should be an important source of such methodologies. The experience obtained by

AGRHYMET (Chapter 57) should also be used. Methods need to be tried and validated locally, in co-operation with the "customers" through extension services.

This implies in practice in relation to weather forecasting that an approach as given by the WMO or as recently proposed as guidance material for agrometeorological services to rice farmers in the humid tropics, should be validated to serve operations in low-input and in cash-crop agriculture in African countries. These forecasts would ideally be in the form of a limited number of ranges, to one of which the forecasted parameter is assigned. This number could then be changed after validation. For specific operations in specific crops, a specific selection of minimally necessary weather elements is required, depending on the needs of the farmer-community. Too much and unnecessarily detailed information can be almost as harmful as insufficient rough information.

It would be advisable to develop such services and test them with research, extension, and weather service personnel in pilot projects in actual farmer's fields prior to establishment region-wide. In these pilot projects, the collection of crop stage and condition information should also be carried out in order to validate functional agrometeorological forecasting (including crop monitoring) methods. Simultaneously, information on trials of actual microclimate modification techniques derived from the farmer's experience should be collected. Advisories, where the second task of weather services lies, must be developed from this knowledge.

The main conclusion is that also at the farmer's level identification of the types of decisions that should draw on weather information limits improvement. Weather information fails to be translated into management because of the absence of operational weather services with and through extension services. Although the knowledge required is different, this is related to the same problems mentioned above as causes for the lack of action towards management in agricultural strategies.

Weather Advisories for the African Farmer

In order to develop better operational and functional weather and agrometeorological forecasting methods, an operational research basis in African agrometeorology must be developed. This would provide a bridge between basic and field research exploring cause-and-effect relationships in farming operations and actual service needs of the farmer. However, management decisions often require more than functional forecasting, and functional advisories are required. Quantitative advice on the optimum benefits of management operations to capitalize, or not to lose too much, as a result of weather events is needed. Such advisories must be derived from the same operational research basis.

Finally there are three additional constraints to the establishment of an adequate research basis for weather services including advisories.

(1) There is need for communication among research workers from different disciplines and between researchers and those close to the farmer's community. For these purposes we established in Tanzania five years ago a National Agrometeorological Committee (NAC) under the National Directorate of Meteorology. The main result of this so far has been much-improved contact between research scientists undertaking work with an agrometeorological component. Although still far from an Agrometeorological Advisory Service, the NAC in Tanzania functions as a catalyst informing its members as to what is going on in the field of agrometeorology in the country. This is a prerequisite for the development of such a service.

(2) From my emphasis on research in Africa into low-input agriculture it follows that such research should be in relation to existing farming systems, something in line with recent trends in, among other examples, the International Agricultural Research Institutes. I believe that what is already being called a second green revolution, with less spectacular but also socially less disruptive results, will be of tremendous importance to Africa. However, the distance between the International Institutes and agriculture in Africa in general is large. When Moyes listed factors which limit the application of results in science and technology by the very poor, his last but not least factor was the absence of local institutions to assist in the transformation of relevant knowledge into packages which can be absorbed by the poorest strata. In this respect, I would like

to make a plea for long-term outside assistance in the establishment of local affiliates of the International Agricultural Research Institutes. This could be achieved by "adoption" of existing rural research institutes or stations. Local affiliated research stations could then follow strategies and tactics developed by the parent institutes and validate them at the local level. Such units would require research and extension staff able to work with the farmers. Experience from Latin America reviewed by White shows the tremendous value of a participatory approach to agricultural research and development.

(3) Meteorology and climatology have so far played only a limited role in the work of the International Agricultural Research Institutes. This is largely due to a lack of attention to the operational urgency towards agriculture from those in environmental research. Harwood has correctly listed the detailed classification of environmental factors among the critical elements largely or entirely omitted in most current systems of agricultural research and development in relation to small-farms.

In summary, agricultural research in Africa should include the agronomical consequences of current and improved techniques of managing and manipulating the microclimate; agrometeorological research should focus there and do so in close relationship to agricultural reality. The education of agrometeorologists and some agronomists also requires redirection and intensification. African weather services need to provide inputs at governmental level and initiate validations of operational weather forecasting for agriculture and agrometeorological forecasting methods with the extension services. These are of paramount importance, but should be accompanied by inputs in and stimulation of operational research by asking appropriate questions; only then can good weather advisories be developed. This is the only means by which national agricultural research and weather services in Africa will have a positive impact on yield fluctuations, amounts, quality and crop protection by the end of this century – through the application of new and improved operational forecasting and management methods. Only then will there be the possibility of more than doubling the food availability when the population of Africa has doubled in 25 years.

I acknowledge useful discussion on parts of this paper with Professor A. Weiss during his visit to my section in Dar es Salaam in early 1983. References are available on request from the author, c/o Mrs. C. Piteo (Secretary), Agricultural University of Wageningen, Department of Physics and Meteorology, Duivendual 1, Wageningen, The Netherlands.

Chapter 63

Remote Sensing for Natural Resources Development in Tropical Africa

I. Availability and Use of Remote Sensing Data

A. B. TEMU

Division of Forestry, University of Dar es Salaam, P.O. Box 3009, Chuo Kikuu, Morogoro, Tanzania.

The Need for Resource Information

The value of any given resource can be expressed as a function of its quantity (or relative abundance), quality and time. In economies dependent upon agriculture, land resources information and climatic data are vital tools of land resource management. In the struggle to meet their immediate needs, many developing countries pay less attention to the development of efficient land resource information systems. The consequence is a threat on their economies and on the long term availability of land-based products. This contribution focusses on the use of remote sensing technology in the identification and monitoring of vegetation resources, with emphasis on agricultural crops.

Background

Remote sensing is defined (Lillesand & Kieffer 1979) as "The science or art of obtaining information about an object, area or phenomenon through the analysis of data acquired by a device that is not in contact with the object, area or phenomenon under investigation". The remote sensing technology developed was initially exclusively for military purposes, but some of its products are now available for civilian use. These include aerial photographs, aerial thermographs, satellite and space imagery (from Multi-Spectral Scanners (MSS), Return Beam Vidicon (RBV) mounted on land resources satellites, and from Apollo vehicles), and microwave imagery (e.g. from Side Looking Airborne Radar*). The various remote sensing devices may be placed near to the earth's surface (e.g. large-scale aerial photography by light aircraft) or thousands of kilometres away (e.g. space shuttles). The images acquired at the various levels are different, but complimentary as far as interpretation is concerned. Aerial photographs are normally taken as the need arises because they are expensive, but satellite imagery is available for any given part of the earth's surface every 18 days; this provides cheap and continuous monitoring of crop conditions.

---

*RADAR is an acronym for RAdio Detection And Ranging.

Utilization of Remote Sensing Products

Visual remote sensing products are normally presented as photographs or facsimiles, known as imagery, on paper prints, transparencies, or television screens. Their main advantage is in providing the possibility to view a large parcel at a glance. The interpreter must be trained to distinguish various land features, natural or man-made, and to do detailed plant and animal studies.

Remote sensing imagery may be used for the sampling of vegetation resources, mapping of land and drainage patterns, land use classification, monitoring land clearing and replanting, monitoring floods, soil erosion and stream/lake sedimentation, monitoring air and water pollution, monitoring fires, detecting crop types (including pastures), monitoring changes in cropping patterns, predicting weather (for meteorologists), predicting crop yields, and detecting and monitoring of the physiological condition of a crop. This last use is a very interesting discovery. Through the use of infra-red (IR) films it is possible to detect the physiological status of a crop long before the unaided human eye could. The cause of stress may be drought, pathological or otherwise, and has to be established from ground surveys. The technique provides advance warning on possible outbreaks of disease or impending drought. Using this technique, diseases such as stem rust in wheat, potato blight, bacterial blight in field beans, and leaf spot in sugar beets, as well as insect damage and drought, have been detected (National Academy of Sciences 1970).

Crop identification and monitoring requires highly specialized skills which can be developed rapidly by those with an agricultural background. Experiments in Arizona using high altitude multi-date IR colour photography produced 80-90% successful crop identification (Lauer 1971); a very high success rate as rarely have interpreters achieved higher than 70% success. The temporal and spatial distribution, level of development and vigour of a crop will normally influence its signature, which therefore changes throughout the life of the crop.

Remote sensing techniques have been applied successfully in forestry in the estimation of areas, and the monitoring of land clearing, forest fires, and reforestation. Large-scale aerial photographs are used to measure tree height and crown diameters, in turn used as variables in the estimation of wood volume. However, small-scale exploitation of forest resources, for instance the extraction of firewood, does not lend itself to immediate detection and monitoring via remote sensing techniques as the changes are small and accumulate over several years.

Discussion and Conclusion

Identification and monitoring of agricultural crops is necessary for the proper crop management at the national and regional levels. The use of remote sensing techniques is one way of complementing and strengthening ground-based crop monitoring systems. Problems which may be encountered when using remote sensing technology in developing countries include: (1) very high variation in crops arising from variabiltiy in seeds and/or mixed cropping (producing highly variable crop signatures); (2) highly variable cropping practices especially in espacement and variability in planting time on the same field; (3) shifting cultivation makes it difficult to monitor crop production; (4) individual farm holdings are often too small to be discernible on small scale imagery; (5) highly variable species composition in forests and woodlands; (6) small-scale extraction of undergrowth or single trees from forests or woodlands is difficult to detect; (7) lack of trained personnel and equipment to do image interpretation; and (8) limited financial resources to buy and maintain a regular flow of imagery.

Despite these shortcomings, there is room to employ remote sensing technology effectively, especially if the products can be made available for multi-disciplinary interpretation. Agricultural, forestry and livestock production all have vegetation as a common denominator. Combined interpretation for the three disciplines would be more cost-effective than when each discipline is taken separately. Remote sensing technology should be adopted and adapted to African conditions for the assessment and monitoring of land resources.

References

Lauer, D. T. (1971) Testing multiband and multidate photography for crop indentification. In *Proceedings of the International Workshop on Earth Resources Survey Systems*, 33-45. Washington, DC; Government Printing Office.

Lillesand, T. M.; Kiefer, R. W. (1979) *Remote Sensing and Image Interpretation.* New York; John Wiley & Sons.

National Academy of Sciences (1970) *Remote Sensing, with Special Reference to Agriculture and Forestry.* Washington, DC; Agricultural Board, National Research Council.

II. Applications to Agro-meteorology

G. DUGDALE

Department of Meteorology, University of Reading, 2 Earley Gate, Whiteknights, Reading, Berks RG2 2AR, UK.

Introduction

In principle, remote sensing is an ideal method for obtaining agrometeorological data. It offers area average measurements over a wide range of space and time scales, and can be applied to the measurement of rainfall, soil moisture and surface water storage, to short range forecasting as well as to features of specialized interest. Remotely sensed data from the ground, satellites and aircraft have been integrated into meteorological data collection system for many years. During the last ten years, there has been progress in applying remotely sensed data to areas of direct interest to agriculturalists. This contribution considers the use of data which are now routinely available from operational satellite systems, concentrating on techniques for using satellite data to monitor water resources. Some of the techniques are ready for use while others are being developed and tested.

Table 1   Operational remote sensing systems useful to agro-meteorology.

Type of satellite or platform	Minimum scale observable	Maximum frequency of observation	Agrometeorological quantity deriveable
Geostationary meteorological satellite	2-5 km	48 day^{-1}	soil moisture, rainfall, short range forecasts
Polar orbiting meteorological satellite	1.1 km	2 day^{-1}	rainfall in conjunction with the above vegetation indices
Earth resources satellites	80 m	1-16 days	surface water extent, land use
Ground based radar	1 km	continuous	short range forecasts, rainfall interpolation

The scales of satellite-sensed data of interest to agriculture range from about 80m with data each sixteen days, to 5-10 km with data many times per day. Radar can give continuous coverage with a resolution of about 1 km at distances up to 150 km. Each scale of measurement has its use according to size of the phenomenon to be monitored or the scale of the planning problem. In general, the better the space resolution the less frequently data is available. The cost of mapped data is usually directly related to its resolution. Table 1 lists types of operational data sources, their characteristics, and uses.

The four main types of sensor which are used in agro-meteorological studies are the thermal infra-red which gives information on the temperature of the viewed surface, the visible which indicates the amount of solar radiation reflected by the surface, the spectrometer which measures the reflected solar radiation in different wavebands and the microwave radiometer (radar) for measuring rainfall and other forms of water. See Barrett & Curtis (1982) for further information.

Rainfall Estimation

Microwave radiometers give direct information on the presence and size-distribution of raindrops. Thermal infra-red and visible data from satellites can be interpreted in terms of cloud cover and this, in turn, related to rainfall.

Microwave methods (including radar)

There are at present no operational satellites which can give rainfall data over land from microwave radiometers, though several experimental systems are being developed. Ground-based radar suitably calibrated and used in conjunction with telemetered rain gauges offers the best possible instantaneous information on rainfall. However, these systems are expensive and require skilled technical support and their use can probably only be justified in heavily populated areas or in zones which can react rapidly and profitably to advice of imminent rainfall events.

Useful rainfall information can be extracted from less sophisticated storm warning radar installed at international airports; this does not give quantitative information on the intensity of rainfall, but the number and duration of events can be monitored and calibrated from existing rain gauges. This method improves interpolation between conventional rain gauges and is also valuable for assessing flood risk where floods are caused by storms in areas remote from conventional networks. The use of existing storm radar for these measurements involves no increase in capital costs, although maintenance and labour charges will increase.

Cloud identification methods

The basis of rainfall estimation from visible and infra-red satellite data is the association of rainfall with cold (high), bright clouds. Orbiting satellites giving 2 images per day and geostationary satellites giving 48 images per day have been used in estimations of rainfall; geostationary satellites are the most promising for the short-lived convective storms which predominate over Africa, although showers affecting less than 25 km² are unlikely to be identified.

Cloud indexing methods have been used mainly with high resolution data from polar orbiting satellites. Each cloud type, as identified in high quality photographic imagery, is ascribed a rainfall probability so that the rainfall over an area can be calculated from the cloud type, its duration in the area, the rainfall associated with that type, and the prevailing synoptic situation. This method has proved effective in areas where most of the rainfall is associated with large meteorological systems, but relies on skilled interpretation. Its utility where most of the rain comes from short-lived localized storms is unproven.

A method which shows great potential in the tropics is the "life history method". The rainfall from convective clouds is associated with the rate of expansion of the top of the cloud as recorded in the digital infra-red and visible band geostationary satellite imagery. This method has given good results over parts of the USA and the tropical Atlantic ocean. The limitation of the method is that the relationships between rainfall and cloud characteristics vary with the climatic zone. At Reading University a group is now developing an operational rainfall measuring scheme based on life history methods and on the known mesoscale structure of West African storms (Wilkinson *et al.* 1983).

The operational application of both these techniques involves the use of satellite data receiving equipment, image processors and computers. However, the large capital expense may be justified on the basis that a single centre can provide coverage for most of Africa, and FAO is investigating the possibility of developing such a system.

Short Range Weather Forecasts

The localized nature of rainfall in Africa makes weather forecasting very difficult and prevents the accurate prediction of storm development. In many areas it is, however, possible to forecast the movement of a storm for a few hours after it has developed. The position and movement of growing storms can be determined from storm warning radar or geostationary satellite data. Aviation forecasters employ these extrapolation techniques, but seldom use the geostationary satellite data. There are many agricultural activities which could profitably respond to accurate warnings of the arrival of a storm, and both the basic data and the forecasting skills are available to provide this service.

Soil Moisture Measurement

Remote sensing techniques for the direct measurement of the moisture content of the upper layers of the soil are under development, but not yet nearing an operational status. Indirect methods using available satellite data are being tested and the indications are that it will soon be possible to estimate the moisture content of the top 10-15 cm of bare or sparsely vegetated soils (Wilkinson *et al.* 1983).

The method uses the thermal infra-red channel of the geostationary satellites to measure the daily change in soil surface temperature under clear sky conditions. Other factors being equal, the amplitude of this temperature change is related to the moisture content of the soil. The relationship varies from one type of soil to another, but the temperature ranges in the extreme conditions of completely dry soil and soil at field capacity can also be established from the satellite. Soil moisture estimates are expected to be most precise in sandy soils as these show a large thermal response to changes in moisture content. With such soils, perhaps a five-point moisture scale could be measured between dry and saturated states. The implementation of this technique requires a capital investment in equipment similar to that for rainfall mapping from geostationary satellite data, but the same centres could carry out both tasks.

Surface Water Monitoring

The LANDSAT satellite with a resolution of about 80 m can be used to give useful data on the variation in surface water storage in lakes and swamps. The maximum frequency of this data is 16 days which limits its utility to measuring the extent of slowly varying water levels. The operational polar orbiting satellites can give daily data but here the resolution is about 1 km which is inadequate except for the most extensive floods. However, the sixteen day data can usefully monitor the seasonal depletion and charging of lakes and reservoirs.

Special Applications: Pest Control

Remote sensing is now being introduced into the desert locust control programme (C. J. Tucker *et al.*, pers. comm.) using spectral red and near infra-red data from polar orbiting operational meteorological satellites. In these wavebands, the solar radiation is reflected very differently by growing vegetation and by dry matter so that areas in which vegetation is developing can be identified. To remove the effects of cloud, data is composited over 14 days when a biomass map is produced. In this way the development of potential breeding sites for desert locusts can be identified. It is hoped to incorporate satellite based soil moisture and precipitation mapping into the same early warning system.

Desertification

The best satellite technique for monitoring desertification is that of land use mapping from LANDSAT data. However, the system described in the previous paragraph can also be used to monitor the extent of seasonal vegetation growth on desert margins. This long-term large-scale mapping is just starting and needs to continue for several years before long-term trends can be separated from normal year to year variations.

Conclusions

Much raw satellite data is routinely available which could be processed to give information on the availability of water for agriculture and rangelands. The techniques for processing the data are available and some of them have been tested and applied locally. When questioned, agriculturalists and planners agree that rainfall and other hydrological data is necessary for their work. However, they are much less forthcoming when questioned on the form in which they would like the data and on what impact it would have on their activities and decisions. A major factor preventing the fuller implementation of the techniques for using remotely sensed data is the lack of a well-formulated demand from potential users. If the requests were clearly made, the methods and funds for providing the answers could be found.

References

Barrett, E. C.; Curtis, L. C. (1982) *Introduction to Environmental Remote Sensing.* London; Chapman & Hall, 352 pp.
Wilkinson, G. G.; Ward, N. R.; Milford, J. R.; Dugdale, G. (1983) Remote sensing for rangeland management in the Sahel. In *Remote Sensing for Rangeland Management*, 93-107. Silsoe; Remote Sensing Society.

Chapter 64

Land, Food and Population in Africa

G.M. HIGGINS[1] and A.H. KASSAM[2]

[1]Chief, Soil Resources, Management and Conservation Service, Land and Water Development Division, FAO, Rome, Italy;
[2]Senior Land Resources Consultant, Echemes Development Services, 5/7 Singer Street, London EC2, UK.

Introduction

The balance, or rather imbalance, between land, food and population is one of the major preoccupations of humanity at the present time. This contribution reports a recent FAO activity (FAO 1982) which sheds some new light on this issue.

The activity, carried out in collaboration with the International Institute for Applied Systems Analysis with the support of the United Nations Fund for Populations Activities, makes a first approximation of the potential population supporting capacities of the lands of the developing world in comparison with their actual and projected populations. In particular, the results for Africa are examined and areas identified where land resources are insufficient to meet food needs.

Methodology

A basic theme in the study is that the ability of land to produce is limited and any attempts to produce food in excess of these limits results in a vicious cycle of degradation and ever-reducing yields. Limits to production are set by soil conditions, climatic conditions and the use and management applied to land.

Accordingly, in attempting to quantify potential productivity and potential food production, it is necessary first to define the use being considered.

In the present study, *three input-levels/use circumstances* have been considered: (1) A *low-input level*, with currently grown mixture of crops, no fertilizers, no application for pest, disease or weed control, no conservation measures and manual labour with handtools. (2) An *intermediate-level*, with half optimum mixture of crops and improved varieties, some pest, disease and weed control, some conservation measures and manual labour with improved handtools or animal traction with improved implements. And (3) a *high-input level*, with optimum mixture of crops and high yielding varieties, optimum fertilizer use, optimum pest, disease and weed control, complete conservation measures and mechanization.

Soil conditions have been taken into account in the assessment by using the Soil Map of the World (FAO 1971-81). Interpretation of this soils data provides an insight to existing soil conditions of Africa with regard to rainfed cultivation potentials. Soils having some form of constraint to rainfed crop production comprise more than 80% of the continent. Coarse textures, shallow depth and fertility limitations are among the major constraints in the region.

Climatic conditions have been taken into account in the assessment through compilation and use of a specifically created climatic inventory characterizing both temperature and moisture conditions through the concept of length of growing period, the duration (days) when both moisture and temperature are sufficient to permit crop growth. Interpretation of the climatic inventory provides a summary of climatic conditions for rainfed cultivation in Africa. The areas which are climatically suitable, in some degree, for rainfed crop production are those with short, long and year-round humid growing periods with no severe temperature constraints. They amount to only 53% of the area of the continent.

Overlay of the climatic inventory on the Soil Map results in an inventory of *land units* each with their own unique soil and climatic conditions, and provides the computerized climatic and soil resource database created by the study and now available for all countries in Africa (FAO 1978-80). This database quantifies the total extents of each soil unit, broken down by texture class, slope class and phase (where present) as they occur in each major climate and in each length of growing period zone, on a country by country basis.

This resource database is the foundation of the population potential assessment and, from it, the potential yield of major food crops has been calculated. Inherent in this methodology is an initial crop and level of input specific determination of agro-climatic suitabilities for rainfed cultivation. These agro-climatic suitabilities are subsequently modified by appropriate and specially compiled soil, slope, texture and phase ratings to provide crop and level of input specific *land suitability assessments* in terms of potential crop yields.

Land areas capable of yielding 80% or more of the maximum yield attainable are classified as very suitable; less than 80 to 40% as suitable; ones less than 40 to 20% as marginally suitable; and less than 20% as not suitable. This classification allows calculation of the total extents of potentially cultivable land as well as the extents of individual crop suitabilities. The results show that the area suitable for maize is extensive (424M ha), but the area suitable for wheat is small (38M ha). The total extent of potentially cultivable rainfed land is nearly 800M ha compared with a presently cultivated area of 185M ha (FAO 1981).

Such basic data (FAO 1978-80), allows calculation of the potential food production and, hence, the *potential population supporting estimates*. The estimates are obtained by analysing separately each area of the land inventory to ascertain which use has the highest calorie-protein production potential under the soil and climatic conditions in each area. Prior to this analysis, deductions are made for land required for non-agricultural use, for irrigation, and for fallow period requirements. Limitations imposed by degradation hazards are also taken into account.

After selection of the use giving the highest potential calorie-protein production, appropriate results are totalled to arrive at calorie-protein potentials for each length of growing period zone. Once the maximum potential calorie-protein production combination is ascertained, including the present and projected contribution from irrigated areas (FAO 1981), application of country-specific per capita calorie-protein requirements allows computation of the potential population supporting capacities, in each zone in each country, computed as potential density in terms of persons ha^{-1}, and is subsequently compared with present and projected population densities. Critical zones are identified where, according to the level of input envisaged, potential production from land resources is insufficient to meet the food needs of the populations either at present and/or projected to be living in these areas.

Two time-frames are employed, namely, the "present" (1975), and the "projected" (2000). The choice of 1975 is dictated by the need to employ population and crop mix data available at a sub-national level.

It is important to bear in mind some *limitations* (FAO 1983) of this first assessment, particularly that the results imply the use of all potentially cultivable land for 16 food crops and grassland for livestock production. The 16 food crops provide 81.5 % of the total dietary intake of populations in 90 developing countries.

The study has not yet taken account of fish production, cash crops, nor fully accounted for fuel wood requirements or timber requirements, and neither does it deal with specialized crops and techniques, which might be more productive in specific environments than those considered. Additionally, it does not assume major land improvements such as flood control measures.

Results

The smallest unit of analysis in the study is the individual country/length of growing period zone. If the results from the individual zones are *aggregated on a regional basis*, it depicts the situation of Africa acting as one entity with massive and unrestricted movement of surplus potential and labour throughout the entire extent of the region.

Under this extreme assumption, the results appear promising (Table 1) and show that the lands of Africa could, in total, meet the food needs of their year 2000 populations.

Table 1   Present and projected populations compared with potential population supporting capacities in Africa.

Year	Population (M)	Population density (person per hectare)	Potential population supporting capacities (density – person per hectare)		
			Low Inputs	Intermediate Inputs	High Inputs
1975	380	0.13	0.39	1.53	4.47
2000	780	0.27	0.44	1.56	4.47
2020	1 542*	0.54	?	?	?

*UN (1982)   Land area 2 878.1M ha

These lands, if used in their entirity for food production, could produce food sufficient for thrice the populations they had in 1975 and for one and a half times (1.6) the projected population for the year 2000 at low levels of inputs. At intermediate levels of inputs, they could, in total, feed more than five times their projected year 2000 populations. These aggregated results, however, involve the extreme assumption of massive movements of surplus potential throughout the continent.

However, for each of the many *individual zones* within a country attempting to attain food self-sufficiency from its own land resources, the situation is drastically different. Under this equally extreme assumption, there are vast areas of critical zones where land resources are already insufficient to meet the food needs of populations presently living on them. Under the low-input level, Africa represents a most serious challenge, for critical zones extend right throughout almost the whole the Sahel through southern Sudan into the drier parts of Ethiopia, Somalia, Kenya, Tanzania, the highlands of Rwanda and Burundi, and into the drier parts of southern Africa. The total area involved is no less than 1.3 billion ha where more than 183M people, one-half of the total African population, are living at the present time. Over 100M of these people cannot be fed from the land resources of these areas, should cultivation with low level of inputs be practised.

Under an intermediate-input level, the extent of these critical zones reduces, but large parts of semi-arid Africa remain critical and some parts stay so even under high-levels of inputs. The area affected with intermediate inputs is 1.0 billion ha (73.5M people) and 881M ha with high inputs (44.3M people). The excess populations are 42.5 and 26.5M respectively.

A more practical guide to the true impact of the problem is provided by results aggregated to the country level, i.e. assuming movement of surplus *within invidividual countries*. These results show that, if low-input levels would prevail by year 2000, the number of critical countries in Africa that could not feed themselves from their own land resources, will rise from 22 in 1975 to 29 in 2000.

Increasing the level of input is vital to reducing the number of critical countries by 2000, but even at intermediate input levels, the study assesses that 12 countries, with a total population of 110M, would remain in a critical situation as compared with 7 (25.5M people) in 1975.

Policy Implications

This broad brush assessment brings to light important considerations in the land, food and population issue:

(1)   That land resources and their potentials are very unevenly distributed both within and between countries.

(2)   That there are very considerable areas where land resources are insufficient to meet the food needs of populations presently and projected to be dependent on them by the end of this century.

There is no doubt that the assessed food deficits can be met from the resources of other areas and countries. However such solutions involve major considerations on aid, trade liberalization, price stabilization and world food security, subjects beyond the scope of this contribution. The present discussion centres on means for attainment of the objectives of food self-sufficiency and self-reliance.

First in *areas which are critical with a low level of inputs*, horizontal expansion of production is not possible and the priority must be to raise input levels, involving not only fertilizer use but also mixtures of crops, conservation and reductions of post-harvest losses. Intermediate input use, estimated to equate to a fertilizer use figure approximating to around 80 kg nutrient crop^{-1} ha^{-1}; some estimates (FAO 1984) of current fertilizer use on food crops in Africa are some 6% of this requirement. Increased mineral fertilizer use should be combined with increased use of organic materials to provide an optimal package. Changes in crop mixes, to concentrate on crops most suited to particular environments anticipated in the increased input use; increased areas of upland rice and soybean would contribute to raised productivity in environments suited to those crops currently growing sorghum. "Farm" conservation measures would prevent assessed average rainfed crop productivity losses of 30%, rising to maxima of over 60% in hazardous environments.

Implicit in the results are contributions due to irrigated production; the importance of attaining projected increases in irrigated area and production cannot be over emphasized.

Coupled to necessary input increases, and of particular importance in *areas which would be critical with intermediate and even high levels of inputs*, are the implementation of major improvements to enhance the land resource base, such as increased irrigation, reclamation of saline and/or alkaline lands, drainage, and flood protection measures. The economics of major land improvement schemes require special economic consideration vis-a-vis development of other resources as foreign exchange earners with which to buy food. The development of other resources, for example tourism and minerals, is of particular importance in areas estimated as critical even under high level inputs, and can include development of areas suited to high-value cash crops.

For *areas which are not critical with low levels of inputs*, there is a wider choice for obtaining increased productivity, expansion of the area under cultivation as well as increased yields. However, much of the land remaining for horizontal expansion is in the humid tropics with special clearing, fertility, and conservation requirements; a lack of infrastructure, services, and poor health conditions, also need close attention, but the potential for such crops as rice, cassava and trees is high.

The most important, but insufficiently recognized, need for increasing productivity in Africa, is clearly long-range sound land use planning with respect to food requirements on a country by country basis. Ecologically sound plans need to be formulated to meet future needs and should include provision for the use of limited resources and inputs into those areas, and on the appropriate uses, most responsive to them, and price incentives and policies designed to ensure their use by the farming communities. Monitoring should be an integral part of the plans.

Prerequisite for the production of long-range plans and the most efficient use of available resources is an inventory of the physical resource base and its potential. To know what we are dealing with is perhaps the most important conclusion to be reached by policy makers in the agricultural field.

References

FAO (1971-81) *FAO/Unesco Soil Map of the World*, 10 vols. Paris; UNESCO.

FAO (1978-80) *Agro-ecological Zones Project*, vols 1-4. [World Soil Resources Report no. 48/1-4.] Rome; FAO.

FAO (1981) *Agriculture: Toward 2000*. Rome; FAO.

FAO (1982) *Potential Population Supporting Capacities of Lands in the Developing World*. [Technical Report on FAO/UNFPA Project no. FPA/INT/513.] Rome; FAO.

FAO (1983) *Land, Food and Population*. [Conference Document 83/18. Twenty-second Session.] Rome; FAO.

FAO (1984) *Land, Food and People*. [Popular version of report of FAO/UNFPA Project no. INT 75/P13] (in press).

UN (1982) *Population Bulletin of the United Nations*, no. 14. [ST/ESA/SERN/41.] UN.

Chapter 65

Irrigation in Africa: Present situation and prospects

C.L. ABERNETHY

Hydraulics Research, Wallingford, Oxfordshire, UK.

Introduction

The first thing to note about African irrigation is that there is little of it. Statistics on this are few and suspect, but it seems likely that in 1984 less than 12M ha in Africa will be under irrigation in formal schemes of various types. This is about 4% of all the irrigated land in the world (Framji *et al.* 1981). Even in Europe, with less than a quarter of the potential arable land that Africa possesses and with much smaller natural water deficits, $2\frac{1}{2}$ times as much land as is irrigated in Africa.

If we consider this question in the light of Africa's total resources, the picture is the same. Perhaps 2% of the land in Africa that could benefit from irrigation is in fact irrigated. Africa irrigates about 0.3% of its total land area, and about 5% of its total cultivated area. In no other continent are the relevant figures so small. However, even these macro-statistics disguise the situation as within Africa there are great inequalities. The Nile Valley states, Egypt and the Sudan, account for more than half of all Africa's irrigation activity. Elsewhere, therefore, it is at a substantially lower level than the foregoing figures suggest.

Africa requires the benefits of irrigation technology more than other continents. Its geomorphological and climatic circumstances make it the driest continent, apart from Australia; even where annual rainfall is adequate it often occurs in concentrated wet seasons whose duration is rather shorter than crops require. Africa's rate of population growth, on the other hand, is especially high; and its per capita production of food appears to have been declining through the 1970s (World Bank 1983). It has not as yet shared significantly in the boost to food production experienced in the past decade in south and south-east Asia due to new, water-responsive cereal varieties.

For these reasons, it appears that the potential to improve conditions in Africa by means of irrigation is great. In this review, I attempt to consider the present constraints upon successful irrigation in Africa, and to distinguish those factors most conducive to its improvement.

Present Situation

Distribution of irrigation

The greater part of existing irrigation activity is in countries at the northern and southern extremes of Africa. 55% of all irrigated land in Africa is in the main Nile Valley states, Egypt and Sudan. A further 15% is in the Mediterranean countries Morocco, Algeria, Tunisia and Libya; and 14% is in South Africa and Zimbabwe. If we add to this the 5% in Madagascar, only the remaining 10% of all Africa's irrigation is in the vast area between the Sahara and the Zambesi. Most of this is either in Somalia or along the major rivers of West Africa: the Niger, Senegal and Gambia. The World Bank (1981) reported:

> "Irrigated agriculture has a small place in sub-Saharan African economics, except in Sudan and Madagascar..... In Madagascar, irrigated land occupies 50% of cultivated area, in Sudan 75%; in all other sub-Saharan countries the figure is below 10% and in most cases below 5%".

In short, irrigation is a special, unusual way of life, for all but a handful of African countries.

Comparisons with Asia

To place the above information into perspective, comparisons may be made with two groups of south Asian countries: the Asean states, and the Indian sub-continent (Table 1).

Table 1

	% of all land that is irrigated	% of cultivated land that is irrigated
Indian sub-continent	12.8	27.5
ASEAN	3.1	19.3
Africa	0.30	5.1
Africa excl. Egypt & Sudan	0.16	2.5

There are physical, as well as human reasons, for the lower level of irrigation development in Africa. Rainfall, on average, is less in Africa. Further, infiltration rates on African catchments are generally higher than in Asia, so run-off in the African river systems is correspondingly less. Some reasonably typical figures show the order of these differences (Table 2).

Table 2

River basin	Average run-off (mm year^{-1})
Brahmaputra	760
Ganges	510
Indian E. Coast rivers	345
Benue	191
Niger	50
Chad basin	22
Tana	47
All Kenya	25
Nile (at Aswan)	28

Moreover, in most African basins there is a prolonged season with very little rainfall, notably in West Africa where its duration can exceed 10 months each year. Therefore, there are comparatively few rivers in Africa capable of supporting substantial run-of-river irrigation systems. The building of storage dams on the head reaches, in order to achieve some regulation of seasonal variations, has in general been a prerequisite of substantial irrigation development; this necessarily involves a Government or parastatal body. There is therefore little in Africa (apart from in Madagascar, where run-off is at the high level of 570 mm y^{-1}) that can be compared with the large, communal, run-of-river systems of, for example, Java.

Organization of the state sector

In considering the ways in which African irrigation is at present organized, it is necessary to distinguish two very different categories. I refer to these as formal and informal sectors, but others might equally well refer to them as the state and private sectors, or large-scale and small-scale, or technical and sub-technical systems.

The development of the formal sector is commonly entrusted to some parastatal organization, or a ministry of irrigation; and it is also usual for that organization to take a centralized and paternalistic view of its role. Often it gives little freedom of decision to the farmers; indeed the central organization may develop policies that imply that the farmer is almost a robot with limited skills. In the Sudan, for example, the Government has endeavoured to regulate crop choices, planting dates, water quantities, etc. This attitude is linked to the high cost of financing state irrigation developments. New developments commonly cost US $10 000 – 20 000 ha^{-1} and we are beginning now to hear of development costs up to $30 000 ha^{-1}. There is usually no expectation that the farmers will generate incomes that enable them to repay such costs. Inevitably it appears that such farmers are fortunate beneficiaries of a heavy investment by their Government, and governments may therefore feel obliged to control farm activities. In general, Government bodies have demonstrated that they can not do this satisfactorily, and so there is now a search for organizational arrangements suited to the requirements. Paternal attitudes towards farmers are, however, likely to remain for some time.

Informal sectors

In several countries, especially in West Africa, there exists a small but energetic informal sector of irrigation. Generally this is not assisted by Government, and sometimes it is frowned upon. Statistics of the extent and production of this small-scale or micro-irrigation sector are difficult to obtain, but its vigour impresses observers with the opinion that the economic rewards it offers to its participants are superior to those of state-controlled systems.

Studies of Nigeria's endeavours in this regard (Carter *et al.* 1983) are interesting. These suggest that in the past two decades the growth rate of informal irrigation may have been around 35 000 ha y^{-1}, whereas formal irrigation has expanded at only 1000 – 2000 ha y^{-1}. Such data lead many people to the view that the small-scale, informal sector, with greater farmer participation and more direct rewards for farmer effort, is inherently a more successful pattern of development.

We should, however, note some of the constraints upon this class of development. It is labour-intensive, and therefore much more attractive in labour-surplus economies such as Somalia or Ghana, than in countries where farm labour is short as in the Sudan. The informal sector can usually only develop land that is near a water source; in some countries that restricts its usefulness severely. Land tenure is a considerable difficulty; often the legal rights of those farming in informal systems are dubious, which may reduce their willingness to invest in improvements.

We know almost nothing about the effectiveness with which water is utilised in the informal systems. In countries where the availability of water is a major constraint upon agriculture, as it is in nearly all parts of Africa, we must choose systems of water use that give good productivity in terms of kg food m^{-3} water. The informal sector probably produces widely differing performances on this criterion, and it would be valuable to know which systems of organisation offer the best results.

Constraints upon formal development

In Nigeria the rate at which formal irrigated land has been brought into production has been less than originally planned. The Federal Department of Agriculture in 1971 gave projections equivalent to developing about 30 000 ha y^{-1}; in 1979 the World Bank reflected very similar predictions. However, the rate of new, formal irrigation actually experienced is not much greater than 1000 ha y^{-1}. The reasons for this failure to meet expectations are numerous, and the balance among them is not entirely clear, but there have been similar experiences elsewhere. There are manpower shortages in the implementing agencies and amongst the farming community, where familiarity with the techniques of irrigated farming is in short supply since (outside Egypt, Madagascar, and the Sudan) irrigated farming is a specialisation of a tiny fraction of the population.

Until this basic human resource exists, development costs in excess of US $10 000 ha^{-1} will probably be found unjustified by results. To justify such capital outlay, production well in excess of 3 t ha^{-1} (taking rice to be the crop) is required as an average over the whole developed area. Recent experience does not suggest that this will be widely achieved.

Summary

Africa has, by world standards, very little irrigation; there are some physical reasons for this, but they can by no means account for all the deficiency. The formal, state-directed sector tends to inhibit farmers' initiative and motivation. This sector is struggling, with (in most places) disappointing yields, disappointing rates of new development, high capital costs, and low-grade maintenance. In some countries a vigourous informal sector is coming into existence, but only in Madagascar does this have a strong tradition; however, it is offering a low-capital alternative in some other countries.

Prospects

The lessons from recent experience provide indications as to the future directions of irrigation development in Africa.

First, I think we must emphasise the enormous need for research. We have at present very little clear, numerate information as to the performance of existing systems. This is particularly true of the informal sector; but the formal sector is also very poorly quantified. In those few countries, such as Egypt and Sudan, where irrigation research institutes do exist, they have not, until very recently, made it a part of their job to investigate patterns of water distribution, water availability, and water productivity, within irrigation networks.

Even before research, there is a more basic need for simple measurement. In Egypt, with by far the strongest irrigation tradition in the continent, the smallest unit of water measurement is ordinarily around 10 000 to 20 000 ha. There and elsewhere, the measurements that are made are usually poorly calibrated, and often play no significant role in guiding irrigation operations, because they are so inadequate. Within the Egyptian systems, there is evidence that supplies to individual farmers can vary by a range as wide as 5 to 1.

Neither the system manager nor the individual farmer can derive any benefit from, for instance, research on crop water requirements, or their relationship to fertiliser response, if he has no means of knowing how much water he is applying. Water measuring structures must therefore be built into new systems at a higher frequency than in the past.

Secondly, I think we have to end the paternal, centrally directed attitude to irrigation management. Irrigation authorities must be perceived as bodies providing a service to farmers; a reliable, predictable service, such as we have a right to expect of any public utility. Against this background, farmers must be helped to develop their own occupational skills and to make their own decisions about when, how, and what to irrigate.

The ultimate objective should, I believe, be to provide water on demand, up to some pre-defined annual or seasonal quota, and to levy a charge from the farmer (or perhaps from small farmer groups) which, even if it is not sufficient to recover development costs, is at least proportionate to the amount of water he has taken out of the system. Only when the farmer knows how much water he is getting, knows that he can get it at times of his own choosing, and knows that in taking more water he is going to have to pay more either in money or personal effort, can we expect his attitude towards water to change. He will then see that it is to his advantage to learn how to optimise his use of water. It is worth noting that on the informal systems, and on the individually owned pumped-groundwater systems that have been such an important element of recent successes in north India, the three pre-requisites of good water control more or less exist already. It is on canal-supplied state systems that they do not.

Thirdly, I think we must look for development schemes at lower cost. Especially in those countries which lack a strong tradition of irrigation, it is over-optimistic to expect farmers to quickly learn the skills necessary to produce levels of yields that justify US $10 000 or $20 000 ha^{-1} investments. State irrigation authorities should therefore search for projects with lower capital costs and actively seek ways to assist the informal sector. Techniques of land and water resource planning to these ends do exist, but are not yet being fully utilized. Satellite imagery and aerial photography can help the identification of potentially irrigable land; hydrological networks (too often being allowed to decline) can quantify the available, dependable water resource.

Stimulation of the small-scale sector needs a many-pronged attack. Extension officers must have appropriate training so that they can give sound advice; more design engineers are needed, because far more design decisions are called for; lands departments must be able to resolve tenure difficulties quickly; systems of licensing of abstractions must be kept up to date so that the interests of down-stream users are protected; forms of co-operative organization must be assisted until the most satisfactory are identified. I doubt that these can be achieved in a co-ordinated way unless a special department exists, within ministries of agriculture, to promote this class of development.

Fourthly, I consider that irrigation authorities must ensure that they are getting the best that they can from the systems that they already have; that means improved attention to operation and maintenance. Policies in regard to weed control, desilting, and the overhaul of small control structures need to be continuously reviewed and better supervised. Good operational performance requires good communications within the system, so that operating decisions are taken on the basis of adequate knowledge. Often that condition is lacking; farmers then find the system delivers water to them erratically which in turn reduces farmers' willingness to take risks, for instance to experiment with water-responsive crop varieties. The work of Farbrother in the Sudan has shown very clearly how erratic is the performance of a large, centrally-managed system, once its internal communications have deteriorated.

Lastly I think there must be a great increase in the use of ground-water for irrigation. This may not be the most welcome idea, with the energy crisis not yet behind us; but we must note the crucial role played by ground-water in recent times in north India. Africa is far behind south Asia in this respect. The aquifer has to be looked on as an alternative reservoir, and the costs of wells and of pumping energy has to be set against dam construction and (sometimes) long delivery systems. Organizationally, the main attraction of ground-water is that it places control, and expenditure, with the individual farmer, or of small farmer groups.

In summary, I consider that the main principles of future strategy should be:

(1) To introduce much more research, measurement and performance monitoring, on both formal and informal sectors, so that future planning decisions can be more soundly based.
(2) To develop the skills of irrigated farming, especially by maximising the freedom of decision available to the individual farmer, within realistic constraints.
(3) To seek (in the formal sector) schemes of more modest capital cost, and to increase practical help for expansion of the informal sector, and for ground-water exploitation.
(4) To improve operation and maintenance policies.

References

Carter, R.C.; Carr, M.K.V.; Kay, M.G. (1983) Policies and prospects in Nigerian irrigation. *Outlook on Agriculture* 12, 73-76.
Framji, K.K.; Garg, B.C.; Luthra, S.D.L. (1981) *Irrigation and Drainage in the World*. New Delhi; International Commission on Irrigation and Drainage.
World Bank (1981) *Accelerated Development in sub-Saharan Africa*. Washington DC; The World Bank for Development.
World Bank (1983) *World Development Report 1983*. Oxford; Oxford University Press.

Chapter 66

Management of Terrain, Soil and Water in Humid Regions of Sub-Saharan Africa

R. LAL

International Institute of Tropical Agriculture (IITA), Oyo Road, P.M. Box 5320, Ibadan, Nigeria.

Introduction

Per capita food production in Sub-Saharan Africa declined by about 20% during the two decades ending in 1980 (USDA 1981). In about 18 countries, the per capita calorie availability is less than 90% of minimal requirements. If similar trends continue, bringing diet up to minimal calorie consumption would require some 18.5M t of additional food grain by 1990, with about 12M t of this needed in West Africa alone (USDA 1981). Average yields of most crops in Africa are very low. For example, Cummings (1976) found that the average rice yield in Africa was 1387 kg ha^{-1} compared with 5703 kg ha^{-1} in Japan and 5117 kg ha^{-1} in the USA; the national average rice yield for Zaïre was only 731 kg ha^{-1}. National yield averages rarely show an upward trend despite available improved cultivars.

Rainfall

Many ecologists believe that the most important climatic factor in tropical Africa is rainfall. The success or failure of crops depends on rainfall amount, distribution, regularity, reliability, and above all on its effective utilization. Annual rainfall in the humid regions of sub-Saharan Africa with equatorial climates, the Guinean zone, ranges from 175 to 300 cm, falling in one or two distinct growing seasons (de Vos 1975). If not properly utilized, this excessive rain is a liability, causing accelerated soil erosion and leaching. In spite of heavy rains, crops suffer from droughts because the water storage capacity of most soils is low. Drought stress limits crop growth even in a humid climate.

Soils

Most soils of humid tropical Africa are sandy, highly weathered, low in organic matter content, and mainly kaolinitic. They are characterized by low effective rooting depth, low available water and plant nutrient reserves, are susceptible to erosion and compaction, and are ecologically stable as long as the forest vegetation cover is undisturbed. Using these soils for intensive food crop production requires careful management. Fertile and stable young soils of alluvial and volcanic origin are proportionately few.

Attempts to increase food production generally mean the replacement of labour-intensive and ecologically stable traditional systems of landuse, displacement of forest cover, the introduction of exotic plants and animals, use of agrochemicals, and heavy farm machinery. These factors transform the ecosystem's stability and often produce discouraging results. Modernising the traditional landuse system requires a thorough understanding of the social system and of biophysical and ecological constraints.

Soil erosion

The humid regions of tropical Africa are particularly vulnerable to soil erosion. Even before many land development schemes were implemented in the forested zone, Fournier (1972) estimated a soil loss of about 715 t km^{-2} y^{-1}, about nine times that of continental Europe. The resultant rapid deterioration of soil physical, chemical and biological qualities led to a decline in yields. This decreased production due to the loss of fine clay and organic matter cannot be profitably compensated for by the increased use of fertilizers.

Production Potential of Humid Tropical Ecology

There is no doubt that considering the moisture regime, the radiation levels, and the length of growing seasons, the potential of crop production for tropical Africa is greater than for the temperate latitudes (Table 1). The potential cultivable land area in sub-Saharan Africa is about 7.10^8 ha (de Vries & de Wit 1983), about 30% of the total area. Assuming that the potential crop production of rainfed fields in these areas ranges from 10 to 60% (average 40%) of that under optimal conditions, the potential productivity is 1.10^{13} kg grain y^{-1}. However, development of this potential through scientific means is lacking. Research supports the conclusion that most soils can be intensively cultivated and produce high economic yields without severe degradation of the soil and environment, provided that the delicate soil-vegetation-climate equilibrium is not drastically disturbed and the soil is continuously covered and protected from torrential rains. Production can be dramatically increased if a cautious and conservation approach is adopted.

Table 1   Comparison of radiation levels and potential productivity at Samaru, northern Nigeria, and Rothamsted, UK (Kowal 1972).

	Samaru			Rothamsted	
Season	Radiation (cal cm^{-2})	Dry matter (kg ha^{-1})	Season	Radiation (cal cm^{-2})	Dry matter (kg ha^{-1})
6 months rainy season May-October	88 000	22 000	7 months growing season March-Sept.	64 000	16 000
6 months dry season Nov.-April	90 000	22 500	5 months winter Oct.-March	12 000	3 000*
Annual total		44 500			16 000

* Production limited by low temperatures

Deforestation and bringing New Land under Cultivation

In comparison with 1961–65, the arable land area increased for the decade ending in 1976 by 12.8% in West Africa, 9.3% in Central Africa, 33.7% in East Africa and by 18.7% in Southern Africa (USDA 1981). There is a large potential of increasing food production by clearing additional land in tropical Africa. In the period 1970–74, for example, the cultivated land with annual food crops in Zaire was only 1.8% of the total land area; out of 1.5M km² of moist tropical forest, Zaire has some 1M km² as forest reserves. Conversion of tropical moist forest to arable land requires caution and proper planning. Attempts at rapidly increasing food production have already resulted in the conversion of more than 1M km² of African forest to arable landuse (Salati & Vose 1982), most of which has been severely degraded by mismanagement. The agricultural potential of the forest zone is 20–40% less than the savanna or the Soudanian zone (Kassam & Kowal 1973) as in the forest zone, gross photosynthesis is lost through the high rate of respiration; average economic yields of crops in the savanna were higher than in the forest by 45% for maize, 86% for cotton, 63% for rice, 33% for soybean, 76% for groundnuts, 25% for cowpea and by 162% for tomato.

Further, the most suitable arable land has already been developed, and much of the remaining forested land is only marginally suitable for food crop production. With these limitations in mind, the First International Symposium on Land Clearing and Development (IITA 1982), concluded that improving productivity on existing lands offered the most economical and viable means of accelerating agricultural production in tropical Africa. Deforestation for arable landuse disrupts the bio-physical, hydrological and chemical balance of tropical soils and to minimise the adverse effects, there is a need to: (1) increase production from land already developed; (2) use proper methods of clearing land; and (3) restore eroded and degraded land. All new land development projects need to be carefully reviewed, and properly planned and executed. In addition to the adverse effects of incorrect land clearing methods, the post-clearing management of soil and water resources has a major role in preserving soil productivity.

Increasing Production on Existing Farmland

There are few quantitative examples from tropical Africa relating the loss of productivity to soil erosion. In the Niangoloko region of Upper Volta, however, an increase in the annual rate of erosion from 143 to 1318 t km^{-2} represents an annual increase in loss of clay and humus from 27 to 50 t km^{-2} and resulted in a decrease in the yield of millet from 727 to 352 kg ha^{-1} (Fournier 1963). High water runoff increased the frequency, duration and intensity of drought stress. Severe siltation of the Msalatu reservoir in Tanzania reduced its water capacity by one-third over 30 years (Rapp *et al.* 1972). For soils in south-western Nigeria the rate of decline in maize grain yield caused by natural erosion was 0.26 t ha^{-1} mm^{-1} of eroded soil (Lal 1983).

The selection of appropriate landuse and soil management practices should drastically curtail and even prevent accelerated soil erosion; it is important to avoid clean seedbed preparation, the use of heavy farm machinery causing soil compaction, using cultivation practices that increase runoff, fire, and increased stocking rates leading to excessive grazing.

Soil Surface Management for Erosion Control

Good soil surface management practices are crucial in controlling runoff and erosion. In general, good farming practices are also soil and water conserving measures; these include:

(1) *Residue mulch.* Maintaining a layer of crop residue mulch on the soil surface is a valuable means of maintaining the soil's capacity to accept high intensity rainfalls. This has been effectively used to prevent erosion in tea plantations in East Africa, in pineapples in Ivory Coast and in arable lands in Ghana and Nigeria. In Kumasi, Ghana, mulching reduced runoff by 11–35 times and erosion by 188–750 times, and in Zanzibar mulching decreased runoff by 1000 times and soil erosion by 27 times (Khatibu *et al.* 1984).

(2) *No-tillage.* Mechanical tillage should be kept to an absolute minimum as soil disturbance and exposure to rain causes erosion. In addition to preventing erosion, the no-tillage method has been demonstrated to create favourable soil temperature and moisture regimes and soil structure. Field experiments at IITA indicate that a no-tillage system can control runoff and erosion on slopes of up to 15%. With the no-tillage system of seedbed preparation, it is often unnecessary to use other erosion control measures such as terraces and diversion channels, as long as there is an adequate quantity of crop residue mulch.

(3) *Cover crops.* The frequent use of cover crops in rotation is recommended to protect steep slopes from accelerated erosion. Grasses and low-growing legumes are good soil builders, and are also effective in the restoration of already eroded and degraded lands. Cover crops have long been used in soil conservation in tropical Africa, and their advantages are well documented. When necessary cover crops can be suppressed by chemical or mechanical means so that seasonal crops can be grown through them. Improvements in soil structure and fertility by fallowing with grass and legume covers also occur. Although infiltration rates may increase during the fallow period, most of this improvement may be eliminated by the ploughing and mechanical seedbed preparation for the first cropping season. It is important, therefore, to adopt a no-tillage method of crop production following the use of fallowing with legume covers. Deep-rooted woody perennials can also improve water acceptance by compacted or degraded soils, and growing woody perennials in

association with seasonal crops (alley cropping) is widely recommended for fertility maintenance in tropical soils (Kang *et al.* 1981).

Runoff Management

(1) *Terraces*. If properly constructed and regularly maintained, terraces provide an effective obstruction to water flow and serve to decelerate erosion. Their usefulness is, however, a controversial issue, originating because of the lack of information needed to evaluate adequate spacing in relation to soil properties, slope, and soil management techniques. Even if the terraces are adequately constructed and maintained, inter-terrace soil management is very important for effective conservation. Millington (1982) recommended stone and stick bunds constructed with native materials in Sierra Leone and although soil losses were low, construction costs were high. Terraces are prohibitively expensive in some developing agricultural areas, and 10–15% of the land area is generally taken out of production by them. Buffer strips of grass or herbaceous vegetation, or hedges of woody perennials, may be more effective and economical than terraces for controlling erosion and reducing runoff velocity.

(2) *Contour ridges*. Seedbed preparation with a contoured ridge/furrow system allows more time for water to infiltrate the soil; adjacent ridges are sometimes tied together to develop a series of small basins permitting infiltration where the rain falls. As with terraces, ridges are effective only for soils with a stable structure and for gentle slopes. Tied ridges can reduced soil loss considerably, but if the ridges are up and down the slope runoff and soil erosion can be more than from unridged land. Crops grown on ridges may also suffer from higher soil temperatures and more frequent drought stress than those on flat mulched seedbeds.

Fire as a Tool for Land Development and Residue Management

Fire is widely used as a principal clearing tool in Africa and is an important environmental factor (de Vos 1975). Burning saves labour, it is a cheap herbicide, and adds neutralizing ash to low pH soils; it is also used for what is believed to be an improvement of grassland and pastures. Widespread and uncontrolled burning, however, leaves the soil bare and the result is usually severe and accelerated soil erosion with the onset of rains. In addition to the lack of residue mulch, ash in some soils may decrease wettability and so increase runoff. It also causes substantial losses of nitrogen and other nutrients. An incomplete and controlled burning can, however, serve some useful purposes without unnecessarily exposing the soil to the harsh tropical climate, but as with other tools, careful planning and rational use with adequate controls is necessary to avoid adverse effects.

Restoration of Eroded and Degraded Land

The soils of humid tropical Africa are easily degraded, and productivity declines rapidly with cultivation because of the deterioration in soil physical, chemical and biological characteristics. One factor necessitating deforestation and bringing new land into arable use is declining productivity in existing lands; the natural resource base is shrinking. A considerable portion of the global annual rate of soil degradation of 5–7M ha occurs in the tropics (Kovda 1977). This degradative process can be stopped by good farming practices and steps should also be taken to restore the productivity of those lands that have been rendered unproductive by mis-management; the technology required is often available. Although slow and often expensive, techniques are available to restore and reclaim the productivity of degraded lands. These include the use of planted fallow, establishing deep-rooted woody perennials, adding organic matter, and learning sub-soil management.

Irrigation Development and the Management of Wetland

Water management is crucial in the humid regions. If not managed properly, rainfall is a severe liability to crop production, yet many agricultural development projects and land settlement schemes overlook the importance of water management. Agro-chemical inputs and improved cultivars are often wasted because of the frequent drought stress.

Planning for land development in the humid regions must, therefore, make provision for: (1) maximising rain water retention in the soils' rooting zone; (2) safe disposal of excess water runoff at gentle velocity; (3) storage of excess water; and (4) recycling the stored water for supplementary irrigation. Small reservoirs for storing surplus runoff can be easily constructed even on 1–2 ha farms and used for supplementary gravity-irrigation. Dry-season farming on small irrigated plots can be particularly profitable; double cropping, for example, has proved very successful in the subhumid region of the Ivory Coast (de Vos 1975). They can also increase the efficiency of livestock production and it is clear that the farm pond is a most desirable approach to present irrigation in the humid tropics. Water management, soil fertility and logistic problems associated with large-scale irrigation projects are often too formidable to overcome successfully.

Presently under-utilized wetlands and hydromorphic soils can be intensively utilized for double or even triple cropping provided that the problems of their development, water management, and health-hazard-based taboos are overcome. In the subhumid and humid regions of West Africa, an area of 2.2M km², the total wetland area ranges from 12.6–28.5% (Hekstra et al. 1983). At least one-third of this wetland is occupied by streamflow valleys and their development for arable landuse should be given priority.

Conclusions

The soils of the humid tropical regions of Africa can sustain intensive economic productivity with judicious land development and soil and water conservation techniques. To do so, it is necessary to prevent or control erosion, maintain adequate levels of organic matter content, prevent surface sealing, increase aggregate stability, and improve water-use efficiency through appropriate agronomic practices. Although the possibility of expanding the cultivated land area exists, priority should be given to increasing the production from existing farmlands by adopting conservation-effective systems.

Soil and water conservation can be achieved through biological methods such as mulch farming, no-tillage, cover crops, and buffer strips, and incorporation into the crop rotation of woody legumes with a deep root system can alleviate compaction and improve water infiltration. It is more beneficial to leave residue mulch on the soil surface than to plough it under, and mechanical devices for the disposal of water runoff should be adopted only after a careful appraisal because they are expensive and are often less effective than biological measures. Some large scale land development schemes in the region have failed because of the severe and rapid degradation of the soil surface structure with increased erosion and resultant decline in soil fertility. The productivity of vast tracts of degraded soils can also be restored. Land evaluation criteria can indicate when soil should be taken out of production and put under a restorative and ameliorative phase before it enters that of irreversible degradation.

Although fire as a management tool has some benefits, its disadvantages outweigh its merits. Multipurpose small-scale farm ponds appear to be particularly useful for storing excess water runoff, and for supplementary gravity irrigation. If packages of agronomic practices can be developed, vast areas of now under-utilized wetlands can also be used for food crop production.

References

Cummings, R.W. (1976) *Food crops in the low-income countries: the state of present and expected agricultural research and technology.* [Working Papers.] New York; Rockefeller Foundation, 103 pp.

Fournier, F. (1963) The soils of Africa. In *A Review of the Natural Resources of the African Continent*, 221-248. Paris; UNESCO.

Fournier, F. (1972) *Aspects of Soil Conservation in the different climatic and pedologic regions of Europe.* [Nature and Environment Series.] Council of Europe, 194 pp.

Hekstra, P.; Andriesse, W.; de Vries, C.A.; Bus, G. (1983) *Wetland Utilization Research Project in West Africa.* Wageningen; ILRI.

IITA (1982) *First International Symposium on Land Clearing and Development in the Tropics.* Ibadan; IITA.

Kang, B.T.; Wilson, G.F.; Sipkens, L. (1981) Alley cropping maize and *Leucaena* in southern Nigeria. *Plant and Soil* 63, 165-179.

Kassam, A.H.; Kowal, J.M. (1973) Crop productivity in savanna and rainforest zones in Nigeria. *Savanna* 2, 39-49.

Khatibu, A.I.; Lal, R.; Jana, R.K. (1984) Effects of tillage methods and mulching on erosion and physical properties of a sandy clay loam in an Equatorial warm humid region. *Field Crops Research* 7, in press.

Kowal, J.M. (1972) Radiation and potential crop production at Samaru. *Savanna* 1, 89-101.

Kovda, V.A. (1977) Soil loss: an over-view. *Agro-Ecosystem* 3, 205-224.

Rapp, A.; Berry, L.; Temple, P. (1972) Erosion and sedimentation in Tanzania. *Geografiska Annaler* 54A (3-4).

Salati, E.; Vose, P.B. (1982) *Deforestation Environment Research and Management Priorities for the 1980's.* Stockholm; The Royal Swedish Academy of Sciences.

USDA (1981) *Food Problems and Prospects in Sub-Saharan Africa: The Decade of the 1980's.* [Foreign Agricultural Research Report no. 166.] Washington DC; USDA, 293 pp.

de Vos, A. (1975) *Africa, the Devastated Continent.* The Hague; W. Junk.

de Vries, P.F.W.T.; de Wit, C.T. (1983) Identifying technological potentials for food production and accelerating progress in productivity. In *Accelerating Agricultural Growth in Sub-Saharan Africa, Victoria Falls, Zimbabwe, 29 August – 1 September 1983.*

Chapter 67

Management of Terrain, Soil and Water in Seasonally Arid Regions

V.V. DHRUVA NARAYANA[1] and P. COOPER[2]

[1]Director, Central Soil & Water Conservation Research & Training Institute, Dehra Dun 248 195, India; [2]Acting Leader, Farming Systems Programme, ICARDA, Syria.

Introduction

The seriousness of soil erosion in India can be gauged by the fact that out of a total of 328M ha of the geographical area, nearly 175M ha are undergoing intense soil erosion processes. The most serious soil erosion is sheet erosion in red (69M ha) and black soils (67M ha). Lateritic soils, associated with rolling and rugged topography and high rainfall (occurring in only 2 – 3 months) also have serious erosion problems; eroded gullied lands (ravines) along the banks of the Yamuna, Chambal, Mahi and other rivers occupy nearly 4M ha. The Himalaya and lower Himalayan regions are in a highly deteriorated condition through intensive deforestation, large-scale road construction, mining, and cultivation on steep slopes. The shifting cultivation system practiced in the north-east produces serious soil losses in that region also (Narayana *et al.* 1983). A comparison of the estimated annual soil erosion within the drainage basins of rivers of several countries of the tropics (El-Swaify *et al.* 1982) shows that the drainage basins of India and South America have extremely high erosion values.

Soil Erosion

The soil erosion hazards in India, as in many parts of Africa, are high because of a lack of sustained vegetation cover during the dry months. When the intense monsoon rains come, the land does not have adequate protection from vegetation. The monsoon season in India is characterized by high rainfall intensities, erosivities and soil erodibilities; combined with faulty cultural practices this leads to both high soil and water losses and a loss of productivity in different regions. Lal (1979) indicates that water erosion is a serious problem in those regions, where the "R" factor is in the range of $700 - 1400$ t year^{-1}. The soil erosion map prepared by Das (1977) shows the areal extent of the different erosion problems of India, and Fig. 1 the seriousness of soil erosion in Africa. Soil erosion (excluding that from landslides, road construction, mining and geologic factors) in India is taking place at a rate of 5.3×10^9 t y^{-1} and that about 500M t of soil are deposited in the surface reservoirs and about 1.6×10^9 t discharged into the sea; in the Sholapur region about 50 cm is reported to have been lost over the last century. Qualitative accounts of the incidence of erosion and its impacts given for the Uluguru mountains of Tanzania (Temple 1972) indicate that the loss of fertility was so great that the soil fertility could not be recouped even after 40 years as fallow.

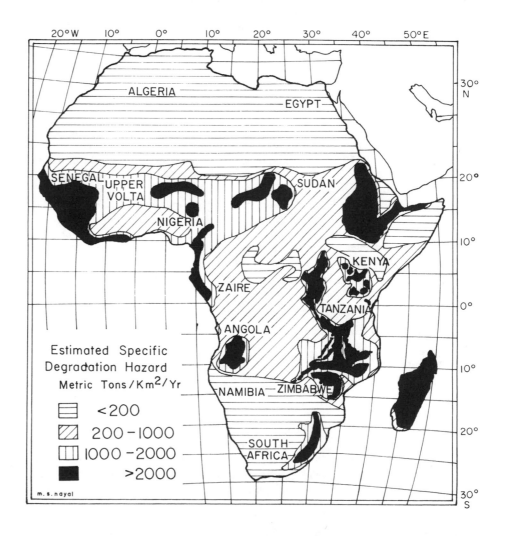

Fig. 1.   The normal erosion risk in Africa (adapted from Fournier 1962).

This loss of productivity should not be allowed to continue in the developing countries. In the last three decades, increased food production in India has been obtained through the improvement of irrigation facilities, introduction of high-yielding varieties, and the application of chemical fertilizers. Future production advances require better soil and water management. The following objectives could be the basis for such management: (1) to minimize or control surface runoff and soil erosion from various land uses; (2) to ensure an adequate moisture supply for crop production either by maximizing infiltration and storage in the soil profile or by manipulation of runoff water; and (3) to store excess runoff water locally for subsequent irrigation or domestic uses.

Soil and Water Conservation Strategy

The basic unit of development in an area will be the watershed, which is a manageable hydrologic unit for the simultaneous development of the entire watershed consisting of the area starting from the ridge-line to the outlet at the end of the *nalah* or natural stream. Appropriate ameliorative measures need to be implemented on barren hill slopes, marginal and privately-owned agricultural lands, and badly eroded *nalahs* and river courses. This contribution concentrates on soil and water conservation measures required in agricultural lands. These include conservation tillage such as contour farming, surface and vertical mulching on gently sloping areas, mechanical measures such as contour and graded *bunding*, conservation ditching on moderately sloping areas, and bench terracing on steep slopes. Appropriate measures can be designed on the basis of precipitation, soil type and crop requirements:—

(1) *Where precipitation is less than crop requirements* these include: Water harvesting through land treatments to augment runoff. Fallowing the land for water conservation for the subsequent crop in rotation. The use of accumulated ground water (if available) for supplementary irrigation. The raising of drought-tolerant crops with suitable management practices. Potential production in these areas is greater than that obtainable by traditional methods.

(2) *Where precipitation is equal to crop requirements.* Generally, in such situations, all precipitation should be locally conserved to the extent possible for current crop use. Depending upon the soil type, land slope, and storage capacity of the soil profile within the rooting zone, conservation measures should store as much rainfall as possible, with excess runoff being stored in small farm ponds and recycled for supplementary irrigation.

(3) *Where precipitation is in excess of crop requirements*, conservation measures are designed to reduce the impact of erosive rains, drain excess runoff, and store this for use during dry interludes and irrigating post-rainy season crops.

Conservation Technology

Fallow cropping

Fallow cropping, whereby land is cropped in alternate seasons to store water, is traditionally practiced in many dry areas. However, in the tropics with high evaporation rates, the proportion of rainfall stored is low, although the practice may be worthwhile if it permits crop production not otherwise possible.

Contour farming

Contour farming and the various practices of mulching, in addition to practices such as cover- and inter-cropping with legumes, will generally be adequate for gently sloping lands. Contour cultivation is found to reduce runoff and soil loss from agricultural watersheds even in high rainfall regions. In an experiment conducted on vertisols with slopes of about 3%, the sorghum crop raised with contour tillage had yields 35–60% over the traditional method of up-and-down cultivation (Rao & Rao 1980).

Vertical mulching

Vertical mulching, with stalks of sorghum placed up to 30 cm deep and at a spacing of 4 m, gave nearly 85% more sorghum than in the control (Rao & Chittaranjan 1983).

Surface mulching

Mulching with vegetation trash protects the soil from raindrop impact. Stubble mulch or trash-forming has potential, especially if accompanied by dense planting. Surface mulches of cut weeds and other waste materials at 4 t ha^{-1} can give about 35 % more yield of sorghum due to better soil and moisture conservation (Rao & Chittaranjan 1983).

Mechanical soil conservation

Mechanical soil conservation works consist of earth banks, terraces, and channels constructed across the slope. These break up the slope, intercept runoff before its volume and velocity become sufficient to cause serious erosion, and give more time for infiltration, so conserving water for agriculture. Water is diverted into channels down safe gradients to suitable discharge or outlet points which carry away water in such a way as to minimize erosion damage. Mechanical measures must precede crop management.

Contour bunding

Contour bunding is found to be suitable for soils having adequate infiltration rates and with rainfall less than 600 mm. An agricultural watershed (22.3 ha) in Agra region was treated with contour *bunds* and the runoff value was brought down from 15.1 % to 0.3 % and the soil loss from 3.7 to 0.1 t ha^{-1} (Sharda *et al.* 1981). In a study on the effect of contour bunding in an agricultural watershed, Sastry and Narayana (1983) have shown the favourable effect of bunding in reducing the runoff.

Graded bunds

Research studies, conducted at the Bellary Centre of CSWCRTI on montmorillonite clay soils (Rao & Chittaranjan 1983) have shown that graded bunds of 0.8 m² cross-section and at a vertical interval of 0.7 m with a channel on the upstream side have reduced the soil loss from 12 t ha^{-1} to 1t ha^{-1} besides improving the yields in comparison to the values from control and contour bunding treatments. These yields are relatively low because sorghum varieties used in this study were not the high yielding ones.

Conservation ditching

Conservation ditching is being tried at CSWCRTI's Research Centre at Bellary. These ditches serve as drainage sinks as well as mini-reservoirs at the field level. In normal and above normal years, the ditches held about 70 to 90 % of the annual runoff. Soil loss occurred from the treated areas only on two occasions when there was overflow. Otherwise no soil loss occurred. The sorghum grain yields, in areas treated with conservation ditches, were 2.85 t ha^{-1} as compared to 2.22 t ha^{-1} from untreated areas.

Tie ridging

Tie ridging, where adjacent ridges are joined at regular intervals by barriers or ties of the same height, allows water to infiltrate and prevent runoff except during intense storms. Except on very steep slopes, this method is sufficient on its own, especially in areas of moderate rainfall. Beneficial effects are recorded for cotton, sorghum and groundnut (the latter two with mulching). Yield increases of 128 % in bullrush millet, 57–87 % in sorghum, 59 % in groundnut, 39 % in cotton, and 15 % in maize, with tie ridging as opposed to contour ridging, have been obtained in Tanzania (Peat & Brown 1960). The broad bed and furrow system in vertisols in Hyderabad gave high yields of sorghum and pigeonpea, a better utilization of rainfall and reduced soil loss (Kampen 1983).

Bench terracing

On steeply sloping land, intensive farming can be practiced only with bench terracing. However, in rainfed areas, this can be recommended only on slopes up to 30 %.

Runoff harvesting, storage and recycling

Traditionally, runoff was allowed to flow into a natural drainage system and was collected in medium sized tanks or a major reservoir. Otherwise, the cropped area would suffer water stagnation. The harvesting of runoff at the micro-level for storage and recycling, which is often neglected, is necessary and possible for the better utilization of rainfall, the control of erosion, and providing essential irrigations during dry spells (Narayana 1979). Seepage control techniques for farm ponds are still in the experimental stages, but in comparison with the cost of water resource development by means of large reservoirs or minor irrigation works, the cost of lining small farm ponds (with bricks and cement and cement concrete) appears to be justified, particularly in areas where there is no other source of irrigation water.

Crop Management

Soil and water conservation measures are often costly. Thus, having optimized erosion control and crop moisture supply, maximum economic benefits will only be obtained when crop management is tailored to maximize water-use efficiencies. Plant growth can be fitted in to existing rainfall conditions through simple water balance approaches (Dagg 1965). Crop management in risk-prone areas involves the following considerations:−

Growing season

(1) The crops can be chosen by matching the duration of the dependable rainy season with appropriate crops and crop varieties.

(2) In selecting crops for an area, the stability of yields over a number of years is a better indicator of suitability than occasional high yields. For example, in the Anantapur region of Andhra Pradesh (a chronic dryland area), pearl millet is reported to be relatively more stable than sorghum and so pearl millet is a better choice for that region.

Crop management during growth

(1) Amongst crop management factors, the timely sowing, or preferably early sowing, of crops has always given higher yields in Indian conditions.

(2) If crops completely fail due to drought stress during the early periods of growth, either the same crop can be resown or a suitable alternate crop grown depending upon the length of the remaining growing season. For example, coriander and safflower can be sown in October in the Bellary region if sorghum fails by the end of September (Rao & Chittaranjan 1983).

(3) On the other hand, if moisture stress occurs at a later stage of the growing period, other means such as ratooning (a high management technique) or thinning are suggested (Venkateswarlu 1982).

Cropping intensities and rainfall

(1) In regions receiving $500 - 600$ mm rain y^{-1}, only a single crop can be raised in any year.

(2) If the rainfall is in the range $700 - 900$ mm y^{-1}, inter-cropping systems can be adopted to increase cropping intensity and crop production per unit land area. Inter-cropping systems include sorghum and green gram in the Akola region, pearl millet and pigeonpea in the Sholapur region, and groundnut and pigeonpea in the Bangalore region.

(3) In regions with rainfall above 900 mm y^{-1} and with a soil water storage capacity exceeding 200 mm, double or sequential cropping can be practiced (Balasubramanian et al. 1982). This involves using a short-duration crop during the rainy season, and a relatively long-duration crop in the post-rainy season.

Performance Evaluation

Soil conservation programmes are in progress in about 31 river valley project catchments and many other small agricultural and non-agricultural watersheds in India. The annual sediment production rates (250 t ha^{-1}y^{-1}) from these areas under untreated conditions are naturally very high, but with implementation of soil conservation measures in the catchments of river valley projects sediment production rates have declined considerably (Sastry & Narayana 1984) and this trend has also been noted in runoff plots.

Inspite of such promising trends, and their economic benefits, progress in the implementation of soil and water conservation programmes is slow in many countries. Increasing attention should, therefore, be directed to developing social attitudes and policies to hasten the implementation of such beneficial measures on a watershed basis.

References

Balasubramaniam, V.; Gangadhara Rao, D.; Hanumantha Rao, C. (1982) Crops and cropping systems for dryland. In *A Decade of Dryland Agricultural Research in India 1971-80*. Hyderabad; All India Coordinated Research Project for Dryland Agriculture, 252 pp.

Dagg, M. (1965) A rational approach to the selection of crops for areas of marginal rainfall in East Africa. *East African Agricultural and Forestry Journal* 30, 296-300.

Das, D.C. (1977) Soil conservation practices and erosion control in India. A case study. *FAO Soils Bulletin* 33, 11-50.

El-Swaify, S.A.; Dangler, E.W.; Armstrong, C.L. (1982) *Soil erosion by water in the tropics.* [Research Extension Series no. 024.] University of Hawaii, College of Agriculture, 173 pp.

Fournier, F. (1962) *Map of Erosion Danger in Africa south of the Sahara.* Paris; EEC Commission on Technical Co-operation in Africa.

Kampen, J. (1982) An approach to improved productivity on deep vertisols. *ICRISAT Information Bulletin* 11, 1-14.

Lal, R. (1979) Erosion as a constraint to food production in the tropics. In *Proceedings of the Conference on Priorities for alleviating soil-related constraints to food production in the tropics, Los Banos, June 04-08, 1979.*

Narayana, V.V. Dhruva (1979) Rainwater management for lowland rice cultivation in India. *Journal of the Irrigation and Drainage Division, ASCE* 105, 87-98.

Narayana, V.V. Dhruva; Ram Babu (1983) Estimation of soil erosion in India. *Journal of the Irrigation and Drainage Division, ASCE* 1983 (Dec.).

Peat, J.E.; Brown, K.J. (1960) Effect of management on increasing crop yields in the Lake province of Tanganyika. *East African Agricultural Journal* 26, 103-109.

Rao, M.S. Rama Mohan; Chittaranjan, S. (1983) Moisture conservation and erosion control in deep vertisols of semi-arid regions. In *Proceedings of the National Symposium on Watershed Management.* Bellarly; CSWCRTI Research Centre.

Rao, V. Ranga; Rao, M.S. Rama Mohan (1980) *Rainfed Agriculture. 25 years of research on soil and water conservation in semi-arid deep black soils.* Bellary; CSWCRTI, Research Centre.

Sastry, G.; Narayana, V.V. Dhruva (1984) Evaluation of conservation practices in small watersheds in Doon Valley (Idia). *Journal of the Irrigation and Drainage, Division, ASCE,* in press.

Sharda, V.N.; Bhushan, L.S.; Srivastava, M.M. (1981) Hydrological studies on ravine watersheds under mixed landuses at Agra. In *CSWCRTI Annual Report 1981,* 190. Dehradun; CSWCRTI.

Temple, P.H. (1972) Soil and water conservation policies in the Uluguru mountains, Tanzania. *Geografiska Analer* 54A, 110-123.

Venkateswarlu, J. (1982) *Dryland Agriculture Problems and Prospects. A decade of dryland agriculture in India, 1971-80.* Hyderabad; All India Coordinated Research Project for Dryland Agriculture, 252 pp.

Chapter 68

Limitations imposed by Nutrient Supply in tropical African soils

P.H. LE MARE

Department of Soil Science, University of Reading, London Road, Reading RG1 5AQ, UK.

Limitations to agricultural production imposed by nutrient supply occur because soils do not supply enough of one, or more than one, nutrient; and sometimes because too much of a nutrient is freely available to a crop. To understand nutrient limitations we need to know the mechanisms by which soils control the supply of nutrients, which may be from the soil's reserves or indirectly from fertlizers, and how much a crop needs. Our knowledge of soil properties and characteristics has progressed in recent years so that some empirical relationships established in the past are now better understood, and we have a sounder basis for improving soil management. Also, advances in soil classification and survey have helped to develop a basis for transferring knowledge from one area to another.

A very large proportion of tropical Africa is covered by oxisols, ultisols and the less weathered alfisols; nutrient supply problems lessen in that order. The clay minerals of these soils are largely kaolinite, and oxides and hydrous oxides of iron and aluminium. Soils with these clay minerals have poor nutrient characteristics because they have small negative charge and small cation exchange capacity. Furthermore, the surfaces of the clay minerals are amphoteric so that their charges vary with the pH of the soil: with increasing acidity the soils develop positive charges so that cation exchange diminishes. With increasing positive charge important anions, especially phosphate and molybdate, are adsorbed strongly so that their concentrations in solution diminish.

The cation exchange capacity of soil organic matter is especially important because it is much greater than that of the clay minerals so that it can contribute greatly to the cation status of oxisols, ultisols and alfisols. Organic matter is also an important cycling agent of nitrogen, phosphate and other nutrients. Nevertheless, the nutrient supply potential of soils dominated by 1:1 layer silicates and by oxides of aluminium and iron is generally small, so that fertlizer management of them is very important and often complex.

In tropical Africa there are relatively few extensive areas of soils that are inherently very fertile, although there are important small areas. The more fertile soils have clay minerals that have 2:1 silica: alumina ratio, with isomorphous substitution of ions causing permanent negative charge so that cation exchange is much greater, and anion retention less severe, than in soils with 1:1 clay minerals. The vertisols in the Gezira of Sudan comprise the major area of these more fertile soils, but there are important smaller areas throughout central Africa.

Phosphate is taken up by plants as $H_2PO_4^-$ and as HPO_4^{2-} but the latter is important only in neutral soils. The problems of phosphate nutrition of crops grown in acid soils, which comprise the majority in tropical African agriculture, arise because the anion is absorbed onto the positively charged surfaces of clay minerals, and because phosphate reacts with iron and aluminium to form sparingly soluble compounds. The uptake of phosphate is further complicated by its very slow movement in soils, a consequence of its reactions with soil constituents. Depth and distribution of roots are important and in some crops, especially cassava but also upland rice, maize, sweet potato, yam and some legumes, vesicular-arbuscular mycorrhizae improve the transfer of phosphate from soil to crop. Research is in progress to see whether inoculating soils with mycorrhizae can lead to improved crop nutrition.

Field experiments throughout tropical Africa have demonstrated that crops respond to phosphate fertilizer; in the past thirty years many papers have reported responses to small or moderate amounts of fertilizer phosphate and to their residues. Residual effects should always be included in assessments of the economic returns from phosphate fertilizers.

For major nutrients response curves are important to provide data for the assessment of economic returns from fertilizer. The common and most satisfactory curve is one with negative curvature which rises steeply with small dressings and shows diminishing returns as more nutrient is applied. However, in some acid soils responses to successive increments of fertilizer increase to a point of inflexion beyond which they diminish, so that the curve is sigmoidal. Liming raises pH and diminishes exchangeable aluminium, so that yield and response to small dressings of phosphate increase, and the whole response curve then has negative curvature.

A more complex sigmoidal response sometimes occurs: small dressings of triple superphosphate diminish yields but larger amounts increase them. Curves of data from Uganda, Ghana and elsewhere in the tropics were of this form. The cause in Uganda was excess manganese and too little calcium taken up with phosphate from manganese-rich soil; when adequate calcium was given the harmful effects of manganese were eliminated.

Sigmoidal response curves are not often reported from Africa but they may occur more widely if soils become more acid with intensive cropping and use of acidifying nitrogen fertilizers. The two types of sigmoidal curves have different causes: in the first by reaction between soil constituents and phosphate, and in the second by improper nutrition of the plant caused by an excess of a cation taken up with the phosphate anion. Careful liming is likely to correct both conditions and make phosphate fertilizer more effective.

Crops take up most of their nitrogen as the nitrate ion, although some, especially rice, utilize ammonium. Problems of soil nitrogen are therefore largely concerned with the supply and movement of nitrate. The primary source of soil nitrogen is the atmosphere but reserves are held in the organic matter. Some atmospheric nitrogen enters the soil in rainfall but the major route is fixation by micro-organisms, which maintain the supply of nitrogen to uncultivated vegetation. Legumes also acquire adequate nitrogen by fixation but non-leguminous crops rely upon reserves in soil organic matter, unless manure or fertilizer is applied. Unless organic matter is maintained, by cropping with fertilizers and manures, and by growing legumes in the system, nitrogen supplies soon become too small for acceptable yields. Legumes are especially valuable in fodder crops if the animal manure is returned to the land, and if pasture and arable crops form a rotation. Leguminous tree crops, e.g. *Leucaena leucocephala*, are promising for some areas because pruned branches and foliage provide nitrogenous mulch and animal feed, and the wood can be used as fuel and building material.

Mineralization of organic nitrogen and the movement of nitrate ions are the major processes that affect uptake of soil nitrogen by crops. Mineralization is rapid when dry soil is wetted, so that in many parts of Africa there is a flush of nitrate when rain falls after the dry season. Similar flushes occur when soil is rewetted after it has dried during a short period without rain; these intermittent flushes may cause greater total mineralization than steady moist conditions.

Leaching of nitrate and associated cations has been investigated widely over many years. Loss of nitrate may be limited by adsorption onto positive charges in the textural B horizon of some soils and the nitrate held may be available to crops as their roots grow downwards. However it is important that other nutrients, especially phosphate, are adequate to ensure prolific roots. In very wet areas leaching under arable crops may be very severe but not under perennial crops as, for example, under oil palm in southern Nigeria.

Large numbers of field experiments, mostly on previously unfertilzed land in tropical Africa, showed that non-leguminous arable crops respond to nitrogen fertilizer and that generally its interaction with phosphate is positive, thus emphasizing the importance of phosphate to ensure optimum response to nitrogen. Often when phosphate and other nutrient deficiencies are corrected the response to nitrogen becomes very large.

Responses to nitrogen, and other nutrients, may be affected by pests and pathogens. At Mwanhala, Tanzania, if the very mobile insect *Calidea dregei* was not killed by insecticide, nitrogen fertilizer decreased yields of cotton because the insects preferred to feed on plants grown with added nitrogen; but when the crop was sprayed with carbaryl insecticide, nitrogen fertilizer caused large increases of yield.

Until recently sulphur was rarely a primary limitation of crop yield because, in the absence of fertilizer or manure, sufficient sulphur was mineralized from organic matter; other nutrients, especially nitrogen and phosphorus, were more likely to limit yield. Where fertilizers were applied they often contained enough sulphur. Now, many fertilizers that supply major nutrients contain little or no sulphur so that deficiency is more common. Where organic manure is used sulphur may be adequate but increasing vigilance will be necessary to ensure crops receive enough sulphur.

In general little calcium is removed in the harvested parts of crops but some, especially broad leaved crops, take up large amounts during growth so that leguminous forage crops may remove large amounts. In soils that lack sufficient calcium, acidity and hence excesses of aluminium and manganese, as well as calcium deficiency, cause crop failures; liming corrects the toxicities and ensures adequate calcium for crops.

In many African soils weatherable minerals may supply enough potassium for most crops for a number of years, even when large yields are grown with nitrogen and phosphorus fertilizers, but then, as large yields are removed, fertilizer potassium becomes necessary. Potassium deficiencies are most likely in and following root crops, plantains, sugar cane and oil palm, in which large amounts of potassium are removed in the harvested parts. In some experiments in southern and western Tanzania potassium chloride (muriate of potash) decreased yields; in western Tanzania harmful amounts of manganese were taken up in the presence of potassium chloride in acid soils, but liming eliminated the effect.

There are few reports of magnesium deficiency except where it is induced by an unsatisfactory ratio with another nutrient, especially potassium. This occurred in Kenya and Uganda in coffee mulched with *Pennisetum purpureum* which added much potassium to the soil.

Micronutrient deficiencies are widespread and become increasingly important as yield levels increase with better crop management, but toxicities also limit yields. The following are a few examples of micronutrient disorders.

Boron deficiency occurs widely and is perhaps the most serious limitation imposed by micronutrient supply. Deficiency was reported in cotton in Zambia, Tanzania, Nigeria, North Cameroon and probably occurs elsewhere. Boron deficiency also occurred in sisal, coffee, eucalypts, pines, maize, tobacco, citrus and vegetables. Sometimes recognition of boron deficiency has been delayed when a pest or a pathogen was thought to be the cause of a plant disorder because boron often causes die-back, and breakdown of cell tissue, which allows infection with pathogens. Boron deficiency is easily corrected but the range of amounts tolerated by plants is small so that care is necessary not to give harmful amounts.

Molybdenum deficiency was reported in groundnuts in Ghana, Senegal and Nigeria. In very acid soils in Zimbabwe molybdenum deficiency occurred in maize; it was ameliorated but not eliminated by lime which also increased availability of phosphate, indicating that molybdate and phosphate anions were absorbed strongly by positive charges.

Zinc deficiency was reported in maize in Zimbabwe, Nigeria and elsewhere; it may be induced by excess of lime.

Manganese deficiency is less likely than excess in African soils; some examples have been mentioned. In general, monocotyledonous plants seem less susceptible to manganese toxicity than broad leaved plants, especially legumes. Adequate calcium is important to counter effects of excess manganese, which inhibits translocation of calcium.

Conclusions

The characteristics of many African soils are such that, unless they are carefully managed, their productivity diminishes rapidly. To remove limitations caused by nutrient supply we must first ensure that the properties of a soil are suitable for crops to take up nutrients. Acidity and organic matter are, perhaps, the two most important variable factors which control satisfactory nutrient supply. Large areas of African soils with nutrient limitations are inherently acid and tend to become more so under cultivation, especially in heavy rainfall areas and if nitrogen fertilizer is used. Acidity is easily corrected and, as pH need rarely be raised greatly, effective amounts of lime may not be large. Any programme that promotes greater productivity should consider the need for lime, or prepare for a future need; otherwise farmers may become disillusioned and rightly feel misled by their advisers if use of fertilizer fails to increase yields economically.

In soils with suitable pH, fertilizers provide the immediate means of supplying nutrients; which nutrients and how much fertilizer to apply must be determined by local experiment and experience. Soil testing can help but in tropical soils, as in temperate soils, relationships between analytical data and crop responses in the field are generally very variable, less because the analytical methods are poor than because many environmental factors affect yields of crops; we must identify and modify these. Nevertheless, a recommendation for fertilizer, especially phosphate, based on soil analysis of an individual field has a large margin of error and the economic risk to the farmer may be too great for him to bear alone. However, when averaged over a large number of farms, soil analysis data are important to guide fertilizer programmes, especially if a co-operative or public organization bears some of the risk of individual farmers.

Although lime and fertilizers provide the means to replenish nutrients removed by crops, and by leaching, methods that conserve and cycle nutrients are essential to sustain improved economic well-being. As larger crops are grown with improved pH and nutrient status, organic matter levels increase and reach a stable level so that soils become better buffered and able to maintain adequate crop nutrition. Organic manure is very valuable; it supplies many nutrients, and helps maintain satisfactory pH and organic matter. Pastures should be improved as part of the overall development of productivity. Some improvement of pH, phosphate and other nutrient status will be necessary to introduce forage legumes and these, in grazed leys, will improve the nitrogen and organic matter status of the soil.

Deficiency of nutrient supply causes serious limitations to production in many tropical soils. More research will be needed but enough is already known to improve yields in most areas. Other characteristics of environments and agricultural systems limit yields also but a multi-disciplinary group of agricultural scientists can, within a short time, often define the major limiting factors and recommend improved methods of farming. It is more difficult to devise the most appropriate combination of methods that farmers can adopt within the limits of their resources. This can be done only by the careful co-operation of agricultural scientists and extension workers with sociologists, economists and politicians to establish the social and economic conditions within which the farmers can adopt new practices and find them profitable.

Copies of a fuller version of this paper, together with references from which data is cited, can be obtained from the author.

Chapter 69

Human Carrying Capacity in Existing Farming Systems

A.J.B. MITCHELL

Land Resources Development Centre, Tolworth Tower, Surbiton, Surrey, UK.

Introduction

The rate of population increase is higher in Africa than in any other major region of the world. It is over 3% in many countries, which means that the population virtually doubles every 20 years. As a result some areas are beginning to become overpopulated.

It is not easy to identify the point at which the population in a given area exceeds the capacity of the land to support it, but by the time it is noticed, it is often too late to avoid serious problems of land degradation. Human carrying capacity is a means of attempting to define this point, and the purpose of this paper is to describe its use in a land use planning project in Tabora Region of Tanzania.

Definition of Carrying Capacity

Human carrying capacity can be defined as the number of people that can be supported by a particular area of land under a given system of management and with a given expectation of self-sufficiency or level of living.

For rural land use planning it is most important to be sure that there is sufficient land to provide a living for the farming families. They must have land for their food and cash crops, additional land for fallowing, grazing for their livestock, and woodland for fuel and building materials. They also need water for domestic use, livestock and sometimes for irrigation. Such a soundly-based rural population is the best guarantee of food security for the nation.

Before one can say how many people can support themselves one has to answer certain questions: (1) What standard of living should the people expect? (2) What crops will they grow? (3) What yields can they expect? (4) If they need manure or fertilisers to maintain yields, will these be available? (5) Will the people be physically able to grow the acreage of crops needed to provide the intended standard of living? (6) How much fallow will be needed, for weed control or to restore fertility? (7) If livestock are kept, are their owners looking for production, subsistence by use, or simply an inflation-proof investment? And (8) If the area under cultivation is increased, will this mean that fewer livestock can be kept, and if so will the resultant lower production of manure reduce crop yields?

In other words, one is trying to identify or project a complete system in which the various parts are in balance, and where changing one part will change the balance of the whole system.

Techniques of Carrying Capacity Assessment

Before describing the use of carrying capacity assessment in Tabora Region, I should like briefly to refer to the work done in Zambia during the 1930s and 1940s. Here Allan, Trapnell and others developed a technique for estimating the amount of land cultivated by an average family, and the amount of additional land needed to allow for restoring fertility through fallowing. They related these to soil-vegetation associations recognised by the local people. The early work was reported in two surveys by Trapnell (Trapnell 1953, Trapnell & Clothier 1957), and Allan later extended the principles (Allan 1967) to provide a basis for examining the relationship between agricultural systems and the environment over the whole of Africa. The factors which formed the basis for their assessment of carrying capacity were:

(1)   The cultivable percentage of land; the proportion of the "total" land area which was available for cultivation

(2)   The land use factor; the length of the cultivation and fallow periods

(3)   The "normal surplus" requirement of subsistence agriculture; the average surplus to their annual requirements grown by most producers to insure against shortage through drought, about 25 % of the food requirement. In good years this surplus may be sold, or used for beer-making, or for hospitality or bartering. The extent of these activities could be used to indicate whether the people were cultivating enough land.

(4)   The cultivation factor; the actual area cultivated per head of population, measured by field sampling. This was influenced primarily by climate and soils, and taken together with the normal surplus was used to calculate carrying capacity.

The cultivation factor was remarkably consistent over much of Central Africa at a little more than 1 acre (0.4 ha) per head, while in East Africa where there are two rainy seasons per year it was about half this area.

Carrying Capacity Assessment in Tabora Region of Tanzania

Description of the region and its main problems

The Land Resources Development Centre of the UK Overseas Development Administration has had the responsibility of running a land evaluation and land use planning project in Tabora Region since 1978 (Mitchell 1982). Important problems in the region are:

(1)   Severe overpopulation in the north-east, due to the immigration of people and their livestock from less-favoured and even more over-populated areas.

(2)   Declining agricultural production over much of the region, which has a number of causes including the creation of settlement patterns which prevent effective use of the land.

(3)   Tsetse infestation of the natural woodlands, which appears to be encroaching into areas that were formerly in the cultivation cycle but are reverting to bush.

The Land Use Planning approach

It became clear during the early part of our project in Tabora Region that agricultural improvement could not be based initially on any technological package that might significantly raise farm production. Such packages will eventually bear fruit, but they depend on the results of research, improvements to the provision of inputs and marketing of crops, and education; all of which take time and on past record are uncertain in their outcome. We concluded that the first priority was to get the framework right, to ensure that the farmers had access to enough land to apply the farming systems that they understood, so as to provide as far as possible an assured food supply and a marketable surplus that would give them an acceptable income.

Our village planning thus depended on assessing the carrying capacity of the land for the farming systems that already existed in the region. We sought to identify very much the same things that Trapnell and Allen had identified in Zambia, with the additional requirement of trying to define a single "living" standard, compatible with modern expectations in Tanzania. There was also the problem of population increase and the recent administrative changes that had disturbed the farming systems.

We developed a series of models for carrying capacity assessment (Corker 1982) based on the following criteria (in relation to those to Trapnell and Allen).

(1)   The proportion of land that can be used for cultivation; which we established by means of natural resource surveys.

(2)   The length of cultivation and fallow periods; based on an assessment of soil fertility and the traditional farming practices related to it. Cultivation and fallow periods could not be estimated directly from observation of present land use, due to the changes brought about by villagisation. Although there would be no return to shifting cultivation, we assumed that manure and artificial fertilisers would continue to be unavailable to most farmers.

(3)   The area that should be cultivated annually by each family (Trapnell's "normal surplus"). Because of the recent changes to farming patterns we tried as far as possible to produce a standard basis for determining what an average family was, what they grew, and what they needed. We used the following criteria:

(a)   standard farming systems related to climate; based on an agro-economic survey the most important part of which was a detailed recording of the daily activities of selected families in villages representing the zones of the most important farming systems. We identified three basic farming systems, all based on rainfed production of staple food crops and cash crops.

(b)   possible intensification by growing rice in mbugas (valley grasslands). While rice is an important part of some farming systems, it has certain features which made it preferable to treat it separately from the upland farming systems, primarily because its feasibility depends on the seasonal wetness and seepage flow in the local valleys (mbuga) which is difficult to estimate in village surveys.

(c)   Standard crop yields for the constituent crops of the farming systems on "class 1" soils were based on data from a programme of simple crop trials, the agro-economic survey, and other trials and demonstrations in the region. Yields varied enormously due mainly to rainfall and management.

(d)   Physical and chemical limitations in the soil were used to create four soil classes. The agronomic data were not sufficient to correlate these limitations, apart from soil depth, with specific yield reductions, and our estimates are therefore subjective.

(e)   The size and composition of an "average" family were found from the 1978 census data for the four districts of Tabora Region.

(f)   Subsistence food requirements were based on estimated calorific needs for the different age and sex groups of the family. Since a comparatively high proportion of legumes is grown, and other green leaves are used as relishes, protein supply was assumed to be in balance with energy in the standard farming systems. Storage losses were assumed to be 25% of total food production.

(g)   A minimum cash income based on crop sales. Two levels of cash income were proposed. The lower level was the cash needed to obtain basic requirements such as salt, soap, clothes, tools, grain milling etc. The higher level was intended to give a purchasing power equivalent to the standard wages of an urban labourer.

(4)   Separately from the criteria listed above, we estimated the carrying capacity for livestock, and the capacity of existing woodlands to meet the people's fuel needs. Livestock carrying capacity was based on measurements made by us of standing forage on land in different categories of current land use and vegetation. Existing and projected needs for fuelwood were estimated, together with the capacity of natural woodlands to supply them, and an estimate of the area needed for plantations.

The assumptions we have made in developing the models are the best that we could make on the information we had. They could not in any way claim to be statistically valid; to obtain such figures would require years of research. Our justification for using them was that we had to have a consistent means for comparing one area against another, and a yardstick against which to compare the performance of individual villages.

Our next step was to build the models in the form of a series of simple algorithms for calculating carrying capacity, which can be used by people without a high level of technical skill or knowledge, to arrive at a single figure for viability or carrying capacity.

Important advantages of algorithms are: (1) By imposing a standard discipline and seeking answers to specific, objective questions, they reduce as far as possible the variation between operators; and (2) if one or more of the variables are changed (standard yield of a crop for example, or prices) the algorithm can be changed accordingly, and ones already completed can be re-worked.

Application of the carrying capacity algorithms to land use planning

The carrying capacity algorithms that we have used are developed from the Village Viability Assessment algorithm developed by the Overseas Development Group of the University of East Anglia for Iringa Region (Coleman *et al.* 1978). They enable land use planners to use reconnaissance or detailed survey information to estimate the carrying capacity of selected areas, or the viability of existing villages, and to identify constraints that will require to be overcome.

One of the final intended products of our project is the proper land use planning of all the villages in the region. There are some 400 villages, and our land use planning teams could reasonably expect to plan about 16 to 20 villages a year. Therefore there must be a selection process. The villages most in need of planning are those where people are having difficulty in maintaining self-sufficiency. Since planning would involve some emigration from overcrowded villages, others with spare capacity would have to be identified also.

The algorithm for regional assessment of human carrying capacity is called the *Carrying Capacity Assessment Model*. It makes use of the reconnaissance soils information, overlaid on a map of administrative boundaries, to find the carrying capacity of the wards.

Comparing human carrying capacity with the 1978 population, we found that 30 of the 74 wards (a small administrative area comprising up to 4 or 5 villages) in the region were overpopulated if the people depended solely on dryland cultivation, while 10 would still be overpopulated if all the suitable land were also brought under rice cultivation. As expected most of these wards were in the north. Other wards with potential for new settlements were mainly in the west.

The second stage of the planning process was to discuss our finding with the District authorities, and draw up a list of priorities for village surveys and planning. These discussions generally confirmed our findings, but also threw up some anomalies. Some villages that we thought would have plenty of land were barely self-sufficient, while other overpopulated villages were apparently quite wealthy. When we began to do detailed village studies some causes for these anomalies were revealed:

(1)   The pattern of human settlements was important. Centralised settlement patterns meant that part of the village area was not used for cultivation, and also that there was little use of manure on the land.

(2)   Farmers in some of the most densely settled villages had intensified their crop production by

growing vegetables such as tomatoes or onions and sugar cane, using water from small dams or seepage areas. Some of these crops fetched high prices due to seasonal scarcity.

We did not find that these anomalies detracted from the value of the initial carrying capacity assessment. On the contrary, by indicating the capacity that should be expected, it helped to focus attention on the reasons for the anomalies.

The final stage in the land use planning process is the detailed survey and planning of the village, including both a physical survey of the natural resources, land use and settlement patterns, and a detailed questionnaire covering all aspects of village life including agriculture, livestock, sources of fuelwood, water, services and infrastructure, and sales of produce. The data are used to complete three algorithms:

(1)   Availability of agricultural land compared to the human population.

(2)   Availability of grazing land compared to the livestock population.

(3)   Potential fuel supply from existing sources compared to the village requirements, and the area of plantation that would supply those requirements.

The algorithms show whether the village is short of farming land, grazing, or fuel. The questionnaire survey throws up other problems such as water shortage, and also shows the pattern of crop sales. The land use map shows the distribution of settlements and cultivated land.

Experience in the field with the use of algorithms

A new team has now taken over at Tabora, and they have approached our ideas with a fairly open mind. They have had some problems with the practical use of carrying capacity algorithms, particularly at the detailed level of the individual village. They have also produced a number of more philosophical queries, for example:–

(1)   *The area that should be cultivated* has raised several questions, first of which was whether such a detailed series of assumptions was justified, given the very limited or uncertain information on which many of them are based. It was felt that there was a danger of giving the algorithms a spurious appearance of being more scientifically objective than they really are. Specific questions were:–

    (a)   *The standard farming systems.* We dealt with the *average* family cultivating an *average* area of land. In practice there is a great range of performance among different farmers, and the systems adopted by the better farmers may be significantly different from the average.

    (b)   *Possible intensification by growing rice.* This is a comparatively high-risk crop, and is therefore dangerous to use for predicting carrying capacities. In practice we only used it for assessing the potential viability of existing villages in comparison with their performance. We did not use it as a basis for planning future settlements.

    (c)   *Standard crop yields.* Our standards were taken, somewhat subjectively, as being what the majority of farmers should achieve, usually without fertiliser, in a normal year. The crops included both staple foods and non-food cash crops. The main problem in using the standard yield is that the data on which they are based show such variations that it is difficult to justify them on any statistical basis. It would be desirable to be able to apply a risk factor to a yield figure and so, for example, allow for a lower risk of failure in food than cash crops.

(2)   The minimum cash income has been a source of problems.

    (a)   Crop prices are Government-controlled. Relative movement between prices of individual crops can upset the equation.

    (b)   Consumer-goods prices have far outstripped crop prices, as well as labour wages. Our original "lower income" minimum was based on what we regarded as basic essentials. Our "higher income" was equated to a labourer's wage and was about $2\frac{1}{2}$ times as much. Now, thanks to the rise in consumer goods prices, the "lower income" is higher than the current "higher income"!

Obviously a better measure has to be found. Carrying capacity figures should not have to be revised annually due to price movements.

The question in the final analysis is, can we really justify the relative complexity and apparent objectivity of our algorithms in view of the highly variable and often uncertain information we base them on? By including several such variables in succession, do the errors compound each other and so lead to highly misleading results? Would it not be better to go back to the simpler formula of Trapnell and Allan and base our targets on what the better farmers are actually achieving under the present system, or some similar measure (i.e. the "normal surplus")?

Such an approach would have definite attractions, but could bring its own problems, especially lack of flexibility. If a new crop is introduced successfully, for example, farmers may respond by changing their cropping patterns, or greater returns may be possible from a smaller area. If the original carrying capacity assessment was not based on clearly identified assumptions, how will you change it?

We feel that although our approach of successively applying a series of defined criteria may be more complex, it does allow for any or all of the criteria to be changed easily, and carrying capacities re-assessed, as our knowledge improves. Additionally, it requires its users to seek to obtain information in a systematic, repeatable way. Where the algorithms *should* be simple is in the field, when they are used by people with comparatively little training. Our *regional* assessment now consists of three steps, using existing natural resource and population data. More complex algorithms are used if any fundamental revision is required. It would be desirable to simplify the *village-level* assessments in the same way.

Conclusions

Tabora Region is faced with a problem of overpopulation in one part and underpopulation in another part. One of our tasks as a land use planning unit was to identify which parts really were overcrowded, and which really had spare land for resettlement.

We have found that a simple technique of carrying capacity assessment enabled us to perform this task. At the same time it has given us a sensitive means of comparing the performance of a village with a standard set of criteria, and thus of identifying its priorities for land use planning.

There are inevitably problems in the practical application of the technique, but none should be insuperable.

References

Allan, W. (1967) *The African Husbandman.* London; Oliver & Boyd.

Coleman, G.; Pain, A.; Belshaw, D.G.R. (1978) *Village assessment in the framework of regional planning in Tanzania.* [UNDP/FAO Project URT/75/076.] Norwich; University of East Anglia, Overseas Development Group.

Corker, I.R. (1982) *Land use planning handbook, Tabora Rural Integrated Development Project, Tanzania.* [Project Record 66.] London; Land Resources Development Centre, UK Overseas Development Administration.

Mitchell, A.J.B. (ed.) (1982) *Land evaluation and land use planning in Tabora Region. Tabora Rural Integrated Development Project, Tanzania.* [Project Report 116.] London; Land Resources Development Centre, UK Overseas Development Administration.

Trapnell, C.G. (1953) *The soils, vegetation and agriculture of North-Eastern Rhodesia. Report of the ecological survey.* Lusaka; Government Printer. [Reprint, originally published 1943.]

Trapnell, C.G.; Clothier, J.N. (1957) *The soils, vegetation and agricultural systems of North-Western Rhodesia. Report of the ecological survey.* Lusaka; Government Printer. [Reprint, originally published 1937.]

Chapter 70

Potentials and Constraints in the Management of Vegetation for Development in Agriculture and Rural Areas

B.N. OKIGBO

International Institute of Tropical Agriculture (IITA), Ibadan, Nigeria.

Introduction

The importance of vegetation lies in its function as the most vital life support component of the biosphere for all organisms, and its constituting one of the most vital renewable resources available to man.

The management of vegetation involves various policies, strategies, techniques and activities used in conservation and a rational utilization of vegetation to satisfy the multifarous uses of vegetation to man and other organisms. Management must be based: (1) on the understanding of man as a component of ecosystems; (2) on the realization that man is not only dependent on the ecosystem but exerts considerable influences (adverse or favourable) to either components or the ecosystem as a whole; and (3) on continuous research and generation of technology to ensure that encountered constraints are eliminated in satisfying the increasing human needs under changing environmental conditions and attitudes.

This contribution reviews African vegetation with respect to changes that it is undergoing as a result of human activities involving the use of indigenous and introduced technologies, and pressures of modernization. The potentials of African vegetation are evaluated and constraints in its management identified as the basis for determining priorities in research and development activitities.

Vegetation Resources of Africa

In any ecosystem, there is a more or less reciprocal relationship between climate, soil (+ parent material) and vegetation. Whenever the earth's surface becomes first exposed, there develops a primary succession of vegetation types which eventually stabilizes to form a *climatic climax* where dynamic equilibrium exists among the constituent elements of the system. When interfered with, a secondary succession occurs. Adejuwon (1976) concluded that the climatic climax vegetation of Africa consisted mainly of (1) *tropical rainforests* in areas where rainfall exceeds 1500 mm in not less than 10 months, (2) *tropical deciduous forest* where rainfall is between 900 – 1500 mm in 5 – 10 months, (3) *tropical xerophytic woodland* in areas of 200 – 900 mm rainfall in 2 – 5 months, and (4) tropical desert in areas of less than 200 mm annual rainfall in 1 month or less. Prevailing rainfall regimes here are based on Le Hoeurou & Popov (1981).

The present vegetation zones in Africa are now regarded as ecosystems of largely anthropic origin, resulting from modifications of the climatic climax vegetation by clearing, cutting, burning, farming, grazing, lumbering, road construction, urbanization, etc.; in almost all parts of Africa the original primary vegetation has been replaced by secondary vegetation only approaching the primary climax after long periods of minimal disturbance. The resultant vegetation zones are best exemplified in west Africa and especially Nigeria where (1) the tropical rain forest has given rise to the mangrove forest and associated coastal swamp vegetation, freshwater forest, forest regrowth and derived savanna; (2) the tropical deciduous forest resulted in the southern Guinea savanna and northern Guinea savanna; and (3) the tropical xerophytic woodlands gave rise to the Sudan savanna and the Sahel savanna (Adejuwon 1976).

On the African continent today, the main vegetation zones consist of mangrove and/or coastal swamp vegetation, tropical rain forest, forest/savanna mosaic, tropical deciduous woodland, savanna grassland, grass steppe, sub-desert steppe and desert, temperate grassland, cape machia and montane vegetation (grassland, forest and bare snow cover).

Potential and Utilization of Africa's Vegetation Resources

Land Area and Cultivated Land

Africa has a total land area of 29.7 of 31M km². The prevailing landuse in Africa in 1980 consisted of 181.2M ha (6.1%) cultivated land, 784.2M ha (26.4%) permanent pasture, 696M ha (23.5%) of forest; woodland and remaining land amount to 1304.9M ha (44.0%), in 1975 the arable land in use in Africa amounted to 30% and 0.64 ha per capita.

Primary Productivity

This depends largely on the prevailing rainfall and climatic conditions. Areas with the highest rainfall and tropical rainforest vegetation in parts of west Africa and the Congo basin have net annual productivity values of over 4000 g m^{-2}, while areas close to deserts with desert steppe or sparse vegetation cover have less than 500 gm^{-2}. Primary productivity also depends on the vegetation cover or dominant species in each ecological zone. UNESCO/UNEP/FAO (1978) provide biomass production observations for various ecological zones in Africa.

Utilization and Ethnobotanical Importance

The most important use of vegetation to man and his domestic animals is food. At first most food was eaten raw, but following the discovery of fire about 50 000 y ago a greater range of products were eaten cooked with less danger of the adverse effects of toxic substances and pathogens. Following the neolithic revolution about 10 000 y ago in the fertile crescent and Egypt, and later in many independent sites in tropical Africa at about 3 – 4 000 BC, there was enoblement of various species of plants in either the seed agricultural complex of the savanna or the vegecultural complex of the forest zone and the forest/savanna ecotone. This resulted in the domestication of several cereals (African rice, sorghum, millets and teff), root and tuber crops (yams, Hausa potato and piasa), grain legumes (cowpea, African yambeans, bambara groundnuts, pigeonpea and Kerstings groundnuts), tree crops such as the oil palm, cola, and African pear, shea butter and locust bean, and miscellaneous vegetables nuts and seeds. These indigenous African domesticates were later enriched by Asian crops introduced via the Sabaean lane or the Island of Madagascar and later by American crops. The Asian crops included wheat, rice, banana/plantains and coconut while the American crops included cocoa, groundnuts, cassava, sweet potato and papaya; some of which became of major importance as staple food crops or as cash crops.

In addition many semi-wild protected and wild plants continue to be harvested. Plants which make up the African vegetation are of ethnobotanical importance and have a remarkable variety of uses. This is especially true for timber, but it should be noted that while there are about 90 commercial timber species from west and central Africa (Jay 1972), less than 25% of the potential commercial woods are currently being exploited.

Human Influences, Strategies and Technologies

Major changes in the relationship between man and nature have occurred, in chronological order: (1) deliberate use and making of fire; (2) deliberate cultivation of crop plants and rearing of animals leading to settlement in villages; (3) urbanization; (4) an industrial revolution based on harnessing sources of energy other than those of man and beasts; and (5) a scientific revolution which resulted from the interaction of science and technology in finding solutions to practical problems. The green revolution which ushered in an era of phenomenal increases in yields of "miracle" seeds (rice, wheat and maize) in certain developing countries of the world might also be added.

In Africa the dominant pattern of land and natural resource utilization involves agricultural production systems ranging from extensive shifting cultivation and nomadic herding to specialized intensive horticulture and livestock production. Burning involved in traditional farming systems and hunting causes a loss of plant nutrients tied up in vegetation. Deforestation and clearing with increased frequency of cultivation causes rapid decomposition and loss of soil organic matter and fertility, erosion, and decreased water retention capacity (Okigbo & Okigbo 1979).

Strategies and Technologies

Traditional and "modern" strategies and technologies are used in the management and utilization of vegetation and associated soil and animal resources. Traditional strategies and technologies of resource management and utilization have been found to be by and large ecologically sound, efficient, culturally relevant and adapted to environmental conditions and human needs and circumstances. Yet many African traditional production systems, practices, and natural resource management and utilization systems are often regarded as "primitive" or scientifically "unsound". This is often the result of a lack of knowledge about the relationship of man and his environment in the tropics and a lack of appreciation of the socio-economic and cultural origins and basis of prevailing traditional technologies.

In developing countries, the desirability of conserving natural resources, preserving habitats for wildlife and minimizing pollution of the environment cannot be evident to people who lack knowledge and the life-style needed to appreciate these concepts. Conservation measures are often viewed by local populations as random oppression by elites making it unlikely that such measures will be effectively practiced unless preceded by education. There is evidence of conservation within traditions of native peoples based on an accurate knowledge of the exploited biota. These may be in the form of taboos coinciding with breeding times of fish, clearing, planting and harvesting of various crops. The best methods of management and developing of natural resources should be those that enhance the tapping of natural resources in a judiciously and ecologically modulated manner so as to significantly satisfy demands without adverse effects on the environment. Scientific conservation should enhance the preservation of existing species while enhancing their utilization.

UNESCO/UNEP/FAO (1978, 1979) reports on tropical forest and grazing ecosystems and reviews prevailing ideas about the evolution of human societies and traditional natural resource management and utilization systems in the tropics. Traditional resource management and utilization systems are the culmination of long experience and intimate association with environmental niches used for livelihood and trial-and-error experimentation. These can only be defined as deliberate efforts to improve and/or protect the value of life-supporting resources and insure long-term viability.

Characteristics of African resources management systems

These include some plant species (1) occurring in East African coastal forests and elsewhere in the region also present in West Africa (Zaïre and/or Guinea) forests, (2) represented by different subspecies in the East African coastal forests and West African forests, (3) of African genera represented approximately by equal numbers of species in the East African coastal forests and West African forests, and (4) of African genera with only one or two species in East African coastal forests but with their principal distribution in the West African forest. Examples are given in Lind & Morrison (1974). It should, however, be noted that the coastal rainforests of Kenya and Tanzania are separated from the easternmost edge of the Zaïrean rainforest by more than 1100 km and from the easternmost extensions of that forest type (the Kakemega forest in Kenya and Kigoma and Bukoba forests of Tanzania) by 650 km in Kenya and 850 km in Tanzania.

UNESCO/UNEP/FAO (1978) was devoted to an interdisciplinary evaluation of traditional African natural resource management and decision-making with an emphasis on the environmental impact consequences of decision making and reactions and attitudes of the human societies concerned. Traditional natural resource management decision-making was compared with that in recent development projects in the forest and preforest, Sudan and Sahel ecological zones. Examples of the failures and partial failures of modern decision-making and strategies are to be seen in development projects in this region. The following conclusions have been drawn: (1) Most modern decision-making occurs in offices in capital cities and is designed to achieve short-term economic gains. (2) Interventions by public authorities have important repercussions, much can be learnt from the way people concerned felt and reacted. (3) Sudan savanna farmers gave the highest priority to food security. (4) Progress made with agricultural equipment can result in a regression of collective work and decline in the solidarity of communities. (5) Farmers of the Sudan savanna were quick to learn if proposed techniques were financially feasible and profitable. (6) Finally, modern decisions and strategies have increasingly tended to initiate individual reactions or responses and the promotion of personal interests rather than traditional ones which affect communities. Similar experiences are also encountered in the forest – derived and Guinea savanna areas of west Africa.

Constraints and Problems of Vegetation Management

Considerable areas of African vegetation have undergone drastic modification by man and deforestation continues at an alarming rate that poses a major threat to environmental quality, and the future survival of man, livestock, and wildlife. In efforts to develop policies and strategies for rational management and utilization to satisfy increasing multifarous needs of man and other animals the following constraints and problems are encountered:–

(1) A lack of ecologically sound and socio-economically acceptable policies for rational and integrated management and utilization of vegetation.

(2) Limited knowledge of Africa's vegetation and related renewable resources.

(3) Deficiencies in the effectiveness, timeliness and scope of research in relation to ecosystems and their response to various multi-use, and the ethnobotany of various cultures in Africa, as a basis for integrating traditional and modern attitudes and uses of plants, improving knowledge of existing systems and problems, human social organizations in different ecosystems and their adaptations to prevailing environmental conditions, and the development of technologies that enhance planning, policy formulations, management conservation, processing and utilization of resources.

(4) Problems of land availability to satisfy the needs of a rapidly increasing population which results in increasing intensity of deforestation and removal of vegetation cover through clearing, farming, burning, grazing, industrialization, urbanization, etc., causing soil degradation, erosion, desertification and environmental deterioration.

(5) Lack of financial resources required for research and development activities.

(6) Deficiencies in human and technical resources in relation to lack of trained manpower.

(7) Presence of human parasites and diseases such as malaria, trypanosomiasis, schistomiasis, onchocerciasis, loasis, yellow fever, helminthiasis, etc., that are detrimental to health and limit the exploitation of resources in certain ecosystems. Related to these are various animal diseases and insect pests such as locust, weeds, graminiferous birds, etc., that reduce crop yields or damage vegetation.

(8) Rapid population growth and pressures of modernization resulting in higher rates of demand on resources.

Coupled with the above are various physical (climatic and physiographical), biological and socio-economic constraints to increased agricultural production. In addition, essential factors for agricultural developments are effective marketing and pricing systems, incentives to farmers, research linked with training and extension, and suitable rural infrastructure such as roads.

Conceptual Framework and Guidelines for Management of Vegetation Resources

Strategy

Knowledge of traditional resource use strategies and technologies in relation to physical and socio-economic environmental conditions is crucial in determining success in agricultural and rural development projects. This calls for a strategy in research and development that is based on the integration of traditional and modern technologies in the development of appropriate technology components of modern technology that are ecologically sound, economically viable and sociologically acceptable. This usually enhances rapidity of adoption while minimizing adverse environmental effects.

Conceptual framework

Policies, planning, conservation and rational utilization of resources must be based on a conceptual framework of integrated vegetation and ecosystem management and utilization strategy. Such a framework should give due consideration to alternative landuse categories and include such landuse options as conservation, special reserves, forestry for timber and other products, tree crop plantations, tree/shrub/herb fallow systems, special agroforestry systems, short-term herbaceous perennials and annual fallows, grazing lands, annual or staple food crops, industrial landscape and special uses of plants, mining and human settlements. For details of priorities and strategies by landuse category see Okigbo (1983).

Guidelines for management

Programmes for agricultural and rural development should consider the following strategies and priorities:–

(1) Vegetation constitutes a vital component of man's support system in the biosphere and any effort directed towards the realization of the full potential of vegetation must on a continuing basis involve vegetation surveys, inventories, classification, evaluation and utilization.

(2) Strategies that aim at striking a meaningful balance between protection, conservation and development goals are to be preferred. Here conservation is used as defined by UNESCO and FAO as "the rational use of the earth's resources to achieve highest quality of living for mankind" (Dassman *et al.* 1973).

(3) Management and utilization should on the macro-level be based on ecological principles in which each agricultural and/or forestry production system should ideally be based on the management of natural, modified or artificial ecosystems in ecological zones, and at the micro-level on integrated watershed mutli-use management strategies.

(4) Consideration should be given to soil conservation; protection from erosion by the maintenance of adequate soil cover, elimination of excessive tillage and making maximum use of biological processes in maintaining soil fertility.

(5) Advantage should be taken of integrated approaches that are based on the complementary and/or synergistic effects of various components in the system, as is involved in game ranching, the integration of livestock and crop production in mixed farming and related agri-silvopastoral systems, and integrated pest management systems (see Chapter 39).

(6) The most crucial constraint is often the lack of capabilities for executing research of sufficient scope and in the various disciplines required.

(7) Use of indigenous plant species that have high potentials for use as sources of food, oil, fibre, drugs, timber, and miscellaneous industrial and other products should be developed.

(8) Rapid deforestation requires the establishment of special reserves and germplasm collections to ensure the range of genetic diversity of potentially useful plants and relatives of already exploited plants is not lost.

(9) An ecologically sound national vegetation management must involve the controlled use of fire so as to minimize adverse effects on vegetation.

(10) Incorporation into educational programmes knowledge of plants and their uses in the human ecosystem.

(11) Governments could find it advisable to formulate policies, development plans, legal and enforcement interventions to ensure sound management, conservation and utilization of Africa's vegetation.

(12) Resources allocated to manpower development, research and training will contribute to sound management, and so rational and efficient utilization of vegetation. Communication between researchers and policy makers and politicians can ensure that national plans include appropriate allocations of resources rather than force reliance on foreign aid programme.

(13) More use can be made of universities not only in teaching but especially in basic and applied research on Africa's vegetation.

Need for Co-operation at all Levels

African countries vary in the resources at their disposal for supporting research and development programmes related to vegetation management and utilization. In addition to interdisciplinary co-operation, there is need for interinstitutional, interministerial and international co-operation in research and development, formulation of policies, exchange of information and funding of relevant programmes. Not only universities but United Nations agencies, such as UNEP and FAO, scientific and professional organizations and non-governmental organizations all have vital roles to play.

References

Adejuwon, J.O. (1976) Human impact on African environmental systems. In *Contemporary Africa: geography of change* (C.G. Knight; J.C. Newman, eds), 140-158. Englewood Cliffs; Prentice Hall.

Dassman, R.F.; Milton, J.P.; Freeman, P. (1973) *Ecological Principles for Economic Development.* London; John Wiley.

Jay, B.A. (1972) *Timbers of West Africa.* Hieghenden Valley; Timber Research and Development Association.

Le Houerou, H.N.; Popov, G.F. (1981) *Agro-climatic Classification of inter-tropical Africa.* [FAO Plant Production and Protection Paper no. 31.] Rome; FAO.

Lind, E.M.; Morrison, M. (1974) *East African Vegetation.* London; Longman Green.

Okigbo, B.N. (1983) Plants and agroforestry in land use systems of west Africa. In *Plant Research and Agroforestry* (P.A. Huxley, ed.), 25-43. Nairobi; International Commission for Research in Agroforestry.

Okigbo, B.N.; Okigbo, L.R. (1979) Soil fertility maintenance and conservation for improved agroforestry systems in the lowland humid tropics of Africa. In *Soils Research in Agroforestry* (H.O. Mongi; P.A. Huxley, eds), 41-78. Nairobi; International Commission for Research in Agroforestry.

UNESCO/UNEP/FAO (1978) *Tropical Forest Ecosystems* [Natural Resources Research no. 14.] Paris; UNESCO.

UNESCO/UNEP/FAO (1979) *Tropical Grazing Land Ecosystems.* [Natural Resources Research no. 16.] Paris; UNESCO.

Chapter 71

Mixed Systems of Plant Production in Africa

P. J. WOOD

International Centre for Research in Agroforestry, Box 30677, Nairobi, Kenya; and Commonwealth Forestry Institute, South Parks Road, Oxford OX1 3RD, UK.

Introduction

The sowing of mixed annual crops has been a common farmer response to risk from climatic uncertainties, to the need for diversity of food crops, and to pragmatic observations of beneficial interactions between species. In the humid tropics, sophisticated tree-crop mixtures have been developed over centuries using many species and where a wide variety of benefits is obtained. In the drier parts of Africa trees have formed important components of grazing systems, or have been left deliberately for their benefits in cultivated land. The use of agricultural crops in the establishment of forest plantations (*taungya*) is widespread, and in Africa as elsewhere shifting agriculture provided, in the past, a sustainable land-use system which is only now breaking down under the pressure of increasing populations.

This contribution considers all multiple cropping systems, including agroforestry, agrisilviculture and taungya as "mixed systems"; i.e. cropping patterns where more than one crop is grown on a piece of land in a year. Agroforestry itself has been defined (Lundgren & Raintree 1983) as:

> "Agroforestry is a collective name for land-use systems and technologies where woody perennials... are deliberately used on the same land management unit as agricultural crops and/ or animals in a spatial arrangement or in temporal sequence. In agroforestry systems there are both ecological and economic interactions between the different components."

Terms such as intercropping, relay cropping taungya (where the ultimate objective is the establishment of a forest plantation) are included in one or other of the above.

Background to Multiple Cropping

Until recently the most striking advances in agricultural production were mostly obtained through the high-input monocrops of the green revolution. However, the African husbandman has continued to use his traditional multiple cropping systems, improved where possible by the adoption of new crops and techniques, despite the efforts of expensive extension services to induce him to adopt monocropping practices (Steiner 1982). The protracted struggle by the Wachagga of Kilimanjaro to be allowed to grow coffee during the Colonial period may have really been a fight for the recognition of the validity of the multiple cropping as opposed to the plantation approach (Bowers, in Macdonald 1981). The renewed interest in farming systems research (see e.g. Collinson 1982) is a reflexion of the recognition that major changes from multiple cropping are unlikely to occur in Africa. The extent of multiple cropping systems continues to be very high, e.g. 80% of areas in southern Nigeria in 1970–71. For further detail on West African intercropping see Steiner (1982). These levels are unlikely to fall due to the increase in population and the fact that small farmers tend to practice multiple cropping more than large farmers. Steiner (1982) reports that an average of over 70% of farms in Côte d'Ivoire, Ghana and Nigeria, are under 5 ha; similar figures probably apply to other parts of Africa with high or rising populations.

Traditional Mixed Cropping Systems

Most traditional systems reveal a thoughtful and sophisticated approach to the problems of sustainable production, whether referring to settled or shifting practices.

In shifting agriculture, examples of farmers' awareness of declining yields following intensive cropping are known (Braun 1974). Variations in the patterns of shifting agriculture are seen depending on soil fertility. The "Citimene" system of Zambia utilizes stored nutrients of a considerably larger area of Miombo woodland than was actually cultivated, on soils of low fertility, whereas complex multiple-species gardens were developed by the Wachagga of Kilimanjaro on their eutrophic brown forest soils.

A deep knowledge of the site requirements of many crops is particularly evident in the wetter zones, for instance in south-east Nigeria, Zaïre, and the East African mountains. The Medge of Zaïre were known to be growing 80 varieties of 30 species of food crop in 1911 (Okigbo & Greenland 1976), and cropping patterns indicated that local topography was cleverly used, with wide-crowned trees suitably dispersed to control weeds and erosion, and light-demanding and shade-tolerant plants appropriately located. In eastern Nigeria great complexity is shown in the placing of large numbers of different crops, sometimes more than 50 species, at various levels on artificial mounds (some 2.5 m high) on hydromorphic soils. In southern Nigeria, also, planted fallowing is practised, using, for instance, *Acioa barteri*, *Anthonotha macrophylla* and *Gliricidia sepium*. Such agroforestry and multiple cropping practices occur in many parts of Africa.

In many drier areas also crop mixtures abound, as in the Hausa lands where 156 crop mixtures have been recorded (Agboola, *in* Macdonald 1981). These areas in the Sudan Zone are also notable for the retention of naturally occurring trees in the arable systems for fruit, oils, medicines, fodder, etc. Animals are an essential part of many mixed systems involving trees. Silvopastoral systems range from traditional nomadic grazing systems of the drier zones to more synthetic systems developed from plantation agriculture. Examples include cattle under coconuts in Tanzania, and under oil palm in Côte d'Ivoire (Lazier *et al.*, *in* Macdonald 1981).

Taungya systems for the establishment of planted forests were first developed in Burma and India in the nineteenth century. Their extent has varied with the need felt by Governments to develop *compensatory*[1] plantations in national forest reserves. Successful taungya schemes are found in Nigeria, Kenya, Tanzania, Uganda and the Congo. 160 000 ha of plantations (Spears 1980) were raised in the forest reserves of the Kenya Highlands under the "shamba"[2] system where licensed farmers grew maize, beans and potatoes before and after the trees were interplanted for 4–5 years; conditions in the East Africa volcanic highlands are unusually favourable, however. In 1975–76 there were over 24 000 "traditional" taungya farmers in 20 000 ha of forest reserves in Southern Nigeria (Ball 1977).Taungya is an imposed rather than an indigenously developed system, although it is an older practice in Africa than many systems developed by peasant farmers. Its success in some places has undoubtedly depended on land hunger and poverty; social aspects have been examined critically leading to the word "agrisiliviculture", intended to be a kind of "taungya with a human face".

Analysis of Multiple Cropping Systems

Mixed cropping is adopted by farmers to increase overall production per unit area, reduce the incidence of pests and diseases, enable planting to take account of soil variations, exploit the different mature heights of various crops, ensure a continued and varied supply of food, provide soil cover against weeds and erosion, and to even out the demand for labour during the year. Phased planting is adopted to even out the demand for labour during the year, reduce land preparation costs, minimise risk (particularly from climatic uncertainties), provide phased harvests, and provide ground cover.

The disadvantages of mixed cropping systems, perhaps more apparent than real, are difficulties of mechanisation, difficulties in applying inputs (e.g. fertilisers) and more complex experimentation (Okigbo & Greenland 1976). Analyses of traditional systems generally show overall increases in crop yields. On dry zone farms incorporating *Acacia albida* trees (to c. 40 trees ha^{-1}), for example Charreau & Vidal (1965) recorded striking effects on the yield of millet. Not only was the quantity of grain increased the quality was also improved; soil analyses in the same cropping zones showed improved conditions in comparison to soils outside the canopy.

The usual measure of improvement in yield resulting from multiple cropping is the Land Equivalent Ratio (LER), the relative land area under sole crops required to produce the same yields from a unit area of intercropping. Very substantial improvements in yield have been recorded in field studies (Beets 1982). There is also mounting evidence of the positive effects on soil conservation, and thus on the sustainability of the systems (Steiner 1982).

The socio-economic basis on which the African farmer makes decisions has been studied. He is not particularly impressed by an LER obtained from on-station experiments, however impressive in percentage terms these may be. The actual increase may be very small on a small field or farm, and farmers are more likely to adopt multiple cropping for reasons other than the LER, for instance a more efficient use of labour, particularly in intensity of cultivation and weeding, despite the fact that the labour input per unit area is often increased. Equally important is erratic rainfall.

---

[1] i.e. compensating for a lack of natural forests, for a lack of merchantable species in natural forests, or for a lack of expertise to manage exploited natural forests for a sustained yield of merchantable wood.

[2] The Kiswahili word for the taungya system in East Africa.

The "contingency mixing" of maize and sorghum in Haute Volta is a good example of a response to this risk.

In some societies there is a traditional division of responsibility for crops between the sexes; women in parts of Cameroun for instance plant "their" crops in their husbands fields, resulting in a complex multicropping system. In many areas near to a powerful market the opposite, adoption of low-labour monocropping, is the farmers' response.

The economic evaluation of taungya has been studied in both West and East Africa, and from the point of view of the national Government generally shows substantial economic advantages for the establishment of plantations. Ball (1977) gives realistically estimated internal rates of return (IRR) for various methods of raising plantations of *Tectona grandis* (teak) and *Gmelina arborea* in southern Nigeria. In the Congo the raising of *Terminalia superba* with bananas is an extremely successful taungya system in which the now maturing trees have been underplanted with cocoa (Koyo, *in* Macdonald 1981). Here, however, the principles of agroforestry mixtures were insufficiently applied, and although the (unthinned) plantation trees continued to grow reasonably well, the yields of cocoa fell over 1970–73, giving 49, 67, 31 and 16 kg ha^{-1} respectively.

The Future of Mixed Cropping Systems

There is no doubt that mixed cropping systems will continue in Africa, and that the complex and expensive external inputs needed to change traditional farming systems will only be available to the few. More research is needed on systems, by both biological and social scientists, on:— understanding the current systems and identification of constraints; estimating national Government's present and future proposals for infrastructure for agricultural development; the development of new, adapted cropping systems; and adaptive, on farm, research (Norman 1976). There is nothing new here, but few resources are allocated to it. In the specific situations of agroforestry, ICRAF has developed a "Diagnosis and Design" methodology with a very similar approach. "There is no substitute for good design" (Raintree 1983). Farmers have pragmatically designed their own systems, whereas modern agroforestry designs are able to draw on scientific knowledge as well as farmers' experience. Such a system involving *Acacia senegal* (for gum production) in combination with grazing, cereal and fuelwood production has been described by von Maydell (1978). Another system becoming widely applied is alley cropping, where trees (often nitrogen-fixing leguminous species, e.g. *Leucaena*) are grown in widely spaced rows with agricultural crops between (e.g. ter Kuile 1983).

There is scope for a great deal of site-specific research on topics such as optimum spacing, choice of species, and technologies on both these systems and for those in the humid tropics (Watson, *in* Macdonald 1981). Spears (1980) has identified it as "truly sustainable tree and agroforestry cropping systems", and Lundgren (Budowski, *in* Macdonald 1981) as: "... economic and nutritional output from land must not only be sustained at the present low levels, but be substantially increased ..."

The problem is how to increase productivity whilst retaining the stability of traditional systems. Research should thus focus on:—

(1) The development of new genetic strains for use in multiple cropping and agroforestry systems (in contrast with most current breeding which is directed towards monocropping). This is especially true for multipurpose trees.
(2) The development of no-tillage, green manure/mulching systems, incorporating trees.
(3) The development of Diagnosis & Design (D & D) methodologies for steady improvements to meet perceived farmer needs.
(4) The application of agroforestry technology to forest fallows.
(5) The development of land use systems in forest reserves that produce a wider variety of products than wood only, and that safeguard soils, water supplies, land ownership and crop ownership.

Lanly (1983) recorded that 7.2 % of the total land area of tropical Africa was under forest fallow at the end of 1980. If a single focus for future effort is needed, this could be it.

References

Ball, J.B. (1977) *Taungya in Southern Nigeria*. [FO No. NIR/71/546.] Rome; FAO.

Braun, H. (1974) Shifting cultivation in Africa (the evaluation of a questionnaire). In *Shifting Cultivation and Soil Conservation in Africa*. [FAO Soils Bulletin no. 24.], 21-36. Rome; FAO.

Charreau, C.; Vidal, P. (1965) Influence de l'*Acacia albida* Del. sur le sol, nutrition minerale et de rendements des mils *Pennisetum au Sénégal*. *L'Agronomie Tropicale* 20, 600-626.

Collinson, M.P. (1982) Farming systems research in East Africa. The experience of CIMMYT and some National Agricultural Research Services 1976-81. *MSU International Development Paper* No. 3.

Lanly, J.P. (1983) The nature, extent and developmental problems associated with shifting cultivation in the tropics. In *Papers from the Expert Consultation on the Education, Training and Research Aspects of Shifting Cultivation, 12-16 December 1983*. Rome; FAO.

Lundgren, B.; Raintree, J.B. (1983) Sustained agroforestry. In *Agricultural Research for Development: Potentials and Challenge in Asia*. The Hague; ISNAR.

Macdonald, L. (ed.) (1981) *Agroforestry in the African Humid Tropics*. Ibadan; United Nations University.

Norman, D.W. (1976) Developing mixed cropping systems relevant to the farmer's environment. In *Intercropping in Semi-Arid Areas* (J.H. Monyo; A.D.R. Ker; M. Campbell, eds), 52-57. Canada; IDRC.

Okigbo, B.N.; Greenland, D.J. (1976) Intercropping systems in tropical Africa. *American Society of Agronomy Special Publications* 27, 63-101.

Raintree, J.B. (1983) *Guidelines for Agroforestry Diagnosis and Design*. [Working Paper no. 6.] Nairobi; International Centre for Research in Agroforestry.

Spears, J.S. (1980) Can farming and forestry co-exist in the tropics? *Unasylva* 32 (128), 2-12.

Steiner, K.G. (1982) *Intercropping in Tropical Smallholder Agriculture with special reference to West Africa*. [Schriftenreihe der GTZ No. 137.] Eischbon; GTZ.

ter Kuile, C.H.H. (1983) The nature, extent and development problems of shifting cultivation. In *Papers from the Expert Consultation on the Education, Training and Research Aspects of Shifting Cultivation, 12-16 December 1983*. Rome; FAO.

von Maydell, H.J. (1978) Agroforestry to combat desertification. *Agroforestry Bulletin* 303, 3-6.

Chapter 72

Timber and Fuel Needs in African Nations and how they can be met

K. OPENSHAW

Field Director (Eastern and Southern Africa), Energy Initiatives for Africa, P.O. Box 39002, Nairobi, Kenya

Introduction

Wood has been used since time immemorial for a multitude of purposes. The woodlands and forests have also been cut down to provide arable and pastoral land, either on a permanent or shifting basis. Seeing that trees are a renewable product, relatively little attention has been paid to their conservation, particularly in Africa. However, trees are only a conditional renewable resource, conditional on them not being over exploited. In many parts of Africa, particularly where the population density is great, trees are being felled faster than they are growing, and so, many countries are faced with a problem of ever diminishing tree resources and therefore in need of conserving what they have. Much of the increase in food production in Africa has come from expanding the areas under crops and animals by clearing forest; this cannot continue if the requirements for wood products are to be met, if soil and desertification are to be prevented, catchment areas preserved, and soil fertility maintained. Unit agricultural and silvicultural output have to increase if the needs of the present and future populations are to be met. In theory this can be achieved by the use of improved seeds, the application of fertilizers and fungicide, better machinery, and improved management. However, most African countries, faced with shortages of foreign exchange, are unable to import these means to increase agricultural production.

Consumption of Wood Products

Most African forest services have in the past emphasized the exploitation of so-called commercial species from the natural forests and woodlands, mainly for the export market, and growing plantations of pine, cypress, teak and eucalypts, to meet demands for sawn wood, panel products and pulp. At the same time, they have endeavoured to protect areas of natural forests that are in watersheds or on erosion-prone soils. Scant attention has been paid to meeting fuel wood requirements, especially of the subsistence farmer, and yet fuel wood and charcoal are by far the largest end use of wood. These fuels are the most important source of energy in developing African countries south of the Sahara. Fuelwood and charcoal accounts for 80–90 % of the wood consumed, and 65–80 % of all energy consumption (Table 1).

Table 1   Consumption of wood and energy for a typical African country.

Wood Consumption	Roundwood equivalent (%)		Energy Consumption[5]	Energy Input (% joules)	
Woodfuel	80	– 90	Woodfuel	65	– 80
Fuelwood	(68)	(80)	Fuelwood	(55)	(70)
Wood for charcoal	(12)	(10)	Wood for charcoal	(10)	(10)
Poles[1]	7	– 4	Other biomass[3]	1	– 1
Sawnwood	8	– 4	Other renewables[4]	0	– 0
Panel products[2]	2	– 1	Oil	27	– 16
Pulp for paper	3	– 1	Hydroelectricity	5	– 2
			Coal	2	– 1
Total	100	100	Total	100	100

[1]Poles mainly for house construction and fencing in the subsistence sector.
[2]Veneer, plywood, particle board, fibre board, etc.
[3]Crop residues, grass, dung, etc.
[4]Wind, water (excluding hydro), biogas, solar, etc.
[5]Excluding animal and human power (≃ 1-3 %).

Fuelwood, and to a lesser extent charcoal, are high bulk low cost products that have to be consumed near where they are produced; unlike products such as sawnwood and oil, they cannot be transported over long distances and still sold at competitive prices. Also, much fuelwood and poles do not enter the commercial market but are freely collected and so to be of use must be within walking distance of the home. Approximately 70 % of all woodfuels are collected, and only 30 % is bought or grown specifically for industry. Consumption depends on availability, elevation, eating habits, and custom, but if wood is reasonably available, average per capita consumption for all wood products is $1-1.2$ t y^{-1} (air dry 15 % moisture content), or approximately $1.4-1.7$ m^3, of which $0.8-1.0$ t $(13-16$ GJ 10^9 J) are burnt as fuel. These figures exclude forests and woodlands cleared for farming purposes.

Future demand will depend on supply, but the present desired per capita demand of $1-1.2$ t y^{-1} (roundwood equivalent), may not change significantly although there may be a slight switch to sawnwood and pulp. In $15-30$ y, there could be a reduction in per capita demand if more efficient conversion methods are introduced and better end use devices, principally stoves, accepted; this could be around $0.8-0.9$ t y^{-1} per capita, but the population will also have increased by about 60 % in this time so that overall demand for wood products will at best be about 30 % greater than today, and at worst 60 % greater.

Supply of Wood Products

The supply of wood products can be divided into "commercial wood" (sawn logs, peeler logs, pulp wood and board wood but excluding wood for charcoal), principally coming from natural forest[1] and plantations; and fuelwood, charcoal and poles mainly from the farm trees, hedgerows, riverine trees and woodlands[1], and branch wood and waste wood from forests and plantations. Forest services of many African countries have made provision to meet the needs of commercial wood even up to 2000, but have neglected the demand for fuel and poles. The trees for fuel and poles should be grown as near to the consumer as possible, i.e. on or near the farm or, for urban people, near towns. To meet rural requirements, a well-developed extension service to assist the farmer is required, and species suitable for growth on the farm. Urban demand could be met from peri-urban plantations which also act as recreational areas. In areas of high population density there are generally shortages of woodfuels which cannot be met adequately from surplus wood areas because they are not within walking distance or outwith the economic transport radius; the wood is often not in the right place. There is also increasing pressure to clear forest and woodlands for agricultural use; although this may prove ecologically and economically disasterous in the long-term, it may have short-term advantages or be the only option of an individual to survive.

Meeting Current Deficits and Future Demands

Population pressures will mean that some forest areas will be converted to agricultural production, therefore there will have to be increased production from a decreased area. This can be achieved by better management of natural forests, and converting some of the forest areas to plantations, while releasing other tree-covered areas for agriculture. This does not exclude taking some land out of agriculture for plantation forestry, as for peri-urban plantations, or for industrial use such as tea-drying and tobacco-curing. Farmers must also be encouraged to plant trees on their own land in a way that does not interfere with agricultural production to any great extent. With a judicious choice of trees and suitable management, arable and pastoral agricultural production may be increased and sustained over a longer period with minimal inputs.

Better tree management will lead to an increase in the production of wood, but should also result in an increase in the utilization of the raw material. Instead of having a single end-use such as saw logs, with tops and branches being discarded, whole-tree utilization could be practised; the first log for plywood, the remaining logs for sawnwood, tops for particle- or fibre-board, and branches for

[1] A "forest" is defined as an area of dense trees which normally has at least 80 % canopy closure with little herbaceous understory; and a "woodland" one of scattered trees with $20-80$ % canopy cover, with a moderate to dense herbaceous understory.

fuel. Wastes from industrial wood production, particularly sawmilling, could be used for board making, or as fuel to burn, or convert into charcoal, or even as a feedstock for producer gas engines. Improved wood drying will make available more energy when it is burnt, and improved charcoaling techniques will lead to increased output. The utilization of woodfuels could be improved by more efficient cooking stoves, wood-fired boilers, tobacco barns, fish curing units, etc.

Increased production from natural forests

Many forest services only exploit the natural forests for so-called commercially valuable species such as mahogany, iroko (mvule) and camphor, and the management of these forests is rudimentary, usually limited to cutting trees above a minimum diameter of about 60–80 cm. Occasionally, natural forests are also exploited for medical products, resins, and foods, but the potential for such resources is generally much greater. All trees have some use, and most, including shrubs and bamboos, as well as ground vegetation, have many uses. Systematic management of natural forests could at least double the wood raw material output from 2–4 m^3 ha^{-1} y^{-1} (1.4–2.8 t airdry) to 4–8 m^3 ha^{-1} y^{-1} (2.8–5.7 t airdry).

Conversion of natural forests to plantations

An even greater increase in production should be achieved by converting some natural forest to plantations. This may also be a way of saving areas of the remaining natural forest which are potentially under threat as the demand for wood products increase. Average annual production of stem and branch wood from a well manaaged plantation in a medium to high rainfall area is 20–30 m^3 (14.3–21.4 t); about four times that of a well-managed natural forest. However, the potential production is greater as breeding programmes are producing tree clones that give consistently high yields. Plantation output may be doubled and the rotation shortened in the forseable future.

Many trees provide other valuable products, such as rubber, resins, and nectar for honey, but many are also nitrogen fixers and are an actual or potential source of animal feed and/or nitrogen rich mulches. *Leucaena* species can yield up to 2 tonnes y^{-1} (airdry) of animal feed ha^{-1}, as well as about 15 tonnes (airdry) of wood, on a two- to three-year rotation. Similarly, *Prosopis* species in low to medium rainfall areas yield about 1 tonne (airdry) of protein-rich pods and 10–12 tonnes of wood per year on a 3–4 y rotation. An animal feed industry could be established alongside such trees, creating employment and boosting pastoral production.

Farm trees

Trees need not necessarily have a long rotation, indeed some may be cropped annually as *Calliandra calothyrsus*, a vigorously coppicing tree harvested annually for more than twenty years in high rainfall areas of Indonesia, giving a wood yield of about 20 tonnes (airdry) y^{-1} plus up to 3 tonnes of foliage for animal feed. It is not always realized that certain trees give quick returns, and that trees may enable farmers to maintain or increase production by: (1) acting as a nutrient pump; (2) decreasing wind velocity and sun scorch; (3) increasing soil friability and the water carrying capacity of the soil; and (4) decreasing erosion, water runoff and the velocity of rain.

Trees planted on farms may increase the incidence of pests and diseases, and if not managed properly can significantly decrease the agricultural output. However, if complementary trees, such as many leguminous species, are planted at wide spacings or in rows on a short rotation, the farmer will be able to maintain the agricultural output and at the same time provide the household with fuelwood, poles, fodder/mulch and perhaps honey. Agricultural production may even be sustained longer, if not indefinitely, with a minimum of outside inputs such as fertilizers (K, P but not N, and perhaps Mn, Ca and S). In one experiment in Rwanda (I. Neumann, pers. comm.) *Grevillea robusta* was planted at 300 ha^{-1} on a six-year rotation; the trees were root-pruned at a radius of 20 cm, and the branches were also pruned. Fifty trees were cut each year, yielding 7.2 tonnes (airdry) of wood, and 2.5 tonnes of green leaves which were used for mulch. A 1% decrease in agricultural production was recorded, it is thought due to the shading effect of *Grevillea*, but crop production was maintained using a fallow period of one year in six.

Many forest services are not geared to providing suitable farm trees. Eucalypts, pines and cypresses may be suitable on farm areas that are inappropriate for agricultural production, but they are not generally satisfactory in fields or along boundaries. If the Third World energy crisis is to be overcome, this must largely be by the main users, i.e. subsistence farmers. By planting trees on their own land, they can have wood on hand and so save time and effort in collection, and help maintain soil fertility.

Peri-urban and industrial plantations

In order to save energy and expense in transporting fuelwood and charcoal to the consumer, it is better to grow it close to the demand centre; the logic of peri-urban plantations and industrial plantations. These may require land that was or is under agriculture, but from a national standpoint the overall balance may favour such plantations. If multi-purpose trees are grown, animal feed could be produced at the same time.

Conservation

Wood can be saved by whole tree utilization, the use of sawmill waste, etc., wood drying, better charcoal conversion and improved end use design for industrial and household use (see above). Savings of around 30 % can be made, which could be significant, especially in areas of large end-uses of wood such as for cooking. However, if efforts are not made to ensure a sustainable supply of wood, conservation will only delay the acute shortage and not eliminate it.

Energy and Agriculture

As the amount of new fertile land is limited, some forest areas will undoubtedly be converted to agriculture. Marginal lands can only be brought under sustained production by applying techniques such as irrigation. The greatest increase in production will have to come from existing agricultural areas, which implies a greater use of energy either to produce fertilizers and/or through more intensive cultivation techniques. Crop storage and processing, if they are improved and expanded, also require a greater energy input. Energy is one of the key factors to increasing agricultural production.

Direct wood energy can only be a marginal source of energy. It is possible to use producer gas units to run vehicles, tractors, and stationary engines (for pumping), but such engines have to be well maintained and serviced and an adequate and constant supply of wood particularly has to be at hand. However, farm trees particularly, can indirectly increase agricultural production by improving the microclimate and providing nitrogen and other essential elements through their leaves and pods. If draught animals are used for ploughing and transport, trees can supply protein-rich fodder to feed them. An adequate and sustained supply of wood energy for household and industrial use will also mean that petroleum products will be more readily available for the agricultural and transport sectors. The preferred fuel of the subsistence sector is wood; when this is in short supply there is a switch to crop residues, grass, twigs, leaves and dung, removal of which has an adverse effect on the land. Eating habits may alter and a less nutritious diet be adopted, so affecting working capability. The urban population, in contrast, will tend to switch to fuels such as liquid petroleum gas, paraffin (kerosene) and electricity; increasing the demand for foreign exchange to purchase these fuels and their appliances.

The use of energy in agricultural processing is critical, especially if the crops processed earn foreign exchange. Tea and tobacco require a considerable amount of energy to dry or cure them, and in most cases wood is the cheapest energy source. In Uganda, the cost of curing 1 kg of tea with oil is of the order of US $0.20, where as when wood is used it is US $0.01; the latter being almost all local costs and the former almost all foreign exchange costs. Substituting for indigenous energy is not one of the best allocations of foreign exchange.

The Cost of Ensuring an Adequate Wood Supply

There is a definite cost attached to the strategy outlined above. However, the benefits of an assured supply of indigenous energy and wood products, the saving of foreign exchange, and the creation of (rural) employment outweight the costs. For example the Ministry of Energy and Regional Planning in Kenya has just completed a study (O'Keefe, *et al.* 1983) to determine present energy consumption and possible future alternatives; the study concluded that wood was the cheapest energy source and that Kenya should ensure that a sustainable supply of wood is guaranteed. It was proposed that the equivalent of 1.4M ha should come under forest management in the form of farm tree planting (0.5M ha forestry equivalent; or 3M ha actual farm land), managed natural forests (0.4M ha), replanted natural forest (0.3M ha), and peri-urban/industrial plantations (0.2M ha). The last would have to be switched from other (agricultural) use, but at the same time, it was estimated that at least 0.3M ha could be converted from natural forests and woodland to agricultural use. The cost of such a programme, including conservation measures and the provision of other wood products such as poles and sawlogs, is about US $300M over 17 years, most of which is local costs (Openshaw 1984). The average return on capital investment for all the options is in excess of 10%. Such a programme would yield an additional 27M m³ (18.0M air dry tons) by the year 2000, enough to satisfy 35% of Kenya's energy requirements by that time and create 100 000 jobs in the growing, management and felling of the new wood resource. In contrast, the Kiambere hydro-electric power station with a rated capacity of 140 Mw, will cost an estimated US $500M, practically all foreign exchange, and supply only about 0.5% of Kenya's energy requirements by 2000. The cheapest alternative to wood is paraffin (kerosene), but if this were imported or refined, the cost would be about six times that of the wood programme, nearly all of which is foreign exchange, and the employment opportunities would be small.

The wood option is the cheapest, uses the least foreign exchange, and provides the most employment. It is also an insurance against soil degradation and ensures that the rural population is capable of increasing and sustaining agricultural production.

References

O'Keefe, P. *et al.* (1983) *The Energy Development in Kenya. Problems and opportunities.* Stockholm; The Beijer Institute (Royal Academy of Sciences).

Openshaw, K. (1984) *Costs and Benefits of the proposed Tree Planting Programme for satisfying Kenya's Wood Energy Requirements.* Stockholm; The Beijer Institute (Royal Academy of Sciences).

Chapter 73

Plantations for Timber Production in the Agricultural System

R. S. W. NKAONJA

Department of Forestry, P.O. Box 30048, Lilongwe 3, Malawi.

Introduction

The silvicultural and management techniques for industrial timber production in forest plantations are well-established and founded on properly researched information in many African countries. The same is true in agriculture, including animal production, where advances in crop and animal husbandry, disease and pest control and breeding research, have increased output many times over. The challenge before foresters and agriculturalists, however, is the one set by the theme of this conference, which calls for closer co-operation between the two disciplines and hence, an integrated approach to agricultural production for the social and economic development of Africa.

This call is urgent in Africa where the economy is primarily agro-based, the rate of population growth is very high and the amount of arable land is rapidly diminishing. Malawi, for example, with a population of 6M, growing at the rate of 2.96% per annum, and a total land area of only 94 400 km² is one of the most densely settled countries of Africa. Agriculture is the backbone of the country's economy, accounting for 95% of the total annual foreign earnings and 37% of the Gross National Product. Firewood accounts for 86% of the total annual energy consumption and has traditionally been provided from the indigenous forests. This response is, however, diminishing rapidly at an approximate rate of 3.5% per annum in response to agricultural expansion, steadily increasing population and increasing wood energy demands. Similar situations prevail in many other African countries. For example, in Tanzania and Kenya agriculture is also the backbone of the economy. Most of the agricultural production is by small farmers who account for 90% and 80% of the total populations respectively as compared to 75% for the whole of Africa. There is, therefore, a great need to develop integrated farming systems that would maximise both agricultural and forestry outputs from the same piece of land.

Tree Production in Agricultural Systems

Forestry and agriculture have traditionally been considered as mutually exclusive activities. Foresters have always confined their afforestation activities to designated forest reserves and have rarely, if at all, considered it as an aspect of agriculture. Similarly, agriculturalists have developed independently from forestry and have abhored the sight of trees in crop gardens. In fact, it was a punishable offence in the 1950s in Malawi even to mix the planting of maize with beans or groundnuts in the same garden. Consequently, monoculture plantations of pines or eucalypts in forestry or of maize, beans, millet, etc. in agriculture have been developed and extended independently. Today, however, the concept of agroforestry is being vigorously advocated by the same agricultural and forest scientists, much to the confusion of the small farmer who had known about the merits of the system much earlier by trial and error.

This contribution discusses several ideas, from a forester's view, as to how tree-planting in farmlands could be approached to enhance agricultural production.

Land Classification Silvicultural Zoning

In many African countries, forest reserves are created and gazetted on mountainous areas or on unstable steep slopes which are not suitable for agricultural production. Their main functions are to conserve the soil and to promote the water-yielding capacity of the catchment areas. Forest plantations have been established on the gentler slopes for saw timber or pulpwood production, thereby increasing the range of benefits from the reserves. The most commonly planted trees on the highlands are *Pinus*, mainly *P. patula*, *P. elliottii*, *P. caribaea*, *P. kesiya* and *P. radiata*, and *Eucalyptus*, primarily *E. grandis*, *E. saligna* and *E. microcorys*. These species cannot tolerate the harsher climates of the lower altitudes where agricultural activities are concentrated.

Silvicultural zoning is needed so that suitable tree species for each zone are properly identified. In this regard, the use of relevant meteorological and soils data or land satellite imagery is necessary, and hence the need for co-operation between foresters and experts in these fields.

This approach has been made in Malawi using rainfall, temperature and, to some extent, soils and altitude as the main variables (Hardcastle 1977). A total of 11 Silvicutural Zones were delineated. Seven of these were identified as the most needy in terms of agricultural production and yet the most needy in terms of wood for construction and fuel for domestic use and for curing farm crops, mainly tobacco. Species and provenance discrimination research was therefore launched in

1978 to identify the species most adapted to these zones, where a shortage of forest products is becoming an important social problem and a constraint on improving standards of living of the rural people. The results of these trials (Nkaonja 1982) form the basis of a follow-up research project in agroforestry starting in 1984.

Management and Utilization of Indigenous Forests

Indigenous forests have a proven record of successful survival in their original ecosystems. They have supplied man with his basic necessities for life such as energy, fruit, timber, medicines, etc., and have promoted other biotic communities such as honey bees, mushrooms, wild animals, caterpillars, etc., which man has utilized in various forms for his survival. Agricultural production must therefore be understood to include the intrinsic and cultural values of indigenous forests.

Unfortunately, the common practice in Africa has been to promote monocultures of exotic tree species in afforestation projects at the expense of the indigenous trees. A similar approach is also adopted when recommending tree species for agroforestry purposes. In fact, with the exception of *Acacia albida*, very few indigenous species have received wide publicity in agroforestry; *Leucaena leucocephala* leads the list among the exotic species.

It is, therefore, being advanced that foresters must put sufficient emphasis on indigenous silvicultural, management and utilization research in order to accentuate the importance of the indigenous forests in development activities. Furthermore, wise use and proper management must be advocated, particularly now in the face of rapid expansion, to conserve the genetic base of the endangered and rare species.

Choice of Species and Provenances

Selection of tree species for agroforestry purposes must be carried out, at least, to ensure that: (1) the species are adapted to the intended planting sites; (2) the species will satisfy the objectives of management, for example, soil enrichment, fodder, erosion control, live fencing or fuelwood and poles; and (3) farmer's preferences to particular trees are duly included.

The first consideration is straightforward. Particular trees will be more productive in those sites they are most suited to in terms of soil and climatic factors. Silvimeteorological data must, therefore, be obtained and matched with tree species to enhance productivity. Similarly, seed source or provenance is a proven variable which influences adaptability of a given species to a new site, in terms of survival and biomass production. Therefore, there is a need for very close co-operation between meteorologists, tree-breeders and silviculturists in the area of species and provenance research for increased biomass production in forest trees.

The second consideration entails that silvicultural and management techniques must be developed for particular agroforestry trees. For example, *Cassia siamea* or *Gmelina arborea* develop aggressive root systems and preclude the growth of farm crops under their shade. On the other hand, *Acacia albida*, an important shade, fodder and nitrogen-fixing tree, enhances crop yields under its crown. Thus, if *Cassia siamea* or *Gmelina arborea* are to be planted together with farm crops, they must be planted at a very wide spacing and must be pruned or pollarded frequently. With *Acacia albida*, the consideration would be on how densely the trees must be planted in order that the total output of the tree and farm crops is maximised. To the extent that agronomic and silvicultural techniques are different, the need for co-operation between foresters and agronomists cannot be over emphasized.

As the largest contribution to the agricultural economy in Africa is made by the small farmers, it follows that the success of any agricultural system will depend on how well the small farmer understands and accepts the system. Even more important, it will depend on how much the farmer has been involved in *developing* the system. Therefore, the co-operation needed to enhance agricultural production in Africa must not be limited to "experts" in agriculture, forestry, soils or meteorology but must be extended to include the rural farmer who finally implements the recommended system. Thus, in agroforestry, the farmer's preferences for particular species must not be overlooked in preference to species such as *Leucaena* which have been widely publicised in international journals.

Collaborative Research in Agroforestry

Agroforestry is a multi-cultural farming system which is a departure from monoculture systems in forestry or agriculture. As such, the system has its own problems which may necessitate special research to provide solutions. Such research cannot be carried out independently by foresters or agriculturalists lest the recommendations derived may have serious implications to the other field.

In Malawi, for example, the growth of some trees in agricultural lands, such as *Melia azedarach* and *Leucaena leucocephala*, was severely hampered by snails of the family Streptaxidae which nibble the shoots and ring-bark the stems and branches. The dryland eucalypts such as *Eucalyptus camaldulensis* and *E. tereticornis* suffer severe attacks from subterranean termites, particulary *Macrotermes* species. The recommended chemical control method of these pests is the application of aldrin or dieldrin suspensions to the trees. However, these chemicals are not recommended for use in tobacco fields because they contaminate the tobacco leaf.

Numerous other examples exist which necessitate collaborative research in agroforestry, particularly those concerning the interactions between crops and trees, such as the allelopathic effects or various spacing requirements for crops and trees when planted in panmixia. These examples have excluded animal husbandry which is included in the definition of agroforestry by many authors.

Conclusion

Timber production in farming systems has a great potential for increasing agricultural output in Africa. Testimony to this are tea and coffee farms in East and Central Africa where trees such as eucalypts, pines, *Grevillea robusta*, banana trees, etc., planted initially as nurse-crops or windbreaks, provide timber, fuel, fruit, and other products, in addition to the main crops. There is need, however, for this system to be extended widely. In this regard, foresters must intensify identification of the potential agroforestry trees and match them with the climatic and soil conditions in the intended planting sites. Close co-operation with meteorologists is necessary at this stage to ensure that correct data is used in silvicultural zoning.

The proper management and utilization of the indigenous forests must be given due emphasis in promoting agricultural production in Africa, and research in these fields should be intensified.

Collaborative efforts of agriculturalists and foresters in agroforestry research and development should be initiated very early in the planning stages. Even more important, the rural farmer must, as much as possible, be actively involved at every stage of development to ensure the success of the farming systems developed to enhance agricultural production.

References

Hardcastle, P. D. (1977) *A Preliminary Silvicultural Classification of Malawi.* [Research Record no. 57.] Forestry Research Institute of Malawi.

Nkaonja, R. S. W. (1982) *The Rural Fuelwood and Poles Research Project in Malawi.* Zomba; Forestry Research Institute of Malawi, 79pp.

Chapter 74

Bees as a Development Resource in Sub-Saharan Africa

J. CORNER

Apiculture Rehabilitation Project, CARE-Uganda, 15 Mackinnon Road, P.O. Box 7280, Kampala, Uganda.

Introduction

Africa includes the largest tropical land area on earth. All of the African woodlands and forests are well suited to beekeeping and even in the dry savannah regions of Africa south of the Sahara, the tough and adaptable honeybee *Apis mellifera adansonii* is able to survive and store surplus honey. Beekeeping has been practised by Africans in all parts of the continent for many thousands of years. To this day, the collection of honey, and beeswax from the nests and hives of these industrious insects, contributes, in no small way, to the nutrition and well being of Africa's people. There are few other livestock husbandry endeavours in Africa where the stock is available at no cost and is continually self renewable. In addition to abundant honeybee stock there are a wide variety of plants, shrubs and trees valuable to bees and man, as a rich source of nectar, pollen and beeswax.

Honey Bees as a Development Resource in Africa

In the context of classifying and evaluating agricultural products; beekeeping and the production of bee products is usually assigned a very low priority. When compiling agricultural statistics beekeeping and honey production are usually combined under the heading "other crops".

During the last 100 years, the art of beekeeping has spread from Africa and the old world to the new world where honey bees are kept as far north as Alaska and the Scandinavian countries and south to Argentina and Australia.

World production of honey in 1975 was estimated at 630 430 t harvested from 50M colonies of bees (Crane 1975); Africa produced 82 700 t or 13.1 % of this total. Such statistics do not account for honey and wax harvested from all feral colonies.

The important honey producing countries of tropical Africa are Tanzania, the Malagasy Republic, Angola, and Ethiopia (Crane 1975). Other countries in Africa with high potential for honey and beeswax production are, Uganda, Kenya, the Central African Republic, and Zaïre. More information is needed on many other countries in Africa south of the Sahara where possibilities exist for the development and expansion of the beekeeping industry.

The third International Conference on Tropical Apiculture will be held in Nairobi on 5 – 9 November 1984. It is hoped that discussions there will centre on the potential and benefits to be derived from the development and expansion of beekeeping in Africa. Direct contributions of beekeeping to African families and communities would be: (1) production of honey and beeswax; (2) production of a nutritious food with good keeping qualities; (3) local village and urban market potential; (4) production of beeswax which can be easily stored and transported and a product which has a proven market potential; (5) other bee products, i.e. pollen and propolis; and (6) beeswax candles.

In many African countries honey is valued for its medicinal properties, especially in the treatment of colds and as an antiseptic for wounds. This is especially true in Uganda but there very little beeswax is salvaged for the market. Honey is also used extensively in the brewing of local beer.

Internal Trade

Very few African countries south of the Sahara have developed a substantial trade in the export of honey. There is some indication that honey as a commodity does move across country borders in Africa, but this is not so much the result of surplus production in the exporting country but rather the proximity of ready markets to areas of high production.

Export Markets

Honey has in recent years become a very competitive product on world markets. Honey in bulk containers is the most common method in which it is exported commercially. However, retail packs in glass and plastic containers are preferred by some importing countries. China has also increased production of high quality white honey at very competitive prices. As the production of honey in African countries increases, it will be necessary to actively seek out and develop export markets. It may be sometime before this happens, but as hives and management improve in Africa so will the volume and quality of both honey and beeswax. Many African honeys have interesting and unique flavours and if properly handled from the producer to the packer level, there will be opportunities to market surplus honey and beeswax and so attract much-needed foreign currency to the country of origin.

Pollination

The most important contribution of beekeeping to agriculture throughout the world is in pollination. In many of the developed countries mono-culture of agricultural crops is practiced; vast acreages of single varieties of fruit and seed crops. Pollination by bees is often essential to success and thousands of hives of bees are consequently rented to pollinate these crops when they are in bloom. In Africa some of the most important and valuable export crops need, or benefit from, pollination by bees. A few of these are coffee, cotton, tobacco, pyrethrum, cashew, macadamia, and sunflowers (Crane & Walker 1983).

However, the pollination of agricultural crops in Africa is not usually given high priority in spite of its importance. *A. m. adansonii* is an excellent pollinator, and as more research on the beneficial effects of crop pollination by honey bees is carried out, the importance of beekeeping in Africa will be increasingly recognized as having a valuable role in the quantity and quality production of crops requiring pollinator services.

Bee Breeding

There has been virtually no attempt made to select and breed *A. m. adansonii* in sub-Saharan Africa. The reason for this is that bee breeding currently requires an unusual amount of colony management and handling; however, systems could perhaps be devised whereby management could be kept to a minimum. *A. m. adansonii* does not always react well to handling and disturbances, frequently responding by aggressive behaviour, absconding, and the destruction of queen cells. Nevertheless, this honeybee orginated in Africa and there are strains of *A. m. adansonii* which are reported to be of a gentler disposition. Africa is therefore the most logical place to carry out selection and breeding programmes using local bee stocks. It is to be hoped that research on this problem will begin in the near future. Although the selection and breeding of honeybee stocks is a complicated and long-term project, any improvement would be of great importance.

Beekeeping Aid Programmes

Developed countries have been generous in their support of beekeeping programmes in developing countries and several of those in Africa have been very successful. Tanzania and Kenya have expanded their beekeeping programmes, which are now well developed and entirely run by local staff. Too often, aid programmes mix research and practical training, with the research receiving a higher priority than extension programmes. The need in most African countries is for the training of national staff in the art of good beekeeping extension procedures and programmes, aimed at training and encouraging the interest of young bee farmers.

Basic problems needing solutions include improvements in local hives; improved harvesting methods; improvements in the quality and grade of both honey and beeswax; and more frequent contact with bee farmers.

For the most part, African attempts at converting and adapting to the use of expensive and unsuitable beekeeping hives and new world systems of management, have proved impractical, expensive, and have usually failed. Behavioural differences between exotic and African bees dictate the type of hives and equipment to be used in Africa as well as management and harvesting methods. There is a strong tendency to send African personnel to developed countries to acquire experience and advanced training in apiculture. Some training of this kind is desirable and useful. However, the most productive training can be carried out in African countries, working with African bees, modified local hive equipment, and solving local problems under local conditions with local people.

African countries interested in developing their enormous beekeeping potential must focus their efforts on strong beekeeping extension programmes aimed at teaching and training extension staff and farmers. Productive research will naturally follow as management improves and production increases.

Bee Diseases

Although there are a variety of predators on honey bees and their nests, bees in Africa are almost entirely free from American foulbrood (*Bacillus larvae*) and European foulbrood (*B. pluton*).

Nixon (1983) provides additional information about the distribution of two very serious pests of the honeybee, the mites *Varroa jacobsoni* and *Tropilaelaps clareae*. These are spreading rapidly throughout the world honeybee populations. *V. jacobsoni* in particular is a serious threat to beekeeping in Africa. This mite is now present in Algeria, Tunisia and Libya, and it is to be hoped that the barrier created by the Sahara will prevent its spread into sub-Saharan African honeybee populations. African countries south of the Sahara that have not already done so would be well-advised to enact legislation prohibiting the importation of honeybees into their countries. If *Varroa mites* were introduced into sub-Saharan Africa, they would probably spread quickly throughout the honeybee populations; the absconding and migratory behaviour of *Apis m. adansonii* would result in its rapid dissemination.

Conclusion

The African honeybee is industrious, adaptable and ideally suited to the ecological and environmental conditions in Africa. It is fortunate that Africans have this efficient insect in abundance for it can and does contribute much to the economy of tropical African countries and to the Africans' well-being and quality of life.

References

Crane, E. (1975) *Honey. A Comprehensive Survey.* London; William Heinemann, 608 pp.
Crane, E.; Walker, P. (1983) *The Impact of Pest Management on Bees and Pollination.* London; Tropical Development Research Institute, 73 pp.
Nixon, M. (1983) World maps of *Varroa jacobsoni* and *Tropilaelaps clareae*, with additional records for honeybee diseases and parasites previously mapped. *Bee World* 64, 124-131.

Chapter 75

Management of Wildlife in the Future of Africa

G.S. CHILD

Wildlife and National Parks Officer, Forest Resources Division, FAO, Rome.

Introduction

Over one million km² of Africa are designated as National Parks or equivalent reserves (IUCN 1982). Such areas are set aside exclusively "... for the propagation, protection, conservation and management of vegetation and wild animals..." (Anon. 1969). In addition extensive areas are established as wildlife management areas, game controlled areas, hunting reserves, partial reserves or similar categories in many countries by national legislation, where the management of wildlife resources is regulated in varying degrees. In most forest reserves hunting is also controlled. Various forms of wildlife management are evidently seen as potentially viable land-use options either in themselves or as components of multiple-use systems.

The categories of areas constituted envisage two broad concepts of resource utilization, protection ("non-consumptive utilization") or harvesting a proportion of the resource ("consumptive utilization").

In order to attempt to predict the future of wildlife management in Africa, it is necessary to appreciate the factors which have influenced its evolution.

Foundations for Wildlife Management

Generalizations often oversimplify which can result in conclusions having a questionable basis. With this reservation clearly recognized, the following analysis is presented.

Objectives of management

Traditional management systems have had the production of meat and other animal products in subsistence economies as their primary objectives. Harvesting technologies employed consisted of methods of hunting and trapping, and management was often influenced by cultural factors. Examination of the latter have demonstrated resource conservation elements in patterns of traditional utilization (Sale 1981). A further interaction between rural communities and wildlife revolved around the protection of crops and livestock from the depredations of wild animals.

The only long-term and extensive commercial utilization of wildlife was to be found in the ivory trade; elephant hunting was not regulated until the turn of the century.

With the advent of the colonial era, European and later North American sport hunting enthusiasts were attracted by the variety of large wild animal species. This resulted in the development of safari hunting. Mention should also be made of game control and elimination exercises undertaken in accordance with contemporary thinking and practice related to animal disease control, crop protection and similar operations.

By the end of last century, concern was being expressed at what was seen as the destruction of Africa's native biota. The designation of reserves and attempts to regulate hunting followed. By 1901 a draft International Convention for the Protection of Fauna and Flora had been drawn up, but it was not until 1933 that the London Convention was signed and received the required number of ratifications, coming into force in 1935 (Anon. 1936). Under this treaty, species were designated for varying degrees of protection, procedures for export and import of certain species and their products were laid down, and categories of protected areas defined. The latter included the concept of the "national park", although in 1925 an area had already been established as such in what is now Zaire.

Legislation

An examination of wildlife and protected area legislation promulgated during the first half of this century shows that of the above elements, traditional utilization was at best ignored. It usually fell within the definitions of poaching, and the technologies employed were declared unlawful methods of hunting. Furthermore, possession of meat or wild animal products, let alone disposal of them, could often constitute a crime. Most legislation was directed at the institution and regulation of sport hunting as perceived in Europe. Concepts such as game animals, hunting seasons, bag limits, trophies, hunting reserves and royal game, which were associated with traditional systems of managing game and hunting in Europe, were adopted in legislative texts. The biological validity of such approaches to management in tropical conditions was apparently not questioned.

Legislation also sought to remove possibilities for exploitation and extensive commercialization of wildlife and its products, presumably because of the conflict of interest with the requirements of sport hunting and fauna preservation. To cater for the protection of livestock and crops, the concept of vermin was often adopted.

The London Convention provided further significant ingredients for legislation relevant to the protection of species and their habitats, including a first attempt at internationally recognized definitions for categories of protected areas. In some countries parts of the convention were adopted without modification in national legislation.

Management Authorities

As game and hunt management were aspects of forestry in Europe, responsibility for wildlife and protected areas was seen as a function of forestry administrations in most African countries. This was so even in the arid parts of West Africa as forestry essentially covered the management of vegetation. However, in many eastern and southern African countries, the tendency was to link game to other resource areas. Thus some countries established Game and Fisheries departments, elsewhere the emphasis was on Tsetse and Game Control and where wildlife was regarded as being of little or even negative significance, it became a minor responsibility of Agriculture or Veterinary Departments. This picture was complicated in that over time institutional arrangements were redefined.

Recruitment of Managers

Where responsibility for wildlife and protected areas fell within the mandate of forestry administrations, the management of these resources and areas was in the hands of foresters. Although, at the time, foresters perhaps came nearest to meeting the requirements of wildlife and protected area managers, they were trained in Europe to manage forests, and such exposure as they had to wildlife was primarily related to temperate species, in terms of recreational hunting or damage to trees.

Elsewhere, wildlife and protected area management was seen as mainly concerned with law enforcement and the control of large and dangerous animals. Thus, an ability to organize armed guards and scouts along disciplined paramilitary lines, for policing functions in respect of protected species and areas, and to hunt and destroy dangerous animals. Ex-military officers with an interest in hunting and nature were considered to be well-qualified for such work. Individuals from the forestry, agricultural and veterinary fields, and ex-policemen with an interest in wildlife or hunting, were also recruited.

The Basis of Wildlife Management

By about 1950 a general pattern had been established as the accepted approach to wildlife management in Africa. Utilization of the resource was based on the tradition and ethics of hunting in Europe. It catered essentially for visitors on trophy hunting safaris and expatriate residents hunting for the pot. There was the feeling that indiscriminate slaughter of wildlife was morally wrong, nevertheless only minimal justification was needed to embark on game elimination programmes. Certain species were legally protected as royal game, and areas had been established as reserves. The objectives of reserves were not always clearly defined but concepts of having stocks of game that would serve to replenish hunting areas or could be utilized as sources of meat in times of drought were becoming apparent, alongside complete protection to enable nature to take its course.

The implementation of this approach was primarily a policing operation to enforce hunting and protection legislation. The control or elimination of wildlife considered to be in conflict with other forms of land use or human interest was also important. However, the role of wildlife in subsistence economies or in terms of nutrition in rural communities was usually ignored.

Wildlife: A Renewable Natural Resource

Wildlife management as a scientifically based subject is very new, originating in North America as game management in the 1930s where the classic work of Leopold (1933) made a major contribution to its definition. It is an established discipline there, with associated institutions, literature, training, and research. The first national park was designated in North America last century, and was the forerunner of the extensive system of protected areas that now exists.

Following World War II, many of the principles and approaches developed were applied to African wildlife and national parks management. This was initiated by American specialists working in Africa and sustained by Africans trained in North America. As a result, the concept of "game" was replaced by "wildlife" with all that this implies, and a static "fauna preservation" approach gave way to the more dynamic "wildlife conservation".

Wildlife and National Parks

Eastern and southern Africa were in the vanguard of these changes, although developments in Zaïre should not be overlooked. An increasing involvement of naturalists and biologists with game administrations paved the way for the reorientation. Early signs that Governments were beginning to treat wildlife more seriously included efforts to implement the London Convention. Of particular significance was the establishment of national parks in several countries. Game authorities were strengthened and given a resource management mandate and in some cases, separate National Parks administrations were created. That the fate of wild animals is ultimately linked with that of their habitat was recognized and research activities were launched.

At the international level, the provisions of the London Convention were examined and proposals for amendments made at the Bukavu Conference in 1953. Subsequently, following technical meetings and preparations, the African Convention for the Conservation of Nature and Natural Resources was adopted by the OAU summit at Algiers in 1968. Perhaps the single event which has had most impact on the development of Wildlife and National Park management over the past two decades was the Arusha Wildlife Conference of 1961, held under the auspices of CCTA and IUCN, the forerunner of OAU/SRTC.

In the wake of this upsurge of interest in wildlife two problems became apparent, (1) a shortage of appropriately trained personnel, and (2) the need to promote public acceptance of management programmes. A milestone in overcoming the former was the establishment of the College of Africa Wildlife Management at Mweka, Tanzania in 1963, which was followed by its sister francophone institution, the Ecole de Faune at Garoua in the Cameroon. In addition, wildlife courses were introduced at a number of African universities, and several countries have developed training facilities for guards and rangers.

Promotion was seen in terms of conservation education and information. Programmes have been initiated in schools (often through the media of wildlife clubs), education centres and interpretive programmes have been established at national parks and information disseminated through the media.

Wildlife utilization

The two approaches to wildlife utilization are best examined separately. The adoption of one of these, or both, in any given situation may ultimately be dependent on policy decisions.

(a) Protected areas: non-consumptive utilization

The acceptance of the principle that national parks and similar protected areas should be used for the benefit of people precipitated changes in the concept of their management. From being something approaching private preserves of game authorities, visitors began to be tolerated, if not welcomed. Initially, expatriates resident in Africa took advantage of this situation, but with the upsurge of international tourism in the 1960s there was a dramatic increase in the numbers of overseas visitors. Tourism based primarily on wildlife viewing and photography became established.

The philosophy underlying the management of National Parks was a subject for considerable debate. There were two major schools of thought, those who advocated letting nature take its course, and those who supported active management to maintain a desired state. In the latter case, management intervention could include the controversial cropping of large mammals.

(b) Game harvesting: consumptive utilization options

As the concept of game management gained ground, efforts were directed towards scientifically regulating the offtake of animals by safari and other licenced hunting. This possibility of taking a sustainable quota from wild mammal populations stimulated interest in examining the potential of harvesting wild ungulates for meat. Such ideas were given impetus by various contemporary lines of research. These included the recognition that a spectrum of wild herbivore species made more efficient use of many vegetation formations than a limited range of domestic stock, various wild species had trypanotolerant, heat-tolerant and drought tolerant attributes, and their reproduction and meat production characteristics were favourable (Ledger & Smith 1964, King & Heath 1975).

Efforts to realize this potential took the form of cropping schemes, game ranching and game farming activities, some launched as Government projects and others undertaken on a commercial basis (Dasmann 1964, Mossman & Mossman 1976, Bindernagel 1968, Parker & Graham 1973, King & Heath 1975). Possible approaches to varying management intensity were examined from systematic harvesting of free-ranging wild animals, through various degrees of confinement, to attempts at "domestication".

Research

Historically the accumulation of knowledge on African fauna relied heavily on the personal interests and dedication of individuals, together with the field work and support of museums based in Europe and North America. The national parks of Zaïre were amongst the earliest to launch research programmes. A more systematic approach became possible with the appointment of biologists by game authorities and the sponsorship of wildlife researchers associated with universities and foundations. In East African National Parks, programmes were launched by the Nuffield Unit of Tropical Animal Ecology (forerunner of the Uganda Institute of Ecology), the Serengeti Research Institute and the Tsavo Research project. Research units were also established in Government organizations and some African universities became active in the field.

Certain studies undertaken had clear relevance to the needs of wildlife management. With regard to techniques, work was carried out on methods for the assessment of populations and the development of systems for the immobilization of free-ranging animals. Other studies culminated in the formulation of prescriptions for the management of species or the rationalization of protected area boundaries. Other activities were of less relevance and at worst designed to assemble data for academic advancement.

As in other areas of applied science, there was a certain lack of communication between research workers and managers. Problems of management were not analyzed and defined for adoption as research projects, and research results were not always translated into implications for management and presented in a form that managers could use.

Current Status of Wildlife Management

Wildlife and protected areas as resources and their management (as a profession and recognized function of Governments) have made much progress, uncertainty remains as to their niche and role in policy-making. It is not difficult to find reasons for this if the apparently incompatible viewpoints on wildlife are examined. These range from extreme non-governmental preservationist lobbies, through international conservation and environmental organizations, to the promotors of active management, utilization and recreational hunting. Decision-makers are so faced with conflicting advice and the situation is compounded by residual influences of the past.

Institutional arrangements

Uncertainty as to the role and linkages of wildlife and protected area management is illustrated by the variety of ways in which they are accomodated in Governmental structures. In many West African countries, they remain incorporated in or closely associated with forestry organizations. In countries of eastern and southern Africa with significant wildlife and national park resources, independent departments or divisions have been created. Elsewhere, they may be grouped in ministries of natural resources (e.g. with forestry, fisheries), in ministries of agriculture, in ministries with tourism, in the ministry of commerce, or with environmental agencies. Wildlife and National Parks administrations can be separated. In a few countries National Parks have been historically administered by parastatal organizations.

Legal aspects

At the international level, the Fourth Meeting of the Convention on International Trade in Endangered Species of Flora and Fauna (CITES) was held at Botswana in 1983. The number of parties to this Convention is now sufficient for it to have a significant impact on wildlife trade. For management, important developments have been the definition of wildlife farming and ranching for the purposes of implementation of its provisions.

Many countries have undertaken comprehensive revisions of national legislation to incorporate provisions for the adoption of modern approaches to wildlife and protected area management but it has proved difficult to break away from past formulations. Constraints are still placed on possibilities for management by now inappropriate legislation.

Wildlife in land use and development

Wildlife and protected area management are seen as viable landuse options that should be examined and, where appropriate, be incorporated into landuse plans. They should also be considered in designing multiple-use systems and included in the formulation of rural development programmes. Such ideas are generally accepted at wildlife conferences and symposia. Unfortunately, planners and other resource managers do not necessarily routinely take account of wildlife and protected areas in the planning process. There may be a mention of wildlife programmes in broad terms at the national level (in development plans and similar documents) and at the local level specific activities may be instituted by wildlife authorities in areas designated by legislation, but integration into the intermediate levels is frequently lacking.

A further problem in relation to development planning concerns the inability of those responsible for wildlife management to translate their requirements into economic planning terms. This results from a weakness in training, which often had an ecological rather than planning and management orientation.

People's participation in wildlife management

It has become increasingly realized that there can be no long-term future for wildlife or protected area management initiatives where people, and especially local people are antagonized. They should be involved and benefit from such schemes, to the extent that they will be persuaded to contribute to the conservation of the resources on which they are based. Initial efforts in this direction were often superficial, but in recent years more realistic approaches have been developed. However, viable and lasting examples remain rare and there is scope for further work in this area.

The role of wildlife in nutrition

The significance of wild protein in the diet of rural communities in Africa has yet to be documented comprehensively. The contributions of subsistence hunting or fishing are not adequately reflected in conventional surveys. However, the indications are that in certain circumstances and situations, wildlife can be the major source of meat (Talbot 1966, Charter 1970, de Vos 1977, Ajayi 1979, Sale 1981). Further, it is often the smaller and less spectacular species that are most important in this respect (Asibey 1972, Ajayid 1979). This aspect of wildlife utilization, which is largely traditional, is generally not accomodated in existing legislative and institutional frameworks.

Game cropping, ranching and farming for the production of meat and by-products occurs in a number of African countries (FAO, in press). Government and private enterprise are involved. Many of these efforts are pilot projects and constraints of a socio-economic and institutional character have to be overcome, including processing, preservation, hygiene, distribution, and marketing of products.

In some ranching situations it has been found that a safari hunting or wildlife-viewing operation is easier to undertake and can be more attractive economically than the production of meat from wild animals. Also, some countries have opted against any form of consumptive wildlife utilization by introducing complete moritoria on hunting.

The Future of Wildlife Management

Wildlife and protected area management are to some extent again in a transitional stage, comparable to that of the post-war period, but today the pressures on resources and environment are greater. Wildlife is also often perceived as a stock source of free food and protected areas are seen as available for conversion to other uses without recognition of their productive, service and environmental roles. Decisions taken over the next few years will have profound effects on the future of wildlife in the next century.

Wildlife Management: appropriate technology

A key factor is the need to ensure compatibility with the socio-economic realities of rural life. In common with current trends in other resource disciplines, a reappraisal of technically and ecologically acceptable approaches to wildlife management is required to ascertain their relevance to rural development.

Pilot projects and schemes in the cropping of wild mammal populations and wildlife ranching have often been conceived as Government implemented programmes or operations carried out in the context of large-scale ranching. Ecological and economic aspects have usually received due attention, but sociological considerations have been seriously neglected. Unless these options can be adapted in the rural development context their future significance must remain in doubt.

Technologies for the production of food from wildlife in Africa exist in traditional systems. These cover harvesting strategies and techniques, processing and preservation of products and, in some communities, conservation. Animal protein derived under such systems makes a significant contribution to the diet of many subsistence farming economies, especially those in humid forested areas or trypanosomiasis zones with limited possibilities for domestic stock. In other situations, it is often the poorest elements of communities that rely on wildlife for their meat (Asibey 1972, Ajayi 1979, Sale 1981). There is scope for the transfer and adaptation of technologies between countries in the region, but the primary requirement is to evaluate traditional systems in biological and ecological terms, with a view to evolving approaches which are sustainable and possibly more productive. Systems compatible with other land-use practices should be identified and developed. In this is the need to recognize that wildlife is not confined to large and conspicuous animals.

Protected Areas: appropriate management systems

The future of National Park and protected area systems at one level will be dependent on the better integration of individual areas into overall landuse planning and rural development programmes. At another level the need will be to re-orientate management systems and practices to ensure that greater socio-economic benefits accrue to local communities. The role of buffer zones requires redefinition with a view to optimizing their contribution to satisfying demands on protected areas which would be incompatible with their management. An aspect of protected area management which needs greater emphasis is compatibility with the requirements and goals of other conservation and multi-use efforts such as watershed management, erosion control, protection forestry and the restoration of degraded land. Similarly, the role of National Parks and equivalent reserves in the field of *in situ* conservation of genetic resources will become increasingly important.

Wildlife in Marginal Areas

Over the next 100 years there could be significant redistributions of human populations in parts of Africa, following similar trends apparent elsewhere in the world. This would include movement away from harsh environments towards urban centres where services are concentrated. In arid zones, an additional factor would be forced movement from areas where productive capacity has been destroyed. This could release extensive areas for wildlife management, but there may be a need to rehabilitate degraded habits and restore wildlife populations. Systems and techniques to achieve these ends will be necessary, including renewed efforts to take advantage of the physiological and ecological adaptations of desert species.

Wildlife in Forestry and Agroforestry

The use of indigenous tree species is being increasingly advocated in African forestry. There is also a body of opinion that favours the concept of multi-species plantations as being more ecologically acceptable than monocultures of exotics. Clearly there would be scope for the development of wildlife utilization activities in such artificially established multi-species formations. To date the integration of wildlife management into agroforestry systems has been minimal. This is an area which requires renewed effort in the future. It could be especially relevant in terms of involvement of rural people and the contribution of wildlife to their protein diet.

Research, Training and Extension

A major shift of emphasis is needed in research and training. Whilst much remains to be studied in the natural science fields, the sociological and economic aspects of wildlife and protected area management are the ones in urgent need. A main thrust of research should be the development of appropriate management systems and techniques. In training, there is likely to be a dichotomy to cater for the specialized requirements of wildlife management and protected area management respectively. Game farming has progressed little beyond the experimental stage in many parts of Africa. An intensification of research effort with the accent on arid zone and trypano-tolerant species is necessary, not only on biology and ecology, but also on husbandry, economics and sociological implications.

With escalating costs of energy stimulating interest in the adoption of more appropriate technologies, the possible future role of certain wild animal species as draught animals could usefully be examined. The Government of Zaïre is currently studying the possibility of reviving its African elephant training centre at Gangala na Bodio, and in Zimbabwe African buffalo are being trained to the plough.

If new approaches and technologies in wildlife and protected area management are to contribute to rural development, adequate provision of related extension services will be essential. Wildlife and National Park personnel must divest themselves of their "police" image and local people must be increasingly involved in development and management programmes. Wildlife managers must actively manage wildlife for the sustained benefit of people.

References

Ajayi, S.S. (1979) *Utilization of Forest Wildlife in West Africa*. Rome; FAO, 76 pp.
Anon. (1936) *International Convention for the Protection of Fauna and Flora (with protocol)*. [Treaty series no. 27.] London; HMSO, 45 pp.
Anon. (1969) *African Convention on the Conservation of Nature and Natural Resources*. Addis Ababa; Organization of African Unity, 46 pp.

Asibey, E.O.A. (1972) *Wildlife as a source of Protein in Africa South of the Sahara.* [African Forestry Commission Working Party on Wildlife Management.] Rome; FAO, 10 pp.

Bindernagel, J.A. (1968) *Game Cropping in Uganda.* Kampala; Uganda Ministry of Animal Industry, Game and Fisheries, 200 pp.

Charter, J.R. (1970) *The Economic Value of Wildlife in Nigeria.* Ibadan; Forest Association of Nigeria, 12 pp.

Dasmann, R.F. (1964) *African Game Ranching.* New York; Pergamon Press, 75 pp.

FAO (in press) *Wildlife Utilization. Proceedings of a Session Symposium of the Seventh Session of the African Forestry Commission Working Party on Wildlife Management and National Parks, Arusha, Tanzania, 19-22 September 1983.* Rome; FAO.

IUCN (1982) *1982 United Nations List of National Parks and Protected Areas.* Gland; International Union for the Conservation of Nature and Natural Resources, 154 pp.

King, J.M.; Heath, B.R. (1975) Game domestication for animal production in Africa – experiences at the Galana ranch. *World Animal Review* 16, 23-30.

Ledger, H.P.; Smith, N.S. (1964) The carcass and body composition of the Uganda kob. *Journal of Wildlife Management* 28, 827-839.

Leopold, A. (1933) *Game Management.* New York; Charles Schriber's Sons, 481 pp.

Mossman, S.L.; Mossman, A.S. (1976) *Wildlife utilization and Game Ranching.* [Occasional Paper no. 17.] Morges; International Union for the Conservation of Nature and Natural Resources, 98 pp.

Parker, I.S.C.; Graham, A.D. (1973) Commercial use of Thomson's gazelle (*Gazella thomsonii*, Gunther) and impala (*Aepyceros melampus*, Leichtstein) on a Kenya beef ranch. In *Third World Conference on Animal Production*, 109-118. Sydney; Sydney University Press.

Sale, J.B. (1981) *The Importance and Values of Wild Plants and Animals in Africa.* Gland; International Union for the Conservation of Nature and Natural Resources, 44 pp.

Talbot, L.M. (1966) Wild animals as a source of food. *United States Department of the Interior, Bureau of Sport Fisheries and Wildlife, Special Scientific Report, Wildlife* 98, 1-16.

de Vos, A. (1977) Game as food. *Unasylva* 29, 2-12.

Chapter 76

Inland and Marine Fisheries for African Development

J.J. KAMBONA

Chief, International Institutions and Liaison Unit, Fishery Policy and Planning Division, Fisheries Department, FAO, Rome, Italy.

Introduction

This paper aims to give a concise review of the importance of the fishery resources available in Africa and the contribution their rational development can make to African economies as a whole and to the food and nutrition situation in particular, both in the short and long term; to describe the strategy developed by FAO to assist African countries in developing collective self-reliance in the exploitation of their own fishery resources; and outline some of the measures required to adjust and reorient the regional and sub-regional arrangements so as to ensure more efficient management and development measures.

The Present State of African Fisheries

Since the late 1950s rapid expansion has taken place in many fisheries in Africa, particularly marine. Fish production from the various fisheries in the oceans and seas and inland waters increased by 64% between 1965 and 1972 but has virtually unchanged since them. The 4.4M tons of fish harvested by African countries (including South Africa) in 1972 and the 4.0M tons in 1981 represented about 7.0 and 5.3% respectively of the total world fish catch, as against 8.3% in 1965. However, this simplified picture hides substantial differences, both in trends and relative importance observed between regions, countries, fisheries or even the participants in the fishing activities.

The inland water bodies of Africa occupy about 250 000 km^2, mainly concentrated in a few large lakes (e.g. Lakes Victoria, Tanganyika, Malawi (Nyasa), Kyoga, Turkhana, Mobutu, Edward Bangweulu, Rukwa, Mweru wa Ntipa) and a few large man-made lakes and reservoirs (e.g. Nasser/Nubia, Volta, Kainji, Kariba, Kafue, Cabora bassa), the rest being dispersed. In addition, rivers total some 12M km in length comprising of some of world reknown such as the Nile, Niger, Congo, Zambesi, etc.

The catch of inland water fish, including brackish waters, coastal lagoons and production from aquaculture, has risen steadily during the decade 1964–74 but has since remained relatively constant at about 1.4M tons y^{-1}, falling short of the estimated potential of about 3M tons y^{-1}. The reason for this is fisheries which remain untapped either because of their isolation, for example the Sudd in Sudan or the Okavango Swamps in Botswana, or because they represent a relatively newly discovered resource which is only exploitable with more sophisticated methods than those at present available, as in the Limnothrissa of Lake Kivu.

Fish Resources

The marine fish resources, now under the control of African coastal nations as a result of the extension of their jurisdiction over the waters surrounding the African continent, might yield yearly catches of about 9M tons. A further 3M tons might come annually from inland waters whereas aquaculture, essentially in inland and brackish waters, may add several hundred thousands to this amount provided appropriate inputs are mobilized. In 1978 production in waters now within the jurisdiction of African countries reached 7.3M tons, of which only 3.3M tons were caught by Africans. Of that amount, 1.9M tons (60%) were marine and 1.4M tons (40%) inland. Four million tons were harvested by long-distance non-African fleets.

These figures show that the prospects for expanding African participation in their fisheries are exceptionally good and economically feasible. However, because of the uneven geographical distribution of the marine resources, these potentials are not equally shared. Most of the marine resources are concentrated in areas of upwelling of cold, nutrient-rich waters off the west coast of Africa in the latitudes of the Tropic of Cancer (between Morocco and Guinea) and Capricorn (off Angola and Namibia), and to a lesser, but still substantial, extent, the north-eastern area off the shores of Somalia. About two thirds of fish catches in Africa come from these areas.

Oceanic tunas have different patterns of distribution. In the Atlantic most is caught in the eastern central Atlantic, in particular from the island and coastal states bordering the Gulf of Guinea.

Exploitation

Over 80% of fish production by African states is by small-scale fishermen. The scattered nature of their activities facilitates the distribution and consumption of fish among communities living in the vicinity. However, it is also an obstacle to the disposal of surplus in resource deficit areas. In the marine sector, artisanal fisheries are well-developed in all coastal areas; they play an important role in the economy of countries such as Ghana and Senegal with landings in excess of 100 000 tons y^{-1}.

Most coastal countries now have fleets of medium-size trawlers, including shrimpers and purse seines, but few have long distance trawlers or tuna clippers.

Prospects for Increasing Fish Production

It is estimated that production of 12M tons y^{-1} could be achieved by African countries, compared to the present 4M y^{-1}. The biggest opportunity for African countries to increase their marine catches arises from extended national jurisdictions making it possible for coastal states to control activities of foreign fleets, and, if desired, to substitute African-based fleets for them.

In the waters off north-west Africa, increases of the magnitude of 2.6M tons have been estimated, and 2.4M tons off the shores of Somalia. The estimated increases in the Gulf of Guinea and south-west Indian Ocean (0.5M tons each) are much lower, and they are modest in all other sea areas.

More than 60% of the overall potential for increasing African fish production is located offshore (i.e. to 20 miles out) in north- and south-west Africa. Increased production can, however, only result from the extension of the offshore fishing operations, but local fishermen have concentrated on the inshore; long-distance foreign vessels have been exploiting more offshore waters. An increase in local fish production will require the introduction of bigger boats and development of handling, processing and storage capacities and markets. The problems have been compounded by rapidly rising energy costs.

In relation to the difficulties of catching and marketing specialized resources (e.g. oceanic tunas), the participation of most African countries in their fisheries has so far remained marginal. However, these offer substantial prospects for island states (e.g. Sao Tome and Principe, Seychelles, Mauritius) and coastal countries which have acquired control over the traditional fishing grounds. Benefits can be obtained from increased participation in catching, processing and marketing or licencing and royalties on foreign vessels.

Lakes and reservoirs account for half the freshwater fish production, and rivers and floodplains for the rest. The potential for increased yields of inland fisheries is moderate and concentrated in eastern and southern Africa. Inland waters now yield about 1.5M tons y^{-1}, but account must be taken of their potential and the possibilities of doubling the catch, particularly for domestic consumption. The importance of inland fisheries is illustrated by the fact that they account for over half of the total African ouput. In 21 countries over half the fish from inland waters, and 13 depend entirely on this source for fish. Development requires the introduction of new methods, improved access, and reduced wastage.

In most parts of Africa fish is recognized as an aquatic resource which is renewable and ever present, but the effects of competing uses of water in modifing the ecosystem may reduce the potential sustainable yield. Rivers are particularly vulnerable to conflicting uses. Floodplains are exceptionally productive and their role in the reproduction, growth and survival of most freshwater species is crucial. These zones are, however, increasingly being used for agriculture or drained for irrigation or flood control schemes. Deforestation and marginal agriculture within river basins lead to silting and changes in runoff patterns, modifying flooding cycles and endangering stocks. Integrated planning in river basin development schemes is therefore necessary.

Estimates of potential are derived from capture fisheries either in natural (marine or freshwaters) or man-made waters and do not include advances which can be made through aquaculture. Increases from the latter are expected to be moderate in the immediate future, but could be significant in the long-term.

Demand

Fish is a highly appreciated food item throughout most parts of Africa and a major source of animal protein, especially in western and central Africa (Table 1).

Present per capita consumption averages 7.3 kg y^{-1}. Demand is expected to grow as a result of population and income increases and the slow progress expected in the production of animal protein. About 600 000 tons y^{-1} of marine fish is used for fish-meal production; a portion of this amount could increase the human food supply and result in increased earnings for fishermen.

Table 1   Relative importance of fish in food supply in African countries (marine and freshwater fish as % of animal protein supply). Based on FAO Provisional Food Balance Sheets Average for 1984.

Over 40%	40–20%	Under 20%
Sierra Leone	Ivory Coast	Guinea-Bissau
Congo, People's Republic	Gabon	Mali
Ghana	Comoros	Central African Republic
Nigeria	Benin	Morocco
Senegal	Togo	Mauritania
The Gambia	Uganda	Tunisia
Malawi	Tanzania	Madagascar
Chad	Guinea	Egypt
Sao Tome and Principe	Cameroon	Djibouti
Seychelles	Mauritius	Zimbabwe
Liberia	Zaïre	Kenya
Cape Verde	Zambia	Rwanda
	Mozambique	Libya
	Burundi	Upper Volta
	Angola	Algeria
		Niger
		Sudan
		Botswana
		Somalia
		Ethiopia
		Lesotho
		Namibia
		Swaziland

Countries listed in order of importance of the contribution of fish to total animal protein.

Distribution and Marketing

Although fish supplies in Africa are largely sufficient to meet demand, their distribution does not coincide with that of human populations, leaving some countries with potential surpluses and others with supply deficits. A considerable amount of trade supplements local production or caters for specific tastes. Self-sufficiency in fish, therefore, implies an expansion of intraregional trade. Fish from rich inland areas such as Lake Chad now reach coastal western central African countries, and ones caught by foreign trawlers off north-west and south-west Africa are commonly traded in the form of frozen blocks in countries bordering the Gulf of Guinea and the hinterland.

Spoilage and waste is particularly crucial in the traditional sector because of insufficient equipment and poor connections with export outlets. Losses due to insufficient cold storage and poor handling may be 10–15%, whilst cured fish, in stock for long periods before sale, becomes infested by insects and losses may reach 40%. Discards at sea are also important, and may exceed one third of total catches. In total, post-harvest losses may well exceed 500 000 tons y^{-1} of fish protein.

Prospects and Constraints

For countries of which long-distance fleets have been active, the new ocean regime represents unprecedented opportunities to expand their own fisheries and to progressively develop their shore-based operations in fish processing and marketing. Full realization of the African fish potential in both marine and inland waters could add an estimated US $1500M (at 1978 prices) to the African gross product, excluding value added in processing and marketing. A substantial time will be necessary for countries with large marine resources to fully exploit these themselves. In the interim period, well-conceived and planned licensing, or joint ventures may considerably enhance foreign exchange earnings. These could, in turn, be used to build up national infrastructures and capabilities to achieve a fuller participation in the exploitation of the resources.

However, African fish production is at present no higher than levels reached in 1972. In the northwestern sector production of coastal countries has not increased in the last six years, whereas the recorded catches of foreign fleets dropped by half from 1977 to 1979, despite substantial investments made in catching and processing facilities by most coast states. The reasons for such stagnation are manifold, and economic, technological, institutional and political. Lack of managerial capabilities, skilled manpower, difficulties and delays in developing distribution networks and export markets, low consumer income, price policies, failure in the formulation and implementation of sound fishery development strategies, insufficient administrative and institutional set-ups, etc., are all serious constraints. In some cases, such as the Sahel, environmental problems have also had an influence.

However, most coastal countries with fish potentials larger than their present production have implemented systems aimed at extracting revenues from the activities of foreign vessels. Despite difficulties in enforcing such schemes, preventing illegal fishing and under-reporting of catches, some already derive substantial benefits. In 1978 Mauritania extracted from fees and fines charged to foreign vessels around US $30M. Such benefits are obviously not reflected in national catch statistics which, thus, cannot accurately reflect the benefits coastal countries already enjoy.

An African country bordering resource-rich sea areas is confronted with the need to assess the fish stocks it now controls, and the means at its disposal for their exploitation, to evaluate and compare development options, to mobilize the capital, equipment and know-how required to implement the selected strategies, to develop its domestic markets and open export outlets, and to design and enforce appropriate management and surveillance systems well-adapted to local conditions and available means. They also need, in collaboration with neighbouring countries, systems to appraise, monitor, exploit and control shared stocks, etc. These tasks require specialist expertise (resource evaluation, fishery development planning, fishery management, fishing companies and fish processing management, legislation, trade, etc.) which is critically lacking in most African countries; this is probably the constraint at present most severely affecting African fisheries with a large potential.

In Africa, the majority of small-scale fishermen earn a precarious livelihood. Their economic, social and technical needs often do not receive the attention, for example with respect to credit facilities, technical assistance and extension services, given to other sectors. Nevertheless, artisanal fisheries in Africa play a dominant role as a source of employment, especially in rural areas and as a major source of food. The new ocean regime also offers Governments the possibility to allocate to artisanal fishermen greater shares of the fish resources now under national control.

Most countries in Africa have now recognized the importance of aquaculture in improving diet, generating employment, providing a cash crop to farmers, and foreign exchange; they are especially important for countries with limited fish resources. Many Governments have accorded high priority to this sector and initiated pilot-scale operations which have demonstrated that fish farming using indigenous species and free material can be highly profitable. However, the failure of some ill-conceived programmes remains a major constraint in convincing farmers and investors of the economic viability of aquaculture. Insufficient appreciation of the basic requirements of an effective aquaculture development programme, and consequent inadequacy of Governmental support activities, have handicapped the development of this industry.

Chapter 77

Natural Resources Management

M. DAGG (Rapporteur)

International Service for National Agricultural Development, The Netherlands.

Introduction

The natural physical resources, and the natural biological resources, where we can sometimes collect desirable products direct from nature, are the starting point for agriculture. More commonly man has to exploit these natural resources to yield the products he requires to sustain and improve his well being. Generally, he would like to do as little as possible to achieve his objectives, but usually he has to put in considerable effort, especially when the resources vary from the perfection he desires, in both time and space. He (and I am conscious that this is very often 'she') would also like his efforts to be as efficient as possible.

We were asked how best can science help to increase the efficiency of the producer's efforts in so far as managing the natural resources base is concerned, so that the production and development of agriculture in Africa can be advanced. This symposium addressed aspects of the primary physical resources of weather and climate; terrain, land and water; and the derived biological resources of vegetation, forestry, wildlife and fisheries.

Description

Throughout we came frequently to the need for better collection and processing of information on the elements of the basic natural resources, so that a steadily improving base or inventory is accumulated. But in order to keep costs down, these measurements should be restricted to key factors, which of course requires a clear understanding of interrelationships between plants, animals and the weather.

Climate and Weather

With 'climate and weather' we started with the very conspicuous impact of drought that featured in most country reports (Chapter 3) as a major, if not the major, constraint to realizing potential production, and a cause of variability and uncertainty that curbs investment in inputs that might feature highly in development plans. The paper on rainfall trends (Chapter 58) concluded that there was no significant climatic change, but that wide fluctuations in rainfall must be expected and planned for, amongst possible ways, by storage of food, and by the use of stored water for irrigation. We looked to a scientific approach to mitigate the impact of drought, which strikes when less rainfall is received then expected. A defence is to grow varieties with lower water requirements or expectations; i.e. to match the crop to a more reliable rainfall regime.

The paper on the principles of plant/weather relationships (Chapter 60) stressed the supreme importance of being able to derive the water balance from rainfall and evaporation estimates and then working closely with agronomist and breeder to fashion plants and operational systems to match less than average rainfall regimes.

An excellent example from FAO of the application of this principle on a broad regional scale was reviewed and acclaimed, as a major step forward in improving planning capabilitiy (Chapter 64). More detailed country analyses are promised that will accommodate some of the objections raised in discussion, such as the neglect of cash crops. Such exercises should be encouraged as a matter of urgency. The interactions of plants with other meteorological aspects of temperature, humidity, rain, wind, radiation, and seasonality (for field operations) are reasonably well understood from international work, but need to be applied with a care for the magnitude of the local parameters.

Experience with the AGRHYMET network in the Sahel (Chapter 57) shows how short-term forecasting can improve the efficiency of man's efforts to protect his crop, for example warning to ridge up maize in anticipation of a line squall. Successful large scale forecasting for agriculture was not likely without considerable changes in the central collection and processing of information. Rapid processing on a regional scale was possible by remote sensing (Chapter 63), yielding useful information on areas of moist soil and development of vegetative biomass with time; an estimate of rainfall from the rate of development of cloud cover shows promise but is not yet perfected. While crop type and condition can be monitored by remote sensing by its spectral signature, characteristic small scale of farms makes it difficult to apply as yet in Africa.

The session specially stressed the need for: (1) more extensive network and data recording and better training for recorders, (2) closer collaboration between agrometeorologists, agronomists and breeders with better quantification of environmental conditions of trials, (3) closer collaboration between agrometeorologists, pathologists and entomologists, and (4) more agrometeorology in undergraduate curricula and in the training of extension staff.

Terrain, Land and Water

Rainfall is the most rapidly varying component of natural resources, and has the most unpredictable effect on yields, but crops are grown on land and productivity can be seriously affected by soil conditions which can change steadily in time as well as space. Some deficiencies can be corrected, but at a cost which many farmers cannot afford; many crops are grown on the intrinsic fertility of the soil resource.

The FAO analysis mentioned above incorporated water balance information with the available inventory of soils to derive the maximum possible yield of different crops on unimproved soil. (It also gives yields for a medium level and a high level of remedial treatment.) From that, broad regional carrying capacities for people have been calculated as a guide for planning. A similar approach to a much more detailed specific situation applied to the possible sustainable population of village settlements was also described (Chapter 69).

One aspect of the FAO analysis showed that at low input levels and under rainfall conditions only, Africa would not be able to support its burgeoning population in 2000 unless as much as 20 % of the cultivated area was irrigated to extend the growing season. However, no irrigated area would be necessary if a high level of inputs was applied to rainfed crops.

A review of the state of irrigation in Africa (Chapter 65) showed that there was probably enough water resource available to support this intensity of irrigation, but that at present only 2 ½ % of the cultivated area was irrigated; more than 70% of this was in north Africa and only 5% in the area between the Sahel and Zambezi. Because settlement farmers in these areas have little tradition of irrigation, parastatal organizations in very costly schemes are attempting detailed control of farming operations. There is an enormous need for more research on water management and irrigated farming under these circumstances. Training of extension agents and farmers in irrigated agriculture is urgently required to reduce the constraint of inexperienced manpower.

In due course more irrigation will be needed to meet the pressure for increased food production and greater stability of production in dry areas. The comparative efficiencies of large formal irrigation schemes and small-scale informal developments should be investigated to guide further development planning. Several participants stressed the Indian experience of the operational advantages of irrigation from ponds and tube wells, and encouraged Africa to follow this path. Irrigation from lakes and swamps was also advocated, and protection of catchments to reduce siltation was also stressed.

The precious soil resource of tropical Africa is vulnerable to erosion under the very intense rainfall of tropical storms. The sandy, highly erosive soils in the humid zone have been protected by forest, but pressure to bring in new land means clearance will have to continue. Experience in West Africa showed that hand clearing or relatively gentle mechanical clearing of trees with a shear blade leaving the roots intact, was necessary to avoid excessive soil loss. Clearing in any environment needs to be followed by appropriate soil conservation measures to manage run off and/or soil surface management to maintain a cover, either living or dead, to break the rainfall impact. Tillage should be kept to a bare minimum. Multipurpose farm ponds can be useful devices for storing excess water.

Given care and planning, even the erosive soils of the humid zone can sustain economic production of annual crops without serious land degradation. However, the great danger of soil degradation is a severe constraint to increasing production by clearing forest.

India has similar erosion problems. In a review of some of the protective measures taken (Chapter 67), planning water management and erosion control on a catchment basin basis was stressed, with a range of appropriate measures matched to different slopes, harvesting of runoff water in ponds and careful use of bottom lands for crops. Detailed benefit/cost ratios for different measures were quoted. Some of the methods might have direct application in Africa and should be tested.

The leached acid soils of much of central Africa are often deficient in the supply of nutrients to crops. This can be corrected by the addition of fertilizers, and recommendations are available in many places. However, the soils have little buffering capacity, and so acidity and nutrient imbalance can result in expensive applied nutrients not being taken up by the plant — with disillusionment on the farmer's side about the value of fertilizers. Sound scientific understanding has been accumulated about these soils and remedial steps, such as liming to correct soil acidity or the adding of organic material, may be very remunerative in making fertilizers effective (see Chapter 68).

The natural physical resources of Africa are substantial, but they are far from perfect for crop production. However, the scientific understanding of the shortcomings is reasonably good, and, with the collection of more data to describe the environment better, the understanding can often point developers and planners to the ways of removing or mitigating the deficiencies, or at least how to avoid disasters in development.

Biological Resources

The management of biological resources for development is a very wide topic. It was treated widely to include all aspects of land and vegetation, animal and fisheries management, and annual crops and domestic livestock utilized in association with trees.

Generally, the management of trees implied a minimal disturbance of the biosphere in which man is an integral dependant part and from which he derives his sustenance (and much of his aesthetic satisfaction). But man is also the the prime mover in upsetting the ecological balance for his own short-term objectives. Despite the current pressure on land due to man's increasing numbers, changes in the vegetational resource should be carried out carefully in Africa as the ecological balance for production is delicately balanced.

Following a review of the vegetative systems in Africa and the multitudious uses of the near-natural state, a set of guide lines were proposed for rational management and utilization of natural vegetation for rural development (Chapter 70). Natural vegetative systems should be described as completely as possible; during changes a well-considered balance between protection, conservation and development goals should be attained; cover should be maintained and excessive disturbance of soil avoided; a fully integrated approach to the best use of crops, livestock and trees should be adopted to improve carrying capacity; and unchanged special reserves for germplasm preservation should be maintained. The major constraint to effective management of vegetation resource is a lack of well-trained research staff and funds to carry out detailed surveys of the full complexity of interactions of the vegetative system.

An extensive review was presented of the benefits of mixed systems of plant production, with and without tree components (Chapter 71). Well chosen mixtures are almost always beneficial under hand tool cultivation; the outstanding disadvantage is that they are difficult to mechanize.

Future research to include trees in mixtures should focus on the use and developement of new genetic strains specifically designed for multicropping; on no tillage, green manure systems; "Diagnosis and Design methodology" to direct experiments to farmers' needs; and on land use systems in forest reserves for products other than wood. Grain crops tend to decrease soil fertility in time, while many trees increase it; balanced mixtures should be sought which can maintain soil fertility in the long-term. A strong plea was made for more collaboration in research between foresters and agronomists towards beneficial goals for the farmers, who should be included at the design stage. Notice should be taken of farmers' preferences for indigenous trees.

The major forest product is fuel wood and the demand for this is growing (Chapter 72). Wood is unlikely to be displaced as the main source of energy in rural areas, and future plans have to include production from fuelwood plantations. In view of the growing pressure on land for food production and reductions in forest area, forests and plantations need to be managed specifically for fuel production. However, the main source of fuel wood in rural areas will continue to be farm trees, and research should be intensified into fuel wood production on small farms. Forest services must be geared to the supply of rapidly growing farm trees that provide minimal competition to crops (Chapter 73).

On a specific topic, illustrative of the multitude of subsidiary forest products, an appeal was made for an intensifying development of bee keeping (Chapter 74), which generates a range of valuable products and welcome supplementary income. New techniques are available for breeding better bees and adaptive research is needed on appropriate designs of beehives. The outstanding constraint is the shortage of trained research officers and extension agents.

As the main traditional user of the wildlife resource has in several countries been regarded as a criminal, it has been difficult to collect accurate information on rates of wildlife consumption. It is now recognized that future management of wildlife must cater for the consumer as well as the tourist, and involve him in planning and management decisions (Chapter 75). Attention must be paid to small as well as conspicuous, large animals. In forests of restricted tree species, research may be neccessary to determine particular animal species for optimum productivity.

In a wide ranging review of the prospects for inland and marine fisheries (Chapter 76), it was shown that there was an exceptionally good opportunity to expand African participation in fisheries, both in inland lakes and in the recently expanded area of marine resources now under the jurisdiction of African countries. African fishermen caught only 1.9M of a total 5.9M tons of fish caught at sea out of an estimated sustainable yield of 9M tons. They harvested 1.4M tons of a possible 3M tons from inland waters. The major constraints to increasing the share of marine production are a lack of suitable boats, equipment and harbour facilities; a need to review the potential; a need for agreements amongst coastal African countries in the main fishing regions; and financing and the development of markets. There is a great need for technical and management manpower. For inland fisheries, better methods and equipment are needed and the means of access to remote water bodies.

Questions were also raised about the relative importance for investment in research and development activites in different sectors of natural resource management, but it was not found possible to resolve such questions at this stage.

Chapter 78

The Contribution of the Commonwealth Agricultural Bureaux

N. G. JONES

Executive Director, Commonwealth Agricultural Bureaux, Farnham Royal, Slough SL2 3BN, UK.

The Nature of CAB

Sir Thomas Scrivenor (1980), my predecessor, started his entertaining history of CAB, *The First 50 Years*, by making two points; that CAB was a Commonwealth body and that it provided three essential services to research workers not only in Commonwealth bodies, but all over the world. Time brings changes, however, and whilst the latter is still true – the first changed in part in 1982 when CAB also became an international body. This was achieved by CAB making an Agreement with the Government of the United Kingdom under which the organization became a corporate entity under English law. Previously, not having had a legal identity, the organization made contracts with staff and other agencies in the name of the Executive Director. The legal implications of this arrangement were somewhat inspiring so the change received a personal welcome. As it also allowed CAB certain privileges and immunities including the avoidance of laws relating to immigration, customs duties and tax, the change was also welcomed more generally.

The point should be made, however, that whilst CAB's legal status changed, the organization remains Commonwealth in nature being owned and managed by 28 Commonwealth countries. I should add though that, not being bound by irrevocable rules, CAB may make arrangements with Governments and other agencies to provide new services for agriculturalists. One such arrangement led to the formation of the International Food Information Service (IFIS) which was brought into being by CAB's association with the Governments of Germany and the Netherlands, and the American Institute of Food Technology. Started in 1969, the Service became self-sufficient in about four years and is now a major world information service.

The Structure and Management of CAB

Conceived in 1927 by an Imperial Agricultural Review Conference, CAB was preceded by the Bureau of Entomology in 1911 and the Bureau of Mycology in 1920. The former, the Bureaux of Entomology, was established following an enquiry among scientists working in Africa. Their replies included reference "to the remarkable role played by insects in relation to all sides of human activity" and continued "It would seem urgent therefore that wider attention should be paid to insect fauna whether it be helpful or inimical to man". Action soon followed, the Bureau being set up, with the following objectives:

(1) To assist in the identification of all injurious insects sent in by officers attached to Departments of Agriculture and Public Health in all Member Countries.
(2) To issue a monthly periodical giving summaries of all current literature, whether British or foreign, dealing with noxious insects, whether agricultural pests, or disease carriers. (The *Review of Applied Entomology* began in 1913 and over the years has included reference to about $\frac{1}{2}$ million papers).

The first Director of this Bureau, Guy Marshal, was instrumental in recruiting Dr B. P. Uvarov who later was appointed the first Director of the Anti-Locust Research Centre when the Centre was separated from the Bureau. Subsequently, this Centre was embraced within the UK Centre for Overseas Pest Research and has latterly become a responsibility of FAO. The Bureau of Entomology, which undertook identification as well as information work, produced another unique service when biological control began as a separate activity in 1927. From this, the Commonwealth Institute of Biological Control has developed over the last 50 years.

Meanwhile, as the value of the entomological service was soon recognized, mycologists were the first to decide they wanted one too. The Bureau of Mycology was therefore started in 1920. Eight main Bureaux were set up by the Conference in 1927 to cover Soil Science, Animal Nutrition, Animal Health, Animal Genetics, Agricultural Parasitology, Plant Genetics, Horticulture, and Pastures and Forage Crops. The responsibilities of these 8, however, were limited to information work, their remit being:-

(1) To collect, abstract and collate information from all sources bearing on the most important problems under investigation.
(2) To make an index of current research in member Countries.
(3) To summarize available statistics where these are important in connection with Bureaux work.

Over the years, three more Bureaux were added, covering Forestry, Dairy Science, and Agricultural Economics, and the Bureau of Parasitology achieved Institute status as those of Entomology, Mycology and Biocontrol. Thus under the aegis of a controlling Executive Council, CAB consists of a group of 14 independent units, each under a Director responsible to the Executive Council for his own sphere of activities. Inevitably, with the passage of time, this structure led to a certain duplication in both scientific and administrative work, and to a certain lack of co-ordination both of which had cost implications.

More recently, member Governments have begun to take a functional view of CAB's activities. A Director of Information Services was appointed in 1976, who, having immediate responsibility overall for this Service, worked towards achieving greater co-ordination of Bureaux activities. To complement this development, a Director of Scientific Services will soon be appointed. He will be required to co-ordinate the work of the three taxonomic Institutes and the Institute of Biocontrol and will have about 45 scientists under his immediate control. A Director of Administrative Services will look after the administration, accounting, and personnel functions, and the whole organization will be led by a Director General.

It is also anticipated that a much wider range of subjects will be covered through collaboration with outside bodies, such as the University of London, the Glasshouse Crops Research Institute, and The Institute of Virology: and, as an illustration of collaboration between member Countries, approval has been given by the Executive Council to provide an identification service on basidiomycetes in culture by an arrangement with the Biosystematics Research Institute, Ottawa, Canada. This is a development which is favoured strongly because as much as the organization has benefitted from close association with research establishments in the UK, additional benefit may come in future from closer co-operation with research establishments in Commonwealth countries.

Financial Objectives and the Current Sites

Coincidental with the changes in management, member Governments have also changed the financial objectives of each Service. Up to 1980, member Governments met the net cost of CAB as they do with similar Commonwealth bodies. From 1980, the Information Service was required to be self-supporting and the Biocontrol Service had to cover at least half of its overhead costs. Since December last, the Biocontrol Service is also required to reduce its dependence on funding by member Governments. These economic targets have introduced a measure of value where only generalization existed before and there can be no question that as accountability has increased, so has the realism with which all work is now undertaken.

Finally, may I say that recent decisions to change the structure and management of CAB's Services are designed simply to enable the organization to play a more active part in the development of agriculture in Commonwealth Countries and in countries throughout the world.

Reference

Scrivenor, T. (1980) *CAB – The First 50 Years*. Farnham Royal; Commonwealth Agricultural Bureaux, 92 pp.

Chapter 79

The Commonwealth Agricultural Bureaux Identification Services

D. L. HAWKSWORTH

Director, Commonwealth Mycological Institute, Ferry Lane, Kew, Surrey TW9 3AF, UK.

Introduction

The correct diagnosis of the cause of a problem is the first step towards the finding of an appropriate solution. In the case of problems due to the ravages of pests and diseases, this means an authoritative identification of the organism involved. Armed with a name, recommended treatments can then be sought in appropriate reference works or by more extensive manual or on-line literature searches. The name is the key to all current knowledge of the organism; a wrong identification can also be disasterous.

The process of identification is basal to the whole of pure and applied biology; it is so fundamental that its significance is all too frequently overlooked. However, this importance was recognized by biologists working in the Empire in the first decades of the present century. The result was the Institutes of Entomology, Mycology and Parasitology which now come under the auspices of CAB.

The need for and primary object of the CAB taxonomic and identification service has been repeatedly affirmed during the last half-century. Its main aim was defined as to:

> "...provide a continuing authoritative and efficient Commonwealth service for the identification
> of organisms of actual or potential importance to agriculture, food, forestry, land management
> and public health; and to provide the service....to non-member countries on appropriate terms." [1]

The CAB taxonomic and identification services were set up in the UK deliberately as history has led to the collections and library requirements necessary to support them being met more fully there than elsewhere. For an authoritative identification comprehensive reference collections are crucial, especially for the numerous groups of tropical insects, fungi, and nematodes where modern monographs are not available. As noted by Mason (1940), the first taxonomist employed at what is now the Commonwealth Mycological Institute, ".... the surest basis of the art of diagnosis ... is the matching of good specimens .... against good specimens that have been correctly named."

Entomology

The Commonwealth Institute of Entomology (CIE), founded in 1913, has its headquarters in Queen's Gate, adjacent to the British Museum (Natural History). Its 14 taxonomists and their

---

[1] Commonwealth Agricultural Bureaux (1981).

support staff are distributed through appropriate sections within the Museum's Department of Entomology. CIE staff have access to the Department's remarkable collections, comprising some 23 million insect and mite specimens, and also are in a position to consult specialists on particular groups within the Museum. These enormous reference collections and the concentration of specialists place CIE taxonomists in a unique position to identify critical material from anywhere in the world.

CIE does not keep separate collections of its own, but contributes 8000 – 12 000 accessions to the Museum collections each year. The Institute provides about 9500 identifications (53 000 specimens) annually. Material received includes pests of crops, some of which are also virus vectors (e.g. *Aphis fabae*), organisms used in the biocontrol of weeds (e.g. *Tyria jacobaeae* on *Senecio*), mosquitoes such as *Aedes annulipes*, relatives of which transmit viruses (e.g. yellow fever), protozoa (e.g. malaria, sleeping sickness), and nematodes (e.g. filariasis).

The Institute was involved in the recognition of the Central American larger grain borer (*Prostephanus truncatus*) in Tanzania (Chapter 12), a case which shows the importance of correct identification. When first discovered in Tanzania in 1976 it was wrongly named (not at CIE) as a *Rhizopethra* species. Only in 1981 was its true identity and significance realized following the submission of material to CIE.

CIE and CIBC (Chapter 80) work in close collaboration; indeed CIBC originated as a subsidiary of CIE. Recent co-operative work has included that on cassava mealybugs (Chapter 19), which also involved Museum staff (Cox & Williams 1981), and the introduction of the weevil *Eiaedobius kamerunicus* into Malaysia in 1980-81 to pollinate oil palms.

Mycology

The Commonwealth Mycological Institute (CMI), the largest CAB Institute, was established at Kew in 1920 to enable staff to have access to the historically important reference collections and libraries there (Ainsworth 1980). In contrast to the situation at CIE, the Institute established its own reference collection (herbarium). By 1930 1500 specimens were available, in 1962 the total reached 100 000, in 1974 it passed 200 000, and the current number exceeds 280 000. This total includes numerous type specimens, probably in excess of 10 000, and has been re-housed in compactors to optimise the use of the available floor space. The CMI herbarium is especially rich in material from the tropics as it has largely been built up from collections submitted for identification. It is now the most important single collection of microfungi in the world. Another advantage is that in 1939-40 all specimens were individually numbered and entered in ledgers, and this practice has been continued. In addition cross-indexes by fungus and host or substratum, and bibliographic citations, are maintained. The herbarium records are a mine of information on host ranges and geographical distributions, and the value of careful accessioning and up-dating as names are revised is evident to all who use this working herbarium.

With a staff of almost 70, including 16 taxonomists, CMI is the largest centre for systematic mycology in the world. About 9000 identifications are provided each year, the material received being in connection with a very broad range of problems. In addition to numerous plant pathogenic fungi, insect pathogens of value in biocontrol (e.g. *Beauveria bassiana*), cellulolytic fungi (e.g. *Chaetomium cellulolyticum*), food spoilage organisms, biodeteriorants, mycotoxin producers, and fungi of industrial and biotechnological importance are covered. It is important to stress that the same fungus can be important to a number of areas of human concern, for example *Penicillium* species able to produce rots in fruit, cause losses in stored products, produce mycotoxins damaging to man and animals, and suitable for exploitation in the pharmaceutical industry. The Institute also has a medical mycologist specializing in fungi of medical and veterinary importance, especially dermatophytes such as *Trichophyton* species causing ringworms.

A small unit within the Institute is the only place in the world providing a wide-ranging service for the identification of tropical plant pathogenic bacteria.

The use of fungi in pure culture is a major aspect of modern mycology, and about 70% of the fungi submitted to CMI for identification are in the form of living cultures. The Institute's Culture

Collection, which since 1947 has incorporated the UK National Collection of Fungus Cultures, maintains over 11 000 isolates of fungi by a variety of the most modern techniques which are continually being improved by CMI staff (Smith & Onions 1983). Its holding of microfungi (other than yeasts) are second only to the Centraalbureau voor Schimmelcultures in the Netherlands, which provides no service for dried material. The Culture Collection is partly supported by the UK Department of Trade and Industry enabling it to provide a wide range of services to industry, and the Institute has also taken over the duties and records of the former Biodeterioration Centre of the University of Aston which closed in 1983. The Culture Collection records are held on computer, together with metabolic and other pertinent physiological data, which facilitates both searching and catalogue production.

CMI is also responsible for the cataloguing of the world literature of systematic mycology. All newly published names of fungi (including lichen-forming ones) are indexed in the twice-yearly *Index of Fungi*, and the systematic literature is compiled into the *Bibliography of Systematic Mycology*, also issued twice-yearly. Another standard work issued by the Institute is the *Dictionary of the Fungi*, first published in 1943 and now in its seventh edition (Hawksworth *et al.* 1983). These services result in CMI having a unique and essential international role in documenting the entire field of mycology.

Parasitology

The Commonwealth Institute of Parasitology (CIP) is relatively new. It only started providing an identification service in 1959 and was given its present name only in 1981 on moving to new premises adjacent to the London School of Hygiene and Tropical Medicine's unit at St. Albans. Its team of five taxonomists undertakes about 2 000 identifications of plant parasitic nematodes and animal helminths of veterinary and medicinal importance each year. Material is received from a wide range of sources and includes, for example: *Dirofilaria immitis*, one of six very similar species causing heartworm in dog arteries, related taxa giving rise to elephantitis and river blindness in man; trematodes causing schistosomiasis infections which affect some 200 million people in the tropics; and of course plant pathogens such as *Meloidogyne incognitis* which can severely affect tomato roots, and *Rhadinaphelenchus coccophilis*, red ring disease of coconut roots (spread by the weevil *Rhynchophorus palari*).

Even in its short period as a taxonomic unit, CIP has accumulated a remarkable collection of over 50 000 specimens of animal helminths, and approximately 3000 bottles and 11 000 slides of plant nematodes. This reference material is of international importance and includes about 2000 type specimens. In addition to material submitted for identification, rich collections made by the staff and collaborating research workers are an important component. The collections also include the extensive material of the London School of Hygiene and Tropical Medicine which was transferred to the Institute in 1972; this includes the personal collections of many of the leading workers of the field. The valuable World Health Organization's Filarial Reference Collection, comprising 516 bottles and 1876 slides of filariad nematodes, is also currently located at the Institute.

Research

The identification services at all three Institutes are backed by substantial on-going research, based both on the collections and materials received, which provides a basis for systematic studies of injurious organisms throughout the world, and there is a continuous flow of papers, books (e.g. Duffey 1980, Sutton 1980) and other identification aids (e.g. *CIH Keys to the Nematode Parasites of Vegetables, CMI Descriptions of Plant Pathogenic Fungi and Bacteria*) from the CAB taxonomists; about 120 research papers are published each year (Commonwealth Agricultural Bureaux 1984). Staff have access to an SEM facility at CMI, useful for looking at surface features of spores and helminth attachment organs. TEM and chemotaxonomic facilities are available by an arrangement with the Royal Botanic Gardens at Kew, and glasshouses are also available at CIP and CMI. Staff also have active roles in refereeing papers, and editing scientific journals, so contributing to the maintenance of scientific standards in their fields. The Secretariats of several professional bodies are also often located at CAB Institutes (e.g. International Mycological Association, World Federation of Culture Collections).

Advisory and Consultancy Services

The Institutes' staff are prepared to make their special expertise available through dealing with enquiries, advisory and consultancy services and also appropriate contract research. Such work is undertaken at the discretion of Directors who must ensure that the effectiveness of the core identification service is not adversely affected by such activities. Particular emphasis is now being placed on developing these aspects of the Institutes' work. Major projects or contracts, defined here as ones individually costed at over £100 000, currently in progress includes ones for the Overseas Development Administration, Department of Trade and Industry, and the Science and Engineering Research Council. The close contact with CAB and other on-line databases, extensive libraries, unique reference collections, and wide range of facilities mean that the Institutes are in a prime position to provide such services in their fields of expertise.

Training

All three Institutes run annual six-week courses particularly designed for workers in the tropics. These were started at CIE in 1979, and have proved consistently successful; they are almost always substantially over-subscribed. These courses are now being complemented by workshops and training courses held in other countries. For example in 1984 CMI staff are also running courses in Kenya and Malaysia, the former in collaboration with the Commonwealth Science Council and the latter with assistance from the Commonwealth Foundation. A more specialised UNESCO-sponsored MIRCEN course was held at CMI for the first time in November 1983, and that Institute is also now providing a series of shorter-courses on topics of major concern, for example one on mycotoxin-forming fungi in March.

CAB Staff are regularly called on to teach in MSc programmes and to act as co-supervisors or examiners for PhD theses. Links with the Universities of London and Reading in particular are now being strengthened with a view to extending the Institute's involvement in post-graduate training. Short courses will never train specialists able to run equivalent services in their own countries. For this reason, in 1984/85 CMI intend to sponsor two students from less-developed countries on the MSc Pure and Applied Taxonomy Course at the University of Reading.

In their training programmes the Institutes work towards a goal where most material can be handled nationally with only the more critical needing to pass to them. This situation already exists with the UK, Canada, Australia and New Zealand. Straight-forward material is screened out by scientists in other counties, in the UK by MAFF's Advisory Service whose staff are CMI-trained, so that only the more critical identifications reach CMI's experts. I am convinced this is the situation to strive for. This goal will take many more decades to evolve in less-developed countries, but some progress is being made now in New Dehli for fungi and in Nairobi for East African region insects (Ritchie 1983). We are most anxious to encourage countries to set up their own services and prepared to assist them in this where we can. National services are to be expected to be more rapid in view of the time now taken to mail material, and so should speed-up prescriptions for treatment. Duplicate specimens, advice on curating reference collections, and our experience of operating identification services can be made available.

Prospects

A major factor setting the CAB Institutes apart from other organizations undertaking identification work is that the provision of a comprehensive world identification service is their primary objective. This is also a vital service at a time when 50% of the food being produced in the world is destroyed by harmful organisms either before it is harvested or during storage. The demand for identifications consequently continues, even though since 1977 the Institutes have been required to charge for work undertaken for countries which do not contribute to CAB. The decade 1970–1980 witnessed a 25% rise in demand from 15 000 to almost 20 000 identifications per year. Increases from Commonwealth countries have risen steadily, but, during 1975–80 material was received from 140 countries; the service is therefore truly international.

Climbing numbers are evidence of the foresight of the founders of the CAB Institutes who developed them to fulfill a world need. Seventy years after, there is no indication of anything but a continued gradually increasing demand for the specialist identification services CAB is able to supply. As recognised by the European Science Research Council (1978: 21–22):

"......foreign aid programmes to the developing world are generating a need for specialised identification services and research, which it is difficult to satisfy. A major centre providing identification services is CAB".

However, limited resources mean that CIE and CMI are increasingly under pressure to be more restrictive about the material they are able to accept and process. This is unfortunate in a service organization and CAB is therefore seeking support to enable it to build on the firm base now established, to extend its services by the employment of additional specialists to cover further groups in depth, to increase the capacity for consultancy and advisory work, to provide more workshops, training courses, and research studentships, and to supply better support facilities for countries to build up their own centres and, for appropriate aspects of the emerging biotechnology industries.

In opening a new building for the Culture Collection at CMI in 1975, Lord Zuckermann said that if the services did not exist it would be necessary to create them, a salutary thought.

References

Advisory Board for the Research Councils (1979) *Taxonomy in Britain.* London; HMSO, 126 pp.

Ainsworth, G. C. (1980) CMI 1920-1980. In *CMI – The First Sixty Years,* 12-27. Kew; Commonwealth Mycological Institute.

Commonwealth Agricultural Bureaux (1981) *Report of Proceedings Commonwealth Agricultural Bureaux Review Conference, London, 1980* [Cmnd. no. 8411.] London; HMSO, 65pp.

Commonwealth Agricultural Bureaux (1984) *CAB Annual Report – 1982-83.* Farnham Royal; Commonwealth Agricultural Bureaux, 60pp.

Cox, J.M.; Williams, D.J. (1981). An account of cassava mealybugs (Hemiptera: Pseudococcidae) with a description of a new species. *Bulletin of Entomological Research* 71, 247-258.

Duffey, E.A.J. (1980) *A Monograph of the Immature Stages of African Timber Beetles (Cerambycidae). Supplement.* London; Commonwealth Institute of Entomology, 186 pp.

European Science Research Council (1978) *Taxonomy in Europe.* [ESRC Review no. 130.] Strasbourg; European Science Foundation.

Hawksworth, D.L.; Sutton, B.C.; Ainsworth, G.C. (1983) *Ainsworth & Bisby's Dictionary of the Fungi.* Seventh edition. Kew; Commonwealth Mycological Institute, Kew; 457 pp.

Mason, E.W. (1940) On specimens, species and names. *Transactions of the British Mycological Society* 24, 115-125.

Ritchie, M. (1983) The National Museum's insect pest identification service. *Kenya Farmer* 1983 (March), 9.

Smith, D.; Onions, A.H.S. (1983) *The Preservation and Maintenance of Living Fungi.* Kew; Commonwealth Mycological Institute, 51 pp.

Sutton, B.C. (1980) *The Coelomycetes.* Kew; Commonwealth Mycological Institute, 696 pp.

Chapter 80

The Commonwealth Agricultural Bureaux Biocontrol Service

D. J. GREATHEAD

Assistant Director, Commonwealth Institute of Biological Control, Silwood Park, Ascot SL5 7PY, UK.

Origins and History

After the first well publicised biological control success in California in 1886, there was for the first time, a means of permanent suppression of major insect pests (see Chapter 37). As a result, an international trade in beneficial organisms began. At first it was haphazard and consequently there were many disappointments. By the turn of the century, the United States Department of Agriculture (USDA) had begun to set up field stations for the detailed survey and study of pests in their areas of origin so as to enable a better choice of potential control agents to be made. After World War I, the need for a similar service to assist British Empire entomologists was met by the Imperial Bureau of Entomology (now the Commonwealth Institute of Entomology). Farnham House was purchased in 1927 and set up as a biological control laboratory by the Assistant Director, Dr S. A. Neave. In 1928, Dr W. R. Thompson, a Canadian, was appointed Superintendent. Thompson was already well-known through his work for the USDA in Europe, his contributions to the taxonomy of beneficial insects, and theoretical papers on the mode of action of biological control agents (Thorpe 1973).

The laboratory was soon busy with field work in Europe and the supply of natural enemies from stocks held at Farnham House. Soon Australian and other scientists were working at the laboratory and projects were undertaken for foreign countries, notably the USA (Thompson 1930). In 1928, Dr J. G. Myers was sent to the West Indies to assess the prospects for biological control, and in 1932, Dr W. F. Jepson was based in Mauritius to work on the cane grub problem.

In 1940, owing to the restriction of field work caused by World War II, Thompson moved to Canada with some of the staff. Ties with the parent Institute were severed and the unit became the Imperial Parasite Service. In 1945, F. J. Simmonds went to Trinidad to undertake a major weed study for Mauritius, and in 1948 work was resumed in Europe from a base in Switzerland. The name was again changed to the Commonwealth Bureau of Biological Control and, shortly afterwards, the Commonwealth Institute of Biological control (CIBC). Stations were built with Colombo plan funds at Rawalpindi in 1957, and Bangalore in 1958. An East African station at Kawanda Research Station, Uganda, opened in 1962. Until 1965, work in North America continued from a laboratory built at Riverside, California. Thompson retired in 1958, and F. J. Simmonds became Director, moving his headquarters to Trinidad in 1962. Thus, by the early 1960s, the CIBC had become, and remains the only world biological control service. Staff have worked for periods based in other countries on specific projects, including Argentina, Fiji, Ghana, Mexico, Japan, Malaysia and Indonesia. Work in East Africa ceased in 1973 because of troubles in Uganda, but was resumed from facilities made available at the Kenya Agriculture Research Institute at Muguga in 1981.

Activities

From the outset, Commonwealth funding has been restricted, and CIBC has been required to work on a contract basis. This has ensured that its principal work has remained directed at achieving practical pest control. Sponsors of research and implementation projects include member Governments of CAB, other Governments, international agencies and commerical undertakings. Increasingly, work for developing countries has been backed by donor agencies which has enabled more effective programmes to be mounted.

This activity is supported and promoted by advisory and information work. Advice is provided in response to enquiries and through consultancy visits. These may concern specific problems or provide more general recommendations on the use of biological control. This work necessarily requires access to the scientific literature and the extraction of facts from it. Consequently, the CIBC is involved, as is the rest of the CAB organization (see Chapter 81), in the gathering, sifting and publication of information.

CIBC also has a training responsibility, both to strengthen the capacity of countries to undertake their own biocontrol activities and to ensure that materials supplied by the Institute are used effectively.

Pest control projects

Projects undertaken by CIBC vary from fulfilling requests for cultures of proven biological control agents to major investigations to find, screen and supply natural enemies of pests not previously studied and, on occasion, implement biological control.

Essentially, the CIBC exists to undertake work which would be difficult, or too expensive, for national teams, as was envisaged when it was established (Thompson 1930). It is able to do this because of its extensive network and experienced staff with knowledge of faunas, floras and working conditions in many countries, and who can therefore respond rapidly to requests. It also undertakes screening work safely in the area of origin of organisms and provides safe, healthy material, so obviating the need for expensive quarantine facilities in the receiving country.

Usage of the service varies considerably from country to country. For example, the major continuing programme for Canada based on the station in Switzerland provides material for use by Canadian personnel, whereas in East Africa CIBC itself has undertaken release programmes in collaboration with national scientists as well as providing natural enemies for use on other continents.

CIBC has participated in many major successes. Those during the 25 year period up to 1979 are summarised by CIBC (1980) and included major programmes covering insect pests as varied as sugarcane stemborers in the Caribbean, winter moth in Canadian forests, potato tuber moth in Cyprus and Zambia, bean flies in Hawaii, white sugarcane scale in Tanzania and Mauritius, and also weeds including thistles in Canada, water fern on Lake Kariba, and cordia weed in Malaysia. In addition, the supply of known natural enemies has benefitted many countries. Where data is available (Scrivenor 1980), it can be seen that the benefits of successful projects far outweigh their cost.

The outlay often bears little relation to the savings when only limited research is required to solve a major problem.

The techniques in which CIBC has expertise can also be used in the introduction of other beneficial organisms. A notable example is the discovery, screening and transfer of oil palm pollinating weevils from the Cameroun to south-east Asia. This work, which cost about £70 000, the time and travel of only one entomologist over $3\frac{1}{2}$ years, has saved oil palm growers US$ 11M per year on hand pollination and increased yield by about 20% in Malaysia alone. Similar results are now being reported from Papua New Guinea, the Solomon Islands, Indonesia, the Philippines and Thailand (Greathead 1983).

Advisory work

CIBC staff are in demand as consultants by the FAO, World Bank, WHO, GTZ and other agencies, as well as national Governments. Some important examples during the last five years for

FAO include rice pests in south and south-east Asia, coconut spike moth in the Philippines, cane borers in Cuba; for the World Bank an assessment of biological control prospects in developing country agriculture and pest control on the Sudan Gezira; for WHO attendance on expert committees; and for GTZ advice to their pest control programme in the Cape Verde Islands.

Information

Thompson began cataloguing records of insect parasites and predators, initially for use by staff. This information is an essential starting point when beginning a biological control programme, and the decision was soon taken to publish it in catalogues; two series are now in print covering the periods 1913-1937 and 1938-1962. The cataloguing work continues and will now take the form of a computer-based data bank which can be more readily updated and corrected, will be accessible on-line, and will provide catalogues and lists as needed.

CIBC also produces a series documenting past biological control efforts. These now cover Australia, Canada (2 volumes), the Afro-tropical region, south-east Asia and the Pacific, and western and southern Europe. An up-date on Canada is now in press, a volume on the British Caribbean is being completed, and a review of work in eastern Europe is in preparation. It also publishes other occasional works, including conference proceedings.

In 1980, CIBC began a new quarterly serial publication, *Biocontrol News & Information*. This includes news items, provides authoritative review articles, and also abstracts the world biocontrol literature. This journal is now essential reading for those who wish to keep abreast of new developments in biocontrol.

Training

CIBC has always welcomed students at its laboratories to undertake research for higher degrees, and scientists wishing to learn new techniques. Staff have been appointed as supervisors or honorary lecturers at universities in India, Pakistan, Trinidad and Uganda, as well as in Europe. Many former students are now practising biocontrol workers in their own countries and overseas, and provide the CIBC with a useful network of contacts and collaborators.

The need for more formal teaching in biological control techniques was recognized with the launching of four-week International Training Courses on the Biological Control of Pests. The first was held in Bangalore in 1980, the second in Trinidad in 1982, and the third will take place in Bangalore in October 1984. These courses, for personnel of pest control departments, of Governments, universities and commerical undertakings, have attracted considerable interest. So far, 38 persons from 20 countries have received training with financial support from the Commonwealth Foundation, Commonwealth Fund for Technical Co-operation, FAO, other donor agencies, Governments, and commercial companies. An additional special course for participants in the FAO/UNEP Programme for Integrated Pest Control in Rice in south and south-east Asia was also held in 1982 in Bangalore, jointly with FAO.

Now, with CIBC staff at Silwood Park, the CIBC is taking a more active role in university education by participating in course teaching and supervising research by PhD and other students.

Future Plans

The CIBC group at Silwood Park is assuming management of the Institute and some administrative functions formerly undertaken by CAB Headquarters, as well as providing a focal point for project preparation. For these purposes, additional staff are being engaged, including a part-time Contracts Consultant, Dr N. W. Hussey, well-known for his pioneering work on biological control-based pest management. The unit assists the station in Switzerland in field work, but also has an important role in support of tropical work as a quarantine and screening centre. For example, natural enemies of cassava pests being transferred between South America and Africa, pass through Silwood Park (see Chapter 19).

CIBC is currently primarily concerned with "classical" biological control, and then almost exclusively with insects (or mites) as control agents for other insects or weeds. However, this is only a part of biological control which increasingly concerns the study of micro-organisms and nematodes, both as control agents and as candidates for biological control. Further, few pests can be controlled in isolation and on most crops, biological control must be integrated with other measures and be compatible with agronomic practice. For these reasons, CIBC is expanding its expertise. An insect nematologist was appointed at the Commonwealth Institute of Parasitology in 1980 and is available for field work with the CIBC, when required. It is itself appointing a plant pathologist, to be based at the Commonwealth Mycological Institute (CMI), to work principally on fungi as weed control agents. Other expertise is being sought through collaboration; formal or informal agreements are now in force with Imperial College (teaching, ecology, protozoology, nematology, pesticide application technology), the Glasshouse Crops Research Institute (insect pathology, breeding of biotic agents), the Institute of Virology (insect virology), and the Tropical Development and Research Institute (tropical entomology). These are UK-based, but discussions are also in progress with the USDA and CSIRO, for mutual assistance, including the sharing of information and facilities. Discussions are also being held with the CGIAR over the provision of services to the International Institutes in exchange for facilities. Closer relations are also being sought with national organizations in countries where CIBC staff are based. In Pakistan, for example, there is now joint management of the Rawalpindi station with the Pakistan Agricultural Research Council, and in Kenya the work programme is agreed in consultation with the Government.

These developments will enable the CIBC to offer more broadly based services, to compete for major integrated pest control contracts, and take a leading role in biological control worldwide. They will also give it greater flexibility and reduce costs.

Current activity for the benefit of agriculture in Africa includes overseas input into the cassava pest control programme, an IDRC sponsored programme on cassava mites in eastern Africa, feasibility studies on the potential of insect parasites of tsetse flies as control agents, and support for Kenya in the development of its own biological control programme. These activities can be expanded and extended in response to national needs. Where additional financial support is required, the Institute can assist in identifying donor agencies and negotiating with them.

References

Commonwealth Institute of Biological Control (1980) *Biological control service. 25 years of achievement.* Farnham Royal; Commonwealth Agricultural Bureaux, 24pp.

Greathead, D. J. (1983) The multi-million dollar weevil that pollinates oil palms. *Antenna* 7, 105-107.

Scrivenor, T. (1980) *CAB – The First 50 years.* Farnham Royal; Commonwealth Agricultural Bureaux, 73pp.

Thompson, W. R. (1930) *The Biological Control of Insect and Plant Pests.* London; HMSO, 124pp.

Thorpe, W. H. (1973) William Robin Thompson. 1887-1972. *Biographical Memoirs of Fellows of the Royal Society,* 19, 655-678.

Chapter 81

The Value of Information in Agricultural Research and Development

J.R. METCALFE

Director Information Services, Commonwealth Agricultural Bureaux, Farnham House, Farnham Royal, Slough SL2 3BN, UK.

Introduction

Agriculturalists, whether they be research scientists, planners or extension workers, are often frustrated at being unable to obtain the information they require. This was confirmed by conference attendees who answered a series of questions posed by the speaker; the answers also revealed that the required information was sometimes found too late to be of use or to be in an unusable form. The result can be duplicated work, time lost and innovations not applied, and the cost, if it could be estimated in economic terms, would be frightening. Information in the right place, in the right form, at the right time has a value; if it is inaccessible, for whatever reason, it is a wasted asset.

The information scene

Information is said to be the fastest growing industry in the world. Several conference contributions indicated that information in the agricultural sector is now receiving close attention; a mass of data and information is accumulating on scientific research, planning, banking, markets, farm performance, and recommended practices. But this is not necessarily an unstructured mass of information. There is a certain general pattern to the flow of information (Fig. 1) which distinguishes the information circulating at an international, scientific level, from that at a more practically orientated, national level. If these two levels of information are put together, and effective links made between them, and between scientists and farmers, there is the makings of an immensely powerful information system. It would provide channels for the unimpeded flow of relevant information to all categories of user, and would ensure that full advantage was taken of farmers' knowledge, experience and habits, which are so often overlooked. This pattern should be seen as the ideal towards which agriculturalists should work, not creating a grand new system, but integrating and developing the elements that already exist.

The role of the CAB Information Services

CAB's aim is to provide a world-wide agricultural information service for scientists and other professional workers; agriculture in this context is interpreted broadly to include, for example, related fields such as forestry and human nutrition. The service is required to be financially self-supporting (turnover approximately £5 million per year) and therefore has to be managed on the same basis as a commerical organization.

The main function of the Information Services is the preparation of the bibliographic database, and from it the range of abstract journals, available in conventional printed or machine-readable form for computer searching (Fig. 2). Of particular interest among these is a new journal, *Wheat,*

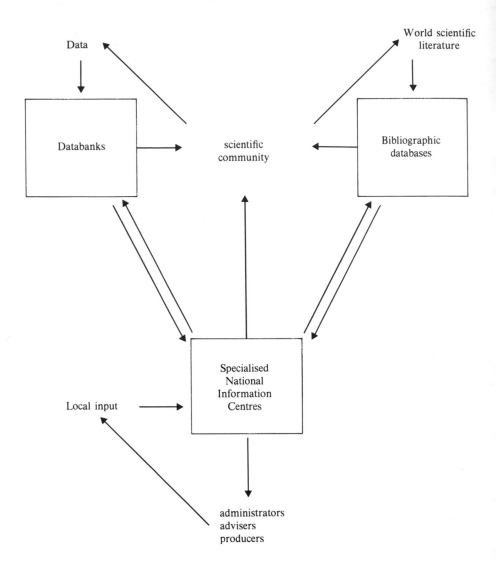

Fig. 1.   The flow of information in the agricultural sector, nationally and internationally.

Barley and Triticale Abstracts, commissioned by the International Maize and Wheat Improvement Center (CIMMYT). Its significance lies in that it is automatically produced through the new CAB computer system and 500 copies are destined for CIMMYT's contacts in developing countries. Thus international funds are being used to channel information where it is desperately needed.

CAB has been producing abstract journals for over 50 years, and the sum total of records, about 5 million, of which 1 ½ million have accumulated in the last ten years, represents a considerable scientific archive, much of it on research in Africa. This, together with collections of data in ministries and departments throughout Africa, constitutes a strong knowledge base, a view the exact opposite of that expressed by Odhiambo (Chapter 6). The value of this knowledge base should not be underestimated, or work already done will be duplicated and the wheel reinvented.

The Information Services objective is to maximise the time the scientist can spend at the bench or in the field. Informative, relevant abstracts of the significant literature save time otherwise spent in perusing the mass of material confronting today's scientists, or sorting through uninformative

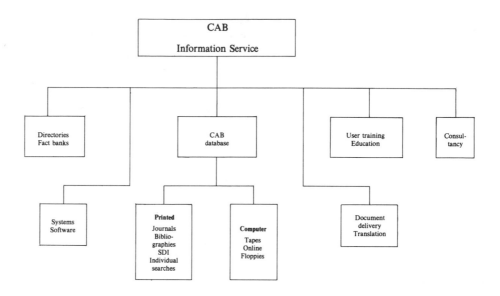

Fig. 2.   The service and products range of the CAB Information Services.

unselective title-only compilations. At the Special Libraries Association Congress in 1976, there was a paper entitled "Information by the kilo". The information product, computer printout, was too much to handle so half was thrown away. This is a poor substitute for informed selection, which is CAB's policy.

The CAB Information Services produce this quality database by by employing about 100 Scientific Information Officers, many of whom are subject and language experts. In certain instances, there is close collaboration with national information centres, for example in China (The Institute of Information for Agricultural Science and Technology, Chinese Academy of Agricultural Sciences) and in Hungary (Agroinform, the information division of the Ministry of Agriculture and Food). Publications for abstracting are sent to CAB from all over the world, and certain centres such as the International Livestock Center for Africa (ILCA) and the International Rice Research Institute (IRRI) provide their accessions lists from which material for abstracting may be selected.

The full range of the CAB information services is shown in Fig. 2. Two of the back-up services are worth particular mention, namely document delivery and training. There is nothing more frustrating to a scientist than knowing about a paper, and not being able to obtain it, so CAB undertakes to provide on request photocopies of most articles abstracted, either from its own holdings or via other organisations' holdings. Training is seen as a highly valued service, particularly for developing countries. There is an annual course entitled "Information in Agriculture" which over the last four years has been attended by 125 people from 40 countries. This year, The Commonwealth Consultative Group for Agriculture (CCGA) has plans to fund two similar courses in 1984, one in Sri Lanka and one in Fiji. By this means, librarians and information specialists appreciate more fully the value of information, the various sources of information, and methods of information transfer.

Limitations on the existing system of information transfer in agriculture

When it comes to information transfer, scientists are in a privileged position. All the tools they need are available: primary journals, databanks, bibliographic databases, document delivery. The only limiting factor, particularly in developing countries, is money. This is one reason why

CIMMYT's bulk subscription to *Wheat, Barley and Triticale Abstracts* is so significant; similarly the sponsorship by ICARDA (International Center for Agricultural Research in Dry Areas) of *Faba Bean Abstracts*. Here international organisations are providing the means for developing country scientists to get the information they need. It is a principle that should be emulated on a much wider scale, by having the international aid organisations directly fund developing country library resources.

However, one form of information that scientists need is not readily available, and that is a rapid alerting service of really new significant developments. Von Kaufmann drew attention to this in introducing the session on livestock information at this Conference (Chapters 52–54). On the other hand, planners and extension workers are not well served. There are several reasons for this:

(1) Scientists are too interested in science for its own sake; they are hobbyists, specialising in the area that delights them most. As an entomologist at the conference admitted, many entomologists are more interested in insects than in people! But this is a gap that can be bridged, for instance with booklets in English and Hausa for sorghum midge control in Nigeria, and as shown by my own experience with the control of sugar-cane leafhoppers in Jamaica. A particularly good example of information services bridging the gap are the publications on rice culture exhibited by the International Rice Research Institute.

(2) Services to end users may be costly, in terms of manpower for preparing the information in suitable form and communicating it, whether personally through the media or in printed form. However, it is necessary to recognize the cost-benefit of such services, for the benefit of well structured services will exceed the cost many times over.

(3) Such services may require input from scientists, economists and sociologists, a combination of specialists that all too rarely meets. It should not be impossible to ensure that workers in these three fields pool their expertise and resources.

(4) The incentives for effective work in agriculture are often lacking. Agriculture is widely looked down upon as an inferior discipline and may not attract top calibre personnel. On the other hand, one speaker showed how, by making advice on correct use of fertilisers part of the retailer's job, an incentive could be introduced (Chapter 31).

How can CAB help in relation to Africa?

In this section, a few ideas are floated. No doubt they can be improved and many others suggested, but they are offered as a basis for discussion.

(1) *Training.* Courses on information transfer, based on the Oxford course, would be immensely valuable to information workers in Africa. The CCGA and other donors could be approached with a view to their providing funds.

(2) Reference has already been made to CGIAR units' sponsorship of specialized abstract journals. There are other abstract journals published by CAB which cover interests of the international research stations, e.g. rice, potatoes, sorghum and millets. These and other titles are open to sponsorship on behalf of developing country scientists.

(3) Simple, but accurate and authoritative handbooks would be welcomed in many fields, as IRRI's series of publications has shown. CAB could undertake or assist with these in selected subjects.

(4) This conference has been unique because of its concentration on the small farmer, as advocated by President Nyerere in his opening address (Chapter 1) and on development (as opposed to research) as required by the Conference Chairman. But there is a risk that the momentum and enthusiasm generated will be lost as soon as the participants disperse unless there is a plan for action or a list of development priorities. We have heard that the potential for increased production in Africa is great, as indicated by the country reports (Chapter 3). Africa was even referred to as potentially the breadbasket of the world (Chapter 5). We must harness this enthusiasm and develop this potential.

On the information side means must be found, possibly through a suitable publication, which would keep conference participants and agriculturalists throughout Africa in touch with the main thrust of agricultural and socio-economic development in Africa. This would help to meet some of the shortcomings in the system whereby information is transferred to the end users.

In conclusion, it can only be restated that information at the right place, in the right form, at the right time has a value. Every conference participant should therefore be committed to the efficient cost-effective transfer of relevant information, and help to create a greater awareness of the value of information. If not, the pace and extent of development in Africa will be far less than they could be.

Chapter 82

Information Services in Advancing Agricultural Production

M. BELLAMY (Rapporteur)

Officer-in-charge, Commonwealth Bureau of Agricultural Economics, Dartington House, Oxford OX1 2HH, UK.

A special session brought together studies of information needs, and of the experience and aims of information providers, all illustrating the vital need for communication and dissemination of research, and awareness of information, and the priority it showed be accorded in research, extension and training programmes.

I. Extension level rice publications in African languages

R. HARGROVE

International Rice Research Institute, (IRRI), Box 933, Manila, Phillipines.

Although IRRI's main language of publication and working language is English, the Institute seeks to overcome the language barrier mainly through co-publication, co-operating with other agencies to translate, publish and disseminate IRRI publications in non-English languages. By late 1983 more than 600 000 copies of 40 non-English editions of IRRI publications had been printed in 30 languages, with another 40 in preparation. Two publications, *Field Problems of Tropical Rice* and *A Farmers' Primer on Growing Rice*, account for almost 90 % of IRRI's non-English publications.

Most co-publication so far has been within Asia, but IRRI is increasingly seeking new co-operative ventures with agricultural agencies and publishers in Latin America. IRRI's advice and co-operation in publishing offers the following: (1) no royalties as payment for non-English editions published in developing countries; (2) complementary sets of artwork without English text; (3) inexpensive printing of certain materials in the Philippines; (4) loan of colour negatives to cut local manufacturing costs; (5) help with promotion of non-English editions; (6) purchase of additional copies of editions in multi-country languages (e.g. Arabic, Portuguese) for distribution in other countries.

Materials in African languages, or co-operative projects, include the following:

(1)   *Swahili*: planned joint Tanzania Agricultural Research Organization (TARO)/IRRI project for an edition of the *Farmer's Primer*.

(2)   *French*: with WARDA, a French edition of the *Primer*, and with IRAT, a translation of *Field Problems*. WARDA is translating the *Technical Handbook for the Paddy Rice Postharvest Industry in Developing Countries*.

(3)   *Arabic*: the *Primer* and *Field Problems* are being published jointly by the Rice Research and Training Project, Egypt, and IRRI.

(4)   *Portuguese*: *Field Problems* is being published jointly by the Ministerio de Desenvolvimento Rural, Guinea Bissau.

(5)   *Other languages*: Translation into Malagasy, More, Wolof, Yoruba, and Hausa are under way or being discussed.

II. Agricultural user population and their information needs: a case study of Badeku Pilot Rural Development Project in Nigeria

C.E. WILLIAMS and S.K.T. WILLIAMS

Department of Agricultural Extension, University of Ibadan, Ibadan, Nigeria.

This contribution (1) examines the needs of agricultural user populations with special reference to Nigeria, using a case study of a rural setting, (2) discusses the methods used in acquiring their needs; and (3) suggests ways of improving the effectiveness of meeting the information needs of the rural sector in general.

Profile of the Nigerian small farmer

Most Nigerians live in rural areas, and are mainly farmers. 69% of farms are less than 4 ha, 25% 4.9 ha, while only 6% are over 10 ha. Farmers produce a wide variety of crops; intercropping has sometimes prevented the adoption of new techniques, e.g. of 528 000 t of maize produced in 1974/75, only 0.78% were produced from holdings using modern agricultural techniques. Most small farmers cultivate only enough land to satisfy basic family needs, producing very little surplus, thereby accentuating the vicious circle of poverty. Soil fertility is maintained by bush fallow (cropping 1 – 4 years, fallow 2 – 10 years). Farmers lack collateral for raising loans and credit.

Badeku Pilot Rural Development Project

This Project is located some 20 miles from the Ibadan University campus, in Ibadan East Division. Its population is about 1300, in 275 households. The village was chosen because of its small size, its easy accessibility, and the receptiveness of its people to change. An initial socio-economic survey resulted in a demonstration of maize and mass vaccination against cholera in 1970. The project expanded to 18 villages in 1973, and 30 in 1974, in two units. One, 18 villages, in the Badeku area of the rainforest, and Fashola, 12 villages in the derived Savanna Zone, Oyo South Division. Data were collected from 140 (74 men and 66 women) respondents at Badeku.

Findings

(1) *Personal characteristics.* All respondents were married, with mean ages of 45 (men) and 40 (women), 76 % males and 88 % females were illiterate. The ratio of full-time male farmers to female was 90: 12.1. Only women (27 %) processed farm produce, and 12 % were traders. 95 % males belonged to farmers' organizations, while 61 % females opted for religious guilds.

(2) *Information needs.* Women sought more information than men. All groups' needs were connected with their occupations. 54 % of men required information about farming, while most women sought information on personal and business matters.

(3) *Sources.* 79 % of women mentioned their husbands as main source, while men (85 %) consulted village elders and close friends.

(4) *Types of Information needed.* These differed with age and occupational preference. Full time farmers (men) wanted to increase their productivity, while women wanted increased financial independence. They mentioned trade and business improvement (60.3 %), loans for trade (50 %), building village market (53 %), good source of drinking water (48 %), child care and pregnancy (48 % and 60 % respectively).

(5) *Form in which information is given.* Village meetings were the preferred method (84.3 %) followed by radio and personal contact.

Conclusions

Information given must therefore be relevant to the needs of the rural population, i.e. to improve farm techniques to increase productivity, and provided in languages and by methods they understand, at specific centres. Women needed information related to their family and community role, especially trading, health and life style.

Implications

Efficient agricultural and home economics extension programmes are needed, which take account of the multiple role of women, and including other income earning activities. Functional literacy programmes based on agriculture and home economics are essential. All programmes must fit with seasonal work pressures. Programmes for women must have consent and cooperation from the male head of household. Improved infrastructure (markets, roads, water supply) cannot be over-emphasized. Both men and women need advice on better use and allocation of time to multiple tasks.

III. Recent Developments in the Transfer of Agricultural Information in India

D.B.E. REDDY

Agricultural Librarian, Andhra Pradesh Agricultural University, Hyderabad, 500030, India.

Introduction

Transfer of technology in agriculture is recognized as a prerequisite for advancing agricultural production in India. The transfer process consists of four important interlinked and interdependent operations: generation, documentation, dissemination and utilization of information. The transfer is in two stages, from research to extension, and from extension to farmers.

Generation of Agricultural Information

A well organized national grid of cooperative research and education is cordinated by the Indian Council of Agricultural Research (ICAR). The 34 Research Institutes, 4 Project Directorates, 54 All Indian Coordinated Research Projects, and 23 Agricultural Universities organize agricultural research, while several other institutions and organizations provide related research.

The National Agricultural Research Project (NARP) was conceived in 1979 to strengthen State Agricultural Universities' capability to conduct location-specific, production-oriented research in agro-climatic zones. Regional committees have been set up in eight agro-ecological regions which review agricultural research and education, and specific agricultural problems. Joint panels consider collaboration research with other agencies, for example ICRISAT.

Documentation

Decentralized library and documentation services attached to the agricultural universities and research institutes cater for the information needs of teaching, research and extension personnel. In addition to book and periodical collections, including secondary sources, there are over 250 agricultural periodicals in some 15 languages, while research institutes produce technical reports. Local indexing and abstracting services cover this literature, as well as theses. The ICAR's Agricultural Research Information Centre, established in 1977, acts as India's AGRIS Input Centre. It also documents on-going and completed agricultural research, and maintains a file of research personnel. A union catalogue of scientific periodicals is produced by the Indian National Scientific Documentation Centre. The Indian Agricultural Research Statistics Institute maintains files and publishes compendia of field experiments. The National Information System for Science and Technology has established discipline oriented centres, such as that for Food Science and Technology.

Information Transfer

A network of extension institutes promotes the adoption of improved techniques, while farm universities, research institutes, state and central governments, voluntary organizations and commercial agencies also disseminate farm information.

Operational Research Projects disseminate technical information direct from research station to farmers' fields.

The national demonstration programme maintains contact between research institutions, farm problems, farmers' training and extension, with emphasis on maximum output per unit area and per unit time, rather than higher productivity per crop.

The "lab to land" programme, in which all ICAR research institutes, agricultural universities and selected voluntary agencies participate, involves their making their research findings and recommendations available to selected farmers, who also receive technical inputs.

Agricultural polytechnics (Krishi Vigyan Kendra) provide young and progressive farmers with non formal education and technology transfer through work experience, as well as training technicians.

Farmers' fairs (Krishi Gigyan Mela) expose farmers to technical developments, and provide conferences and meetings.

The training and visit extension system has been introduced, and a training institute for its monitoring, training and evaluation is to be established.

Extension workers' training is centred on 150 Farmer Training Centres and three Extension Training Institutes.

Mass media are used extensively, through INSAT–18.

Utilization of Information

The Community Development Programme, introduced in 1952, now covers 566 850 villages in 5002 blocks. Each village-level worker covers 10 villages.

Newer programmes include the Intensive Agricultural District Programme, the High Yielding Varieties Programme, and the Multicropping Programme.

Conclusion

Co-ordination of effort is needed between the many agencies and organizations involved in transferring farm technology.

IV. Training Programmes in Agricultural Bibliographical and Information Services

M. BELLAMY

Officer-in-Charge, Commonwealth Bureau of Agricultural Economics, Dartington House, Oxford OX1 2HH, UK.

Introduction

CAB has come relatively recently into the field of providing training for agricultural information personnel. The success and expansion of its courses supports the hypothesis that many existing training courses do not answer the needs of those whose function is to provide information. Unfortunately the users of information, who also frequently control funds for all agricultural training, accord very low priority to information training.

Formal Training

Agricultural information personnel are generally either (1) trained librarians / information scientists specialized in agriculture or (2) diverted research or extension personnel or administrators. Very few institutions or courses provide adequate, relevant programmes; agricultural librarianship is generally neglected. Agricultural librarianship could well become a model for library courses; as is the case in a recent British Council project in Beijing.

Short courses are a potential solution, as they can cater for a wide range of needs: refresher courses for librarians; orientation for librarians and documentalists moving into agricultural information; scientists and agriculturalists moving into information work, and support staff. Such courses are either "once off", or regular, such as those run at Aberystwyth and Pittsburgh.

CAB

Two-week courses have been run for the last five years, attended by 125 participants from over 40 countries. Their aim is to provide a sound background in agricultural information practices, ability to establish and run an information service, knowledge of information sources, instruction in new technology, and a forum for exchange of ideas.

CAB has also assisted at workshops on agricultural information sources (Indonesia and Thailand, mid-1970s) and run 1 – 6 week information services courses for CAB input suppliers (e.g. Beijing in 1982). Some trainees are placed in CAB units, notably five Chinese agricultural information personnel during 1983/84.

On-line workshops are held regularly, mainly in Europe and North America.

Plans for 1984 and beyond include training workshops in Sri Lanka, Malaysia and the South Pacific, sponsored by the Commonwealth Consultative Group on Agriculture.

Funding

Sponsorship for information training courses comes from many sources: British Council, FAO, Commonwealth Foundation, UNESCO and parent organizations and Governments. However, many international organizations and research institutions offer very little support to training their own or national agricultural information personnel, suggesting that agricultural information training has low priority in the allocation of funds.

The Information-Research Gap

Training of information personnel needs to be accorded priority by research institutions. More and better entries in directories of educational courses and sources of funding could assist in this. The "gap" lies between the library and information field and the research field; while research is the main user of agricultural information sources, and the need for information is constantly being acknowledged, research funds and training grants are not easily available for information workers. The information worker should be trained first and not last, as the information unit and library are the infrastructure upon which to build successful research, and this allows cost-effective use of scientists.

Future Emphasis

Better communication is seen as a major role of training in research institutes. Higher priority and funding for information and communication training are vital. "Appropriate" information is required for extension personnel and programmes. An international agency (perhaps coordinating existing ones) is needed to bring together specialists to discuss the problem.

V. A Computerized Agricultural Information System in Africa

M. HAILU

Library and Documentation Services, International Livestock Center for Africa (ILCA), P.O. Box 5689, Addis Ababa, Ethiopia.

Introduction

ILCA has recognized the important role that a well equipped documentation centre can play in collecting, processing and disseminating information relevant to livestock production in Africa. One of its mandates is to retrieve, assemble and make available in English and French all relevant information on animal production in tropical Africa.

Collecting, processing and disseminating non-conventional literature

In 1976 ILCA began to put on microfiche non-conventional literature relevant to the fields of animal production and health, filming such literature in various countries of sub-Saharan Africa. Once identified and retrieved, the literature is publicized through on-line literature searches and printed country indexes, or subject bibliographies. Countries are provided with copies of relevant microfiches, indexes and a microfiche reader.

Initial funding came from ILCA's own sources, and was able to cover Ethiopia, Botswana, Kenya, Mali and Senegal. Subsequent support from IDRC has extended the project to 18 countries.

The project also hopes to increase national awareness of existing research.

Computerized literature search and selective dissemination of information (SDI) services

ILCA has geared its services (1) to answer the requirements of researchers before they start a project, or to deal with a problem (involving an exhaustive literature survey over 10, 20 or 30 years), and (2) for current information.

References on livestock production in Africa are selected from various sources, catalogued, indexed and abstracted. They are entered on ILCA's data base, supported on the MINISIS software package. The inhouse data base now contains over 25 000 items, and is increasing at about 5000 items per year. A new title *African Livestock Abstracts* is planned. ILCA provides literature searches and back up document delivery.

The centre has contact with AGRIS, CAB and AGRICOLA, and searches all three data bases.

An SDI service has recently been made available to key African researchers as well as ILCA staff. Items are selected from monthly AGRIS and CAB tapes.

Conclusion

ILCA's aim is through these services to contribute to and strengthen national livestock research capabilities in tropical Africa. (See also Chapter 54.)

Chapter 83

Some Considerations on the Relevance of Indigenous Food Security Strategies to Development Programmes

H. SIBISI

Agricultural Sociologist, Box A92, Swazi Plaza, Mbabare, Swaziland.

The problem of advancing agricultural production in our continent at this time is much more than one of improving on existing performance. Except in a few of our countries, food production has declined in the past 20 years. According to a recent FAO announcement, there is an actual shortage of food in 24 African countries with a total population of 150M. A food crisis of major proportions has come upon us despite Africa "enormous physical potential to feed itself" (Eicher 1982). It is a crisis unprecedented in our history; for thousands of years our ancestors lived off this same land and with much the same kind of animals without needing any outside assistance. Yet, less than 15 years after the Sahel drought of the early 1970s, we are again trying to recover from a disastrous lack of rain and seeking international help in order to do so.

Much thought has gone into identifying the barriers to tapping our potential, as Eicher (1982) again puts it. He listed the colonial development and training policies, narrow confines of research, low priorities given formerly to investments in agriculture, inappropriateness of assumptions made by foreign (principally western) economic and planning advisers, and so on. In his words "the crisis stems from a seamless web of polictical, technical and structural constraints which are a product of colonial surplus extraction strategies, misguided development plans and priorities of African states since independence, and faulty advice from many expatriate planning advisers".

There is little that could fruitfully be added to this assessment. Nonetheless, I should like to raise a somewhat different question, essentially complementary however to that of the causes of this collapse in our historical ability to support ourselves unaided without recurrent disaster. How was it possible for the calamity to occur at all? If there has been a collapse, just what is it that has collapsed?

To ask this is to try to identify more precisely or in detail just what arrangements or practices made it possible in the past for our peoples to survive in the absence of outside support, especially through bad seasons of the kind which nowadays make us objects of international charity. All I can do here is to focus on one aspect of such practices, taking an example from the southern and eastern parts of the continent, where cattle-keeping and the growing of grain are combined; but I shall add some comments on some other sides of the same question, and then try to show how in general the breakdown has come about.

Anywhere in the world where crops can periodically fail on which people rely for their staple food, they can only hope to survive if *reserves* are available. Today stockpiling of grain for such emergency needs is usually pursued at the national or international level (see e.g. Sarris *et al.* 1979), but Africans formerly needed *local* reserves wherever there was a risk of climatic failure. In other words, lacking means of bringing relief supplies from elsewhere, they had to have devices for long-term storage, which would keep grain especially in good condition for many years if necessary. Moreover, in the absence of means of refrigeration, they had to devise techniques for preserving other products (see below).

An underground repository was, and often still is, used for such long-term grain storage, called in many of the Nguni languages an *ingungu*, located beneath a homestead's cattle byre. Such respositories are usually cylindrical, narrowing at the mouth, and hold about 3 m³ of maize or sorghum, being deep enough to enable a man standing inside to receive a container from above in his upstretched hands; the diameter is about the same as the depth. The lining is of termite earth, familiar all over Africa for making hard floors, and in this case fired for additional impermeability. The top-most layer of grain is covered with several layers of bark, over which a stone slab can be laid if the mouth is narrow enough. The principal sealant however is cattle manure, present in abundance. The repository is hidden under the layer of manure which normally covers the floor of the byre, and of course is renewed by the animals themselves, making an extremely airtight covering.

This is a device for long-term storage, for ensuring that there are always reserves of staple food available in time of need. It is rarely opened and not used for everyday storage, for which there are above-ground receptacles. If a particular harvest is poor, people draw on the underground reserves and fill up the space left with bark so that there is a minimum of air left inside. In bumper years the reserves are replenished, perhaps taking out the existing grain first and putting the fresh supply at the base of the repository.

Grain kept in this way undergoes chemical changes which convert the retained air into a gas known to be lethal, so that anyone removing the cover waits for fresh-air to circulate before decending to bring out the grain. Presumably this also means that pests which getting into a repository do not live long. The chemical changes evidently modify the taste of part of the stored grain, but people became used to this, and in any case different dishes in the indigenous cuisine made use of differently flavoured grains; whatever modifications occur they are not harmful.

As a method of providing reliable grain reserves for a local community without need of outside supplementation, this indigenous African storage technology has not obviously been surpassed by anything known in western practice. It ensured food security at the basic level of subsistence of the homestead, using materials available locally, and so did not entail dependence on trade for the importation of manufactured containers or construction materials.

Since a repository uses cattle manure to make it airtight, however, the method requires cattle. Such a storage technology is an aspect of an economy in which crop cultivation and herding interlock in several ways and support an integrated self-sufficient society. Cattle are the main form of joint wealth for an entire homestead, providing milk and manure for all as needed, as well as latterly draft power for ploughing. Manure of course regenerates the soil more effectively than fallowing or the burning of grass or other vegetation, although these can also be used. Milk in its curdled form provides a stand-by item of diet for all members of a homestead, too. Thus, in addition to the empirical practicality of siting a grain repository in the homestead cattle byre, the location can be seen as having a symbolic value, implying that the stored grain, like the produce of the animals, is available for all who need it, when they need it.

The very effective concealment of the repository is no accident. No doubt in the past when raids by neighbouring peoples were a hazard it was an advantage to be able to turn to such a reserve of grain on returning to the site of one's homestead. Also, concealment of wealth is necessary when the awareness by neighbours that you possess a surplus entitles them to request help when afflicted by scarcity. This double theme of equality (with resentment of undue affluence) and consequent concealment of wealth is familiar in southern Africa and elsewhere south of the Sahara.

As well as requiring cattle, this type of underground storage must also depend on security of tenure of the land where the repository is dug. Contrary to what sometimes has been supposed, the indigenous form of land tenure can be shown to have ensured such security, and continues to ensure it where it has not been undermined by land reform policies.

Maize and sorghum are not the only foods which African people knew how to preserve. There were, and are, methods of storing green vegetables using a technology not of refrigeration but of drying in the warm sun. Several types of vegetables were, and often still are, known which do not require frequent watering or intensive care, and can be grown along with grain crops and harvested before the latter are due. Some also are gathered from the wild. All can be blanched and kept for long periods when dried.

Milk, as previously mentioned, is kept in a curdled form which makes more sense than trying to keep it fresh in a warm climate without refrigeration. It is preserved in large gourds from which it is consumed each day as required, an item available for everyone. As new supplies of milk are brought in they are added to the gourds and thus are themselves curdled.

What is most striking about this, for I am sure that comparably reliable means of survival and self-support can be found in other zones of the subcontinent, is not that such indigenous technologies and practices existed, for if our ancestors had not known how to survive unaided in an African environment most of us would not exist today. Rather, what is notable is that agricultural advisers and other kinds of development expert have done little to acquire knowledge of indigenous practices. Instead of building on what has already been achieved, perhaps looking for better ways of using the same or similar techniques, the consistent preference appears to have been to bring in exotic and unfamiliar methods, evidently on the assumption that they are virtually superior by definition.

Such preferences and assumptions have caused much damage. It is perhaps not surprising if outsiders rely on methods and technologies with which they are already familiar, although the influence of the common colonial view that local and indigenous practices could be no more than the product of ignorance and backwardness can reasonably be surmised as well. What is more dangerous is that our own people, in the course of their training, are led also to neglect and even despise local practices and crops. Hence, highly efficacious traditional storage methods are ignored, and people are encouraged instead to make use of storage tanks which have to be bought and which are apparently not intended as long-term reserves; in using them grain is often affected by mould becomes useless even as animal feed.

Disregard of traditional strategies for survival can be seen as well in the promotion of newer and higher-yielding strains of crops, most markedly hybrid maize. No research has been done, it seems, on whether this can be stored on a long-term basis underground, yet growers were pressured into abandoning traditional maize varieties which can provide such reserves. As hybrids come as a package with fertilizer, costs are entailed which must be recouped from sales; if markets are unreliable the grower may be in trouble, particularly if the hybrid cannot be used for long-term reserves.

As to other crops of exotic origin, especially western vegetables, these normally require more intensive care as well as more water, than most indigenous African vegetables, yet it is not apparent that they are any more nutritious or, indeed, have clear advantages of any kind other than greater marketability. Neglect of African crops stems mainly from a, possibly unstated, view that it is in some way "backward" to grow them at all. It is not wrong in itself to assist people to cultivate vegetables or other crops of western origin or type, but that fostering such changes without making clear, and perhaps even without fully knowing, the consequences of abandoning former crops or techniques. Disasters of development usually stem from either over-confidence in the success of exotic varities or techniques inadequately tested in an African setting, or obliging people to commit themselves to a course from which they can not easily retreat when the outcome is not as expected.

Where drastic resettlement projects are promoted, unless the local way of life is understood thoroughly and it is certain that new methods and arrangements will work in the local setting, it can easily happen that traditional systems are disrupted, people are also prevented from re-establishing it, and at the same time unforeseen difficulties nullify the project's anticipated results.

If people are moved away from cattle byres to newly-constituted settlements, they are not likely to dig new repositories, partly because they may be unsure whether they will be moved again, but also because it is not feasible to construct this type of underground repository at a newly-occupied site. Then if the markets whose availability is usally taken for granted in such schemes prove to be unreliable, the supposed beneficiaries may be left with unsaleable surpluses and no reliable means of storing reserves. Even where the object of resettlement is no more than to enable people to benefit from, say, a better water supply or more accessible roads, at least it is prudent to consider the total costs of the move in addition to expected benefits.

There are also difficulties of a more inherent or built-in type in applying western science and technology to African circumstances. In particular, the implications of the specialization of knowledge and belief in the essential superiority of "market production" over "subsistence". The systematic specialization of knowledge and research is no doubt part of the success of science and technology in the West, since it made possible the concentration of effort on solving detailed problems. Yet it also imposes the penalty of ignorance of most fields other than one's own. If recommendations of specialists in one field are followed without regard to their effects on other parts of an indigenous agricultural system harm can result.

For example, livestock and crop specialists can, by working independently, cause policies to be followed which undermine the close connection between pastoral and arable practices enabling people to operate traditional storage technology. People may be induced or obliged to reduce their cattle in the "interests" of better animal husbandry, and thereby lose not only part of their traditional means of maintaining reserves of grain but also their chief means of restoring soil fertility, without acquiring anything on which they can rely so firmly as they did on their cattle.

Similarly, economics in general and agricultural economics in particular has devoted too little attention to self-supportive systems. The benefits in terms of productivity, and therefore higher living standards, of commercial production are stressed without regard for the benefits of reliable means of ensuring basic long-term security of life. In industralized countries there are ways of protecting ordinary people from the consequences of failure of the market system, as when there is unemployment; in Africa there is no prospect of such Social Security provisions, no alternative to destitution apart from international charity.

There are at least three reasons for not accepting uncritically the assumption that switching to commerical production will provide unalloyed benefits. First, it is foolish to rely on market production for a living if you cannot depend on the market and if what you have to pay for staple foods in bad years is more than your surplus in good ones. If instead you switch from feeding yourself to producing a cash crop, the result can be starvation when the market fails.

Second, as Eicher (1982) and others have remarked, many African Governments have pursued negative pricing and taxation policies for agriculture. Agricultural producers generally have little control over their Governments despite forming the majority of the population.

Third, there is the historical example of the consequences of an unguarded adoption of commercial production in place of basic self-support, India in the latter part of the last century. India experienced the most severe series of protracted famines in its entire history as a result of converting large tracts of grain-yielding land to commercial crops at a time when British demand was high; as during the American Civil War. When American cotton became available again, the Indian market collapsed and millions of peasants weaned from grain production now found themselves riding the world-market economy and unable to convert their commercial agricultural surplus back into food in depression years; the population declined by several millions between 1895 and 1905. Have we reason to believe that world markets for primary commodities have greater dependability now than they had for Indian peasants a hundred years ago?

I am not advocating rejection of market production, trade is after all not unfamiliar in Africa, but querying the assumption that we can benefit from producing for markets only if we jettison the means by which we have ensured basic food security. We need a balance between the two approaches to earning a living. Determining that balance, deciding how far commerical production and no doubt the freehold ownership of land can be taken without gravely weakening our capacity to provide for ourselves and survive bad seasons, is not an easy task. It requires harder thinking and attention to detail. We should appreciate that it is sweeping umbrella recommendations, aggregating together all kinds of important and vital differences, that have caused many of our troubles.

At the level of the individual family or homestead, instances are not lacking of reasonably successful balancing or the combining of traditional practices with responses to modern opportunities. This possibility depends on factors such as the accessibility of urban centres and communications, but it does appear that the indigenous homestead system and land tenure arrangements, in southern Africa at least, lend themselves to various adaptations. On the homestead it is primarily the married women who see to the fields and everyone's livestock; other members, especially young unmarried men, go to seek employment elsewhere or engage in business.

My own research in Swaziland shows homesteads can farm on a regular commercial basis, feeding themselves and producing a surplus which they sell. Although in the traditional system no land is owned outright, in the sense of being freely disposable, a homestead is left in enjoyment and occupancy of the arable land allocated to it; extra land can be borrowed from neighbours in return for services such as ploughing or a share of the crop. All retain their basic rights to land, and the creation of a landless class is avoided. If ordinary people have discovered or evolved ways of participating in the modern commercial and industrial economy without relinquishing the benefits of indigenous institutions, it should also be possible for developers and planners to combine the advantages of the two systems at the policy level.

Any such achievement would have to be based on a thorough understanding of local culture and indigenous economic arrangements. These arrangements, as they help to ensure long-term survival on a self-supporting basis, cannot be ignored without risk of devastation. People should not be induced to abandon what they have and can rely on, for prospects which are a matter of surmise or calculations which can be nullified by unforeseen events. They may end up in much distress, or even dead; the expert who recommended or even designed the unsuccessful programme will probably not even be out of a job.

What is at issue here is not just the desirability of supplementing the techniques of economists by sociological research, rather a full comprehension of rights and responsibilities, in each cultural area, by which people have ordered their lives and ensured their survival over many centuries, coping with a climate they know extremely well. To attain such comprehension we need specialists of diverse disciplines to work together, or perhaps a new breed of specialist, accustomed to a multidisciplinary working environment and mutual consultation.

When we have started replacing the habits of neglect or scorn of indigenous accomplishments by sustained endeavour to understand them, we shall probably find it easier to devise programmes in full consultation with people they are intended to benefit. They in turn will be enabled to assess more adequately the probable consequences of participation, and can then make an informed choice of the degree and kind of commitment appropriate for them. We shall find ourselves building on past achievements, possibly discovering ways of improving them in various particulars, rather than merely discarding them in favour of exotic methods of insufficiently guaranteed reliability. We may be able actually to *revive* devices and practices which have been discouraged or neglected, and discover how to improve them. A local technology indigenous to one part of the continent could perhaps sometimes be introduced to another with profit. Scarcely any systematic attempts have been made to facilitate the cross-fertilization of local knowledge.

The present crisis in food production and supply in Africa appears to have been brought about by a breakdown in the long-established arrangements, social and technological, by which African peoples were able to support themselves without external assistance. One major aspect of such arrangements is the provision for surviving adverse harvest seasons. Eicher's (1982) call for a sustained increase in long-term research, especially in African universities, and the co-ordination of assistance from abroad on the basis of adequate information and relevant models, unquestionably deserves support. Some of the information needed can be found already in Africa, if we care to look for it, and some of the most persistent human constraints on the development of agriculture are to be found in the assumptions and practices of practitioners of development.

References

Eicher, C.K. (1982) Facing up to Africa's food crisis. *Foreign Affairs* 61, 151-174.
Sarris, A.H.; Abbot, P.C.; Taylor, L. (1979) Grain reserves, emergency relief and food aid. In *Policy Alternatives for a New International Economic Order* (W.R. Cline, ed.), 157-214. Praeger Publishing.

Chapter 84

The Culture of Agriculture

T. S. EPSTEIN

AFRAS, University of Sussex, Brighton, Sussex, UK.

Much Third World development planning is based on ethnocentric assumptions. The reasoning being: "If I were an African poor farmer I would do this or that". In fact only the African farmers themselves know the totality of their socio-economic situation and the outside expert may get it all wrong. This was made perfectly clear in Dr. Sibisi's contribution (Chapter 83). Yet scientific researchers still continue to assume that they know what is good and desirable for African farmers. At meetings of agricultural experts it usually soon emerges that the various distinguished scientists present regard "advancing agriculture" as a purely scientific problem; while they consider the practical implementation of the new techniques they develop as a political problem outside their own sphere of interest.

The performance of many of these scientists is no doubt impressive in as much as they effectively diagnose a problem and proceed to work out a solution. For instance, one expert diagnosed Africa's energy problem and suggested reforestation by small farmers as a solution. He went on to say that of course to get African small farmers to plant trees is a socio-political problem, which is beyond his own sphere of competence. It seems that " the basic deficiency now rests in the insufficient knowledge of the *psychosociological* factors, that is of the deep motivation of the African farmer" (Dumont 1969). Without understanding the culture in which African agriculture is so deeply embedded it is unlikely that much advance will take place in improving rural productivity in Africa.

Why Investigate Human Resources?

Since farm populations, rather than scientists or politicians, are at the centre of agricultural activities it is important that more is known about the socio-cultural and economic aspects of farm systems before we can expect technical advances to be widely implemented for the advance of African agriculture. Without reliable information and a sound understanding of existing agricultural practices and the social system of which they are part, it is unlikely that new farm techniques will be designed that will be acceptable to the majority of African small farmers. In this context, President Nyerere (Chapter 1) stressed, "we cannot over-emphasize the importance of the sociological preponderance of peasants in Africa. It has macro-economic implications; for the most part they use far the largest proportion of the arable land....this structure of production underlies the whole traditional way of life, the relations between individuals, families and groups, as well as the ethical base of society and people's sense of belonging".

To ensure ready acceptance of new farming techniques, farmers have to be made part of the development process already in its early stages. They have to be involved in the diagnosis of the problem as well as in the search for solutions. Such "participatory development" promises a much higher rate of agricultural advance than the top-down approach presently practised (e.g. Waddimba 1979).

African farmers no doubt lack the scientific training and know-how to develop new techniques by themselves, but what they can contribute to the scientific quest is a sound understanding of the total impact on their household system of any one new technique. They are better aware than the outside expert of the ripples that run right through the total system even if only a minor innovation is introduced. For instance, the introduction of multiple cropping with its increased and peaked demand for labour affects the customary intra-household division of labour: a male farmer may need to have more wives and more children to work the new cropping technique; this in turn will change intra-familial relations besides tending to increase the birth rate, etc.

Most donor agencies, as well as most newly independent African Governments, operate with a strong in-built urban bias. The majority of the modern services (e.g. hospitals, secondary and tertiary education) are concentrated in the larger towns or cities. Lipton (1977) claims that the rural sector is generally exploited to provide for the urban population.

Education too has a built-in western urban bias and in most cases is totally inappropriate to equip students and trainees with the ability to become more productive farmers. "The legacy of disappointment and frustrations that we are left with today from many well-intentioned literacy projects warns us that these efforts can succeed only when linked with concrete plans to improve the social and economic situation of the participants" (Ahmed 1980). Therefore, the content, format and extent of education available in rural Africa needs urgently to be examined with a view to harmonizing it with the overall development thrust.

Existing education, particularly in rural Africa, usually perpetuates the traditional gender-specific division of labour with a strong male bias. "Rural women form a nation within a nation. Deprived of their right, forced into domestic drudgery, destined to early marriage, subjected to frequent child-bearing, suffering from severe malnutrition and the victims of high mortality rates, their lives are a constant struggle for survival" (Abed & Rahman 1980). Women's life-styles thus differ from men's, which is likely to be reflected in different sets of aspirations. "Part of any process of change and development is the perceptions people have of it and the effect it has on their lives....It is possible that women have different perceptions of development than men in the same society" (Nelson 1981).

The market has assumed such overriding importance in African development that women's activities most of which fall outside it, are generally considered peripheral to the economic system and not defined as economic. The strong urban and male bias is reflected in an emphasis on cash crop development. Farmers with marginal and smallholdings are usually labelled as "subsistence cultivators", which means they remain outside the development thrust pursued by agricultural scientists. The term subsistence implies an economic unit that can produce only enough to meet the basic consumption requirements of its members, without any marketable surplus. However, in the majority of cases this is not so. Almost every household these days is in need of some cash to meet its minimum requirements. We are therefore not dealing with "subsistence" but rather with "sell-to-subsist farmers", who are involved in market transactions not because of a crop surplus but rather because their meagre resources force them to sell (or sometimes barter) their labour and/or a proportion of their crops just to provide for their basic needs. They cannot therefore be regarded as wholly subsistence farmers, nor are they real cash croppers, who cultivate crops to sell in the market. For sell-to-subsist farmers the sale of crops and/or labour often amounts to the equivalent of barter. Many of them presently sell their crops at the expense of lowering their own nutritional levels below minimum food requirements (see Epstein 1984).

The over-emphasis on discovering means to improve the productivity of cash crops, particularly where these crops differ from staple food crops, has resulted in relegating the sell-to-subsist farmers, particularly women, to remaining within the poverty syndrome. "The majority of rural men do not consider themselves responsible for assuring the family's food supply; accordingly increased food production and marketing by men will not lead to early and substantial increases in rural market demand for food. The men will spend most of their incomes on other items and family food needs will continue to be met primarily by women's subsistence production" (Bryson 1981).

As soon as a major staple dietary crop begins to be produced for cash men are generally expected to take over cultivation arrangements. Gambian Irrigated Rice Projects exemplify how difficult it is to convert operations traditionally performed by women into male fields of specialisation. Three outside agencies (i.e. The Taiwanese Agricultural Mission, The World Bank, The Agro-Technical Team of the People's Republic of China) planned to introduce double-cropped irrigated rice in the Gambia between 1966–79, oblivious of the traditional division of labour among people living by the River Gambia, whereby women have been responsible for growing rice in tidal swamps, inland depressions and on hydromorphic soils while men have cultivated the free-draining upland. "In the initial project the Taiwanese Mission established an organisational procedure which was unquestioningly followed by subsequent missions. The Taiwanese technicians assumed that the local subsistence production system was based on a household which was a unified unit of production directed and controlled by a single male head.... Although women were effectively excluded from owning irrigated land and receiving the credits necessary for cultivating irrigated rice on their own account, their labour, particularly for transplanting and weeding, was nonetheless crucial for the success of the projects. Under the customary division of labour, women were under no obligation to work for their husbands....In order to secure female labour, men have been forced to pay village women cash wages....However, in the rainy season women have the alternative of growing their own rainfed and tidal swamp rice. Planners have tended to regard this rice purely as a subsistence crop and have under-estimated the amount of local rice which is sold or bartered, and therefore the importance of this income to women....Many men gave as one of their major reasons for not growing a rainy season crop the fact that they cannot afford to employ labour in the rains" (Dey 1981).

African women are the backbone of African agriculture. With increasing male rural/urban migration an increasing proportion of rural households is headed by women. For instance, already in 1972 it was reported "that a very large number of small holdings in Kenya were worked and possibly managed entirely by women, most obviously when the husband and other male members of the household were in town" (ILO 1972). Yet the agricultural extension service everywhere is heavily weighted in favour of men. There exists as yet only a small number of trained African female extension workers. The male extension officials work together mainly with male farmers. This arrangement assumes that there is a male household head who makes all the farm decisions and/or that the male farmer conveys to his womenfolk the new techniques suggested to him. These assumptions hold good only in rare cases. Frequently, when male farmers are shown new ways of carrying out what are traditionally considered "female jobs" they fail to pass the messages onto their womenfolk. One of the major reasons for this bottleneck in the transfer of information may be that "a man performing a duty which is looked upon as that of a female is ridiculed as is a woman who performs a labour assigned to males" (Ndongko 1976). Accordingly, "the best approach would be to use a female extension staff with the stated purpose of improving the family's food supply while making efforts to ensure that the women have ready access to markets in order to dispose of any surplus....Development which makes use of potential benefits to be derived from enhancing women's activities has the best prospect of rapid and sustainable success" (Bryson 1981). Women ought to be considered the pivot round which African agriculture rotates rather than as appendices to their men. This is not a "feminist" argument, but a realistic appraisal of existing practices in Africa and an important factor in the advancement of African agriculture. Moreover, there is plenty of evidence to show that as soon as a woman's economic contribution becomes generally recognized, instead of labelling her "housewife" the birthrate begins to decline. Therefore, specifically women's economic activities, such as poultry farming and kitchen gardens as well as staple dietary crops, need to receive more attention by agricultural scientists to ensure that development does not make the poor poorer. However, before embarking on such development projects it is important to consult the population they are meant to benefit, so making sure that there is a "socio-cultural fit".

How to Study Human Resources

There are several methods available to examine the political and socio-economic factors which affect African agriculture (e.g. Apthorpe 1979). The two most frequently used methods are the quantitively-oriented questionnaire surveys and the qualitatively-focussed in-depth studies. Each of these methods involves a different data gathering technique with its respective advantages and disadvantages.

Questionnaire surveys
Enquiries based on a random sample (often a stratified random sample). Questions are framed so that the answers can be expressed in a quantitative form. These surveys have the advantage that they lend themselves readily to large scale generalisations. On the other hand, investigators are usually little-trained and lowly-paid individuals who are often expected to collect answers in a robot-like fashion, while informants are inclined to give the answer they think is expected.

In-depth studies
These involve senior researchers conducting participant observation style enquiries. Such studies focus on a small social unit, i.e. a village, hamlet, or sometimes even individual households, and do not approach informants with a pre-conceived set of questions; rather they try to elicit the emic view of things. This means an investigation not only of how informants view their universe but also a holistic approach to the study of socio-economic changes. Such in-depth studies are usually conducted by senior and trained researchers who live in the midst of their informants for an extended period of time (3–12 months) in order to try and apprehend the culture of agriculture in its totality. The resulting analysis helps to illustrate processes of change rather than constitute propositions representative of a larger universe, which is the case with questionnaire surveys.

Frequently advocates of one of these two methods denegrate the other. By contrast I suggest here that, instead of being considered alternatives, the two methods should be regarded as complementing each other. The hypotheses emanating from one of these methods need to be tested by the other method.

To organize questionnaire surveys presents little difficulty; they are being conducted regularly in the context of various kinds of enquiries. Therefore no more needs to be said here concerning such surveys. The main problem with in-depth case studies is the length of time they take. However, since it has taken more than 20 years to develop a weather forecasting system, it should not surprise natural scientists that it may take a number of years before sound predictions can be made in the context of socio-economic changes.

This does not necessarily imply that no qualitative data can be obtained in the short run. It should be possible to have a "social enquiry" division within African departments of agriculture, which could not only conduct questionnaire surveys but also institutionalize the collection of in-depth case studies. Only a close liaison between the different types of natural as well as social scientists concerned with rural productivity can provide a sound basis for "Advancing Agricultural Production in Africa".

References

Abed, T.A.; Rahman, A. (1980) Leadership training for village women in Bangladesh. *Assignment Children* 51/52, 127-136.

Ahmed, M. (1980) Mobilizing human resources: the role of non-formal education. *Assignment Children* 51/52, 21-40.

Apthorpe, R. (1979) Social indicators and social reporting in Papua New Guinea. In *Measurement and Analysis of Progress at the Local Level.* Geneva; UNRISD.

Bryson, J.C. (1981) Women and agriculture in sub-Saharan Africa: Implications for development (an exploratory study). *Journal of Development Studies* 17, 29-46.

Dey, J. (1981) Gambian women: unequal partners in rice development projects. *Journal of Development Studies* 17, 109-122.

Dumont, R. (1969) *Tanzanian Agriculture after the Arusha Declaration.* Dar es Salaam; Ministry of Economic Affairs and Development Planning.

Epstein, T.S. (1984) *Differential Access to Markets and its Impact on Agricultural Development.* Hyderabad; ICRISAT. [In press].

ILO (1972) *Employment, Incomes and Equality.* Geneva.

Lipton, M. (1977) *Why Poor People Stay Poor.* London; Temple Smith.

Ndongko, T. (1976) Tradition and the role of women in Africa. *Presence Africaine* 99/100, 143-155.

Nelson, N. (1981) Introduction. *Journal of Development Studies* 17, 47-58.

Waddimba, J. (1979) *Some Aspects of Programmes to Involve the Poor in Development.* Geneva; UNRISD.

Chapter 85

Advancing Agricultural Production in Africa: a Personal Review

A. H. BUNTING (Conference Chairman)

University of Reading, Earley Gate, Whiteknights Road, Reading RG6 2AG, UK.

In this very exciting and crowded week, we have reviewed many of the things that need to be done, and can be done, to advance agricultural production in Africa, and particularly in sub-Saharan Africa, at the present time. This review is an attempt to assemble what I believe we have learnt into a classified check-list which Governments and agricultural scientists may find useful in deciding how they may best direct their efforts to ensure the progress we so urgently need. It is a strictly personal affair. It does not claim to represent any agreed conclusions or even a consensus among the participants, nor does it necessarily reflect the views of the Tanzanian National Research Council or the Commonwealth Agricultural Bureaux.

The list is no doubt incomplete. Moreover, not all the points in it are necessarily applicable in any one country at any one time. The countries of the continent, and even their individual provinces and districts, have different environments and histories and are at different stages along the path of development. They have different advantages and disadvantages. Nevertheless I believe that the check-list contains at least something for all of us, and so I offer it as a useful basis for future thought and action.

Advancing Agricultural Production

What is it that we seek to do, when we attempt to advance agricultural production? We are not thinking of advancing the output of all crops, all types of livestock products, all forms of timber and fuel. We are really thinking of how to apply our rural resources in order to get the best possible mix of activities and products to suit the needs of each nation and its people.

The contributions of rural people to development

Let us start by reviewing the contributions which rural people make, or can make, to the societies and nations of which they are a part. Their first task is to feed themselves and to produce raw materials for family needs, such as clothing, fuel, structures, household furnishings and equipment (including blankets, mats, baskets and containers), and for technical equipment (including string, ropes and hand tools). In the family and the community, they also help to feed the old, the ill, and the less fortunate; and, if they can, they build up reserves against hard times to come.

Next, they contribute to the supplies of food, fuel and raw materials for more distant people, in towns and cities, who do not produce these things for themselves. Third, they may produce what we still call cash crops (though in many countries food crops are now cash crops also) and animal products to aid the national balance of payments as exports or as substitutes for imports. Fourth, they may produce these and other products as raw materials for industries located in their own country. These are their chief roles as producers.

They have at least three more important roles. Once they are profitably engaged, as most or all of them wish to be, in the market economy, rural people can become a significant part, even the greater part, of the domestic market for the industrial products of their own country. They also create, or help to create, substantial quantities of fixed capital, in agricultural land, feeder roads, bridges, dams and irrigation systems, houses, stores and community buildings, all produced with local materials and labour at times when there is no more important or urgent work to do. Finally, through those family members who leave the rural scene (though they retain their ties with it), they contribute a substantial part of the skilled and unskilled labour force on which development in the non-agricultural sectors depends.

Let us think broadly, therefore, about agricultural advance, taking all these things into account, as a route by which rural societies may contribute more fully, and with more benefit to themselves, to national development in general. Development in the rural space, including development in agriculture and animal production, and non-rural, non-agricultural development, are two inseparable and interdependent sides of the same coin.

Food, money and markets

Though evidently food is not all, many people, in discussions like these, tend to concentrate on "food production". No-one questions that food is important. But we have to think about food, not as some sort of technical abstract, but as a marketed good. That distinguished biochemist, the late Professor Philip Handler, once set out three principles which he felt were essential starting points for discussions of food. The first is that we already produce, in the world, more than enough food to provide an adequate diet for all. Of course it is both produced and distributed unevenly, and much of it is fed to animals, but the total is in fact ample. The second is that nowhere in the world does a man go hungry who has money in his pocket. That is to say, the proximate cause of hunger is not a failure on the production side: it is poverty. Now poverty is a social and political phenomenon of control of resources. An agricultural scientist may help to lessen its effects, but he cannot do much about the social relationships which produce it.

Handler's third principle is of the greatest importance for those of us who seek to induce rural people to produce more. It tells us that nowhere in the world is there a farmer who will produce more than he wants for himself and his family unless someone else gives him something acceptable in exchange. Farmers, and particularly the small farmers about whom there is so much concern, cannot afford to be philanthropists and use their meagre resources to provide free lunches. Evidently, what we offer as scientists will be the more useful if it is linked to market demands and the realities of costs and returns.

Perceiving African problems

We who belong to Africa, like not a few of our non-African friends, seem to have got into a habit, in recent years, of looking inwards. We see African problems with exclusively African eyes, against the background of a distinctly selective view of African experience. We are concerned that our populations are increasing rapidly, as indeed they are. We accept that African tropical environments are "fragile", whatever that may mean, and unable to sustain an intensive or permanent agriculture. We are oppressed by our difficulties with droughts and floods.

Yet those of us who have been able to travel have seen, in other countries and other continents, environments which resemble ours, except that the human populations are very much more dense than anything we have yet even dreamed of. We have much to learn from the achievements recorded in other tropical environments, both wet and dry, including those of the Indian sub-continent, the tropical parts of Australia, the savannahs of Brazil, and the humid forest regions of south-east Asia. In some of these regions human populations are far more dense than ours, though it is true that they are now increasing far less rapidly. India's remarkable agricultural achievements of the past twenty years, which have a long way yet to go, are largely based on the management of terrain, land and water, in environments and on soils not unlike those of Africa, coupled with the widespread use of bullock power and the development of increasingly intensive crop and animal production systems appropriately adapted to the circumstances. In Malaysia, development policy has based rural advance on tree crops producing export products, in systems in which both plantations and smallholders play their parts. Australia has demonstrated both technical and commercial possibilities, particularly in seasonally-arid tropical environments.

Indeed, in our own continent we know that many environments can support highly productive crop and animal production systems. Though each has special features, I need only mention the settled indigenous agricultural systems of Ethiopia, the large-scale semi-mechanized production of cotton, sorghum and other crops in the east-central Sudan, and the agricultural development of the Ivory Coast, Kenya, Liberia, Malawi, Zambia and Zimbabwe, to remind you of what has been technically and commercially feasible in African environments.

Of course, each of the examples I have mentioned, whether in Africa or elsewhere, has attracted criticism, often on social or political grounds. But let us not forget, as some of the more doom-laden critics all too often do, the positive technical achievements which show what can be done.

Factors of Change

This section presents seven main groups of factors which offer opportunities or constraints for agricultural and rural progress. They are: (1) volume of effective demand for rural products; (2) output delivery systems; (3) the resources which producers command; (4) production methods; (5) knowledge systems; (6) the policies and practices of Governments; and (7) international relationships.

Effective demand

Without a sufficient volume of effective demand, output will not increase. The volume of effective demand depends formally on the number of would-be purchasers in the market place and the average demand per purchaser. So the number of purchasers becomes very important. If we can imagine a country where Government policy has somehow managed to hold 80% of the population on the land, then only 20% of the population are available to act as customers for any surpluses the rural majority may produce – export crops excluded, of course. This is not likely to offer a very satisfactory margin for farm business.

The movement of people away from the land can have many positive benefits, whatever negative consequences are perceived by the casual visitor or the development expert who rides past in his Mercedes and sees the shanty towns between the airport and the big city. Though some of the reasons why people leave the countryside may be heart-rending, most of the migrants appear to be able somehow to eat and to clothe themselves. More people seem to find a place on the carrousel than have officially-recognized jobs. But they are there, and they provide not only an important part of the market for those who have stayed behind, but often a supply of remittance money which helps the family back home to pay for modern inputs on the farm.

So while I accept that it is possible for things to go badly wrong, I have never seen any particular virtue in a policy that seeks to hold people on the land. I see many virtues in a policy that seeks to diversify the economy and to make all its parts both larger and more fully complementary to each other. This is far from easy to do, but I suggest that we have to try.

The other component of effective demand relates to the income level of the non-rural population. In Africa, by and large, we are not held back by the class structures which so severely impede progress in some countries of Latin America and Asia. But we do need new economic activities, located so far as possible in the rural space itself, as the rural development policy of Zimbabwe indicates. It is these activities that put money into the non-agricultural parts of the economy and so increase the effective demand for rural products.

The result of all this will of necessity be a society very different from the traditional African societies which many of us have known and admired. Development changes society in many ways, and the process is usually turbulent and untidy. The road of development is hard.

Output delivery systems

Even if there is a sufficient volume of effective demand, rural producers will not be able to respond to it, or even to perceive it, unless what I call the output delivery system is sufficiently effective. The output delivery system consists in part of physical infrastructure: road, railways, waterways; the vehicles that move on them, and the fuel, spare parts, competent operators, and maintenance necessary to operate them economically. It also includes markets, and market masters who can keep track of prices and quantities. Beyond the market it includes storage facilities, processing and packaging plants, and commercial arrangements for wholesaling and retailing – the complete apparatus, usually taken for granted in the richer countries until a forty-ton juggernaut crashes into the front garden, which conveys products from the metaphorical farm gate to the ultimate user or consumer, and the signals of the marketplace from the user or consumer to the producer.

I do not need here to spell out the extreme importance of the storage and processing components of the output delivery system. Storage is essential for all commodities which are harvested not more than once or twice a year but are required all the year round. It is also essential as part of the system of insurance against climatic or political adversity. As to processing, let us remember that almost nothing we eat comes straight from plant or animal to the bowl, and that time spent on domestic processing can often be more profitably used elsewhere.

The output delivery channel must also contain important commercial management features. They include banking and credit, the management of costs and prices, understanding of the home and overseas markets, management of exports and shipping and of insurance. Whether it is publicly or privately controlled, none of this will work well without competent managers and reliable and economical communications and information systems, including telex, telephone, radio, and postal services.

In many, perhaps most, African countries the output delivery system is physically defective and administratively ineffective, even corrupt. I am not talking exclusively about parastatal marketing boards – the private sector impedes output delivery in some countries too.

Resources for production

Given that the farm family feels the pull of the market place through the output delivery system, what resources do they command with which to respond and make some money? The conventional resources for farm production are grouped under the main headings of capital, land and labour. I like, myself, to add time and attention, which I am not easily able to see as components of labour. I like also to think about the weather, climate, water and soil separately, rather than as descriptors of land.

(1) *Weather, climate and water.* Weather and climate were well reviewed in the natural resources symposium. I think there were two important conclusions – first that we do not have enough weather and climate information, and do not make good enough use of what we have. The development of AGRHYMET at Niamey (Chaper 57) was not merely a necessary response to the Sahel drought: it is already an example of what we need and what could be done elsewhere. Weather hazards are not confined to the Sahel.

Second, we make far too much of the uncertainty of African climates. All climates are uncertain, including the seasonally cold and dry climates of much of the temperate zone. The first response is to develop farm systems which minimize the effects of climatic variation – such as are to be found, in Africa and everywhere else, wherever populations have been sustained by agriculture through thick and thin for many generations. These systems may well be imperfect, but there must be much which we can learn from them.

However good the systems, fluctuations in output are inevitable. All African farmers know that, and all of them respond as best they can through traditional crop storage systems, to which their crop varieties are appropriately adapted. There are in fact four classes of things which can be stored – products, animals on the hoof, water, and money, gold or jewels hidden below the bed or under the floor, or if possible in a reliable bank.

We in Africa have a long way to go in the small-scale harvesting and storage of water which is so important in other continents. Yet it is not foreign to African tradition. Small run-off tanks or hafirs are communally controlled along the seasonal migration routes of transhumant graziers in the Sudan. When we cleared impenetrable *Commiphora* scrub at Kongwa, at the southern end of the Masal steppe 250 miles south of here, we found old hafirs at an average of one to every two square miles. They did not seem to be associated with old agricultural sites, so we concluded that they were intended for watering stock, not for irrigation.

If a catchment is equipped from the top down with small scale water harvesting and storage of this sort, any surplus which runs down the valley may be captured by larger dams. Because the smaller dams will capture much of the silt, the larger structure may well accumulate silt less rapidly. If the small dams are under communal control, it may be that the community will remove the silt and put it to good agricultural use.

There are in addition many other ways of handling local water supplies – artificial small-scale catchments filled by local runoff on impermeable clays, pumping from wells, and the fuller use of swamps, marshes, and seasonally wet depressions. We do not do enough about these things. Perhaps, since we are so concerned about the risk of drought, it is high time we did.

(2) *Terrain and soils.* In total, Africa's land resources are sufficient to meet future needs for food. In some areas, land reserves are such that the area cultivated could be expanded to provide future needs and more, even at present levels of technique, let alone the more productive levels which seem certain to be technically possible. In other areas, however, the limits of cultivable land have already been, or soon will be, reached, and land resources are too small to provide for the needs even of the present population, using the present methods of production. It is in these areas, too, that future population growth seems likely to be most rapid. In such areas, most of the increases that will be needed in future will have to come from more intensive and productive use of the land which is already cultivated. These differences in land endowment will have to be taken fully into account in plans to advance agricultural production.

Many of us have been concerned about the losses of soil by erosion, particularly as shifting cultivation systems become permanent. In fact, in many regions of Africa permanent systems have evolved which include important means of control of erosion, even though they may be far from perfect. Minimal tillage and cropping systems which maintain plant cover over the soil are widespread; and they seem to me to be essential if larger scale engineering methods adapted from North American practice are to succeed. Indeed, experience at the International Institute for Tropical Agriculture (IITA) in Ibadan suggests strongly that minimal tillage by itself can protect an erodible soil from damage by severe erosive rain storms even on slopes as steep as 10%.

To produce larger yields per unit of area and of time, as well as to maintain a more or less continuous cover and accumulate the crop residues that are so important in minimal tillage, crops need adequate supplies of plant nutrients. Some observers have suggested that because many farmers in developing countries cannot afford fertilisers at present, they must therefore use other sources of nutrients, such as organic manures and biological nitrogen fixation. These sources are no doubt important, but I see no guarantee that they will provide all that we shall need in the future. As several speakers told us, we have to grapple with the problems, not evade them. As crops are marketed and cash begins to flow towards rural producers in Africa, they will both need, and become able to afford, more fertilizers. Governments and commercial agencies will no doubt be prepared to respond even more widely than in the past. Meantime we must get on with the experimentation, determine the needs of different soils and cropping systems, and find out more about how best to formulate and to use fertilizers in African conditions. There is really no other way out. Fortunately a good deal of help is available internationally for work of this sort, and we can learn from the experience of other continents.

(3) *Fragmentation and tenure.* In some areas of the continent, population pressure has led to holdings too small and scattered to allow the potential of the environment to be realized or to provide sufficient support for the people. In these and other cases access to land may be restricted, for part of the population, by existing patterns of control over land and the social and legal structures that support them. Ways of developing more productive tenure arrangements will have to be found, and this is bound, at least in part, to be a political process which may be have to be supported by the threat or the reality of force. Land reform is not an academic pursuit. At the same time, Governments will wish to ensure that efficient production systems are not damaged or lost to the nation in the process. In not a few instances, in other continents at any rate, politically-attractive "land-to-the-tiller" policies have led to loss of important productive capacity, particularly in large irrigated developments, since the new users or proprietors have lacked resources, skills and organization to make good use of their newly acquired land.

In planning more productive uses of terrain, land and water, the appropriate planning unit is evidently the catchment (both large and small) rather than a purely administratively defined area.

(4) *Labour and power.* In most of Africa at the present time, land is not the main factor limiting output. We are however increasingly short of power at critical times, for sowing, weeding and

harvest. The movement of people away from the land accentuates this shortage. The power of draught animals could help, but like human power it is difficult to concentrate it in order to perform tasks rapidly and at the most suitable times. Many regions of Africa have long since moved to mechanical power. The tractor is with us to stay, but it is a costly power source and can be sustained only by a profitable agriculture. It also requires spare parts, maintenance and skilled people. We may take heart from the large scale semi-mechanized private-sector farming of the eastern Sudan rainlands, which has managed somehow to sustain more than 2M ha of land in profitable production for many years in spite of seemingly impossible logistic and technical difficulties.

This experience, and the success or at least the emergence of larger scale mechanized systems, financed by merchants and other entrants from outside the traditional rural scene, may suggest to us that Africa's agricultural future need not rest exclusively on the progress of small-scale producers. In many countries, we have land enough, and we need the products; and the modern systems are likely to prove more attractive to the educated sons of farmers than the digging stick or the hoe. None of this will lessen inequity, but it may promote development and will therefore no doubt be considered very seriously.

That power is our main limiting factor, while land area is not, suggests that as scientists we should think about yield per unit of energy input at least as much as yield per unit area. Ultimately, of course, yield per unit area must be important, since one of the chief ways of increasing yield per unit of energy is to produce larger yields per unit area.

(5) *Capital, cash and credit.* It is customary to see the question of capital resources largely in terms of credit. Credit is indeed important; and we heard a valuable account of the work of Barclays Bank in providing small loans to rural producers (Chapter 5). Perhaps we should also have arranged for an account of the rural credit arrangements supervised by the Bank of India, which has done so much to enable small producers to participate in the Green Revolution. But most rural people would probably agree that the main source of finance for them should be sales off the farm. They want cash, not debts; and moreover they have no way of repaying debts until cash begins to flow towards them from the market – or from the outposted family members whose remittances can be so important in stimulating rural change.

(6) *Competition for scarce resources.* Many African rural communities are poorly endowed with resources, and they have to allocate them among many competing uses. Labour and cash have not only to be allocated among crops and between crops and livestock, but also between the agricultural sector of the life-system and other sectors. Time and labour spent collecting fuel, fetching water, grinding grain or pounding yams, cannot be used for production, or for looking after the house and the family. The systems in which we investigate resource allocations and constraints have to take account of the whole of this management problem – which rural African families seem to surmount with remarkable competence. I have often wondered whether I could do as well as they do in maintaining life, dignity and cheerfulness with such limited resources and so many competing calls on them.

(7) *The environmental resources of tropical Africa.* Correctly used, the environmental resources of Africa as a whole, whatever their weaknesses, appear to be more than adequate to produce all the agricultural products we need and our export customers require for a long time to come. We have to learn to use them more productively, but there is no room for pessimism. The evidence for this view is in part in the splendid assessment produced for FAO of the human carrying capacity of the continent, and in part in the comparison with more densely populated but otherwise similar regions in other continents. We must look elsewhere than among the environmental constraints for the principal sources of our difficulties.

Methods of production

(1) *Studies of existing systems.* We hear a great deal about the need for appropriate technology in developing countries. Appropriate systems of crop and animal production are in fact to be found everywhere in Africa – on the lands and in the herds of the rural people. They are appropriate, in

the sense that they sustain both life and population growth, though they may not be sufficiently productive to meet the needs of the future. But they are adapted to the environments and to the current, or at least recent past, needs and resources of the people.

The first task, in this domain, is to determine the rationale of the existing systems. The idea is not new: the word "rationale" was used in this context around 1890. To make the existing systems, which are inevitably the starting point, more productive, we have to understand how they work now, and determine what prevents them from working more productively.

Of course, we need novelties. I have at times been so impressed by the beauty of the adjustments and adaptations of an existing system that I have tended to forget that to understand the system is not enough: we have to change it so that the existing resources, and any others we can find at an acceptable price, become more productive.

(2) *Plant and animal improvement.* The first gift agricultural science has to offer to a crop producer is a range of improved varieties, adapted to the environment, including the time tables of cropping which it imposes, and with at least some measure of inbuilt tolerance towards as many as possible of the pests and diseases which are important in his locality. I include here the storage pests and diseases which are all too often neglected by those breeders whose eyes are too exclusively fixed on yield per hectare. With this the breeder may combine greater potential response to more fertile conditions and to irrigation, and/or tolerance to environmental stress, together with culinary and aesthetic qualities appropriate to the producers as well as their more distant customers on the other side of the market place. Breeding for multiple objectives is essential.

However, none of this is likely to be of much use, even in self fertilized crops, without a seed industry and effective seed legislation. It may also be necessary to protect the investment in plant breeding, public or private, by plant variety rights legislation. In the UK, this has the effect that at least a substantial part of public-sector plant breeding pays for itself. As a tax-payer, I think this is rather a sound idea. It interests me to find that plant breeders rights are presented as a device of the multinationals to oppress the developing countries. They may in fact be required in developing countries in order to protect the public investment against the industry, which is otherwise free to take advantage of the public effort virtually free of charge.

I am not qualified to add anything to Mr Ellis' review of livestock improvement in his report on the animal production symposium (Chapter 55).

(3) *Protection.* Much of the two symposia on crop production and on animal production was devoted to protection, which is consequently well covered in the corresponding reports. Some leading topics may however be useful here.

There is no lack of diseases and pests of both crops and livestock; and no doubt all are scientifically interesting and merit study. But the producer is not interested in science or in control so much as in output and profit. In all but the most frankly damaging cases, the producer needs information about the costs of the diseases that affect his crops and stock and the profits, particularly in the short term, of treating them sufficiently to keep losses down to an acceptable level. This makes the design of a protection system very much more than a combination of control recommendations for all the economic organisms at risk. It requires a combination of economics with knowledge of the biology and methods of production of the economic organisms and of the biology, "epidemiology" and possible methods of control of the pests and diseases. Evidently this combination is well advanced for animal diseases and pests but less so for the pathogens and pests of crops.

In important cases we still lack information about the extent of damage and even the biology of the pest or pathogen. Perhaps the most notable instance is that of the nematode pests of tropical economic plants. From those cases which have been adequately studied, we may expect attacks to be widespread and their effects serious, but we know all too little.

We will all, I think, support Professor Hirst's plea (Chapter 40) for more work on the biology and control of weeds, particularly because the control of weeds is one of the most labour-demanding tasks of the crop cycle. However, not all of the control need be mechanical or chemical – shading by economic plants is an important means of controlling weeds in mixed and sequential systems.

I imagine we will all accept that the goals, in protection, are integrated management systems which are based on productive but resistant or tolerant economic organisms and use an appropriate mix of biological and chemical means to achieve the appropriate degree of control. This is easy to write, but far more difficult to put into practice, particularly as no two holdings or producers are alike. Whatever system is devised, it seems likely that it will work better if the producer knows enough about it to participate – as Indian cotton farmers do who scout their own crops and know when, and when not, to spray.

Knowledge systems

The advances in agricultural production which we need, and which we believe are, or can be brought, within our reach, depend substantially on the fuller use of knowledge, new and old. In Africa, as in many other regions, knowledge systems are incomplete and fragmentary. I want at this point to consider the structure of knowledge systems and how they are used.

Knowledge about agricultural production is stored in libraries, records and archives, including the old filing cabinets, and in the minds of individuals. It is increased by enquiry processes, which range from humble investigations to more exalted research. Where the motive for the enquiry is to achieve or approach some practical end, the adjective "applied" is added, leading to interesting but often unproductive semantic debate. Knowledge is delivered through processes which we separate into education, training, advice, extension and so on. All of this ideally forms a single system, but in practice it tends to be divided, in the affairs of nations, between ministries of agriculture, forestry, animal production, animal health, education, science and technology, and rural development. Moreover, the contribution of the rural people is all too often excluded: "research" is seen as the lead agency, and the "farmer" as the not-always grateful recipient of guidance from on high.

Yet rural people have much to contribute; and it is in part their needs that the whole system is intended to serve. (It also serves the purposes of Governments, which have to think about the needs of non-rural people and about the balance of payments, and usually have to pay for the whole system as well). In some African countries, the most effective advisers of rural people are successful producers themselves.

Moreover, at the other extreme, the knowledge system is international; and its links within the nation are provided largely by books, journals and specialised information systems like those of the Commonwealth Agricultural Bureaux (CAB). It includes the international agricultural research centres, which have been so well represented at this Conference, and whose contributions to national research systems are proving more valuable year by year.

Most of our nations have much yet to do to develop national knowledge systems for agricultural development which fully serve their needs. I can do no more here than point to the elements that need to be brought together. The CAB, the Organization of African Unity (OAU), the international agricultural research system, and FAO all have services to offer. The initiatives will have to come from the agricultural science communities of the nations.

The policies and practices of Governments

At the end of the day, it is only the rural producers who can actually advance agricultural production. But the extent to which they succeed depends on the policies and the practices of Governments. Any failure of development is a failure of government. But at the best, the tasks confronting Governments are extremely difficult, and the human resources available, at both the political and executive levels, are usually limited in quality, and often in numbers also. It takes time to build up effective traditions in politics and in government, and few African nations have had enough time.

Nonetheless, it is necessary to record that incorrectly chosen price levels, foreign exchange rates, and foreign currency allocations, and incorrect decisions about investment, all hold back both rural and non-rural development. When to these we add the repeated attempts to manage agriculture, and often to carry out commerical production, by bureaucratic processes, we can discern the roots of not a few disasters. The Groundnut Scheme, which brought me to this country many years ago, was a hard school of experience. It still attracts a long queue of eager pupils.

Finally, I should fail in my duty if I did not refer to corruption, both public and private. Against the power of the corrupt, the rural people are defenceless; and until it is abated, our chances of making rural progress are slim.

International relationships

(1) *Trading relations*. Many African nations have difficulties in exporting traditional agricultural products and in managing their balances of payments. These matters were fully covered in Mr Belshaw's general statement (Chapter 2) and there is no need to add anything here, except to suggest that (as in other continents) part at least of the future of many nations may be with new trading partners, particularly in other developing countries, and with new products. We do not have to be exclusively tied to the arrangements of the past.

(2) *Aid*. Substantial amounts of capital and recurrent aid are allocated each year for development in agriculture and in the rural space in African nations. There are however many difficulties in spending aid funds wisely and effectively, for lack of basic infrastructure and of skilled and experienced manpower. One African country I have visited recently receives aid from more than one hundred agencies, which suggests also that there may be real difficulty in orchestrating so many contributions. However, the advances in physical infrastructure which one sees in so many countries appear to be having substantial effects, reflected in not a few cases in output. One may hope that the benefits will be used in the first place to maintain the new facilities, and not lost in less productive secondary directions.

Substantial amounts of aid have also been provided for manpower development and for education and training institutions. I feel that in some countries training has already run ahead of employment and financial possibilities. In one, staff costs already consume so much of the research budget that research itself has very nearly ground to a halt. Yet more people continue to be trained. This may well lead trained people to emigrate, with substantial financial loss to their own nations in spite of the remittances they send home.

(3) *International agricultural research*. Most African nations have associations with one or more of the international agricultural research centres funded through the Consultative Group on International Agricultural Research (CGIAR). Many of us who are associated with the centres hope that these will develop so that the centres consciously work together, as well as with Governments, to advance the national research systems as wholes, and not merely the national capabilities for research on particular commodities. The International Service to support National Agricultural Research (ISNAR), which is part of the system, could help here. In addition some feel that the donor agencies, many of which support CGIAR, could make fuller use of the centres in their bilateral and regional programmes, and so increase the returns on their investments in the system. I hope also that Governments and national research agencies will increasingly take the lead to ensure that the contributions of donors strengthen the national research system as a whole, whatever the special preferences of particular donors.

Some Suggestions for Agricultural Research

This section summarizes the principal suggestions for research which have arisen in earlier sections of this review. It is surely incomplete, but I hope that it will stimulate new thought about the contributions of our profession to development in agriculture and in the rural space in Africa.

Economics and social sciences

In much of the Western tradition of agricultural research and education, the social sciences and the natural sciences usually stand apart from each other. Evidently a much closer relationship is needed to strengthen the analysis of existing production systems, to identify opportunities and constraints, and to indicate the economic and social costs and benefits of technically feasible development options. This relationship is developing fairly well in the international system: it must advance in national systems also. The task is not easy, because the human and natural sciences differ so much in content and in ways of thinking and working, but many young and lively minds find their encounters across the great divide both exhilarating and productive. The relationship should therefore begin in the universities – in spite of the discomforts it may cause for some of the more staid among the older hands.

The work of the human scientists is not confined to farm systems studies. It should include studies of endowment and competition for resources between competing sectors of rural life systems, of rural/non-rural links of all kinds, of domestic and foreign markets and prices, of demand and of the effectiveness of output delivery systems.

Agro-ecological and farm systems studies

The object of studies of farm and rural life systems is to know and understand the rural scene as it is now, to identify objectives, resources and constraints, and to suggest new needs in research and new paths for development. We are fortunate that much significant work on farming and rural life systems and their environmental, social and economic relations has been done in Africa by such great originators as Trapnell, Gluckman, Audrey Richards, Platt, Margaret Haswell, de Wilde, Collinson, Uchendu, Bruce Johnson and his colleagues, Anthony, Yudelman, Norman and Eicher. We have a sound basis on which to develop studies of agro-ecological zones and recommendation domains. Perhaps the most urgent task will be to test the methods of rapid survey already available and to learn how to adapt them to different circumstances.

These studies should include comparative work in similar environments of other African countries and of other continents, taking into account the historical changes in the recent past and the trends for the future.

Similar remarks apply to animal production systems, as the record of the symposium on that subject shows.

Resource assessment

Work of this sort leads, in one direction, to the preparation and assessment of inventories of rural resources and the identification of priorities in the development of regions and of communications, and in the location of agricultural research. The work of Higgins, Kassam and their colleagues associated with FAO sets an example here (Chapters 8 and 64).

Agricultural climatology and meteorology

Even though everyone knows that variations in output are caused more by variations in weather than by any other single factor, these subjects are underdeveloped in the agricultural science of all countries, and not least in Africa. They offer to national meteorological services, hitherto largely preoccupied with forecasting for aviation, opportunities for new services of great importance for development.

Hydrology

Studies of agricultural climate and weather will be essential for the development of work on the fate of rainfall, and of the management of the national terrain to make the most profitable use of water. Here too comparative studies of other African countries and of countries in other continents will be advantageous.

Timber and fuel

However severe food problems may be in African nations, in many the fuel crisis is at least as severe. We need more studies of species and provenances of woody plants adapted to different environments and patterns of needs, and of systems in which they can be produced. Some of these may be grown in agroforestry systems, and so they may need appropriate adaptations.

Soils and fertility

As populations have become more concentrated, and the fallow intervals of shifting cultivation have become shorter, other means of restoring and sustaining the fertility of the environment have become necessary and have indeed been found by African rural people. We do not know enough about these achievements, and we need to learn more. The methods needed in the future will include measures to control weeds, pests and diseases, but in many cases the most important tasks will be to offset the effects of changes and deficiencies in the chemical and physical attributes of the soil. Studies of responses to added nutrients (including sulphur and "minor" elements as well as nitrogen, phosphorus, potassium, calcium and magnesium) will have to be extended and completed, and associated with inventories and maps of agroclimatological zones. The work of the International Fertilizer Development Center on forms of fertilizers appropriate to tropical soils and crops (Chapter 31) will have to be extended and tested collaboratively in Africa.

Studies of biological nitrogen fixation will no doubt be developed at those few centres in Africa which are appropriately staffed and equipped. It seems likely that additional centres will be needed.

Power

Since labour and power are limiting resources in much of African farming, we need to advance appropriate engineering studies of methods and equipment as well as of prime movers, new and old. Collaboration with international centres, and perhaps especially IITA and IRRI, are likely to be valuable here.

Plant breeding

In collaboration with the appropriate international centres wherever possible, we shall need to develop plant breeding for multiple objectives (pest and disease resistance, including resistance to storage pests and fungi; suitability for transport and storage; suitability for the preferences and requirements of consumers and users; large potential yield and adaptation to stress and/or to irrigation and more fertile environments).

Research on seeds and propagation

Without effective seed and propagation industries the effects of plant breeding are bound to be limited. Most African countries lack such industries. We shall need to develop in each nation or region suitable and economically effective methods of large scale rapid multiplication and of seed production, particularly where a hybrid or population improvement route has been chosen, and we shall need the appropriate procedures for storage, transport and testing. This is development rather than research, but it is science based, and it needs strong professional support.

Storage and processing

Storage of any product on a commercial scale needs professional services for design, management and monitoring. In part, the work may be based on studies of the rationale of traditional storage systems, but it will probably take new research to develop reliable procedures for roots and tuber crops and for fruit and vegetables. As urbanization has advanced, industrial processing (albeit sometimes on a "cottage" scale), before or after storage has become increasingly important: it too requires professional support.

Protection

Methods of protection against pests, diseases and weeds (pest, disease and weed management), designed to attain acceptable levels of loss or damage, need to be built into the production systems. They include resistance breeding, and they may require both biological and chemical interventions. Methods of loss assessment and of forecasting will be needed; and producers may also need scouting methods and advice. Agricultural meteorology has a most important contribution to make. More basic biological and economic information is needed on many pests and pathogens, and particularly on nematodes and weeds, as well as on methods of control.

Knowledge management

Finally we must make far fuller use of existing information about the agricultural environment, economic plants and animals, and systems of protection and production. In most African countries, the older information is usually inaccessible because it has never been assembled or collated into a usable form. So "new research" all too often repeats the old and we make little progress because we continually return to the starting point as staff changes. In part this is a task for individual scientists, but they cannot do it unaided. This may be one of the most valuable future tasks for the Commonwealth Agricultural Bureaux.

Participants

Abassa, Dr K.P, University of Florida, Gainesville, Florida 32611, USA.
Abernethy, C.L., 6 The Mint, Wallingford, Oxfordshire, UK.
Abubakar, A., P.O. Box 6115, Arusha, Tanzania.
Adero, W.E., Ministry of Agriculture & Livestock Development, P.O. Box 30028, Nairobi, Kenya.
Agble, Dr W.K., Director, Crops Research Institute (CSIR), P.O. Box 3785, Kumasi, Ghana.
Akilmal, B.M., National Ranching Company, P.O. Box 9113, Dar es Salaam, Tanzania.
Akobundu, Dr I.O., International Institute of Tropical Agriculture (IITA), P.M.B. 5320, Ibadan, Nigeria.
Alexander, D.M., c/o CDO Office, Town Council of Voi, Private Bag, Voi, Kenya.
Allard, Ms G., P.O. Box 159, N.C.D.P., Zanizibar, Tanzania.
Allen, C.J., Ministry of Agriculture, Hope, Kingston, Jamaica.
Allen, Dr D.J., Centro Internacional de Agricultura Tropicale CIAT, Att. 6713, Cali, Columbia.
Alluri, K., International Institute of Tropical Agriculture (IITA), Oyo Road, P.M.B. 5320, Ibadan, Nigeria.
Anteneh, A., International Livestock Center for Africa (ILCA), P.O. Box 5689, Addis Abada, Ethiopia.
Andersen, B., Coffee Research Institute, Gyamungu, Box 3004, Moshi, Tanzania.
Anthonio, Prof. Q.B.O., Dept. of Agricultural Economics, University of Ibadan, Ibadan, Nigeria.
Atkinson, Mrs A.A., University of Reading, Earley Gate, Reading RG4 2AG, UK.
Attere, A.F., c/o ILRAD, Box 30709, Nairobi, Kenya.

Bals, E.J., Micron Sprayers Ltd, Three Mills, Bromyard, Herefordshire, HR7 4HU, UK.
Barrett, A.T., Planning & Evaluation PMO, P.O. Box 1521, Mwanza, Tanzania.
Baum, Dr H., P.O. Box 30467, Nairobi, Kenya.
Beets, Mr W., International Center for Research in Agroforestry (ICRAF), P.O. Box 30677, NBI, Nairobi, Kenya.
Bekele, D., c/o United Nations Development Program, P.O. Box 9182, Dar es Salaam, Tanzania.
Bekure, Dr S., International Livestock Center for Africa (ILCA), P.O. Box 46847, Nairobi, Kenya.
Bellamy, Miss M.A., Office-in-charge, Commonwealth Bureau of Agricultural Economics, Dartington House, Little Clarendon
              Street, Oxford OX1 2HH, UK.
Belshaw, D.G.R., School of Development Studies, University of East Anglia, Norwich NR4 7TJ, UK.
Bendera, O.M.S., Ministry of Planning & Economic Affairs, P.O. Box 9242, Dar es Salaam, Tanzania.
Birichi, E., Kenya High Commission, 45 Portland Place, London W1, UK.
Bitanyi, Dr H.F., Tanzania National Scientific Research Council, P.O. Box 4302, Dar es Salaam, Tanzania.
Birke, Dr L., Ethiopian Science & Technology Commission, Box 2490, Addis Ababa, Ethiopia.
Boag, Ms C., c/o Land Use Component, P.O. Box 591, Tabora, Tanzania.
Bolam, P., Barclays Bank, 94 St. Pauls Churchyard, London EC4, UK.
Bridge, Dr J., Commonwealth Institute of Parasitology, 395A Hatfield Road, St. Albans, Herts. AL4 0XU, UK.
Brown, Ms L., Mjwara/Lindi Ridep Project, P.O. Box 608, Mjwara, Tanzania.
Browning, L.G., Bayer Zimbabwe (Pvt) Ltd, P.O. Box AY78 Amby Harare, Zimbabwe.
Buck, N.G., 1 Market Place, Bedale, North Yorkshire DL8 1ED, UK.
Budden, M., Tanzanian Seed Co. Ltd, P.O. Box 939, Arusha, Tanzania.
Bujulu, J., Tropical Pesticide Research Institute, P.O. Box 3024, Arusha, Tanzania.
Bunting, Prof. A.H., 7/8 Q Building No. 4, University of Reading, Earley Gate, Reading RG4 2AG, UK.
Button, R.G., Tan-Can Wheat Project, Box 6160, Arusha, Tanzania.

Calverley, D.J.B., Tropical Development & Research Institute, London Road, Slough, UK.
Canhao, Sister J., Dominican Convent, Harare, Zimbabwe.
Caveness, Dr F.E., International Institute of Tropical Agriculture (IITA), P.M.B. 5320, Ibadan, Nigeria.
Chari, A.V., c/o FAO Office, P.O. Box 2, Dar es Salaam, Tanzania.
Chavunduka, Dr O.M., 2 Kenilworth Road, Highlands, Harare, Zimbabwe.
Chelangwa, M.M., May & Baker Ltd., P.O. Box 2111, Dar es Salaam, Tanzania.
Chemponda, S.F., Bank of Tanzania, P.O. Box 3043, Arusha, Tanzania.

Chilagane, A., A.R.I. Arusha (Wheat Project), P.O. Box 6024, Tanzania.

Child, G.S., Forest Resources Division, FAO, Rome, Italy.

Chimphamba, Prof. B.B., University of Malawi, Bunda College of Agriculture, P.O. Box 219, Lilongwe, Malawi.

Chinganga, H., FAO Ifakara, Private Bag Ifakara, Tanzania.

Chipepa, Dr J.A.S., Livestock & Pest Research Centre, P.O. Box 49 Chilanga, Lusaka, Zambia.

Clark, Dr J.S., Agriculture Canada, Ottawa, Ontario, Canada.

Coles, B.O., FAO/UNDP/GRT, Project Tanzania Livestock Development Phase 2, P.O. Box 2, DSW, USA.

Conklin, Dr F.S., Office of International Agriculture, Oregon State University, Corvallis, Oregon 97331, USA.

Corner, J., Care-Uganda, P.O. Box 7280, Kampala, Uganda.

Crosby, Dr D.G., Canadian High Commission, 1 Grosvenor Square, London, UK.

Cross, E., Cold Storage Commission, P.O. Box 953, Bulawayo, Zimbabwe.

Cross, Mrs J., 1 Caithness Road, Hillside, Bulawayo, Zimbabwe.

Cutinha, C.A., Arusha International Conference Centre, Arusha, Tanzania.

Dagg, Dr M., ISNAR, The Netherlands.

Darnhofer, Dr J., International Centre for Research in Agroforestry (ICRAF), P.O.B. 30677, Nairobi, Kenya.

Darvall, Ms A.E., Australian High Commission, P.O. Box 2996, Dar es Salaam, Tanzania.

David-West, Dr K.B., Federal Livestock Department, Lagos, Nigeria.

Davidson, A., United National Development Program (UNDP), 30 Waterside Plaza, New York, New York 10010, USA.

Davis, J.B., c/o United Nations Development Program (UNDP), P.O. Box 9182, Dar es Salaam, Tanzania.

Deleeuw, P.N., International Livestock Center for Africa (ILCA), Box 46847, Nairobi, Kenya.

Devik, C.L.E., P.O. Box 400, Mbeya, Uyule Agricultural Centre, Tanzania.

Dhruva Narayana, V.V., Central Soil & Water Conservation Research & Training Institute, P.O. Dehradun 248195, India.

Dissanayake, Prof. A.B., University of Ruhuna, Matara, Sri Lanka.

Doggett, Dr H., 15 Bandon Road, Girton, Cambridgeshire CB3 0LU, UK.

Donald, R., FAO, Box 2, Dar es Salaam, Tanzania.

Dugdale, G., University of Reading, Reading, Berkshire, UK.

Eames, Mrs S.N., Commonwealth Institute of Parasitology, 395A Hatfield Road, St Albans, Hertfordshire AL4 0XU, UK.

Ellis, Mrs M., c/o Department of Agriculture & Horticulture, University of Reading, Earley Gate, Reading RG4 2AG, UK.

Ellis, P.R., Department of Agriculture & Horticulture, University of Reading, Earley Gate, Reading RG4 2AG, UK.

Epstein, Prof. T. Scarlett, AFRAS, University of Sussex, Brighton, Sussex, UK.

Fatteh, Dr M.A., Agricultural Research Corporation, P.O. Box 126, Sudan.

Filshie, Dr B., CSIRO Centre for International Research Co-operation, Canberra, Australia.

Franco, G., Agronomist, I.A.O., Florence, Italy.

Gessesse, K., University of Dar es Salaam, Faculty of Agriculture, Forestry & Veterinary Science, Box 3022, Morogoro, Tanzania.

Gezahagh, T., Research Assistant, Ethiopian Science & Technology Commission, Box 2490, Addis Ababa, Ethiopia.

Goetz, E., c/o World Bank, P.O. Box 30572, Nairobi, Kenya.

Golden, B., 19 Sandymount Avenue, Dublin 4, Ireland.

Golob, Dr P., Tropical Development & Research Institute, College House, Wright's Lane, London SW8 5SJ, UK.

Gombe, A., CIRO Africa, P.O. Box 6115, Arusha, Tanzania.

Green, J.H., ULG Consultants PLC, c/o A.R.I. Lyamungu, P.O. Box 3004, Moshi, Tanzania.

Greenland, Dr D.J., International Rice Research Institute, Los Banos, Philippines.

Griffiin, J., P.O. Box 318, Chake Chake, Pemba, Tanzania.

Glasford, R.M.D., Mumias Sugar Co., Private Bag, Mumias, Kenya.

Gooch, P.S., Assistant Director, Commonwealth Institute of Parasitology, 395A Hatfield Road, St Albans, Herts. AL4 0XU, UK

Goodchild, Dr A.V., Livestock Production Research Institute, Taliro, Private Bag, Mpwapwa, Tanzania.

Greathead, Dr D.J., Assistant Director Commonwealth Institute of Biological Control, Silwood Park, Ascot, Berks, UK.

Hahn, S.K., International Institute for Tropical Agriculture, PMB 5320, Ibadan, Nigeria.

Hailu, M., International Livestock Centre for Africa, P.O. Box 5689, Addis Ababa, Ethiopia.

Haimanot, T., P.O. Box 2, Dar es Salaam, Tanzania.

Haki, Dr J.M., Sugarcane Research Institute, P.O. Box 30031, Kibaha, Tanzania.
Hamad, A.S., Tanzania National Scientific Research Council, Box 4302, Dar es Salaam, Tanzania.
Hanks, J., VEERU, Reading University, Earley Gate, Reading RG4 2AG, UK.
Haq, Dr A., Dept. of Food Science & Technology, University of Dar es Salaam, Morogor Road, Dar es Salaam, Tanzania.
Haque, M.M., Director, Food Production & Rural Development Division, Commonwealth Secretariat, Marlborough House, Pall Mall, London, UK.
Hargrove, Dr T.R., International Rice Research Institute, Box 933, Manila, Philippines.
Harris, Dr K.M., Chief Entomologist, Commonwealth Institute of Entomology, c/o British Museum (Natural History), Cromwell Road, London, SW7 5BD, UK.
Hatibu, Dr C.G., Box 3084, Arusha, Tanzania.
Havener, R.D., CIMMYT, 40 Londres Avenue, Mexico 6 DF, Mexico 06600, USA.
Haule, K.L., IARO-Mlingano, P.O. Box 5088, Tanga, Tanzania.
Hawksworth, Dr D.L., Director, Commonwealth Mycological Institute, Ferry Lane, Kew, Surrey TW9 3AF, UK.
Henrich, Dr J., 6238 Hofheim − 3, Nussbaumstr. 5, West Germany.
Higgins, G.M., Room B712, Soil Resources, Management & Conservation Service, FAO, Via delle Terme di Caracalla, 00100 Rome, Italy.
Hindmarsh, P., Tropical Development & Research Institute, Slough, UK.
Hirst, Prof. J.M., Director, Long Ashton Research Station, Long Ashton, Bristol, UK.
Hodgson, N.J., Land Use Component, Box 591, Tabora, Tanzania.
Hubert, K., c/o NCDP, P.O. Box 6226, Dar es Salaam, Tanzania.
Hunt, G.L.T., International Potato Centre, C.I.P., P.O. Box 25171, Nairobi, Kenya.
Huxley, Dr P.A., International Centre for Research in Agroforestry (ICRAF), P.O. Box 30677, Nairobi, Kenya.
Hyera, T.M., Crop Monitoring & Early Warning Systems Project, P.O. Box 5384, Dar es Salaam, Tanzania.

Innes, Prof. N.L., National Vegetable Research Station, Wellesbourne, Warwickshire, UK.
Isike, H., Chief Photographer, Arusha International Conference Centre, Arusha, Tanazania.
Issai, S., Wildlife Division, Box 1994, Dar es Salaam, Tanzania.

Jakobsen, H., NORAD, Box 2646, Dar es Salaam, Tanzania.
James, A.D., Veterinary Research Laboratory, P.O. Kabete, Kenya.
Jana, Prof. Dr R.K., c/o Bank of Uganda, P.O. Box 7120, Kampala, Uganda.
Janneh, S.K., Dept. of Agriculture, Cape St. Mary's, Bakau, Via Banjul, The Gambia.
Johnson, J., The Vicarage, Leyburn, North Yorkshire, UK.
Johnston, J.E., Rockefeller Foundation, 1133 Ave of the Americas, New York, USA.
Jones, Dr D., Ford Foundation, P.O. 308 41081, Nairobi, Kenya.
Jones, T., Tropical Development & Research Institute, College House, Wright's Lane, London SW8 5SJ, UK.
Jones, N.G., Executive Director, Commonwealth Agricultural Bureaux, Farnham Royal, Slough SL2 3BN, UK.
Jones, Dr T.A.O.C., Njala University College, University of Sierra Leone, P.M.B. Freetown, Sierra Leone.

Kaaya, Mrs J.E., Tanzania Agricultural Research Organisation Headquarters, P.O. Box 9761, Dar es Salaam, Tanzania.
Kachecheba, J.L., Tropical Pesticides Research Institute, P.O. Box 3024, Arusha, Tanzania.
Kaduma, I.M., Centre on Integrated Rural Development for Africa, P.O. Box 6115, Arusha, Tanzania.
Kajuni, A.R., TALIRO, LRC West Kilimanjaro, P.O. West Kilimanjaro, Tanzania.
Kambona, J.J., FAO, Via delle Terme di Caracalla, 00100 Rome, Italy.
Kamara, Dr J.A., Njala University College, PMB Freetown, Sierra Leone.
Kamvazina, Dr S.S., Ministry of Agriculture, Dept. of Veterinary Services, P.O. Box 30372, Lilongwe, Malawi.
Kapalasula, M.J., Private Bag Ngomeni (Mlingano), Tanzania.
Kapinga, A.M., USAID/Tanzania Mission, P.O. Box 9130, Dar es Salaam, Tanzania.
Kapingu, P., Agricultural Research Institute, Ukiriguru, P.O. Box 1433, Mwanza, Tanzania.
Kasembe, Dr J.N.R., Tanzania Agricultural Research Organisation, P.O. Box 9761, Dar es Salaam, Tanzania.
Kassam, A.H., Echemess Development Services, 5/7 Singer Street, London EC2A 4QA, UK.
Kaufmann, R. von, International Livestock Center for Africa (ILCA), PMB 2248, Kaduna, Negeash, Nigeria.
Kaungamo, E.E., Director, Tanzania Library Services, P.O. Box 9283, Dar es Salaam, Tanzania.
Kayumbo, Dr H.Y., Faculty of Agriculture, Forestry, Veterinary Science, SLP 3005, Morogoro, Tanzania.

Khatibh, A.I., Ministry of Agriculture, P.O. Box 159, Zanzibar, Tanzania.
Kikopa, Dr R., Animal Diseases Research Institute, P.O. Box 9254, Dar es Salaam, Tanzania.
Kinyawa, P.L., National Coconut Development Programme, P.O. Box 6226, Dar es Salaam, Tanzania.
Kirk-Greene, A.H.M., St Antony's College, Oxford University, Oxford, UK.
Kirkwood, P.C., Tanzania Canada Wheat Project, Agriculture Canada, Ottawa, Ontario, Canada.
Kirway, T.N., Uyole Agricultural Centre, P.O. Box 400, Mbeya, Tanzania.
Kitalyi, J., Veterinary Juestication Centre, Private Bag Mpwapwa, Tanzania.
Kombwa, B., Radio Tanzania, Box 1236, Arusha, Tanzania.
Kondela, J., Tanzanian Agricultural Research Organization, Lyamungu, P.O. Box 3004, Moshi, Tanzania.
Kullaua, I.K., Agricultural Chemist, Box 3004, Moshi, Tanzania.
Kundy, Dr D.J., P.O. Box 9254, Animal Diseases Research Institute, Dar es Salaam, Tanzania.
Kusekwa, M.L., Taliro, Livestock Production Research Institute (PLRI), Private Bag, Mpwapwa, Tanzania.
Kuwite, C.A., Tanzanian Agricultural Research Organization Headquarters, P.O. Box 9761, Dar es Salaam, Tanzania.
Kwayu, E.L., State House, Box 9120, Dar es Salaam, Tanzania.
Kyando, D.S., Uyole Agricultural Centre, P.O. Box 400, Mbeya, Tanzania.
Kyomo, M.L., University of Dar es Salaam, Faculty of Agriculture, Forestry & Veterinary Science, P.O. Box 300, Morogoro, Tanzania.

Laing, Dr D.R., Centro Internacional de Agricultura Tropicale (CIAT), Apartado Aereo 6713, Cali, Columbia.
Lal, Dr R., International Institute of Tropical Agriculture, Oyo Road, PMB 5390, Ibadan, Nigeria.
Lamboll, D., Suluti Experimental Station, SLP 355, Songea, Tanzania.
Lawson, Dr T.L., International Institute of Tropical Agriculture, Oyo Road, PMB 5320, Ibadan, Nigeria.
Lema, N.M., Tumbi Agricultural Research Institute, P.O. Box 306, Tabora, Tanzania.
Le Mare, Dr P.H., Department of Soil Science, University of Reading, Reading, UK.
Lewis, R., Box 43233, Nairobi, Kenya.
Libaba, G.K., Director of Fisheries, P.O. Box 2462, Dar es Salaam, Tanzania.
Lilley, L.K., P.O. Box 5047, Tanga, Tanzania.
Liwenga, Dr J.M., Uyole Agricultural Centre, P.O. Box 400, Mbeya, Tanzania.
Lolegrave, Dr A.J., May & Baker Ltd, Dagenham, Essex, UK.
Lugenja, M., LRC, P.O. Box 5016, Tanga, Tanzania.
Lujina, I., Tanzania Investment Bank, P.O. Box 9373, Dar es Salaam, Tanzania.
Lupa, M., Box 9033, Dar es Salaam, Tanzania.
Luzuka, B., AICC, Box 3081, Arusha, Tanzania.
Lyimo, Dr J.J., Tanzania National Radiation Commission, P.O. Box 4302, Dar es Salaam, Tanzania.
Lyimo, S.D., Tanzania Canada Wheat Project, Box 6024, Arusha, Tanzania.
Lyvers, K., Agricultural Development Officer, USAID, USA.

Mabeba, D.M., National Poultry Co. Ltd, P.O. Box 9391, Dar es Salaam, Tanzania.
McDonald, I.R., P.O. Box 47098, Nairobi, Kenya.
Macha, Dr A.M., Tanzania Livestock Research Organization, P.O. Box 6910, Dar es Salaam, Tanzania.
Machaga, J.S.E., Tea Research Institute, P.O. Box 8 Amani, Tanzania.
Machange, Liti Tengeru, Box 3101, Arusha, Tanzania.
Madallali, Dr S.A., Ministry of Livestock Development, P.O. Box 9152, Dar es Salaam, Tanzania.
Madelfy, J., 19 Woodfarm Close, Caversham, Reading, UK.
Madulu, J.D., Box 306, Tabora, Tanzania.
Maeda, D.N., Tropical Pesticides Research Institute, P.O. Box 3024, Arusha, Tanzania.
Magembe, Dr S.R., Ministry of Livestock Development, P.O. Box 9152, Dar es Salaam, Tanzania.
Majahasi, E.N.B., Box 2003, Dar es Salaam, Tanzania
Malima, Mrs V.F., Ministry of Agriculture, Research Division, P.O. Box 9071, Dar es Salaam, Tanzania.
Majok, Dr. A.A., University of Juba, Juba, Southern Sudan.
Manang, E.Z., Uyole Agricultural Centre, Box 400, Mbeya, Tanzania.
Manyanza, D., Tropical Pesticides Research Institute, P.O. Box 3024, Arusha, Tanzania.
Maradufu, Dr A.N.M., Tropical Pesticides Research Institute, P.O. Box 3024, Arusha, Tanzania.
Mardamootoo, Dr P., 15 Boundary Street, Rose-Hill, Mauritius.

Masaki, Dr R.L., Ministry of Livestock Development, P.O. Box 9152, Dar es Salaam, Tanzania.

Masaoa, A.P., Tanzania Livestock Research Organization, Pasture Research Centre, P.O. Box 5, Kongula, Tanzania.

Mascarenhas, Prof. A., University of Dar es Salaam, Institute of Resource Assessment, Box 35097, Dar es Salaam, Tanzania.

Maslen, Dr N.R., Tropical Development and Research Institute, London Road, Slough, UK.

Mavoa, E.M., Livestock Training Office, Liti Tengeru, Box 3101, Arusha, Tanzania.

Maxwell, W.C.M., Commission of the European Economic Communities, Box 1523, Moshi, Tanzania.

Mazev, G.T., Agronomist, P.O. Box 306, TBR, Tabora, Tanzania.

Mazhani, L.M., Agricultural Research Station, Private Bag, 0033, Gaborone, Botswana.

Mbilinyi, Prof. S., Ministry of Agriculture, P.O. Box 9192, Dar es Salaam, Tanzania.

Mbise, Dr A.N., Scientific Officer, Veterinary Investigation Centre, Box 1068, Arusha, Tanzania.

Mbise, T.J., Tropical Pesticides Research Institute, P.O. Box 3024, Arusha, Tanzania.

Mbowe, F.F.A., Tanzanian Agricultural Research Organization (ILONGA), P.O. Ilonga, Kilosa, Tanzania.

Mbwana, A.S.S., Tanzanian Agricultural Research Organization – Maruku, P.O. Box 127, Bukoba, Tanzania.

Mchechu, Dr J.E.U., Ministry of Livestock Development, P.O. Box 9152, Dar es Salaam, Tanzania.

Mchinja, S.J., Principal Research Officer, Tropical Pesticides Research Institute, Box 3024, Arusha, Tanzania.

Mejoqli, Dr A.S., National Artificial Insemination Centre, P.O. Box 7141, Arusha, Usa River Arusha, Tanzania.

Metcalfe, Dr J.R., Director Information Services, Commonwealth Agricultural Bureaux, Farnham Royal, Slough SL2 3BN, UK.

Mgheni, Dr M., University of Dar, P.O. Box 3004, Chuo Kikuu, Morogoro, Tanzania.

Mgonja, Dr A.P., Tanzanian Agricultural Research Organization, Katrin, Fakara, P/Bag Ifakara, Tanzania.

Minja, Miss E.J., U.A.C. P.O. Box 400, Mbeya, Tanzania.

Minja, E.M., Tanzanian Agricultural Research Organization – Naliendele, P.O. Box 509, Mtwara, Tanzania.

Minja, M.D., CIBA–GEIGY, P.O. Box 444, Arusha, Tanzania.

Mitchell, A.J.B., Land Resources Development Centre, Tolworth Tower, Surbiton, Surrey, UK.

Mmari, S.M.J., Kilimo Mkoa, Box 3168, Arusha, Tanzania.

Mmbaga, M.E.T., Box 3004, Moshi, Tanzania.

Mmbaga, M.T., Botany Department, University of Dar es Salaam, Box 35060, Dar es Salaam, Tanzania.

Mnzava, E.M., Forest Division, Box 426, Dar es Salaam, Tanzania.

Mohamed, R.-A., Head of Agricultural Dept., Kibaha Secondary School, P.O. Box 30053, Kibaha, Tanzania.

Mollel, N., P.O. Box 6028, Arusha, Tanzania.

Morley, G.E., Tropical Development and Research Institute, 127 Clerkenwell Road, London EC1R 5DB, UK.

Mosha, A.S., P.O. Box 0024, Arusha, Tanzania.

Mosha, C.J., Tropical Pesticides Research Institute, Box 3024, Arusha, Tanzania.

Moshi, A.O., Tropical Pesticides Research Institute, Box 3024, Arusha, Tanzania.

Moshi, Dr A.J., TARO-Research Institute, Ilonga, Kilosa, Tanzania.

Moteane, Dr M., Livestock and Veterinary Services, P.B. A82, Maseru, Lesotho.

Mpelumbe, Dr I.S., Ministry of Livestock Development, P.O. Box 9152, Dar es Salaam, Tanzania.

Mphuru, Prof. A., Faculty of Agriculture, Forestry and Veterinary Science, P.O. Box 3000, Chuo Kiku, Morogoro, Tanzania.

Mrope, R.A., Tanzania Fisheries Corporation, P.O. Box 4296, Dar es Salaam, Tanzania.

Msabaha, Dr M.A.M., P.O. Box 1433, Mwanza, Tanzania.

Msami, H.M.H., Animal Diseases Research Institute, P.O. Box 9254, Dar es Salaam, Tanzania.

Msanga, J.F., Veterinary Investigation Centre, TALIRO, P.O. Box 129, Mwanza, Tanzania.

Msangi, A., Animal Diseases Research Institute, P.O. Box 9254, Dar es Salaam, Tanzania.

Msangi, Prof. A.S., Tanzania National Scientific Research Council, P.O. Box 4302, Dar es Salaam, Tanzania.

Msolla, Dr P., Dar es Salaam University, Box 3021, Morogoro, Tanzania.

Msororo, R.A., P.O. Box 3084, Arusha, Tanzania.

Mtei, Dr B.J., Veterinary Investigation Centre, Tabora, Tanzania.

Mughogho, Dr L.K., ICRISAT, Patancheru P.O., Andhra Pradesh 502 324, India.

Mukangi, D.J.A., TALIRO, VIC, Box 129, Mwanza, Tanzania.

Mukhtar, H.A.M., Arab Bank for Economic Development in Africa.

Mulder, Dr J.L., National University of Lesotho, P.O. Roma, Lesotho.

Mulilo, J.B., Livestock and Pest Research Centre, National Council for Scientific Research, P.O. Box 49, Chilanga, Zambia.

Munjal, Dr G.S., Uyole Agricultural Centre, Box 400 Mbeya, Tanzania.

Muriithi, I.E., Director of Veterinary Science, Box 68228, Nairobi, Kenya.

Muwila, M.L., Ministry of Agriculture, Box 30134, Capital City, Lilongwe 3, Malawi.

Mwakatunon, Dr A.G.K., P.O. Box 1823, Dar es Salaam, Tanzania.

Mwakilasa, B.A., Planning Officer, Ministry of Livestock Development, P.O. Box 9152, Dar es Salaam, Tanzania.

Mwambene, R.O., Uyole Agricultural Centre, P.O. Box 400, Mbeya, Tanzania.

Mwambuma, J., Secretary for Agriculture, Juwata (Union of Tanzania Workers – Agricultural Section), Box 15359, Dar es Salaam, Tanzania.

Mwandemere, H.K., Ministry of Agriculture, Box 30134, Lilongwe, Malawi.

Ndamugoba, D., Tanzanian Agricultural Research Organization – Maruku, P.O. Box 127, Bukoba, Tanzania.

Nchimbi, W.D., Tanzania Information Services, Box 3054, Arusha, Tanzania.

Ndunguru, Prof. B.J., University of Dar es Salaam, Dept. of Crop Science, P.O. Box 3005, Chuo Kikuu, Morogoro, Tanzania.

Ngana, J.O., Institute of Resource Assessment, University of Dar, P.O. Box 35097, Dar es Salaam, Tanzania.

Nganga, S., CIP Regional Representative, Box 25171, Nairobi, Kenya.

Nguti, H.M., Senior Scientific Officer, UTAFITI, P.O. Box 4302, Dar es Salaam, Tanzania.

Ngutter, L.G.K., Ministry of Agriculture, P.O. Box 30028, Nairobi, Kenya.

Njau, R.J.A., Research Officer, Tropical Pesticides Research Institute, P.O. Box 3024, Arusha, Tanzania.

Nkaonja, R.S.W., P.O. Box 30048, Capital City, Lilongwe 3, Malawi.

Njobvu, C.A., Mt Makulu Research Station, P/B 7, Chilanga, Zambia.

Nsengwa, G.R.M., Veterinary Investigation Centre, P.O. Box 186, Mtwara, Tanzania.

Ntondolo, R.J., Journalist, Box 1069, Arusha, Tanzania.

Nuru, Prof. S., National Animal Production Research Institute, PMB 1096, Zaria, Nigeria.

Nutting, E.R., Radio Tanzania, Box 1236, Arusha, Tanzania.

Nyange, Dr J.F.C., Tanzania Livestock Research Organisation, Veterinary Investigation Centre, P.O. Box 1068, Arusha, Tanzania.

Nyambo, B.T., A.R.I. Ukiriguru, P.O. Box 1433, Mwanza, Tanzania.

Nyassi, P.A., Tanzania Agricultural Research Organisation, Box 9761, Dar es Salaam, Tanzania.

Obedi, W., Horti, Tengeru, P.O. Box 1253, Arusha, Tanzania.

O'Conor, R.G., C/-FAO 10 Box, Dar es Salaam, Tanzania.

Odhiambo, Prof. T.R., ICIPE, P.O. Box 30772, Nairobi, Kenya.

Ogallo, Dr L. University of Nairobi, P.O. Box 30197, Nairobi, Kenya.

Okombo-Ngassaki, Dr V., Service Inspection Veterinaire, Ministere Agriculture Elevage, BP 83, R.P. Congo.

Okigbo, Dr. B.N., International Institute of Tropical Agriculture, PMB 5320, Oyo Road, Ibadan, Nigeria.

Ole-Karyongi, I., CIRDAFRICA, P.O. Box 6115, Arusha, Tanzania.

Olney, G., Land Resources Development Centre, Overseas Development Administration, Tolworth Tower, Surbiton, Surrey KT7 7DY, UK.

Openshaw, K., Regional Director (E&S) Energy Initiatives for Africa, Box 39002, Nairobi, Kenya.

Ovinge, J., Livestock Co-ordinator, P.O. Box 600, Tanza, Tanzania.

Owadally, Dr A.L., 5 Malartic Avenue, Quatre-Barnes, Mauritius.

Page, Dr O.T., Apartado 5969, Lima, Peru.

Papayiannis, C., Agricultural Research Institute, Nicosia, Cyprus.

Parsons, Dr M., Mtwara/Lindi RIDEP Project, P.O. Box 273, Lindi, Tanzania.

Parzer, A., Agricultural Attache, P.O. Box 41537, Nairobi, Kenya.

Patel, K.N., East African Tanning Extract Co., Lonrho (E.A.) Ltd., PO Box 190, Eldoret, Kenya.

Pedersen, G.N., Dept. of Agricultural Engineering, P.O. Box 3003 UDSM, Morogor, Tanzania.

Pedgley, D.E., Tropical Development and Research Institute, 127 Clerkenwell Road, London EC1R 5DB, UK.

Perkin, S., Projects Officer, International Union for Conservation of Nature/World Wildlife Fund, Box 48177, Nairobi, Kenya.

Ramos, A.H., National Agricultural Laboratories, Box 30028, Nairobi, Kenya.

Reddy, Dr D.B.E., Andhra Pradesh Agricultural University, Hyderabad 50030, India.

Rijks, Dr D., 16 rue Gautier, 1201 Geneva, Switzerland.

Rimington, R.P., Manager Circulation Services, Commonwealth Agricultural Bureaux, Farnham Royal, Slough SL2 3BN, UK.

Rimmer, D., Centre of West African Studies, University of Birmingham, P.O. Box 363, BI5 2TT, UK.

Ringo, D.F.P., Registrar of Pesticides, Tropical Pesticides Research Institute, P.O. Box 3024, Arusha, Tanzania.

Ritchie, Dr J.M., National Museums of Kenya, Section of Entomology, P.O. Box 40658, Nairobi, Kenya.

Rossiter, Dr P., Veterinary Research Department, Kenya Agricultural Research Institute, Muguga, P.O. Box 32, Kikuyu, Kenya.
Robertson, I.A.D., Overseas Development Administration, P.O. Box 90950, Mombasa, Kenya.
Rombulow-Pearse, C.W., Land Use Component, P.O. Box 591, Tabora, Tanzania.
Romney, D.H., National Coconut Development Programme, P.O. Box 6226, Dar es Salaam, Tanzania.
Romney, Mrs S., National Coconut Development Programme, P.O. Box 6226, Dar es Salaam, Tanzania.
Rose, Dr D.J.W., Kenya Agricultural Research Institute, P.O. Box 30148, Nairobi, Kenya.
Rudgers, Dr L.A.L., Tanzania-Canada Wheat Project, Box 6160, Arusha, Tanzania.
Rumisha, C., (Press), Shihata, P.O. Box 6028, Arusha, Tanzania.

Saadan, H.M., Tanzanian Agricultural Research Organization, ILONGA, P.O. Ilonga, Kilosa, Tanzania.
Saidi, J.A., Tropical Pesticides Research Institute, P.O. Box 3024, Arusha, Tanzania.
Saka, Dr V.N., Bunda College of Agriculture, Box 219, Lilongwe, Malawi.
Samki, J.K., Director, Agricultural Research Institute, Mlingano, Tanga, Tanzania.
Sampson, M., Ministry of Agriculture and Livestock Development, P.O. Box 180, Voi, Kenya.
Sanders, S., Regional Technical Advisor, BAYER, Agrochem., Bayer (2) PVT Ltd., P.O. Box AY78, Amby, Harare, Zimbabwe.
Sarap, Dr H. International Foundation for Science, GREV Turegatan 19, 11438 Stockholm, Sweden.
Sauwa, P.E.L., National Ranching Co. Ltd., P.O. Box 9113, Dar es Salaam, Tanzania.
Schwermer, W., Protection Officer, P.O. Box 30577, Nairobi, Kenya.
Sekambo, M.M., Botswana Agricultural College, P.B. 27, Gaborone, Botswana.
Sembony, G., Tanzania News Agency, P.O. Box 6028, Arusha, Tanzania.
Semuguruka, Dr G.H., Research Institute Lyamungu, P.O. Box 3004, Moshi, Tanzania.
Serafimosa, Mrs N., Bulgarian Embassy, Dar es Salaam, Tanzania.
Shao, Dr F.M., Director of Research, TARO Ilonga, P.O. Ilonga, Kilosa, Tanzania.
Sheldon, V.L., International Fertilizer Development Center, P.O. Box 2040, Muscle Shoals ae 35662, USA.
Shomari, S.H., Director, Tanzanian Agricultural Research Organization – Naliendele, P.O. Box 509, Mtwara, Tanzania.
Siarra, A.S., Shell Chemical Company of Eastern Africa Ltd, P.O. Box 91521, Dar es Salaam, Tanzania.
Sibisi, Dr H., Box A 92, Swazi Plaza, Mbabane, Swaziland.
Simon, Dr B., Tropical Pesticides Research Institute, P.O. Box 3024, Arusha, Tanzania.
Simons, J.H., Oilseeds Research Project, TARO-Naliendele, P.O. Box 509 Mtwara, Tanzania.
Smith, A.J.L., Delegation of the Commission of the European Communities, P.O. Box 9514, Dar es Salaam, Tanzania.
Smyth, F.P., Training and Visit Extension Advisor, Coffee Authority of Tanzania, Box 732, Moshi, Tanzania.
Speidel, Dr D., Coordinator, National Coconut Development Programme, P.O. Box 6226, Dar es Salaam, Tanzania.
Srivastava, Dr N.C., Senior Research Officer, Ministry of Agriculture, P.O. Box 9071, Dar es Salaam, Tanzania.
Stigter, Prof. C., Physics Department, P.O. Box 35063, Dar es Salaam, Tanzania.

Tapley, Dr R.G., Consultant, CAT/CDP, Box 732, Moshi, Tanzania.
Tarimo, B., Senior Economist, Bank of Tanzania, P.O. Box 2939, Dar es Salaam, Tanzania.
Tarimo, C.S., Tanzania Livestock Research Organization, Tsetse and Trypanosomiasis Research Institute, P.O. Box 1026, Tanga, Tanzania.
Tatchell, Dr R.J., P.O. Box 30470, Nairobi, Kenya.
Taylor, Dr B.R., Agricultural Research Institute, P.O. Box 509, Mtwara, Tanzania.
Tegegn, A., Assistant Dean, Addis Ababa University, College of Agriculture, P.O. Box 138, Dire Dawa, Ethiopia.
Temane, B.K., Administrative Officer (CIRDAFRICA), Box 6115, Arusha, Tanzania.
Temu, Dr A.B., Division of Forestry, University of Dar es Salaam, P.O. Box 3009, Chuo Kikuu, Morogoro, Tanzania.
Temu, A.E.M., Research Agronomist, Uyole Agricultural Centre, P.O. Box 400, Mbeya, Tanzania.
Terry, Dr E.R., International Institute of Tropical Agriculture, PMB 5320, Ibadan, Nigeria.
Thomas, B., c/o Peace Corps, P.O. Box 30518, Nairobi, Kenya.
Thomas, Dr R.J., Dept. of Agriculture, University of Newcastle, Newcastle upon Tyne, UK.
Tibi, S., Principal, P/Bag 0027, Gaborone, Botswana.
Tsakiris, A., P.O. Box 9192, Dar es Salaam, Tanzania.
Tupa, C.K., Director, Finance & Administration, Tanzania Agricultural Research Organization, Box 9761, Dar es Salaam, Tanzania.

Umbima, W.E., International Centre for Insect Physiology and Ecology (ICIPE), Box 30772, Nairobi, Kenya.

Urom, B., Research Officer, TPRI, P.O. Box 3024, Arusha, Tanzania.
Uronu, A., Research Officer, Box 3024, Arusha, Tanzania.

Vahaye, Dr G.A., TARO Mlingano, P/Bag Ngomeni, Tanga, Tanzania.
Voerman, M.P., Chuo Kikuu, P.O. Box 3006, Morogoro, Tanzania.

Walker, P.T., Tropical Development Research Institute, Porton Down, Salisbury, UK.
Waller, Dr J.M., Commonwealth Mycological Institute, Ferry Lane, Kew, Surrey TW9 3AF, UK.
Walyaro, Dr D.J., Coffee Research Foundation, P.O. Box 4, Ruiru, Kenya.
Wangati, Dr F.J., P.O. Box 29203, Nairobi, Kenya.
Watson, M., BDDSA, Box 30059, Lilongwe 3, Malawi.
Way, Prof. M.J., Imperial College, Silwood Park, Ascot, Berks. UK.
Williams, Dr C.E., Dept. of Agricultural Extension, University of Ibadan, Ibadan, Nigeria.
Wilson, A.D., Kilifi Plantations, Kilifi, Kenya.
Wood, P.J., Senior Forestry Adviser, ICRAF, Box 30677, Nairobi, Kenya.
Wodageneh, Dr A., African Inter-Country IPC Programme, FAO/UN, POB 913, Khartoum, Sudan.
Woodford, Dr E.K., Commonwealth Agricultural Bureaux, Farnham Royal, Slough SL2 3BN, UK.

Zan, Dr K., International Institute of Tropical Agriculture, Oyo Road, PMB 5320, Ibadan, Nigeria.